MODERN
POWER STATION PRACTICE

Third Edition

(in 12 volumes)

Incorporating Modern Power System Practice

Authors

Chapters 1 and 2

M. J. Brindle, BTech, CEng, MIMechE

I. Cresswell, BSc, CEng, MIMechE

J. C. Greenslade, BSc, CEng, MIMechE

A. R. Jones, BEng(Mech)(hons), MIMechE

K. Jones

P. C. Martin, BSc, DMS, CEng, MIMechE

A. R. Smith, BSc, CEng, MIMechE

I. Stagg, BEng(hons)

Chapter 3

J. D. Blackhall, BSc(hons), CEng, MICE

P. D. Davis, MSc, MICE

P. M. Emberson, BA, DipLD, ALI

E. J. Forsey, CEng, MIMechE

R. Hawkridge, HNC(Civils)

M. F. F. Hynes, BSc, CEng, MICE

J. Irving, BSc, CEng, FICE

D. J. Mallard, MSc, MICE, FGS

P. A. Moulding, CEng, MICE

R. A. Pope, BSc, CEng, MICE

C. H. Trent, DRTC, CEng, MICE

R. A. Vevers, CEng, MICE

Series Production

Managing Editor	P. M. Reynolds
Production Editor	H. E. Johnson
Resources and Co-ordination	T. A. Dolling J. R. Jackson

MODERN
POWER STATION PRACTICE

Third Edition

Incorporating Modern Power System Practice

British Electricity International, London

Volume A

Station Planning and Design

PERGAMON PRESS

Member of Maxwell Macmillan Pergamon Publishing Corporation

OXFORD · NEW YORK · BEIJING · FRANKFURT

SÃO PAULO · SYDNEY · TOKYO · TORONTO

U.K.	Pergamon Press plc, Headington Hill Hall, Oxford OX3 0BW, England
U.S.A.	Pergamon Press, Inc., Maxwell House, Fairview Park, Elmsford, New York 10523, U.S.A.
PEOPLE'S REPUBLIC OF CHINA	Pergamon Press, Room 4037, Qianmen Hotel, Beijing, People's Republic of China
FEDERAL REPUBLIC OF GERMANY	Pergamon Press GmbH, Hammerweg 6, D-6242 Kronberg, Federal Republic of Germany
BRAZIL	Pergamon Editora Ltda, Rua Eça de Queiros, 346, CEP 04011, Paraiso, São Paulo, Brazil
AUSTRALIA	Pergamon Press (Australia) Pty Ltd., P.O. Box 544, Potts Point, N.S.W. 2011, Australia
JAPAN	Pergamon Press, 5th Floor, Matsuoka Central Building, 1-7-1 Nishishinjuku, Shinjuku-ku, Tokyo 160, Japan
CANADA	Pergamon Press Canada Ltd., Suite No. 271, 253 College Street, Toronto, Ontario, Canada M5T 1R5

First edition 1963

Second edition 1971

Third edition 1991

Library of Congress Cataloging in Publication Data

Modern power station practice: incorporating modern power system practice/British Electricity International.—
3rd ed. p. cm.
Includes index.
1. Electric power-plants. I. British Electricity International.
TK1191.M49 1990
62.31'21 — dc20 90-43748

British Library Cataloguing in Publication Data

British Electricity International
Modern power station practice.—3rd. ed.
1. Electric power-plants. Design and construction
I. Title II. Central Electricity Generating Board
621.3121.

ISBN 0-08-040511-8 (Volume A)
ISBN 0-08-040510-X (Set)

Printed in the Republic of Singapore by Singapore National Printers Ltd

Contents

Colour Plates

(between pp 66 and 67)

Foreword

G. A. W. Blackman, CBE, FEng
Chairman, Central Electricity Generating Board
and Chairman, British Electricity International Ltd

FOR OVER THIRTY YEARS, since its formation in 1958, the Central Electricity Generating Board (CEGB) has been at the forefront of technological advances in the design, construction, operation, and maintenance of power plant and transmission systems. During this time capacity increased almost fivefold, involving the introduction of thermal and nuclear generating units of 500 MW and 660 MW, to supply one of the largest integrated power systems in the world. In fulfilling its statutory responsibility to ensure continuity of a safe and economic supply of electricity, the CEGB built up a powerful engineering and scientific capability, and accumulated a wealth of experience in the operation and maintenance of power plant and systems. With the privatisation of the CEGB this experience and capability is being carried forward by its four successor companies — National Power, PowerGen, Nuclear Electric and National Grid.

At the heart of the CEGB's success has been an awareness of the need to sustain and improve the skills and knowledge of its engineering and technical staff. This was achieved through formal and on-job training, aided by a series of textbooks covering the theory and practice for the whole range of technology to be found on a modern power station. A second edition of the series, known as Modern Power Station Practice, was produced in the early 1970s, and it was sold throughout the world to provide electricity undertakings, engineers and students with an account of the CEGB's practices and hard-won experience. The edition had substantial worldwide sales and achieved recognition as the authoritative reference work on power generation.

A completely revised and enlarged (third) edition has now been produced which updates the relevant information in the earlier edition together with a comprehensive account of the solutions to the many engineering and environmental challenges encountered, and which puts on record the achievements of the CEGB during its lifetime as one of the world's leading public electricity utilities.

In producing this third edition, the opportunity has been taken to restructure the information in the original eight volumes to provide a more logical and detailed exposition of the technical content. The series has also been extended to include three new volumes on 'Station Commissioning', 'EHV Transmission' and 'System Operation'. Each of the eleven subject volumes had an Advisory Editor for the technical validation of the many contributions by individual authors, all of whom are recognised as authorities in their particular field of technology.

All subject volumes carry their own index and a twelfth volume provides a consolidated index for the series overall. Particular attention has been paid to the production of draft material, with text refined through a number of technical and language editorial stages and complemented by a large number of high quality illustrations. The result is a high standard of presentation designed to appeal to a wide international readership.

It is with much pleasure therefore that I introduce this new series, which has been attributed to British Electricity International on behalf of the CEGB and its successor companies. I have been closely associated with its production and have no doubt that it will be invaluable to engineers worldwide who are engaged in the design, construction, commissioning, operation and maintenance of modern power stations and systems.

March 1990

Preface

Chapters 1 and 2

The planning and design development of new power station proposals at first sight appears straightforward, but experience has shown that such a process can involve complex interaction between not only the various engineering disciplines which contribute to the process but also environmental, planning, economical, political and social pressures.

In this third edition, the opportunity has been taken to restructure the information and CEGB experience to provide a logical review of the investigations and engineering design activities which are required to underwrite power station development.

The Engineering studies are outlined in some detail together with examples showing how such activities need to be brought together to fully define the project parameters. Engineers by nature and training prefer to make decisions on factual information but in reality judgement is often required. A sound basis for exercising this need is not only experience but also knowledge, and the revised text attempts to illustrate how the progressive and interactive nature of investigations allows for project evolution from initial conception through to commitment to construct.

The implications of the more intangible 'non-engineering' factors are reviewed and their potential influence on the development process discussed in general terms, but for any particular proposal these aspects will have varying degrees of relevance. It will be the responsibility of the development manager in his own particular circumstances to judge the influence of these factors and the implications they may have for the cost and programme of his project.

There is always so much that is of relevance in preparing a text that the engineers responsible for this chapter have had to be responsible for choosing those aspects which they consider most important. If the material they have chosen is found to be of benefit to the reader and helps to pass on the experience and 'know how' of the CEGB engineers working in this field, then the authors will no doubt draw some satisfaction from their efforts.

P. C. Martin
Advisory Editor — Chapters 1 and 2

Chapter 3

Civil engineering and building in the power industry may initially be seen, in simplistic terms, as the provision of weatherproof containment over the mechanical and electrical plant and the necessary foundations to support it. In this edition of Modern Power Station Practice my contributors and I have tried to illustrate that this truism leaves much unsaid. The civil engineer and the architect are main contributors in every sense throughout the design and construction on any new power station. From the earliest planning and site selection studies through to the final landscaping both these disciplines together with the support of colleagues in estimating, finance and quantity surveying, are fully involved.

Indeed the start of a project presents the civil engineer with perhaps his greatest challenge. Unavoidably he is unable to proceed with his design work until the weights and loads of the plant and their locations are known with some certainty. This late start must then be compensated by design and contract to allow the construction phase to go ahead. Under these circumstances the civil engineer must accept the soil conditions as found, since the proximity of cooling water, fuel supplies and transmission connections are likely to outweigh the poorest ground strata in the overall planning considerations.

Setting up his own 'factory' on the site and providing good access and working facilities for the mechanical and electrical contractors presents its own unique set of problems for the civil engineer. Doing so with a labour force that is mainly recruited locally, or itinerant, and assembled specifically for the project, requires considerable man management skills.

The timely and successful completion of the civil works is the key factor without which no power station project can meet its overall criteria of quality, programme and cost. These seemingly supplementary items to the boiler and turbine plant — the roads, drains, culverts, cooling towers, chimneys, building and structures — are likely to represent at least 25% of the total cost of a fossil-fired or nuclear station and up to 66% of a hydro or pumped storage station.

Beyond cost, the potential for cumulative delay is massively enhanced if the civil engineering and building works run late and hence delay the erection of the largely factory-produced mechanical and electrical plant.

Equally the final stage on site — the landscaping — has its own importance. Hard and soft landscape treatments are essential in leaving a completed station which reflects credit on its designers and builders, inspires pride and dedication in its operators and shows the public and planning authorities that the promises made at the outset have been fully kept.

I hope that this civil engineering chapter of the Station Planning and Design volume is able to convey to the reader a little of the technical skills, imagination, excitement, perseverance and devotion that are always present in any successful civil engineering and architectural team working on a power station project. If it also conveys any sense of the innate satisfaction and fun that so often helps motivate that team, our purpose will have been well served.

I. W. HANNAH
Advisory Editor — Chapter 3

Contents of All Volumes

Volume L — System Operation

Volume M — Index

CHAPTER 1

Power station siting and site layout

1 Planning for new power stations

1.1 Introduction

The construction of a major new power station takes typically about five to six years from the decision to build the station to the commissioning of the first unit. The CEGB's annual plans therefore include the provision for specific new generating stations that are planned for commissioning in the period seven to nine years ahead of need (referred to as the *planning years*). Before it can embark on the ordering of a new power station the CEGB must have received the Secretary of State's consent required under Section 2 of the Electric Lighting Act 1909, together with any related consent and licences, and must separately have received financial sanction from the Government.

The CEGB has to evaluate the need for power stations in the light of its statutory duties. It considers whether there is a need for new capacity in order to maintain an adequate security of supply, or to give greater economy, or to improve the security of fuel supply by allowing the types and sources of fuel or primary energy to be diversified. In addition, it may be justifiable to build a new form of generating capacity in order to develop the ground for a possible future benefit.

1.2 Capacity considerations

Capacity requirement is determined by the need to meet the peak demand of the year. The first step in estimating generating capacity requirement is therefore to forecast the peak demand for each future winter up to the planning years. The forecast presumes that the peak is most likely to occur on working weekdays in December to February during a spell of cold weather of average severity and is thus described as the *average cold spell* (ACS) winter peak demand. ACS conditions are determined by a statistical analysis of past weather data and the variation in demand caused by weather variations.

Each year the Electricity Supply Industry prepares new estimates of the unrestricted ACS winter peak demand, the corresponding values of restricted peak demand after allowing for the expected reduction in peak demand by load management, and the total number of units of electricity (unit requirements) to be produced by the CEGB or purchased from external suppliers. The unit requirement therefore equals the sum of the CEGB sales of electricity to Area Boards and to its direct consumers and the transmission losses on the CEGB system.

After consideration of the various forecasts, recommendations are made to the Electricity Council as to estimates of demand and unit requirement up to the planning years. The Electricity Council then formally adopts these forecasts, together with provisional estimates for the subsequent two years, on behalf of the Electricity Supply Industry in England and Wales.

In order to meet the statutory requirement to provide a continuous supply of electricity except in cases of emergency, the industry has over many years aimed to provide sufficient generating capacity to meet the future demand with a high degree of security. Since it is impracticable to ensure 100% security of supply there will, on occasions, be insufficient generating capacity to meet demand even after the application of load management. In such circumstances, the first action would be to reduce the voltage and/or frequency within permissible statutory limits. This has the effect of reducing the magnitude of demands which are sensitive to voltage or frequency while maintaining continuity of supply to all consumers. In this way the overall demand can be reduced by up to 7.5%, but if the remaining demand still exceeds the generation available, then some consumers must be disconnected.

It is the CEGB's function to ensure that sufficient generating capacity is provided to meet the generation standard and it achieves this by planning a reserve margin of generating capacity called the *planning margin*. This is defined as — *the percentage margin of additional generating plant planned to be in service in the planning years over and above that needed to meet the peak demand.*

The CEGB and Electricity Council make estimates of the expected average availability and of the expected magnitude of variabilities of availability and forecast demand. A simple statistical calculation then gives the size of planning margin that meets, or approximately meets, the security standard.

1.3 Economic considerations

The provision of new capacity to meet the forecast demand is not the only reason which might justify the construction of new generating plant. New construction might also be justified on economic grounds and might allow the retirement of some existing capacity. In principle, a plant is retained in service until it becomes more economic to replace it with new capacity. Evaluations are made for certain economic indicators for existing stations and for the potential new stations that might be built:

- For existing stations, the annual avoidable cost is evaluated on a year-by-year basis of retaining certain stations or parts of stations in an operable condition. This cost is called the *net avoidable cost* (NAC) expressed in units of £/kWpa.
- For new generating station options for commissioning by the planning years, the CEGB calculates the net effect on total system costs of building and operating the station over its lifetime and converts this into an average annual cost, in units of £/kWpa, called the *net effective cost* (NEC).

These indicators allow two economic comparisons to be made. Firstly, the comparison of NEC for alternative new generating plant allows, for given assumptions of input parameter values, the indication of the most economic option, namely the one with the lowest NEC. Secondly, for that option, it is possible to test whether it is economic to install the new plant and decommission existing capacity.

When making an economic appraisal of alternative new generating station options, it is necessary to assess the probable cost of installing and running each station and its impact on other system operating costs, and to ensure that there is likely to be sufficient fuel available at an acceptable price throughout its expected operating life. Some generating plant options may have a relatively short construction time and have the potential of being economic after a short period of generation. However, the planner must consider all options, including ones with a construction lead-time of five to six years and operating life of up to 40 years. Hence the planner needs to take a view of electricity demand, fuel availability and fuel price many years ahead.

Figure 1.1 shows a possible future plant mix as envisaged by CEGB in 1985.

1.4 Future requirement predictions

The interrelationship between estimates of economic activity, fuel prices, energy supply and demand, electricity demand and the implications for electricity supply have been more fully examined through the development of economic scenarios (i.e., imagined sequence of future events).

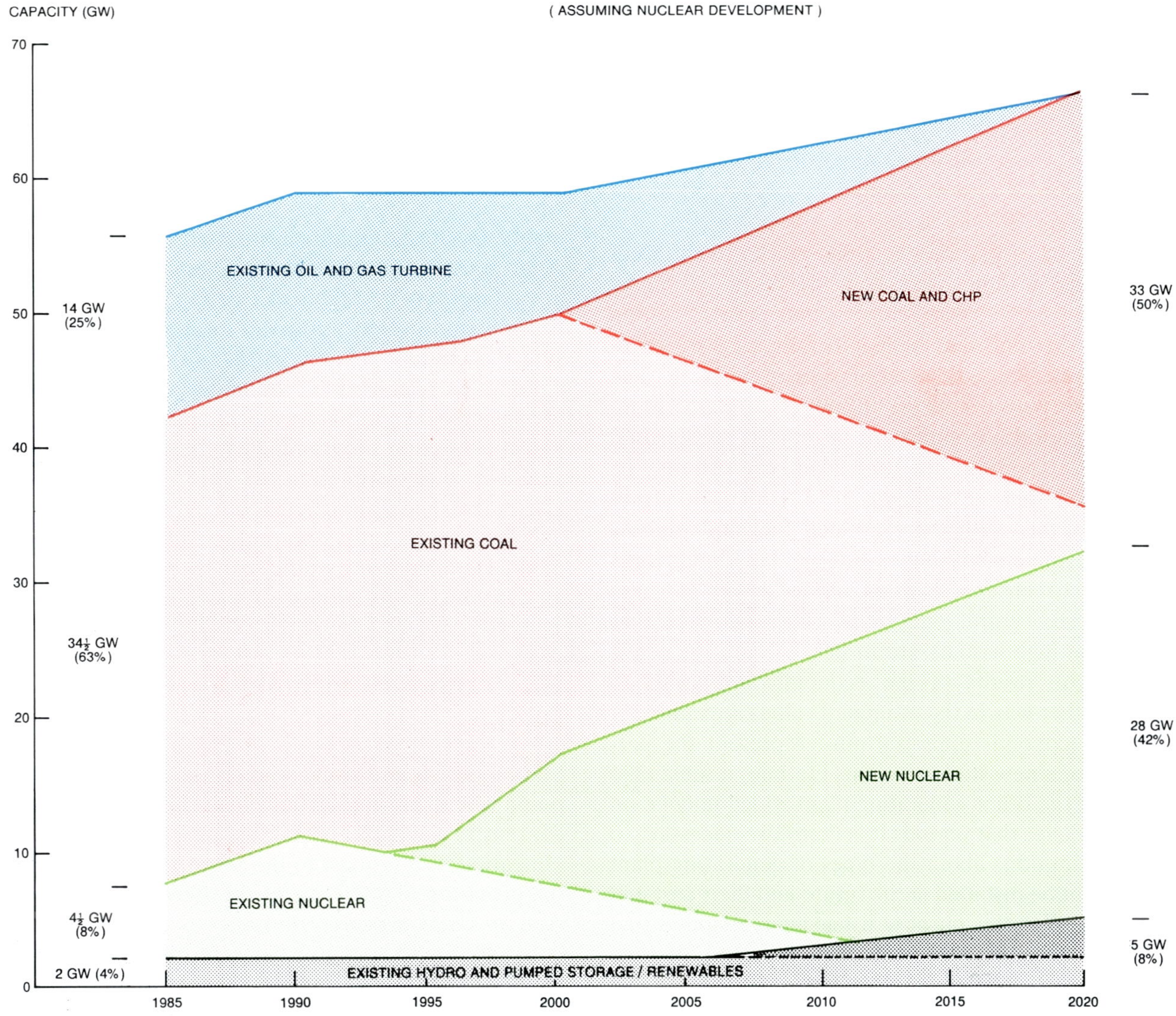

FIG. 1.1 Possible future CEGB plant mix

The scenarios set out a spectrum of possible future developments which can be used in a variety of planning studies. More specifically:

- They form a valuable aid to the judgement of the range of plausible outcomes that should be allowed for in planning, especially with regard to the future extent and composition of economic activity, energy supply and demand, energy conservation, fuel prices and availabilities, and electricity demand.
- The relative economic merits of alternative generating plant types for each scenario are evaluated.
- The implications for economic operation and security of fuel supplies of alternative generating plant development options within each scenario are considered.

The scenario approach does not require the CEGB to estimate specific probabilities of occurrence of the alternative scenarios, but provides a background against which planning judgements can be made for a highly uncertain future. However, each scenario is considered with care when it has been fully developed and is judged whether or not it still appears to be plausible and with a significant prospect of occurrence in real life. Provided the scenarios individually pass this test, the CEGB aims ideally to plan so as to be able to respond to any one of these plausible outcomes. In practice, some reasonable latitude would be acceptable; for example, in the case of a scenario entailing rapid growth of electricity demand, it might be practically necessary to accept a moderately lower standard of security of supply for some interim period before generating capacity could be fully adjusted to the requirement. However, it is planned to avoid a really serious failure to achieve a secure and economic supply for any plausible scenario.

The scenarios allow the CEGB to examine the risks attached to alternative generating plant options which arise, in particular, from variations in future electricity demand or fuel prices. In addition, the analysis of risk covers uncertainties which attach to the alternative options, especially with regard to capital cost, performance factors, lifetime and construction time. The wider strategic aspects are also considered, of which two are worth particular mention, namely the security of fuel supplies and making provision for future investment options.

Figure 1.2 shows the effect of particular scenarios on the estimation of system demand up to the end of the century, using 1979/80 as the base year.

In order to assess the economic merits of different types of generating plant it is necessary to make estimates of capital cost, construction period, station efficiency, lifetime and availability in service, all of which are relevant to the overall value of the plant. The construction period and the incidence of expenditure over that period are important in relation to the total capital investment and the time when a return may be expected on that investment; the lifetime and availability (together with the estimated fuel and running costs) determine what that return will be. Endeavours are made to ensure that, as far as possible, these estimates are central ones (i.e., those which are as likely to be high as they are to be low) and the sensitivity of the results of the economic appraisals to changes either way in the estimated values is examined.

As the economic appraisal must represent the performance of a new station over its lifetime, it must take account of the other generating plant which may be on the system over that period. It is therefore necessary to make assumptions about the types of generating plant which might be installed in the future and their cost and availability. For this purpose it has been assumed that the values of capital cost, construction period, lifetime and availability for later stations would be the same as for the stations being appraised unless there is justification for doing otherwise.

1.5 System planning studies

Early planning work begins with the examination of system load flows and the identification of future generation and transmission needs. This process shows regional requirements and notional locations of generation sites. One important factor which is taken into account is a CEGB policy to develop existing sites wherever possible, if this satisfies the system requirements. By developing such sites to their full capacity, as determined by any technical and environmental limits, advantage can be taken of existing facilities, such as transmission outlet, improved local roads and minimising the amount of new works.

At an early stage the type and size of power station are selected. An initial technical appraisal, capital cost and construction programme can be produced for each site. At the completion of this review, a list of alternative generation sites will have been compiled and they can be ranked in preferred order for development. Detailed siting studies can then be carried out — as described in Section 2 of this chapter.

1.6 Authority to build a new power station

Station design and siting studies are carried out, to the point where an application is made for Government consent to develop a site. This procedure, a statutory obligation, is a request to the Secretary of State, under the provision of Section 2 of the 1909 Electric Lighting Act, to build a power station.

In addition to Section 2 Consent, the CEGB requires planning permission under the Town and Country Planning Act of 1971. Part of this Act empowers the Secretary of State to direct that planning permission

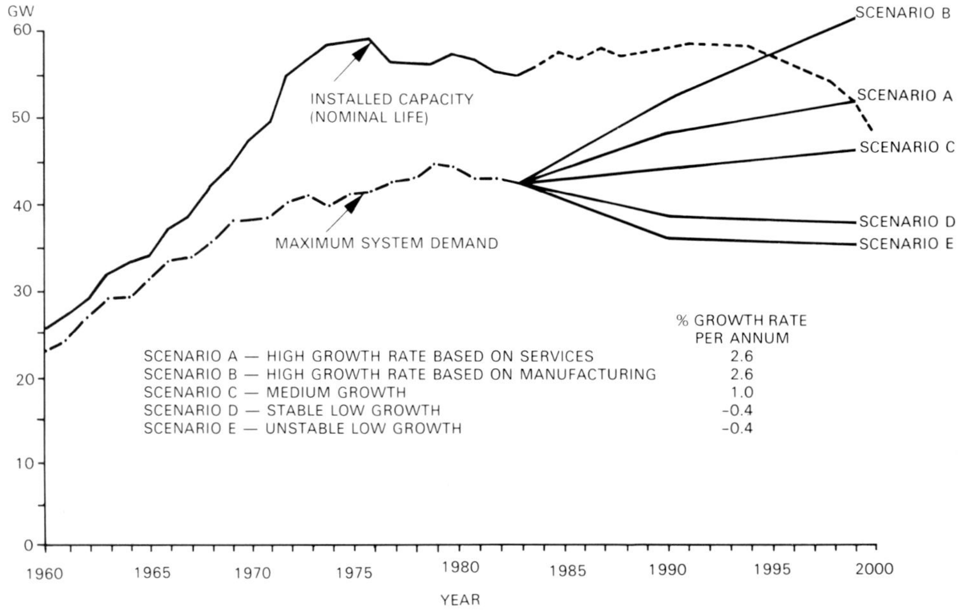

FIG. 1.2 Effects of scenarios on demand

is granted at the same time as Section 2 Consent. However, the Secretary of State may attach conditions, as he thinks appropriate, in regard to the planning approval.

Following receipt of Section 2 Consent and financial sanction, the CEGB proceeds with the design and construction of the project. Figure 1.3 shows the typical timescale for power station planning and early construction.

An important part of the investigation programme is consultation with Ministerial and Local Authorities and other statutory bodies such as Water Authorities. As part of the procedure for ensuring that all parties are fully aware of agreements which have been negotiated and which must be observed during the station design and construction period, a document called 'Station Development Particulars' is issued, which records all discussions and agreements with parties and also contains a schedule of statutory consents which must be obtained.

The Station Development Particulars also contain a technical section dealing with the transmission connections and parameters of the main plant, particularly the generator transformer, so that they are properly matched to the transmission system. The details cover matters such as power factor, synchronous impedance, frequency regulation, the dynamic response of the unit to change in load demand and guidelines on the electrical auxiliary system to ensure that this is a reliable network.

2 Site selection and investigation

2.1 Basic site requirements

A power station is simply a factory for the conversion of the energy stored in the fuel into electrical energy. The basic requirements for a power station are, therefore, similar to those of any other factory:

- A supply of raw material at a competitive cost (fuel).
- Access to the markets for its products (transmission).
- A labour force of the size and quality required.
- Means of disposal for any trade effluent or by-product.
- Land for construction and operation.

The raw material from which electricity is made in a thermal power station can be coal, oil, uranium or natural gas. Electricity, the main product, has its own access to centres of consumption through the transmission and distribution system. By-products are ash or irradiated uranium fuel elements and the economic disposal of the former is often a major consideration. The trade effluents are the large quantities of heat, the disposal of which generally requires very large quantities of water which, for cost reasons, must be available close to the site. The products of combustion, in the

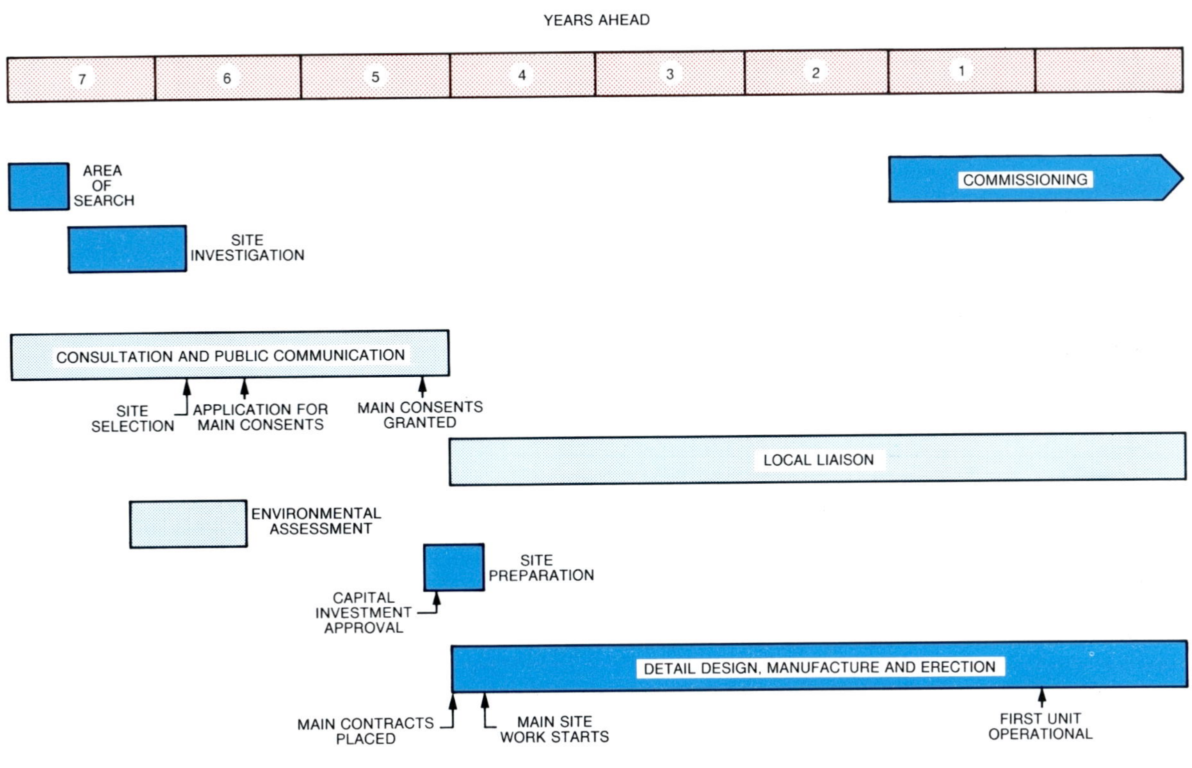

FIG. 1.3 Typical timescale for power station planning

form of large volumes of flue gases, must also be dispersed without contravening the national clean air policy or causing atmospheric pollution.

The main technical requirements of sites for nuclear and coal-fired stations of the size being considered currently are summarised in Table 1.1.

2.2 Area of search for initial site selection

In densely-developed countries like England and Wales, suitable power station sites are difficult to find. Many of the best sites have already been used for one

TABLE 1.1
Technical site requirements

PARAMETER	COAL-FIRED STATION (1800 MW)	NUCLEAR STATION (1200 MW)
Fuel delivery	Economic means of delivery up to 5 million tonnes per year	Low volume and infrequent
Cooling water (a) Direct cooled (b) Tower cooled	52 m^3/s Up to 3 m^3/s abstracted, about 0.75 m^3/s evaporated	52 m^3/s Up to 3 m^3/s abstracted, about 0.75 m^3/s evaporated
Land (direct cooled)	100 ha including construction area but excluding PF ash disposal area	60 ha including construction area
Geology	Ground able to support heavy loads	Ground able to support very heavy loads with virtually no differential settlement
Access (a) Construction materials (b) Abnormal loads	 Road, rail or sea access to deliver up to 2 million tonnes, most in first 3 years Road or sea access to deliver about 80 very large loads	 Road, rail or sea access to deliver up to 2 million tonnes, most in first 3 years Road or sea access to deliver about 80 very large loads (beach landing facility for pre-fabricated PWR parts)
Nuclear siting		Sufficiently remote from population and potential sources of hazard to comply with NII requirements
Flue gas desulphurisation materials	Delivery of about 0.3 million tonnes per year of limestone. Disposal of about 0.5 million tonnes per year of gypsum.	
Waste disposal	Means of disposing of up to 1 million tonnes per year of ash	Near to railhead for transport of flasks of irradiated fuel (about 20 loads per year)
Transmission	Suitable for connection to a point on the grid able to accept output of station	Suitable for connection to a point on the grid able to accept output of station with sufficient security

purpose or another and more and more of the undeveloped areas are being conserved. In fact some 12% is built on, while over 40% is given statutory protection; on the coast, the respective figures are 25% and 60% (see Fig 1.4). Nevertheless the CEGB must be able to meet the need for new stations as foreseen by its estimates of future demand.

The considerable length of time that is required to plan and construct a station and the regular revision of future demand estimates means that it is wise for the

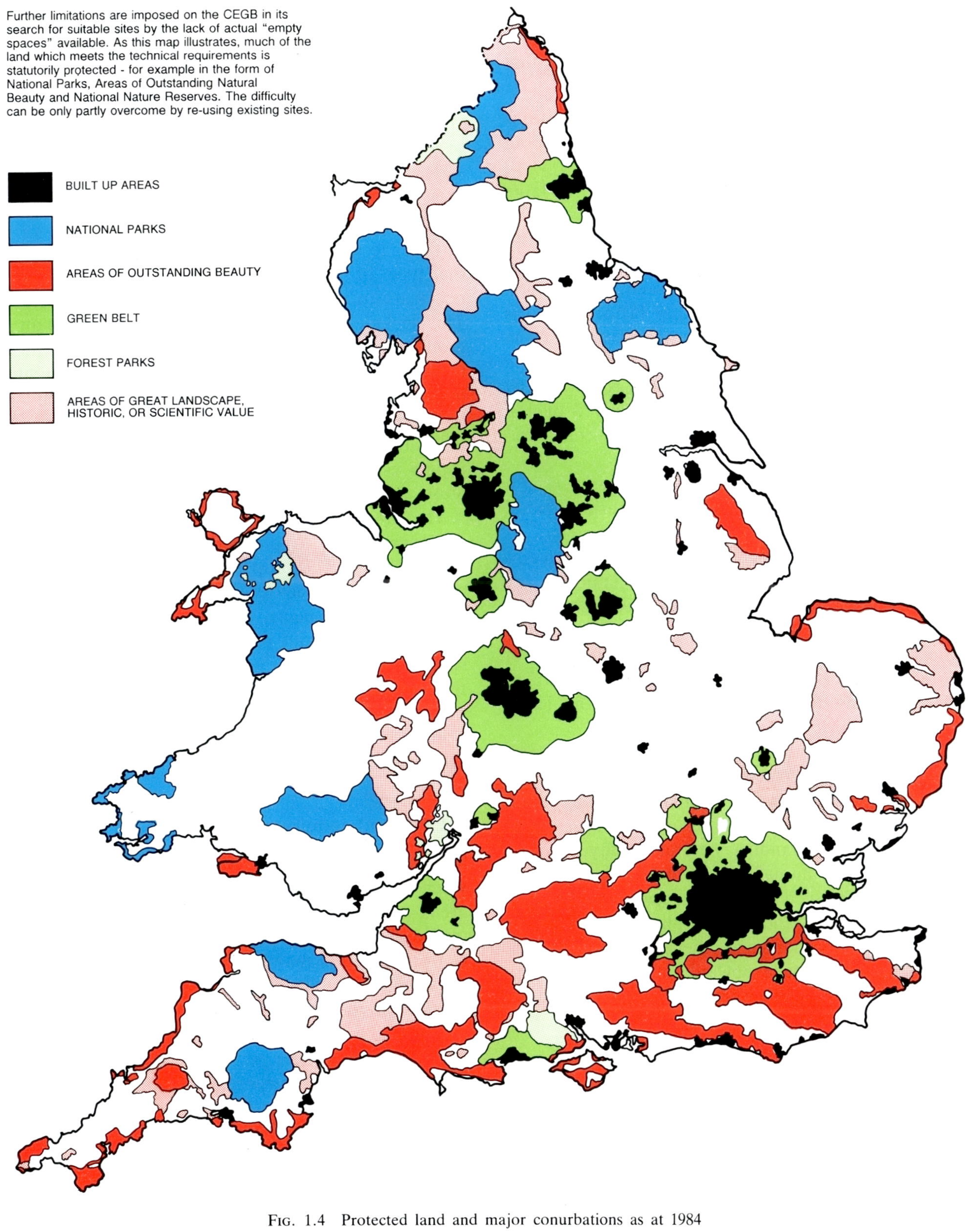

FIG. 1.4 Protected land and major conurbations as at 1984

CEGB to maintain a 'pool' of potential sites from which suitable candidates can be chosen as necessary. This pool is made up of the following three types of site:

- Existing power station sites capable of further development.
- Pieces of land already purchased by the CEGB for future development.
- Pieces of land not owned by the CEGB, but that have been identified as potential sites.

The identification and investigation of potential sites is usually divided into two phases: *area of search* and *detailed investigation*.

Although the Supergrid allows the transfer of large amounts of electricity from one part of the country to another, its capacity to do so is limited by both technical and economic constraints. Therefore, when the need for new generation is foreseen, transmission considerations combined with other factors, such as fuel sources, usually indicate in which part of the country it would be best to locate the station. The type of station required (nuclear, coal or oil) is dictated by such factors as the relative costs, the desired overall balance between fuels, and environmental considerations.

Having identified the need for new generation in a certain region, a large area, perhaps covering several hundred square kilometres, is studied to find out its potential. Any known sites are also reviewed. Government Departments are invited to draw attention to any places of special concern to them. Bodies such as The Countryside Commission and the Nature Conservancy Council, who have responsibility for preserving areas of natural beauty or of scientific value, are also notified and discussions are held with officers of the Local Planning Authority.

Information is gathered and analysed on technical matters such as water resources, geology, population distribution, road and rail system; as well as on environmental aspects such as areas of scientific interest or of outstanding natural beauty, historic features and recreational areas. Much of this information can be obtained from ordnance and geological survey maps, local and county plans, aerial photographs, Admiralty charts and other published material. These studies may take upwards of a year before a shortlist of sites thought worthy of detailed investigation can be prepared.

2.3 Detailed site investigation

Prospective sites may be identified through the *area of search work* or because changes in land-use give new opportunities, e.g., the closure of defence installations. Before detailed site investigations are started, the bodies previously consulted are notified, the owners and occupiers are approached, and announcements are made in national and local newspapers.

It can take over two years to carry out the necessary detailed studies to prove the viability and determine the optimum capacity of each of the alternative sites being considered. During this period consultations take place with the authorities concerned with planning, environmental protection, transport, water supply, flood protection, fisheries, safety and other relevant subjects. A careful study is made of the technical and amenity aspects of power station siting. The main topics covered for a typical nuclear power station site are shown on Fig 1.5. The major aspects of the studies are described as follows.

2.3.1 Preliminary station layout

In order to assess the suitability of a particular site for the type of power station being considered, it is necessary to establish the initial basic station design. This includes the disposition of the major plant or groups of plant in the main station buildings, leading up to the determination of the shape and size of the buildings and then the grouping of the various individual buildings, and external plant items to produce a co-ordinated station design which achieves the lowest capital cost, ease of construction and efficient operation and maintenance of the power station.

The preliminary station layout enables the on-site geological work to proceed and assessments to be carried out on the proposed site level, disposition of construction contractors' plant and storage areas and environmental aspects.

The station layout would be developed during the study period to take the fullest possible advantage of the available site area and of the recommendations of the architectural and landscape consultants. The principles used in the development of the station layout are described in Chapter 2 of this Volume.

2.3.2 Land requirements

Sufficient land will be required not only for the station when it is in operation, but also to provide adequate areas during the construction period.

The area occupied by an 1800 MW tower cooled, coal-fired station may be up to 100 ha (excluding ash disposal areas). The station buildings will take only a portion of the site. The remainder depends on the needs of the coal store and railway sidings.

A 1200 MW nuclear power station will require 16–20 ha for operational purposes.

A considerable area will be required during the construction of both coal-fired and nuclear power stations. Typically 28–34 ha would be required to provide adequate working and storage areas for the contractors and for the construction car and bus parks. In addition, storage space will be required for topsoil removed during excavations (the area required would depend on the particular site) and for excavated material required for backfill.

Most of the temporary construction areas for a coal-fired station could probably be accommodated on the coal store area, although some extra land may also be required. Extra land would be required temporarily for nuclear stations. Therefore a suitable site for a coal-fired station would require about 100 ha and for a nuclear station about 60 ha. Some further off-site land may also be required to provide areas for planting or landscaping to screen the station.

Figure 1.6 shows the typical land requirements for a pressurised water reactor (PWR) station location next to an existing nuclear station.

2.3.3 Cooling water

The total cooling water (CW) required depends on the ultimate station capacity planned. Typically for a coal-fired station a 900 MW turbine requires a main CW flow rate of about 24 m^3/s. For a PWR station a 600 MW turbine requires about 23 m^3/s. Allowing for

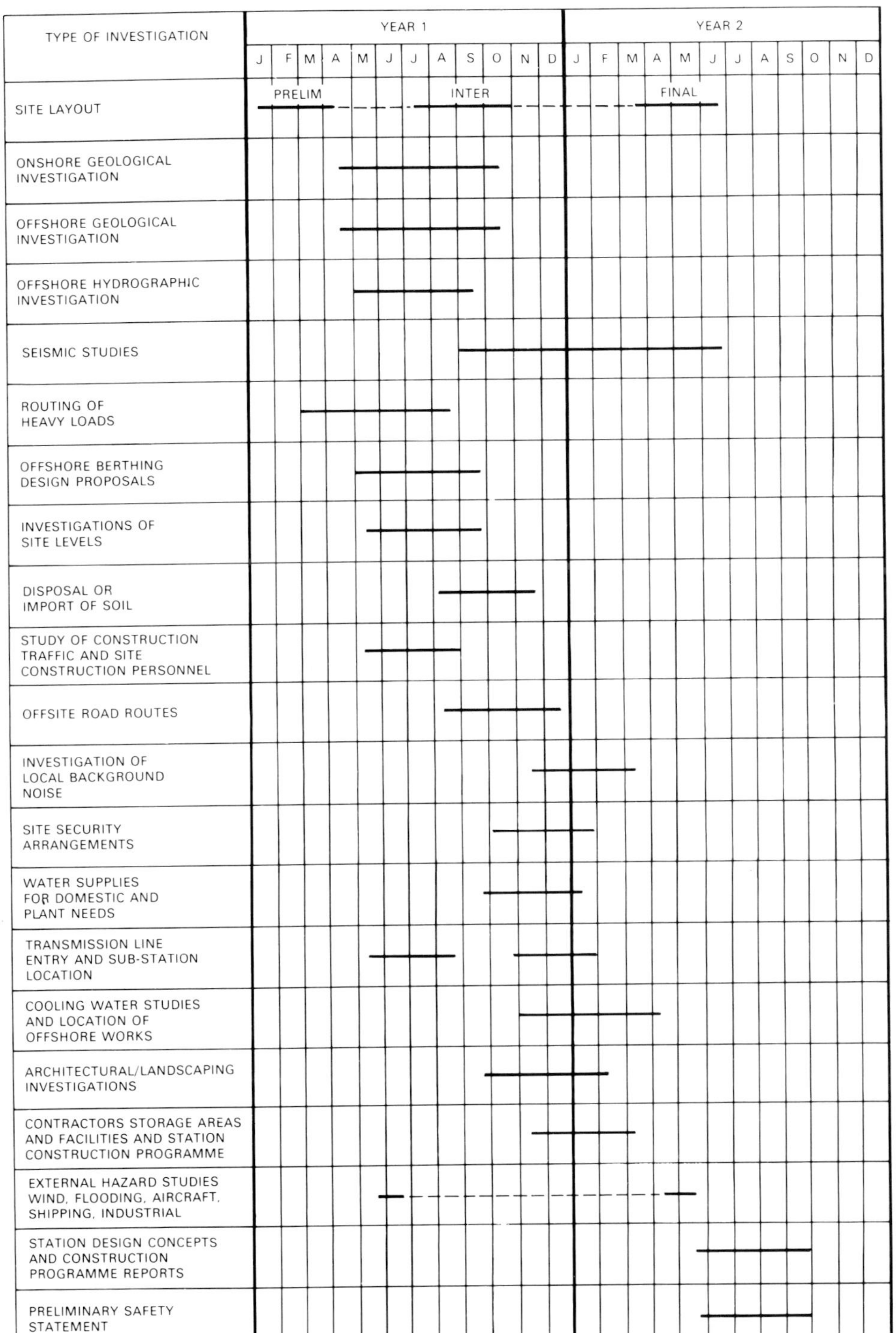

FIG. 1.5 Typical site investigation programme for a nuclear power station

KEY

		AREA (HECTARES)
1	RECLAIMED AREA FRONTING TEMPORARY WORKS AREAS	2.2
2	RECLAIMED AREA FRONTING NEW STATION SITE	1.9
3	NEW STATION AREA - EXISTING OWNERSHIP	15.9
4	NEW STATION AREA - OUTSIDE EXISTING OWNERSHIP	1.0
5	LANDSCAPING WITHIN EXISTING OWNERSHIP	3.9
6	ACCESS ROUNDABOUT SERVING NEW AND EXISTING STATIONS	0.6
7	SITE OFFICES / PERMANENT CAR PARK	2.6
8	TEMPORARY CONTRACTORS' AREAS	27.3
9	COASTAL STRIP TO BE PROTECTED	2.7
10	SITE HOSTEL	11.4
11	CONSTRUCTION CAR PARK	6.4
12	SITE ACCESS ROAD AND LANDSCAPED AREA	10.9
13	LANDSCAPING SUB-STATION AREA	1.5
14	CABLE RESERVE AREA TO NEW SUBSTATION	0.9
15	WOODLAND TO BE PROTECTED	1.0
16	EXISTING NUCLEAR STATION	37.3

EXISTING CEGB OWNERSHIP BOUNDARY
PROPOSED NEW BOUNDARY
AREA REQUIRED FOR NEW STATION

AREAS 1, 8, 10 AND 11 WOULD BE RE-INSTATED AFTER COMPLETION OF CONSTRUCTION AND POSSIBLY RETURNED TO AGRICULTURAL USE

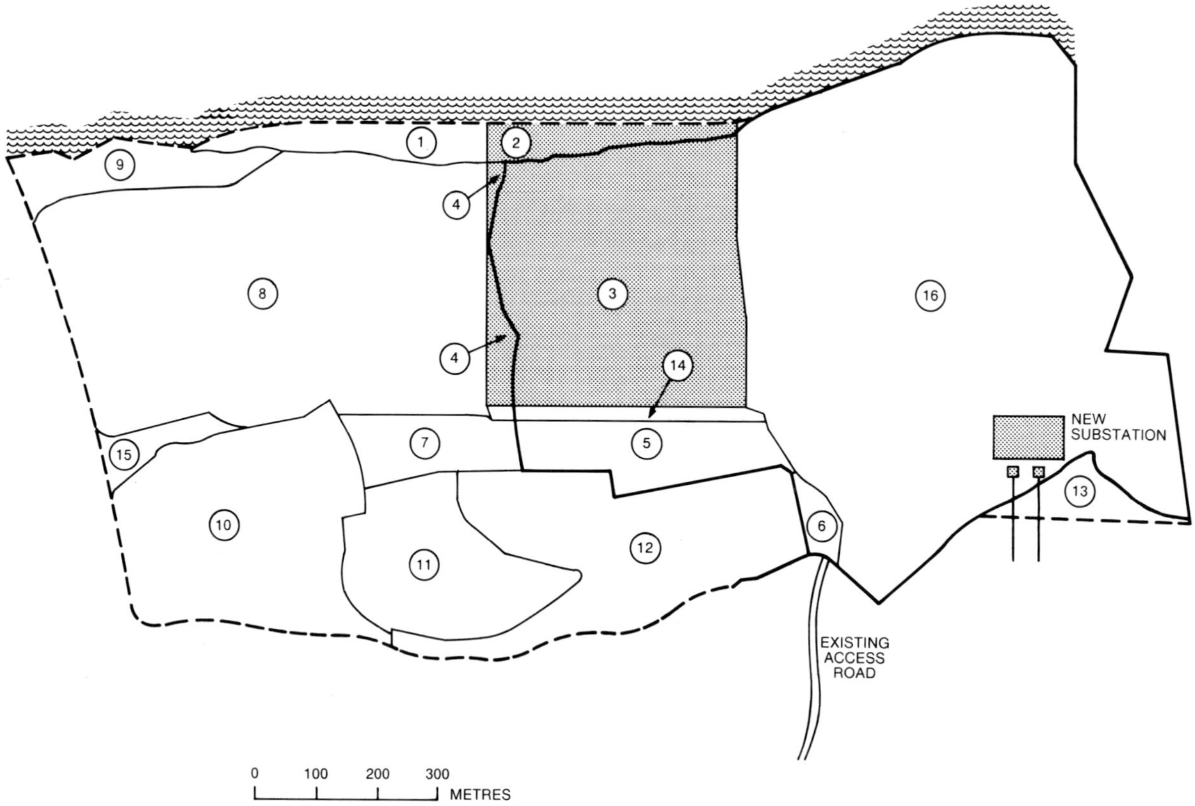

FIG. 1.6 Typical land requirements for a PWR station next to an existing nuclear station

other cooling water requirements this means that an 1800 MW coal-fired station would require about 52 m^3/s and a 1200 MW PWR station about 50 m^3/s.

As the cooling water flows through the condenser tubes, its temperature is raised and this could typically vary between 8°C and 12°C. This warmed water must then be cooled using cooling towers or, in the case of direct cooled stations, by discharge to the water source and be dispersed in such a manner as to minimise its recirculation back into the cooling water intake with attendant loss of steam cycle efficiency.

The use of cooling towers requires that a suitable make-up water supply be identified which would typically amount to 2% to 3% of the total cooling water flow. Whilst the actual flow would be influenced by the site-specific water quality aspects, it is usual for about two-thirds of this abstraction to be returned to the water source as purge to maintain the concentration factor within the cooling system. This water would be about 10°C warmer than the ambient water temperature.

If such a water supply is to be obtained from a river, then studies are required to identify minimum flows and the consequences of the water abstraction and return on the environmental well-being of the river system. In the UK, water authorities often hold long term records of water flows and details of licensed abstractions. A seasonally variable 'minimum prescribed flow' is often applied to rivers, which prohibits abstractions if the actual river flow falls to the specified level.

The preferred location for a power station from the cooling water viewpoint, is near a large river, estuary or sea coast to obtain the large volume flows at lowest temperatures. One of the key problems facing the cooling water system designer is therefore to provide the optimum location and separation between the cooling water intake point and the outfall. Another important requirement is to design a system which has the minimum effect on marine ecology. In this connection it is necessary to ensure that warm water is adequately dispersed to avoid harmful effects on marine life. The acquisition of information on currents and water temperatures over a large area is necessary for these cooling water studies.

With a direct cooled system abstracting from and discharging to the sea or estuary, the eventual loss of heat to the atmosphere is a lengthy process, and in the intervening period the dispersion of warm water discharged from a station outfall can be identified in a number of separate stages. The first, or near-field stage, is represented by the immediate mixing of newly discharged warmed water into the ambient sea.

After a brief transitory period, a second or midfield stage is represented by a buoyant plume of warm water lifting towards the surface and spreading outwards at a rate determined by gravity currents, momentum effects and the action of the tidal stream. A midfield plume can eventually reach several hundred metres in width and can extend in length for 1 km to 2 km in the direction of the tidal stream (see Fig 1.7).

The normal practice is to minimise recirculation by physical separation of the intake and outfall structures. Civil tunnelling costs may limit the degree of protection that can be afforded by this practice, but additional protection can be sought by designing the intake structure to minimise drawdown from an overhead plume, to ensure maximum possible depth of water over the period of coverage, and to minimise the period of coverage.

CEGB surveys have identified a third stage in the heat dispersion process at a number of sites. During periods of calm weather conditions it has been observed that sequential flood and ebb movements of the midfield plume alongshore over a period of several

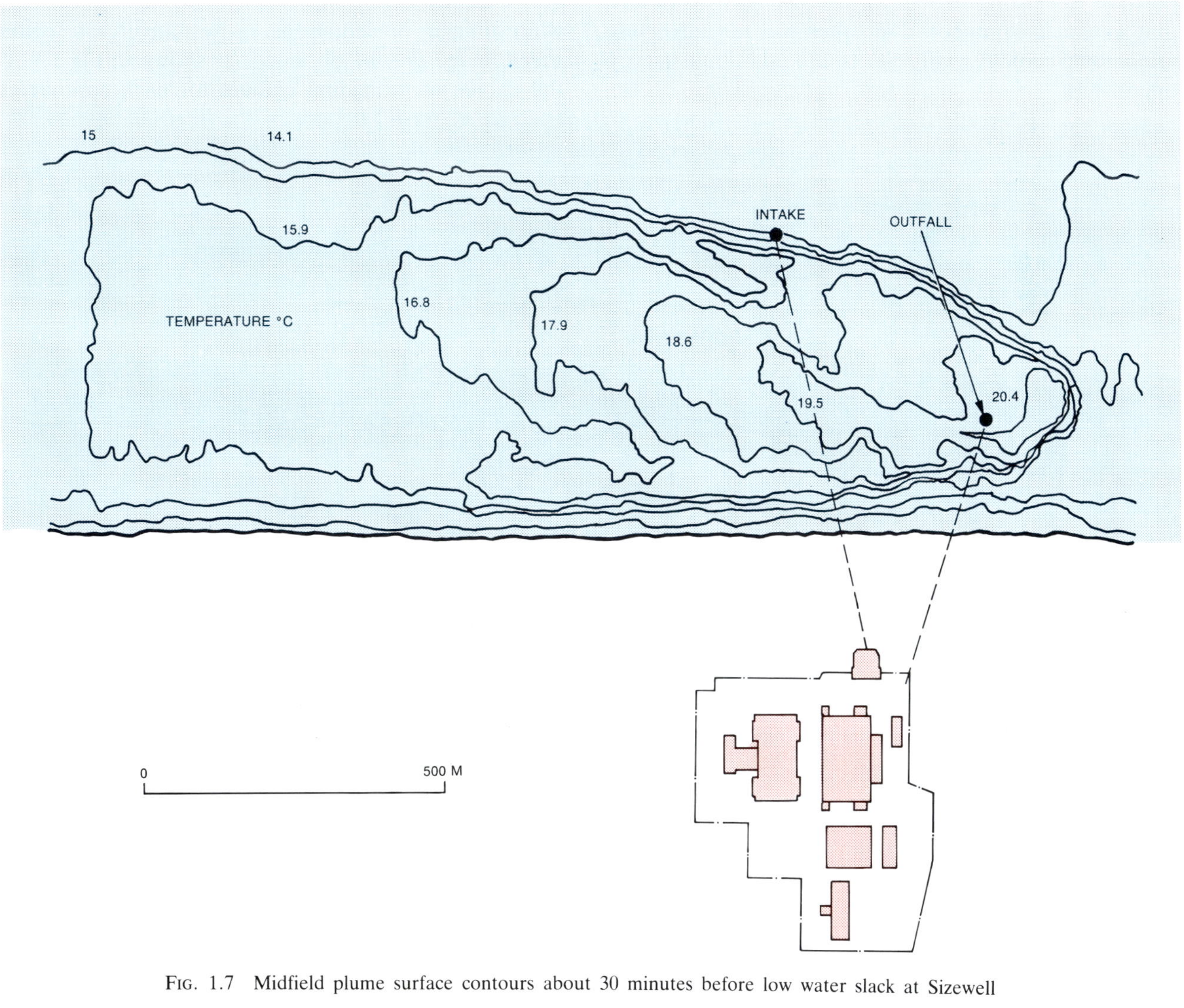

FIG. 1.7 Midfield plume surface contours about 30 minutes before low water slack at Sizewell

days can develop a far larger pool of warm water; Fig 1.8 shows this condition at Sizewell on the Suffolk coast. It can be seen that the spatial spread extended a full tidal excursion alongshore and several kilometres offshore. This far field plume is also moved alongshore by the reversing tidal stream and an amount of secondary recirculation cannot be avoided, in the example shown, as the cost of separating the intake and outfall structures exceeds the recirculation penalty. It has been found that a far field plume is quickly dissipated with increasing wind strengths.

It is important that survey operations should be conducted over a long period to ensure that the eventual design of the cooling water offshore works is founded upon a data base that sufficiently represents the variable meteorological and tidal current conditions local to the site. It is equally important that the survey period should include the calmer and warmer conditions of the summer months when the natural and artificial thermal fields are most likely to reach a combined maximum temperature.

Some hydrographical information will be available for proposed sites near to an existing power station. However, the increase in size of new developments, the need to place the new offshore works in correct juxtaposition to existing structures, and the need to ensure that eventual combined discharges will not adversely affect local ecology, will still require additional survey operations.

A survey will comprise an array of moored instrumentation to record continuous data of flow patterns and water temperature changes throughout the survey period, and a number of individual operations generally limited in time to a single tidal excursion. The moored array can include current meters, tide gauges and thermistor stringers which, together with an onshore automatic meteorological station, provide an overall record of data to improve understanding of the results from individual survey operations. These individual operations can include float tracking, infra red photography from helicopters or satellites, dye release, thermal plume profiling, and temperature/current/salinity profiling. Alongside these activities, which are mainly designed to assist in evaluating thermal plume behaviour, the survey will contain the necessary echo sounding, side scan sonar, seismic work, seabed sediment sampling and wave recording to supply the information required by civil engineers for designing the station structures. The data is also used for assessing the movement of materials on the sea bed and beach under the influence of winds and tides.

Thermal images of offshore coastal waters or estuaries can be obtained using infrared cameras on satellites. Contours of temperature may be marked by different grey tones for each temperature band step on black and white image presentations. Alternatively, a *colour sliced* image may be obtained, as shown in Fig 1.9, by the choice of individual colours for each contour.

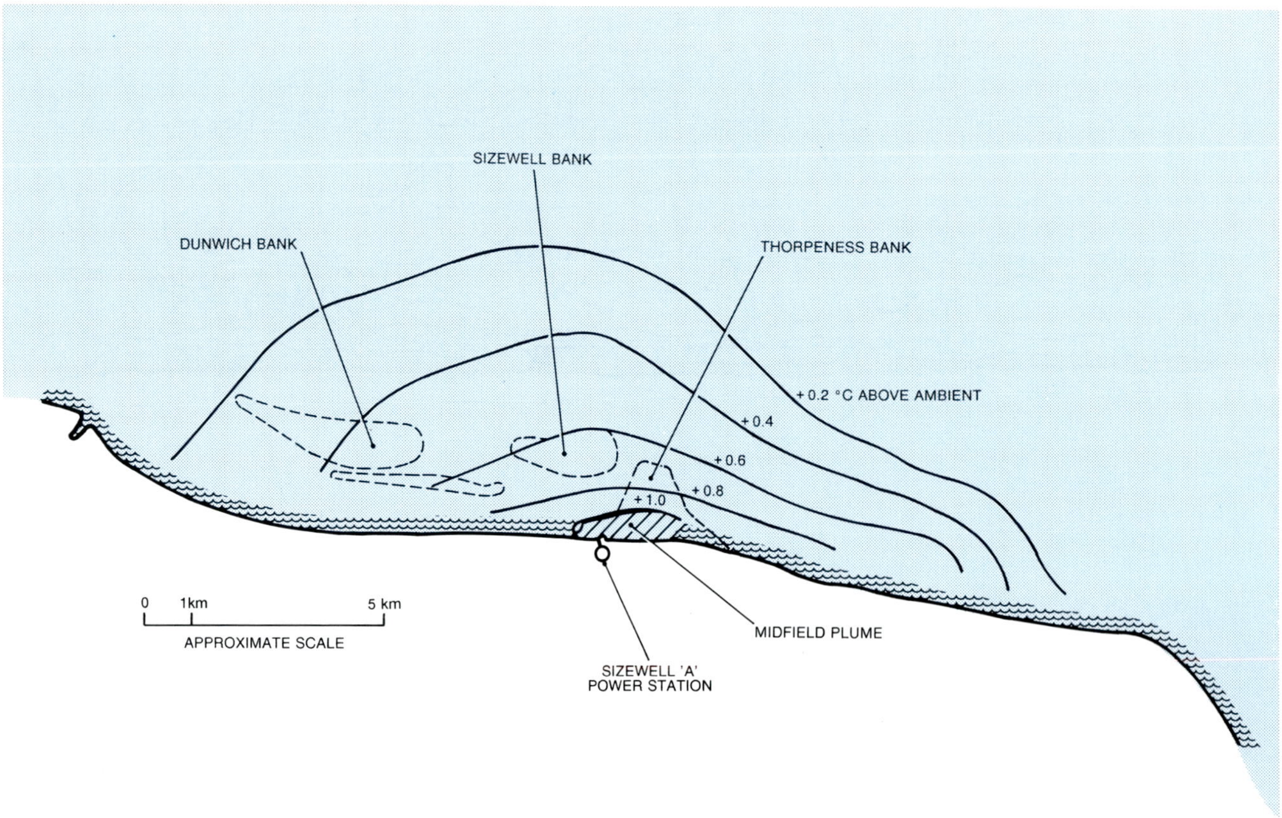

FIG. 1.8 Summer far field at end of flood tide

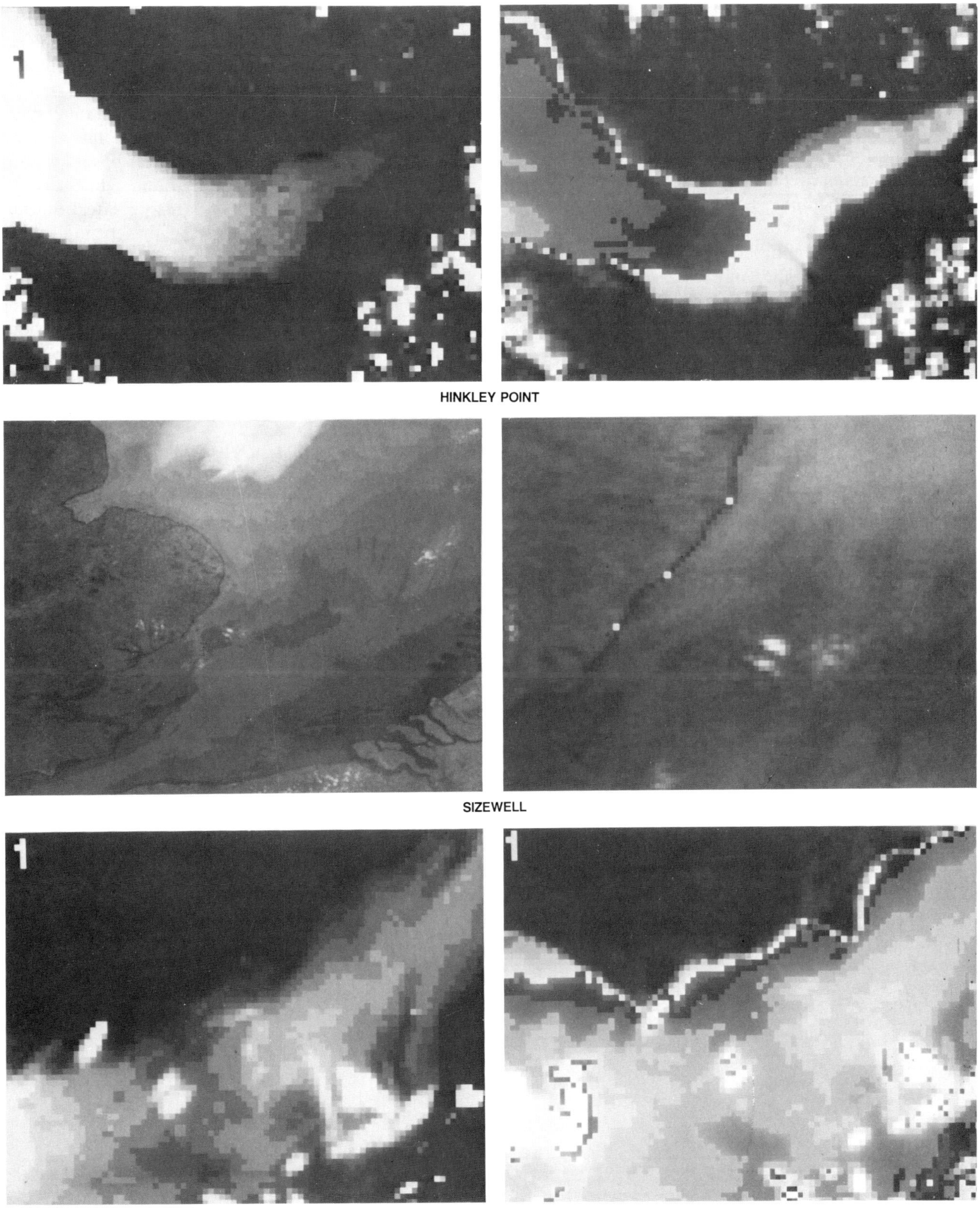

FIG. 1.9 Satellite imagery of thermal discharges from power stations
(see also colour photograph between pp 66 and pp 67)

Improvements in survey operations and measuring equipments have been paralleled by development of mathematical modelling techniques. Experience however has clearly demonstrated the complex problems involved in both modelling the separate temperature fields that make up the thermal structure of a body of water, and of estimating the relative contributions of these temperature fields at different sites and under differing tidal and meteorological conditions. Consequently, despite some good correlation between survey results and model predictions, the CEGB continues to regard the hydrographic survey as the primary tool in present investigations. In particular, it provides a validation source for the flow predictions of a mathematical model and the only satisfactory means of identifying the natural temperature fields. Hydraulic and mathematical models in which conditions such as tides, water and silt movement can be reproduced, are often used when planning and designing the cooling water works and jetties for power stations to ensure that they will not be adversely affected under these conditions.

Figure 1.10 shows a typical prediction of current circulation patterns for a coastal power station location.

The cooling water studies described here, which may take several months, need to be carried out before the final selection of a site can be made.

2.3.4 Transmission

A route must be available for the transmission lines from the proposed power station site to a suitable point on the 400 kV supergrid system or major load point on the Area Board system which can accept the power station's output. Increasing opposition from the public, amenity societies and planners to overhead lines makes routes increasingly difficult to obtain, and sometimes the only solution is to put sections of the line underground. There are tremendous financial penalties and engineering difficulties for underground cabling, however, and so their use can have a major effect on the selection of sites when considering the economics.

Studies of the pattern of power flows which will occur after the commissioning of the proposed station are carried out to discover if any reinforcement of the grid system (such as uprating or addition of transmission lines) will be needed.

2.3.5 Geology

Modern power stations, both coal-fired and nuclear impose very heavy loading on the subsoil which must be able to support it with suitably designed foundations.

The general nature of the soil can usually be obtained from records and maps of the geological strata. Before making a final decision on a site, however, a detailed

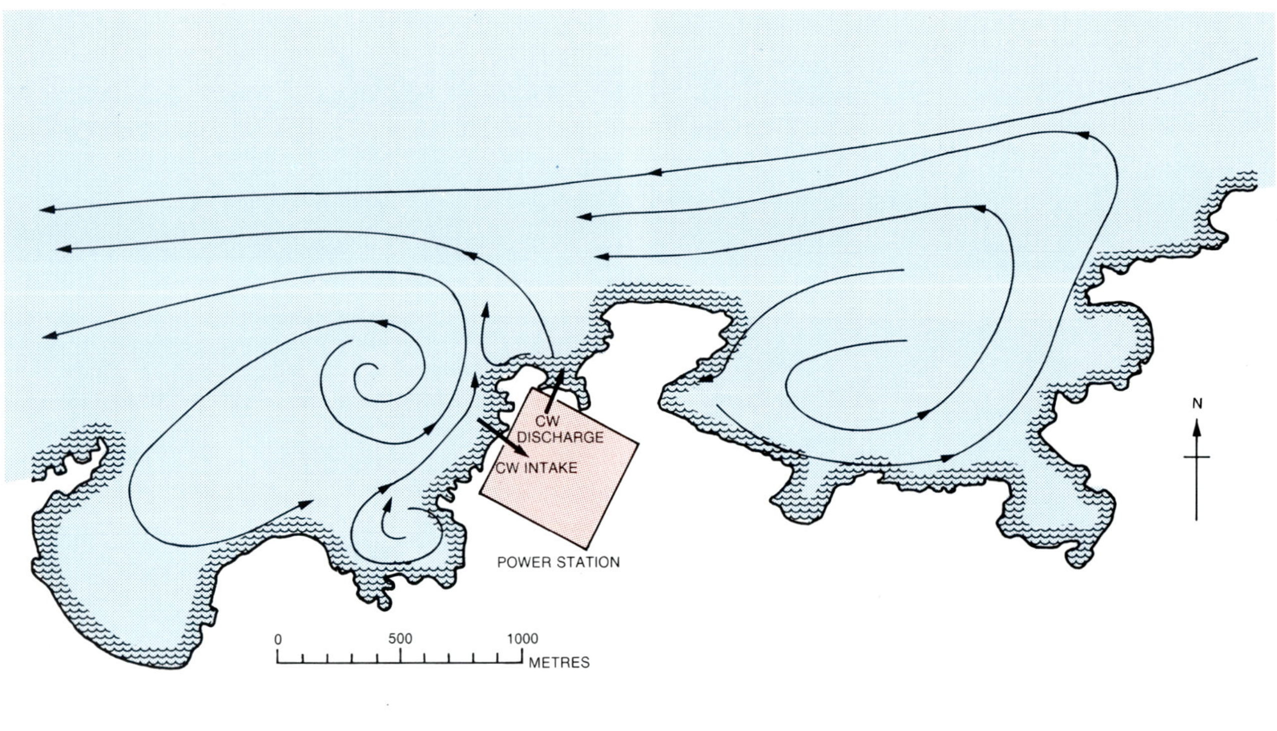

By analysing the results from float tracking over a tidal cycle and from longer term monitoring with moored current meters it is possible to predict the current circulation pattern at different states of the tide. This diagram shows the situation one hour after high tide on a spring tide. Similar complex patterns occur at other stages of the tide.

To minimise recirculation of the warm water discharge without having to discharge into the strong currents 1km offshore which would incur significant costs it is possible to discharge into the small bay to the north of the station. Discharge into the bay to the west would obviously cause problems with recirculation.

FIG. 1.10 Typical prediction of current circulation patterns survey of the subsoil conditions must be carried out to determine the ability to carry the loads; the costs of suitable foundations, which can vary widely, can then be estimated. Where the proposed cooling water works require tunnels to and from the sea or rivers (usually a major item) subsoil investigations are necessary. These could include permeability tests and groundwater tests to enable the feasibility and cost of building the tunnels to be determined.

The subsoil investigation usually requires a number of exploratory boreholes to be sunk, some perhaps over 120 m deep, and trenches dug to expose geological features. Samples are tested both in situ and at soil mechanics laboratories to determine the thickness, strength and other physical properties of the strata under the site. More detailed information about the subsoil conditions between the boreholes can be obtained by the use of seismic reflection techniques. This involves vibrating the ground with either hand-held or vehicle-mounted machines. The resulting shock waves reflect off the various layers of strata and are picked up by instruments set out on the ground. By moving the vibration source along a line and measuring the time taken for the waves to reach the detectors, a fairly accurate picture of the strata along that line can be obtained. In order to build up an understanding of the geology of the site and surrounding area a grid of such lines is traversed.

In industrial areas it is essential to know the previous uses of the land; old foundations, mine workings, or similar features must be located.

Having established the general geology of the area, further boreholes are then sunk to cover the whole site in detail before design and station layout decisions are made (see Section 3.2 of this chapter).

2.3.6 Site and station levels

A site should be reasonably level, not liable to flooding and not so high above the source of cooling water that excessive pumping power is required to supply water for cooling purposes.

A site requiring extensive filling to bring the level above the general flood level, or excessive excavation for the purposes of siting buildings, greatly increases the site preparation costs. However, in order to provide for the major capacities now required, such additional cost can often be justified. For base-load stations, where subsoil conditions are suitable, it can be economic to excavate deeper for the turbine house foundations to save on pumping power.

2.3.7 Access

Access to a power station is required for construction materials and plant, fuel supplies and employees.

Good road access is essential for construction, and rail and sea facilities are useful advantages. Direct access to a main trunk road to bring in heavy loads is desirable.

Modern transformers and generator stators for large units can weigh up to 350 tonnes and, therefore, impose point loads unacceptable to many local authorities, through whose areas the heavy transporters pass on their way to the power station from the manufacturer's works. The CEGB has to bear the cost of any bridge strengthening necessitated by the heavy loads.

A design of trailer with the facility to spread its load when negotiating bridges or unsuitable load bearing roads was therefore an attractive idea. A hovercraft principle of floating the trailer on a cushion of air was adopted for transport over those roads where a weight restriction was in force. This vehicle, which is now in regular operation, is known as *air cushion equipment*. It relieves individual axle loadings on the road and saves the CEGB a great deal of expense reconstructing and strengthening weak links along the route (see Fig 1.11(a)).

An alternative to transporting a load the whole distance by road is to have it shipped from the port nearest to the place of manufacture to the point nearest to the site which special shallow-draught, end-loading vessels can reach. The CEGB owns some specially designed ships of this type known as *roll-on* and *roll-off* vessels. These have a capacity of 1000 tonnes and enable a 600 tonnes single load, inclusive of vehicles, to be shipped (see Fig 1.11(b)). Very large components can also be shipped by large sea-going barges.

The construction of a heavy load berth suitable for barges and roll-on/roll-off vessels adjacent to the site allows very large prefabricated components to be delivered, thereby reducing the amount of on-site fabrication. Road traffic can also be reduced by delivering aggregate by sea.

While a power station is being built, traffic is greatly increased and so local roads adjacent to the site are often reconstructed and re-routed to avoid undue inconvenience or risk to other road users.

The overriding consideration for a conventional power station in full operation is access for its fuel supplies. The site therefore, must be conveniently situated either close to a main railway line to accept rail-borne fuel or, in areas remote from the coalfields or refinery, on an estuary or the sea coast to enable it to take its fuel from colliers or tankers.

For nuclear stations there must be a rail link on or near the station for transport of the flasks of irradiated fuel.

2.3.8 Water supplies for make-up and domestic purposes

Town mains water is used for all services where water has to be fit for domestic purposes, or where clean water is required for make-up or control purposes. However, it is the CEGB's policy to increase the security of supply by providing, where possible, an alternative source. This could be by suitable treatment of seawater or from a river, lake or borehole.

(a) AIR CUSHION TRANSPORTER

(b) ROLL-ON / ROLL-OFF VESSEL

FIG. 1.11 Transport of heavy or abnormal loads
(see also colour photograph between pp 66 and pp 67)

Where water for fire fighting is to be taken from the town mains, allowance should be made either to duplicate the supply or to provide adequate storage capacity to ensure 100% availability; this is most important during the commissioning period of a boiler when the demands on the supply are heavy. Where the Water Authority permits, arrangements should be made for a bypass round the meter so that in cases of extreme emergency, a full flow is available for fire fighting.

If a heavy, uneven demand is anticipated which is likely to exceed the maximum capacity of the water main, suitable storage capacity should be provided to give continuity of supply and to reduce the rate of draw-off from the main.

During the construction period, water consumption depends on the size of the labour force, the nature of the civil engineering works, e.g., aggregate washing, water jetting of piles, concreting, etc., and plant testing. For example, the demand for water during the construction of a 1200 MW nuclear station would usually be about 1100 m^3 per day but could rise to three times this amount during commissioning. The maximum consumption during operation would be about 2200 m^3 per day.

2.3.9 Ash and dust disposal

The CEGB is the major consumer of the lower grades of UK coals which are being produced with an increasing ash content.

During the early years of operation an 1800 MW station, when operating on base load, can produce nearly a million tonnes of ash each year. If this were consolidated it would cover about 12 ha of land to a depth of 10 m in one year. When selecting a site for a coal-fired power station, very careful consideration must therefore be given to the provision of suitable economic ash disposal, either on low-lying ground or worked out mineral workings which can be filled by the creation of landscaped hills, or by the sale of pulverised fuel ash (PFA) to the construction industry.

Figure 1.12 shows ash disposal at the reclamation site which takes ash and dust from the Drax power station. Here, a hill is being constructed of PFA which is progressively top-soiled, grassed and landscaped. Alternative schemes involve reclaiming worked-out gravel pits which can then be used for agriculture or for amenity purposes, including fishing and sailing.

2.3.10 Flue gas desulphurisation byproducts

An 1800 MW coal-fired power station fitted with the limestone-gypsum type of flue gas desulphurisation (FGD) plant will produce about 500 000 tonnes of gypsum per year. This gypsum could be used to make building materials such as wallboard. The amount produced would be sufficient to supply an average size wallboard factory, and it is therefore possible that such a facility could be built adjacent to the site, otherwise the gypsum would be removed from site for use elsewhere or disposal.

The byproducts produced by regenerative processes would have to be transported to a suitable chemical works for use. It is difficult and undesirable to store large quantities of products such as sulphuric acid or elemental sulphur and therefore a regenerative process would only be used where a reliable market for such products existed.

2.3.11 Detailed investigations related to nuclear safety

When considering the suitability of a site for a nuclear power station, additional safety-related aspects need to be studied. These can be divided roughly into the following four categories, the first three of which are potential sources of hazard to the station, and the fourth involves the safety of the public in the event of an incident:

(a) *Earthquakes (seismicity)* The geology of the site and the surrounding area is investigated to find out the local faulting pattern.

Historic research is carried out in order to estimate the location and size of any earthquakes that may have occurred in the general area in the past. For some sites it may also be necessary to place sensitive detectors at various places in the locality of the site to monitor the occurrence of very small seismic events. All this information can be put together to assess the probability of earthquakes of various sizes at the site. Nuclear stations are designed to safely withstand a certain size of earthquake. However, it is necessary to satisfy the Nuclear Installations Inspectorate (NII) that the combination of design and siting is such that the risk of an uncontrolled release of radiation as the result of an earthquake is as low as reasonably acceptable (ALARA) that is, less than once chance in ten million years.

(b) *Other natural hazards* Studies are carried out into the potential hazard from other natural sources such as extreme weather conditions or flood. In this country it is unlikely that extreme weather conditions would present an appreciable hazard, however, the positioning of nuclear power stations on the coast means that the possibility of flooding must be carefully investigated. Not only must extreme sea levels, caused by a combination of tide and wind be evaluated, but the possible erosion of the coast must also be taken into account. From this work the need to raise the level of the site and/or provide coastal defence works can be assessed. It must be borne in mind, however, that raising of the site not only increases cooling water pumping costs but requires the delivery to site of large amounts of fill material.

FIG. 1.12 Ash disposal site
(see also colour photograph between pp 66 and pp 67)

(c) *Industrial hazards* The investigation of the potential 'man-made' hazards generally covers two areas; aircraft crash and transport and storage of hazardous materials. The study starts with a broad survey of potential hazard sources in the area such as industrial installations; rail, road, sea or pipeline routes along which toxic or flammable materials may be transported; airports and flight paths; and military training areas. This review identifies areas of possible concern that need more detailed study, for instance the passage of large tankers of liquid petroleum gas (LPG) past the site.

An assessment of the risk from these various sources can then be carried out to check that it meets the probability criteria. If it does not, design modifications have to be considered. The layout of a nuclear station is such that materials stored on site do not present a hazard. However, where an additional station is planned for an existing site, it is necessary to check that the two stations will not become a source of hazard to each other.

It should also be noted that when considering an area for a coal-fired station, a careful study must be made of airfields (both commercial and military) with their associated flight paths that could be affected by a future power station. Regulations to safeguard aircraft impose stringent restrictions on the height to which the buildings and chimneys and cooling towers can be built. At modern power stations, the heights of boiler houses often exceed 60 m, cooling towers are now 115 m and taller ones (up to 165 m) are currently being considered; chimneys can be up to 240 m in height.

(d) *Population distribution* Information from the national censuses (carried out every ten years), from Local Authorities and from actual house counts is used to determine the distribution of population up to about 30 km from a potential nuclear site. Criteria have been established which place limits on the number of people resident within various distances of the site. Although these provide a set of guidelines against which to measure the acceptability of the site, they are not hard and fast rules. A site will only be licensed by the NII if it is satisfied that the people living near to the site can be evacuated sufficiently quickly in the event of an incident and that the site is not too near to any large concentrations of population. Once a nuclear station has been established, restrictions on development are applied close to the site to ensure that the local population does not increase to an unacceptably high level.

2.4 Environmental considerations

There were two main statutory duties imposed upon the CEGB when it was formed in 1958:

- To develop and maintain an efficient, co-ordinated and economical system of electricity supply.
- To take into account any effect which its proposals would have on the natural beauty of the countryside and on flora, fauna, features, buildings and objects of special interest.

The implications of this dual responsibility are far reaching. The CEGB has to reach a balanced judgement for each development, one that seems the most appropriate in all the circumstances.

Planning a power station involves making an assessment, at a point in time, of many technical, economic and social factors. The CEGB has to draw up a balance sheet of costs and benefits which will range over the life span of the station. In a coldly impartial assessment, these factors would be quantified and reduced to present day values for comparison with price tags placed on the natural resources employed, including all aspects of amenity. But as no-one has developed a way of costing such qualities as natural beauty and tranquility, subjective judgements have to be made in drawing up plans. The CEGB has evolved methods of seeking and proving sites which make the best use of experience and new techniques; this includes a full input from national and local bodies, and there are opportunities for the general public to contribute their views. Following these initial consultations, the statutory planning and consent procedures to which proposals are subjected ensure that they are adequately tested against the opinions, both of experts and of the general public.

For convenience, the environmental considerations may be grouped together as ecological, amenity and socioeconomic effects. In practice there are no definitive divisions, measures taken to reduce one effect can have repercussions on another. For example, the installation of cooling towers would reduce temperature rises in the adjacent waters and changes in aquatic life, but at the cost of a visual intrusion.

2.4.1 Ecological effects

The ecological effects can be defined as those factors which affect flora, fauna and human health. They are due to the physical occupation of land for the power station and for its construction, and to discharges from the site during both construction and operation.

As far as possible, the highest grade agricultural land, nature reserves or sites of special scientific interest would be avoided when selecting power station sites. However, experience has shown that providing the CEGB has taken the correct protective measures, power stations do not have serious effects on the use of adjacent land. For instance, farming can continue up to the station boundary, and the number of birds recorded in the vicinity of some power stations has actually increased over the years.

There are statutory controls over environmental pollution matters such as emission of solid particulates discharged through the chimneys of fossil-fuelled power stations, gaseous and liquid radioactive discharges from nuclear power stations, the temperature of cooling water discharges, and noise. There is routine monitoring of these parameters and regular consultation with the controlling authorities (Government Departments, Statutory Bodies and Local Authorities) to discuss the regulations and any proposed changes to them.

One example of a recent change involves recognition of a possible cause of *acid rain* requiring the installation of flue gas desulphurisation plant on new coal-fired stations and retrofitting on some existing stations, and also the introduction of low nitrogen oxide (NO_x) burners.

Mathematical models have been developed for assessing the ground level concentrations of suspended particulate matter and gaseous emissions (e.g., SO_2, NO_x) derived from specified chimney heights under varying meteorological conditions. Such models allow the estimation of the chimney height required to satisfy stipulated regulatory guidelines for a given station generating load pattern. The models have been validated against the results from field studies of the atmospheric dispersion of emitted chimney plumes. Wind tunnel facilities are also used for assessing the chimney plume dispersion problems which may be induced by neighbouring buildings or local topography.

Figure 1.13 shows diagrammatically the stages of chimney plume dispersion.

The effects of cooling water returned to the sea with increased temperature up to about 10°C above ambient and low residual chlorine, have been examined extensively. No discernible harmful effects have been found, indeed rather the opposite. It is a common feature of cooling water outfalls to find fish attracted to the warm turbulent water. Their presence attracts a wide range of seabirds which in turn attract ornithologists. At one station several million oysters are now flourishing in the warm water area around the outfall.

The effects of the cooling water abstraction are likely to be limited to the immediate vicinity of the intakes. The amount of fish drawn in to suitably designed and located intakes has been found by experience to be not of commercial importance and only in estuaries where salmon are running has there been any justification for

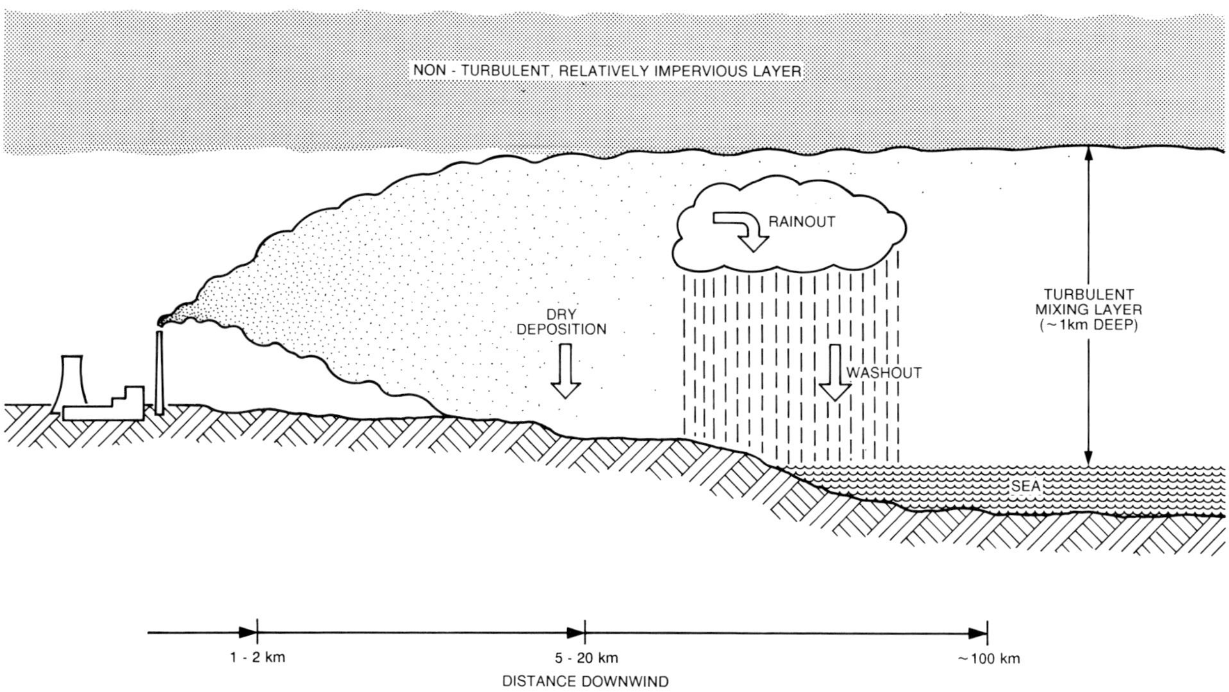

The gaseous oxides of sulphur and nitrogen in emissions from chimneys are dispersed by the wind. At whatever height they are emitted, from tall chimney stacks or from ground-level sources, they become well mixed with the air after a few tens of kilometres. This is because of natural turbulence up to a kilometre or two above the ground. The atmosphere above that level is relatively impervious to emissions.

The gases may be lost by dry deposition as a gas at land and sea surfaces, or by wet deposition where the gases may be dissolved in cloud droplets during the rain-forming process (rain-out) or be scavenged from the lower atmosphere by falling rain (wash-out). Solutions of sulphuric and nitric acids in rain will, of course, increase its acidity.

Ground level concentrations reach a maximum where the plume first touches the ground, usually within 5 - 20 km from the chimney and thereafter they progressively decrease. Mathematical models are used to predict these concentrations under varying meteorological conditions with different chimney heights.

FIG. 1.13 Stages of chimney plume dispersion

any special measures. There has in fact been some success with mechanical means of recovering fish from the intake screens and returning them to the water source.

The location and design of headworks for both intake and outfall has to pay attention to fishery interests and the risk of local erosion or accretion or other changes to the sea bed. Experience has shown that headworks extending above the water level have only limited advantages and pose some structural maintenance problems, as well as being noticeable features. Therefore for future stations it is anticipated that the intake and outfall headworks will be below low water level.

The ecological effects caused by the construction of a power station can be reduced subsequently by restoring land which was used temporarily for construction storage areas. The CEGB has also set up nature parks and trails at some of its power stations. Figure 1.14 shows the layout and gives details of the Trawsfynydd nuclear power station nature trail in Snowdonia.

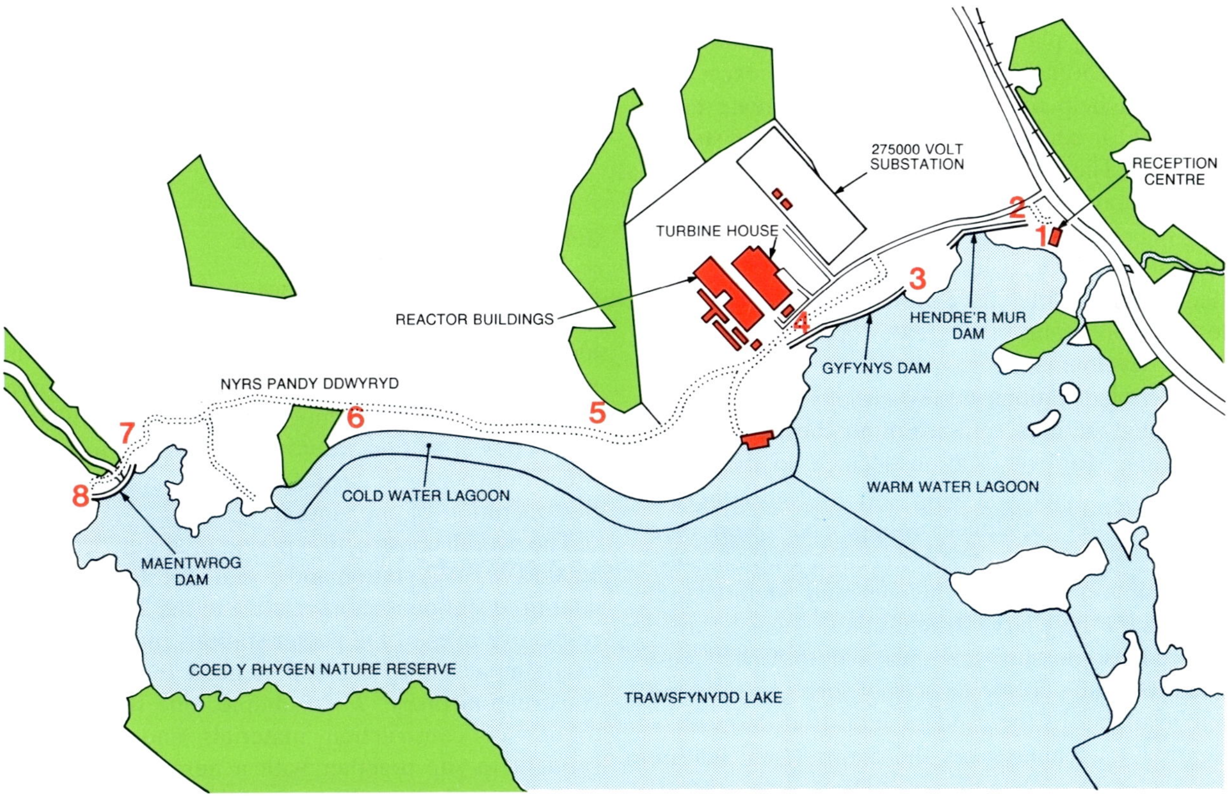

1. Reception Centre amongst oak trees: views over the Llyn Trawsfynydd, a man-made reservoir which provides cooling water for the power station and water for Maentwrog hydro-electric power station.

2 & 3. Hendre'r Mur and Gyfynys dams: great care was taken to ensure that their architecture harmonised with the station's surroundings and many different species of tree were planted to give variety of form and colour to the grounds of the power station.

4. Rainbow trout hatchery: the trout reared in the warm water discharged from the station are used to stock fisheries in the reservoir.

5. The bund wall forming the cold water lagoon and the islands within the warm water lagoon provide nesting areas for birds such as herring gulls, lesser black-backed gulls, great black-backed gulls, black headed gulls, terns and common sandpipers.

6. Nyrs Pandy Ddwyryd is a stand of beech and Scots pine where sparrow hawks, spotted flycatchers, green woodpeckers, willow warblers, carrion crows and buzzards have been recorded. There is an important oak wood in Coed y Rhygen Nature Reserve.

7. Maentwrog dam provides views down the ravine which contains many species of plants, mammals and birds such as mosses and ferns, redstarts, warblers, dippers and barn owls.

8. Panoramic views of countryside in the Snowdonia National Park.

FIG. 1.14 Nature trail at Trawsfynydd power station in Snowdonia National Park

2.4.2 Amenity considerations

Amenity can be defined as *the conditions in the landscape that the public as a whole wishes to see and enjoy*. It should be the aim of amenity planning to preserve or create these conditions. This is a daunting task, power stations and transmission lines will always be seen from somewhere; most people dislike changes to the landscape to which they have become accustomed, even if it is not of great value.

Buildings over 60 m high are impossible to hide and so become new features in the landscape, but imaginative architectural treatment and careful grouping of building and structures can help make what is seen, positive and acceptable. The preservation of existing woodlands, new planting (both on and off-site) and sometimes creating mounds from surplus excavated material can help to both conserve and create landscapes, and to reduce the areas from which the station can be seen. The CEGB commissions leading architectural and landscaping consultants to advise on these issues.

Nuclear stations have less visual intrusion than fossil-fuelled stations as they have no tall chimneys, no large scale fuel stores, and offer greater opportunities for a compact development.

Direct cooled stations are less visually intrusive than tower cooled stations, and are often cheaper to build and operate. However, stations located on rivers and upper estuaries require cooling towers because there is insufficient water for direct cooling. The towers then represent a major visual feature because of their bulk. It is possible to develop layouts and architectural treatments for these large structural forms which can be considered as pleasing (and has led to Civic Trust Awards) by using all the devices of form, texture and colour available to the modern industrial designer.

Figure 1.15 demonstrates the major visual feature presented by the cooling towers at Drax power station.

In order to reduce the visual bulk of the cooling towers it is possible to use assisted draught rather than natural draught towers. This can result in lower or fewer cooling towers while still providing the same cooling performance as the natural draught towers. However, the resulting improvement in visual appearance is not achieved without cost; there would be increased capital and running costs of the station mainly due to the provision and power consumption of the fans.

Figure 1.16 compares the relative sizes and quantities of the two types of cooling tower for a 2000 MW station.

The plume of water vapour that emerges from the cooling towers forms another problem of visual amenity that is more difficult to solve. There are no technical measures available which would significantly reduce the length of the plume. The length of the visible plume is determined by the prevailing meteorological conditions; the plume is short and wispy on warm sunny days, and long and dense on days of high humidity. It is fortunate that the latter condition usually coincides with dull overcast weather so that the plume tends to be lost from view against a cloudy background. The plume from an assisted draught tower would be denser than that from a natural draught tower. Dry cooling schemes would have no plumes; some such schemes have no towers, but they have the potential disadvantages of additional capital and operating costs and noise nuisance.

The CEGB also considers the effect of noise both from the power station and construction site as well as from local roads. Where appropriate, measures are taken to reduce noise nuisance, for example by providing double glazing for affected houses.

The CEGB avoids valued features such as buildings of historic or architectural importance and conservation areas. Their existence can be established during the site investigations.

2.4.3 Socioeconomic effects

More interest has taken place in recent years into the effects of large scale industrial developments on the local people. Studies are now carried out on the nature, magnitude and direction of the various socioeconomic effects on local communities which the construction of a new power station would have. These effects would occur in two stages; during the construction and during the operation of the station.

The overall construction period of about 7 years for a 1200 MW PWR station and 5 years for a twin 900 MW coal-fired station is short relative to the 30 and 40 years respective lives of the power stations, but the sharpest local impacts may be experienced during the early construction phase. For example, one to two million tonnes of construction materials and fill may be brought to site together with a number of large and heavy loads. In the UK, most of the materials and plant required will be moved by road.

Other traffic will arise from the movement of workers to and from the site. It is therefore necessary to arrange with the Highway Authorities to improve roads, and on occasions build new roads to bypass urban areas at an early stage; usually with substantial financial contributions from the CEGB.

The total employment for a 1200 MW nuclear station is about 40 000 man years, of which about a quarter will be on site. While a substantial part of the site labour force may be within daily travel distance, there will be others who will move into the area. Studies are carried out to establish the availability and suitability of local skilled and unskilled labour. Predictions can then be made of the likely number required to be brought into the area, accommodation required and so on. Figure 1.17 shows the projected labour force for construction of a PWR power station.

Both the numbers of the labour force coming in and the local impacts from their arrival are difficult to

FIG. 1.15 Cooling towers at Drax power station
(see also colour photograph between pp 66 and pp 67)

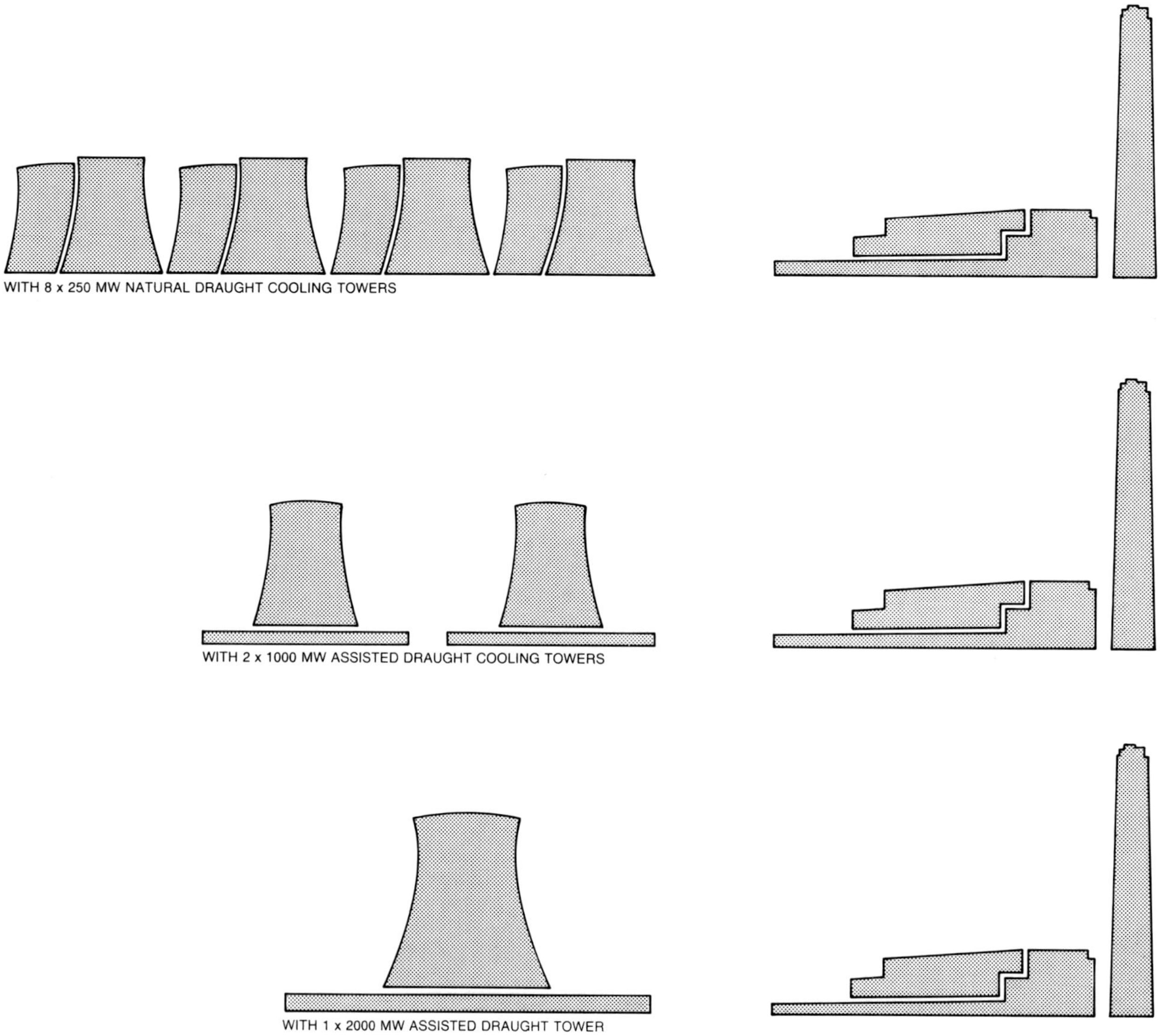

FIG. 1.16 Comparison of cooling tower sizes

quantify. Experience shows wide variations from site to site depending on local conditions, on contract procedures, on personal preferences, and on whether a site hostel and caravan parks are provided. In planning to meet possible demands for accommodation and other services, including education for children, it is essential to maintain close contacts with the Local Authorities and to provide for staged development so that provision can be matched to demand. Lodgings and rented accommodation are commonly found within daily travelling distance of sites, but contractors have brought labour in daily from distances up to 50 km. Permanent power station staff tend to live close to the station but this depends on the relative availability of accommodation.

Other facilities provided during construction, such as improved potable water supplies, 11 kV or 33 kV electricity lines, and in at least two cases, new churches, may remain for the use of local people, or they are removed if not wanted.

The long term effects of an operating station concern several areas; housing demands, travel patterns, expenditure distribution and effects on local employers.

The expenditure effects can be considered as money spent by the power station on goods and services, the rates paid by the power station and wages spent by employees. Revenue expenditure at power stations on goods and services, excluding fuel purchases, covers items ranging from plant spares to canteen food and carbon dioxide to telephone bills. Many items can only be purchased nationally or regionally and only a small percentage of the total can be purchased locally. Nevertheless, even these small expenditures can be significant in the local economy and some local businesses can benefit. There can also be some stimulus to the private house market as many power station staff buy their homes, but these are distributed over a wide area.

The pattern of expenditure by power station staff in rural communities may be such that only a proportion of this income remains within the region, partly because more people have their own transport and do much of their buying at major shopping centres some distance away.

While local labour may be recruited it would usually be only a proportion of the total number employed at the station. Although some local employers would lose employees, studies have shown they would generally be spread over a range of businesses and would probably not cause any local employer to curb his business activities. Some pressure may be brought on wage levels in rural areas; this may be seen as a good thing by the employees concerned.

When applications for stations have been submitted, fears have been expressed about the effects of the development on tourism and recreation. There is no evidence that this has a harmful effect, rather power stations of all types attract visitors. As far as public recreation is concerned the CEGB has taken steps to preserve existing facilities, and where possible to enhance them. Examples are: the beach and grass covered sand dunes at Sizewell for public recreation; the creation at various stations of nature reserves, field study centres, bird sanctuaries; the licensing of fishing facilities and the establishment of fish farms for re-stocking purposes, as distinct from the commercial fish farms making use of warmed cooling water discharges.

2.4.4 Communication with local people

From the outset of site investigations the CEGB maintains a policy of open communication. The proposed development is initially discussed with the relevant Statutory Authorities and other organisations with specific interests, to determine local conditions and matters for investigation. During consent application these consultations are repeated in greater detail. Figure 1.18 shows the authorities with whom the CEGB consults on the local impacts of power stations.

In addition, public announcements are made of the intention to carry out investigations and later to apply for consents. This publicity with the information available, enables both individual members of the public, local societies and national bodies such as the Council for the Protection of Rural England and the Royal Society for the Protection of Birds to contribute

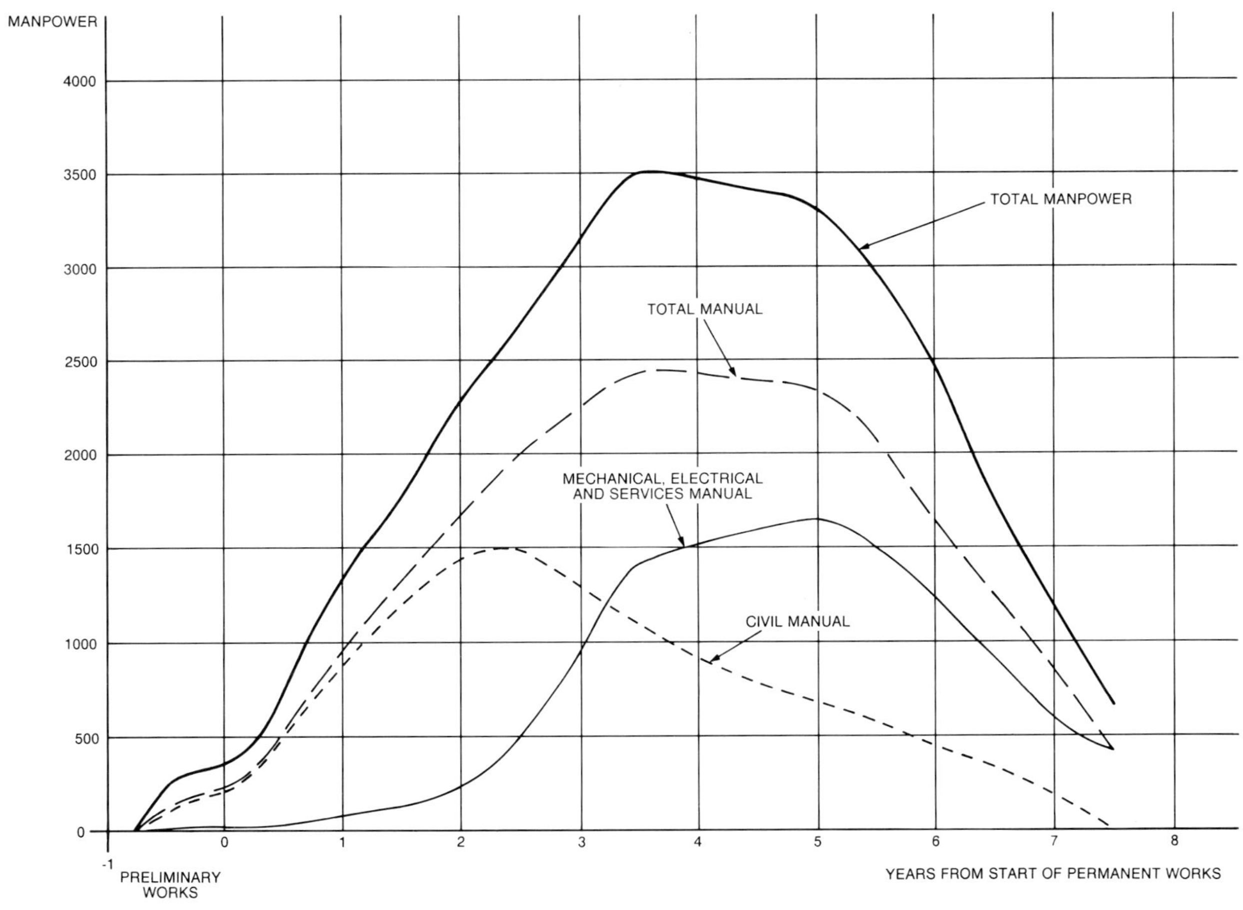

FIG. 1.17 Projected construction labour force at a typical PWR station

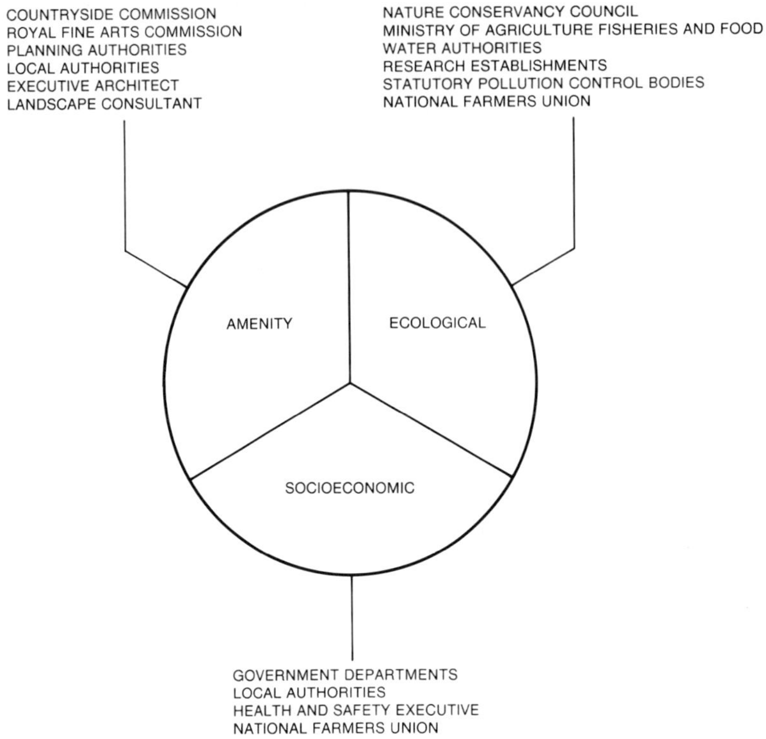

FIG. 1.18 Consultations on local environmental impacts

and comment if they wish to the CEGB and/or to the Planning Authorities. The CEGB has in fact undertaken to produce formal environmental statements for all future power station proposals. If it is necessary, a public inquiry will be held. The CEGB provides evidence and witnesses and the case is given the most critical scrutiny by the objectors, as shown by the inquiry for Sizewell *B*. Local meetings are held during the planning, construction and operation stages.

2.4.5 Assessment of environmental effects

In order to predict the changes to local conditions brought about by a power station, it is necessary to find out what the existing situation is; to identify those factors which are sensitive and to predict how they will be affected.

The CEGB employs its own technical and environmental specialists as well as having planners and designers who are practised in seeking better solutions based on information fed back from monitoring the effects of previous stations.

The CEGB has worked for more than 25 years on fundamental and applied research into environmental impacts, particularly into the ecological factors. The wide range of problems set by a diversity of sites with different environmental conditions, has resulted in the accumulation of knowledge which can be used to estimate, with some precision, the effects of further power stations on existing or new sites.

Equally, the CEGB is alert to new techniques or methods that may be adopted, and to relevant work being done elsewhere in the world. Recent examples are the use of mathematical models for predicting cooling water dispersion and the efforts to set up satisfactory ecological models, the use of computer graphics for the presentation of hydrographic information and in visual intrusion studies. Satellites have been of only limited use to date in giving information on geological faults when considering seismic conditions. However, they might provide more information in future on such matters as the surface patterns of warmed water. Spray cooling as an alternative to cooling towers has been tried in the USA but experience suggests that this is unlikely to be applicable in this country.

Experience of considering environmental impacts and consultation with the Statutory Authorities helps in the early recognition of where significant effects may occur and enables resources to be concentrated on studying them. Experience also guides efforts to improve and to create better conditions; but perhaps nowhere is experience more useful than in the very difficult task of forming a balanced judgement on what combination of various environmental impacts would be acceptable. Some can be estimated with precision, such as the loss of agricultural land. Others such as noise levels, can be scientifically predicted; but people's perception of noise nuisance differs. Factors like visual intrusion are largely subjective and even more difficult to balance. There is no easy way to consider all the different environmental impacts together to determine what measures are justified to reduce them, or to rank the various proposed power station sites. The search for better aids to judgements is essential but unending.

Figure 1.19 shows the Dinorwig pumped storage power station built deep inside a mountain in Snowdonia and which is the largest in Europe. It represents a supreme example of a generating station in harmony with the environmental and ecological needs of a national park.

2.5 Site selection

After the sort of investigation described, some sites may prove to be unsuitable on technical or environmental grounds and will therefore be rejected. Sites found to be acceptable but not already owned by the CEGB may be purchased for either immediate or later development. However, if the site is likely to remain available for use, then the CEGB may decide to delay purchase until development is required.

When detailed investigations have been completed, the overall cost of constructing and operating a particular size and type of station on the site is calculated. Alternative sites can then be evaluated and a recommendation made to the CEGB Executive as to the site or sites which give the best balance of technical and economic considerations on the one hand and impact on the environment on the other. Sites which have been found to be suitable for development but which are not required immediately are placed in reserve as potential sites for later reconsideration.

3 Site layout — thermal power stations

3.1 General

It is not possible to say that there is a recognisable site layout pattern which is adopted at most stations, but like the design of the station itself, the problem of site layout is to find an optimum arrangement which results in minimum overall cost consistent with ease of erection, ease of operation and good appearance within the limits of the available site (see Fig 1.20). There are always conflicting requirements to be reconciled and alteration of one feature often has repercussions on others so that every major aspect of the design must be considered in relation to the others. The main factors influencing site layout are discussed in the following paragraphs.

3.2 Foundations

When selecting a site the general foundation conditions are determined by sinking exploratory boreholes. The boreholes must then be extended to cover the whole site in detail before foundation design and site layout decisions are made. The tests determine the most economic location of the main buildings and indicate any geological features which might limit the area in which the power station can be built. All other things being equal, a situation would be chosen for the main power station buildings where foundation costs would be at their lowest, as long as a good layout is obtained.

It is also necessary to establish whether there are any geological faults that could interfere with the location of the power station. A fault is the result of some long previous disturbance of the earth's crust where there has been a vertical sliding movement; along the line of this the underlying strata at each side of the fault line will be found at two different levels. Where such a condition exists the station should be constructed clear of the fault line and not across it, as this could give rise to a subsequent differential settlement in the foundations. In the case of nuclear stations, the seismic safety requirements insist that the safety related buildings are at a minimum distance away from any fault. This is assessed for specific sites depending on the significance of the particular fault.

A piled foundation is usually adopted where the subsoil is inadequate to carry the required loading, and it is necessary to penetrate to some depth to reach a load bearing strata.

A raft foundation is adopted where ground conditions are suitable to carry the required loading. The raft, which can be of solid concrete or of cellular construction, that is with cavities or tunnels within the mass concrete, is used to distribute the point loadings equally over the ground.

Assuming that there is no marked preference because of cost, the cellular raft has the advantage in that some accommodation is available for running pipes and cables below basement level.

Also associated with the strength of the ground is the question of coal mining. Where underground coal mining is expected to be carried out in the vicinity of a power station during its lifetime, it is necessary to prevent subsidence of the main components on site by refraining from mining within a closely defined volume of ground below the power station. The shape of this volume is known as a *pillar of support* and varies according to the surface area which is to be supported and the geological characteristics of the ground. This shape is formed by extending outward sloping lines, below ground level, at approximately 35 degrees to the vertical from the extremities of the foundations on the pillar. The power station designer should aim to keep this area to a practical minimum. The CEGB has an agreement with British Coal to restrict this supported area and so reduce the quantity of any coal which becomes unobtainable. Since some compensation to British Coal is payable under the agreement, the area is usually limited to the main buildings and cooling towers (if any) and sometimes the switch compound, but not the coal store and rail sidings.

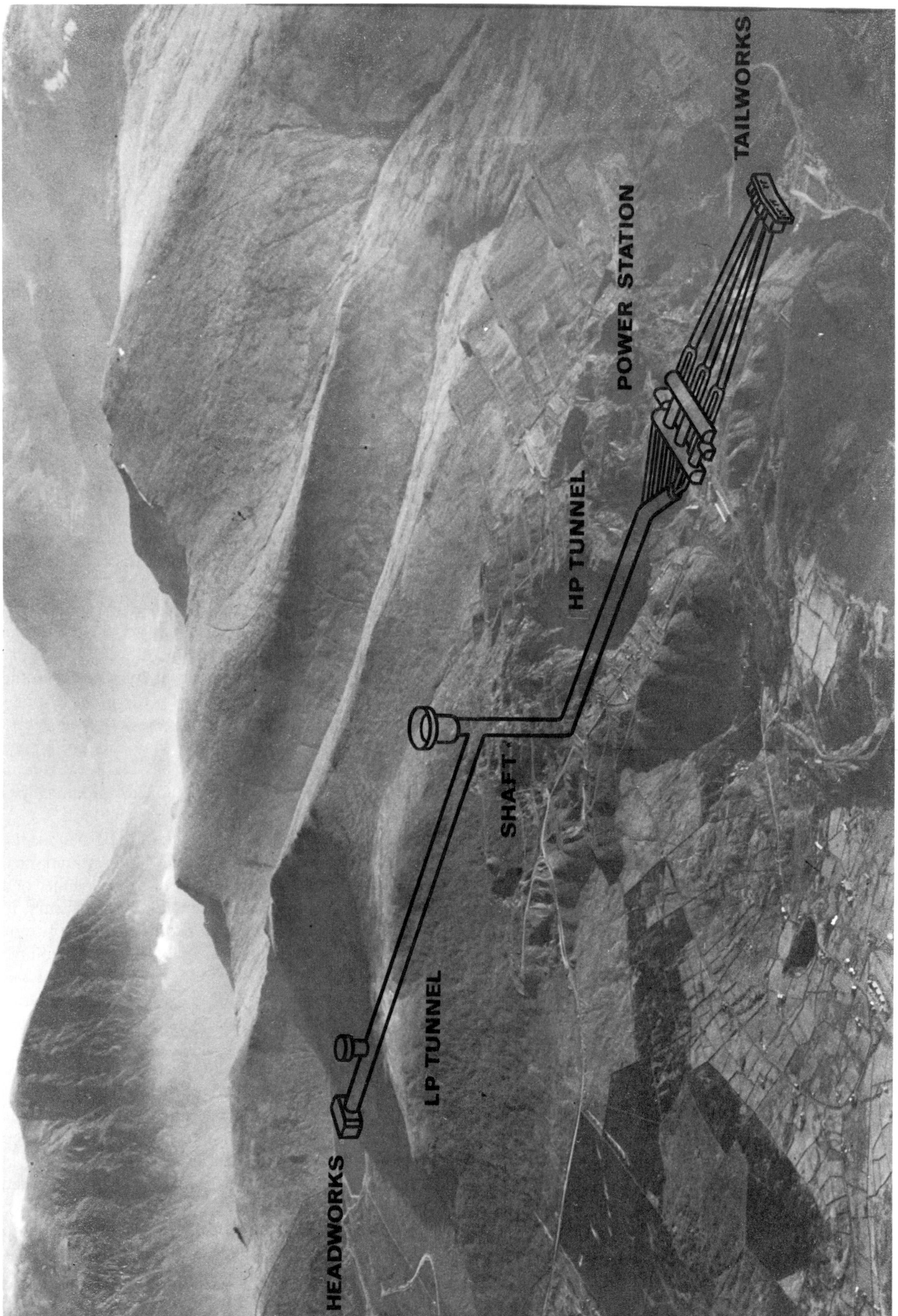

FIG. 1.19 Hidden power: Dinorwig pumped storage power station (see also colour photograph between pp 66 and pp 67)

FIG. 1.20 View of Drax power station nearing completion
(see also colour photograph between pp 66 and pp 67)

3.3 Site and station levels

The main factors governing the choice of power station and site levels are:

- The need to protect the power station against the risk of flooding.
- Capital cost of civil works.
- Cooling water pump running costs.
- Ease and speed of construction.

For economy and general convenience on the site, the power station basement, roads and rail siding are generally constructed at the existing ground level. This avoids the necessity of extensive excavation and removal of soil, or the importation of filling material. If a site is above the predicted maximum sea or river level, the necessity for protection against flooding does not arise. If the site is below flood level there are two main methods of providing protection. The surest way is to lift the power station basement and all other installations essential to the safe operation of the power station, above flood level. This method is adopted where the cost of the required filling material is cheap and cost of pumping cooling water is not materially increased.

Where the filling material and pumping costs are expensive, flood banks are relied upon along those sides of the site below flood level.

The level of a power station with a closed cooling tower system is not dependent on the absolute sea or river level and, in consequence, the choice of power station basement level is little influenced by pumping costs.

However, with direct cooled systems (see Section 3.8.1 of this chapter) the water has to be pumped from the adjacent sea or river; in these cases the height between the power station level and water level is very important. This water level is itself constantly changing because of tidal variations; the maximum pumping head necessary is associated with the lower water level.

Because of high pumping costs, advantage is taken at direct cooled power stations of the principle of siphonic action. Atmospheric pressure acting on the surface of the sea or river is sufficient to lift the water to a height of about 10 m; this means that the cooling water pump has to generate a head sufficient to overcome only culvert and condenser friction before entering the outfall system.

There are sites, however, with the power station basement at site level, where the top of the condenser would be considerably more than that equivalent to 10 m head of water above the lowest level from which the water had to be pumped. If this is the case, one of two alternative arrangements is usually adopted. Either the condenser is lowered relative to the site level to reduce the height, or it becomes necessary to construct a seal pit or weir to restrict the length of the siphon leg. The choice is usually made by carefully examining the advantages and costs of both schemes (see Fig 1.21).

3.4 Main buildings and orientation

New power stations within the CEGB are now planned from the outset to be completed with a given number of units all of the same rating and over the years the layout of major thermal stations has developed into a general standard pattern. The boiler house and turbine hall arranged side by side can be regarded as a central feature. Very often between these two a mechanical annexe accommodating auxiliary plant items is interposed. The building is usually completed by an auxiliary switch bay with generator transformers on the side of the turbine house and a bunker bay on the side of the boiler house. The other items required to complete the concept of this standard pattern are the electrostatic precipitators (on coal-fired plant), the induced draught fans, the flue gas desulphurisation (FGD) plant (where fitted) and the main gas flue and chimney. These features are illustrated in Figs 1.22 and 1.23.

It is currently proposed that future coal-fired stations should be based on the use of 2 × 900 MW(e) units, and the details of a *reference design* have now been established. Similar standardisation is also occurring on nuclear power station layouts, and it is intended, for example, that future PWR station buildings would be based on the station design used for Sizewell *B*. It can be seen therefore that it is intended that future site layouts for conventional and nuclear plants will be based on the use of reference design arrangements within the restrictions of site-related characteristics. As mentioned previously in Section 3.1 there are always conflicting requirements to be reconciled when deciding the best location and orientation for the main buildings. It is almost always impossible to satisfy every requirement perfectly. If a cooling water system is as short as possible, then the connection between the generator transformers and the switchgear may be increased in cost; if coal conveyors or fuel pipelines are reduced to minimum length then some other services may suffer, and so on. The emphasis which can be placed on each factor in terms of money, operational convenience and amenity value depends on the type of power station, the site and its environment.

Within these restrictions the orientation of the main station complex is chosen to give the best compromise of the following factors:

- Making the most economic use of the ground conditions by placing the heavier conventional boiler or nuclear reactor loads on the best ground.
- Minimising generator transformer cable and transmission routes.
- Minimising pressurised cooling water culvert and outlet culvert routes.
- Minimising fuel supply and ash and dust removal routes.
- The adequacy and convenience of construction areas.
- The ability to place the administration buildings in a reasonable environment and close to the main access to the site.
- Meeting the requirements of the appointed Architect with respect to the important off-site views of the complex and its relation to other developments, particularly any existing power station on the site.

Figure 1.24 shows the layout of a proposed coal-fired reference design station.

3.5 Ancillary buildings

Ancillary buildings can be broadly categorised into two groups, those that are directly related to the main plant

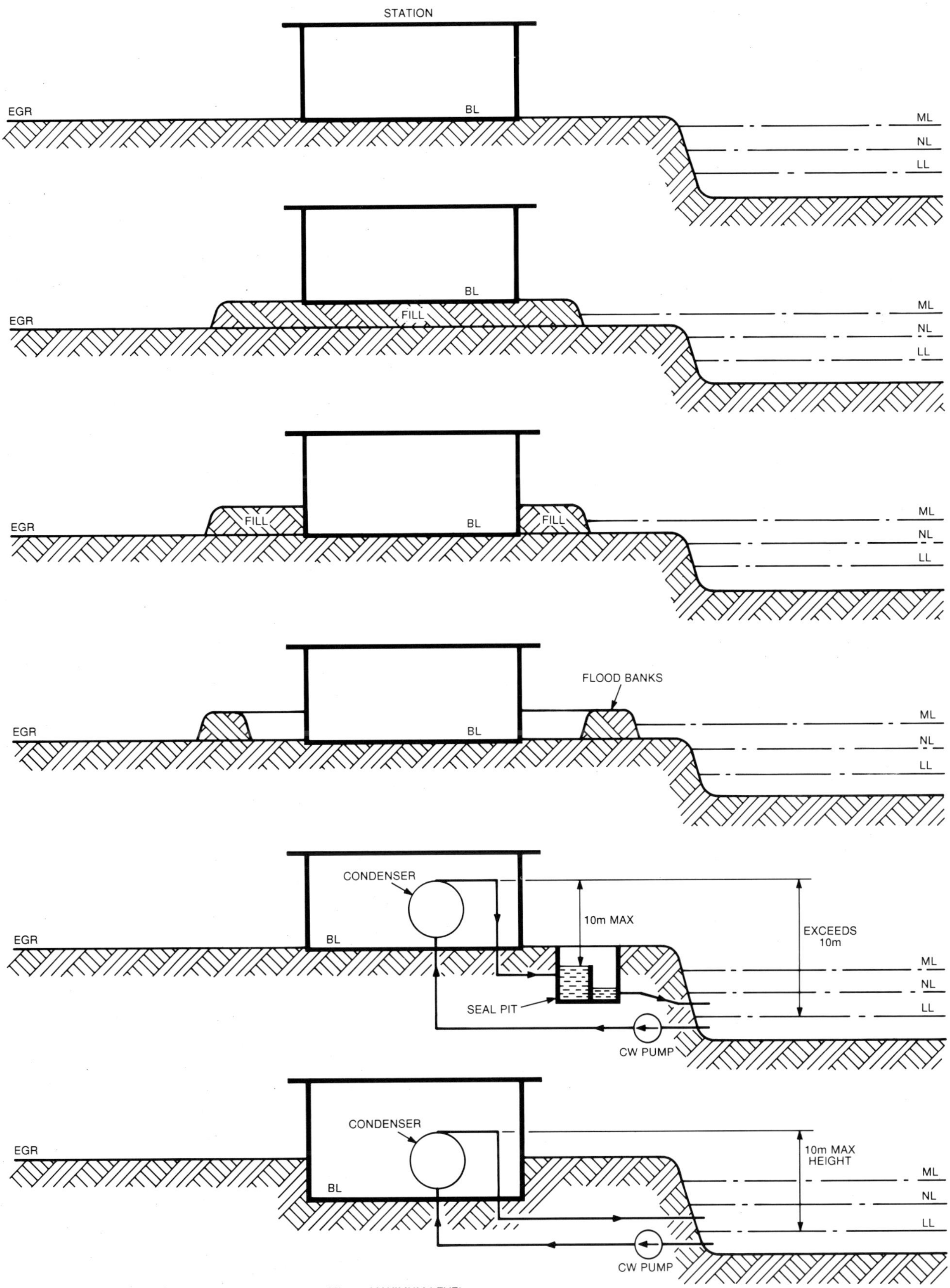

ML - MAXIMUM LEVEL
NL - NORMAL LEVEL
LL - LOW LEVEL
EGR - EXISTING GROUND LEVEL
BL - BASEMENT LEVEL

FIG. 1.21 Site and station levels

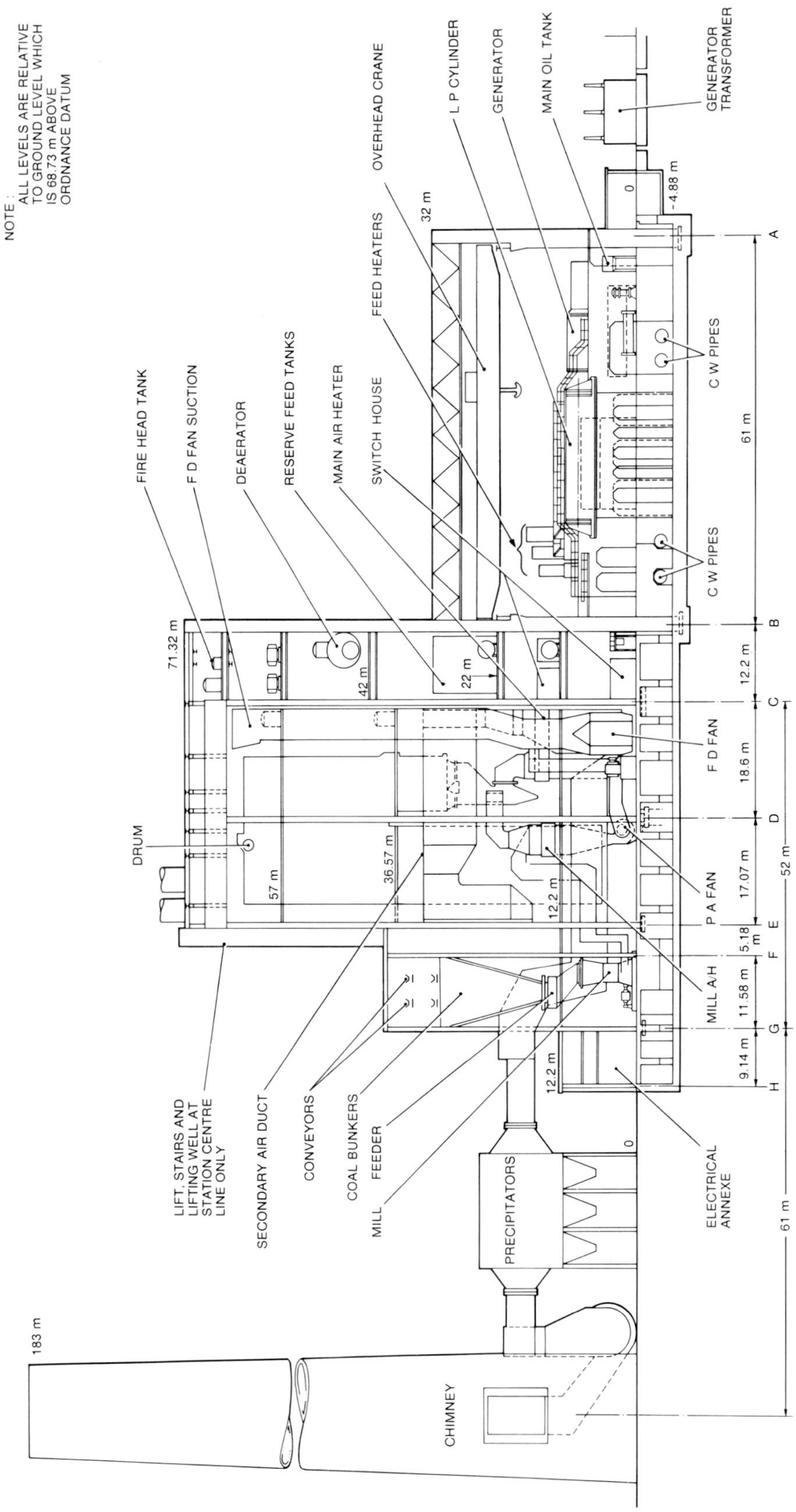

FIG. 1.22 Section through 500 MW unit coal-fired power station

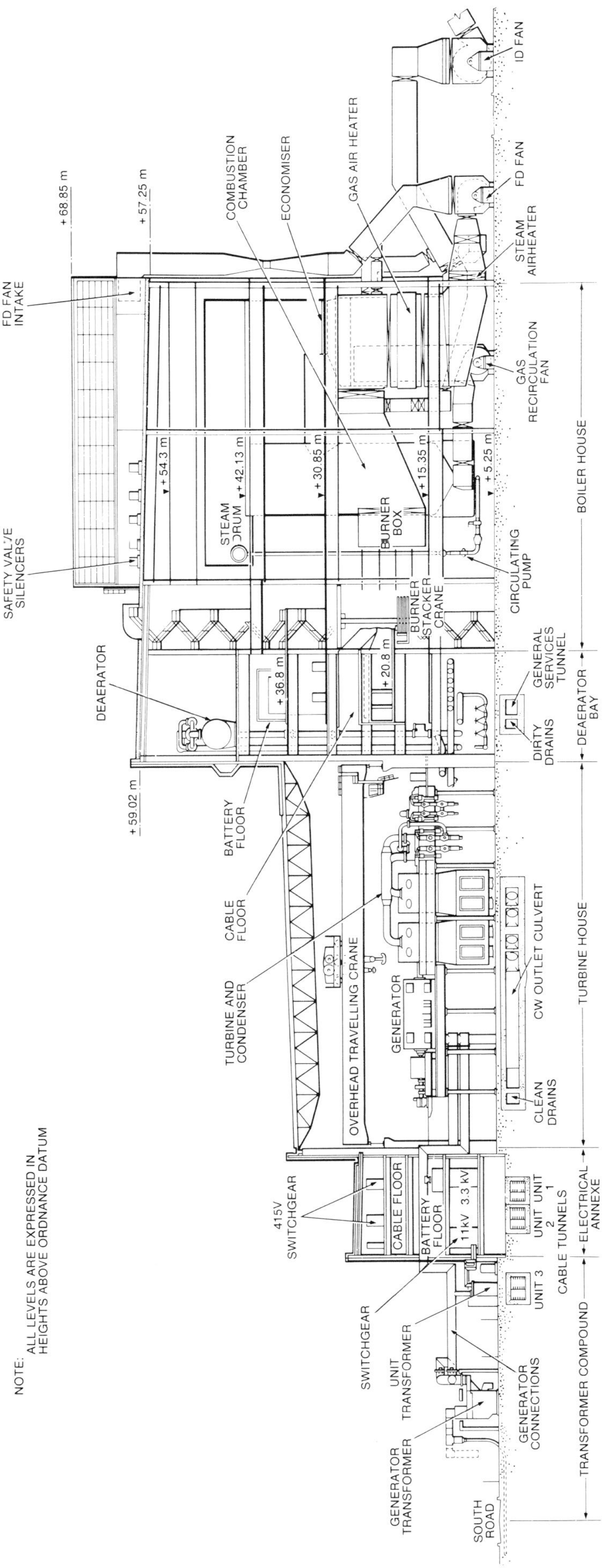

FIG. 1.23 Section through 660 MW unit oil-fired power station

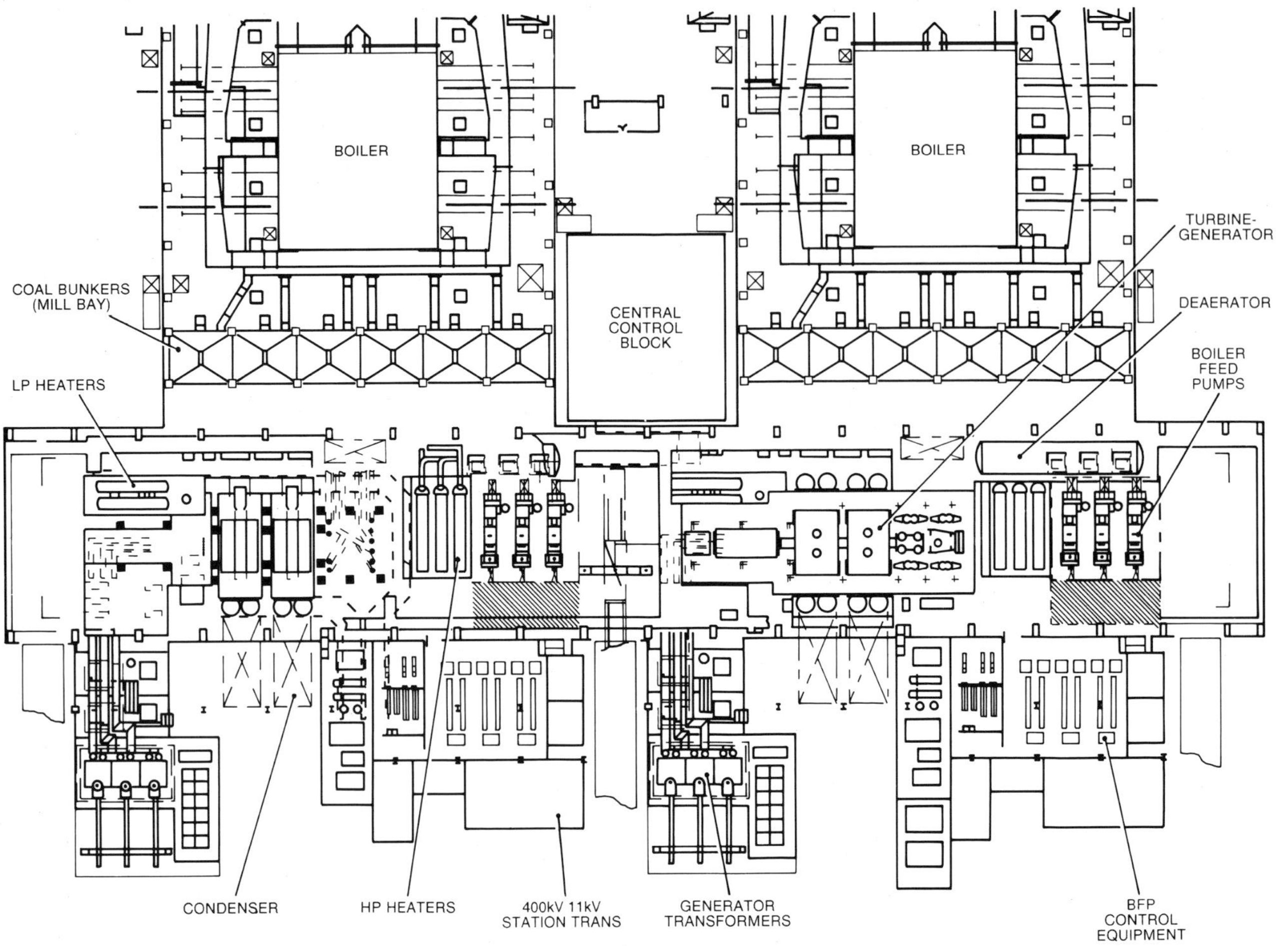

FIG. 1.24 Layout of proposed coal-fired reference design station

operation and those that are needed for accommodating personnel, or providing a service function. The first group would typically include:

- Town water reservoirs.
- Fire fighting pumphouse.
- Make-up water treatment plant.
- Bulk chemical stores.
- Hydrogen production plant.
- Hypochlorite production plant.
- Gas stores.
- Auxiliary boiler house.

Additionally on a nuclear site they would include:

- Diesel generator buildings.
- Active waste management building.
- Decontamination workshop.
- Radioactive fuel store.
- Reserve ultimate heat sink.

The locations of these buildings are to a large extent dictated by their functional relationship to the main plant. As far as possible they are positioned to minimise the length of pipework and cable connections and to provide good operational access routes.

As the reference designs for the PWR and new coal-fired stations are developed, it is the intention that the relative locations of these buildings will as far as possible be fixed within the constraints of a particular site. This applies particularly to the buildings associated with safety aspects of the PWR.

The second group covering personnel accommodation and services includes:

- Administration building.
- Welfare building.
- Canteen.
- Fire station.
- First aid and medical centre.
- Workshops.
- Heavy and light stores.

Consideration of the routine and emergency requirements for the power station as a whole suggests that the location of the above items relative to each is significant, and they should be sited adjacent to each other, and collectively as near as possible to the centre of the plant.

The administration building houses the majority of the personnel who require office accommodation, but a small number of personnel will be housed in offices adjacent to working areas, e.g., control room, coal handling plant and FGD plant. The administration building quite often accommodates the canteen and welfare facilities.

The workshops are provided in order to carry out maintenance work on the plant, and special care should be taken to ensure that the various workshops are sited to facilitate the duties they perform. They are usually associated with the main buildings and with the stores building which provides the materials necessary for effecting repairs. It was frequently found satisfactory on large stations to provide an island of ancillary accommodation which contained administration, welfare, workshops and stores functions and had easy access to the main building.

The current coal-fired reference design stations are planned to have the workshop and stores complex attached to the end of the turbine hall. This naturally lends itself to the administration building being in a similar location because of the functional links required. However, the location of the office building will depend on the overall station layout and on any special conditions which may affect the disposition of the ancillary buildings, but should be located where the following conditions can be fulfilled:

- Freedom from noise.
- Freedom from dust.
- Easy access to other ancillary functions and to the main building.
- Easy access between the maintenance engineers and the maintenance workshops.
- Easy access between the personnel concerned with operation and the control room.
- Reasonable aspect.
- Close to the main access road and car parks.

3.6 Main access and on-site roads

Good road access to a power station is required during the construction and operational phases.

The station should be so arranged on the site that if possible the access road from the nearest main public road is as short, direct and as level as possible. However, this external access is difficult to influence as the positions of the main building and ancillary buildings are often dictated by the shape of the site and the general topography.

Where a new development is proposed for an existing site then careful consideration needs to be given to the road access provisions for the operational station during the construction of the new station.

The roads within the site boundary should satisfy the following criteria: main access roads would be designed to be suitable for heavy loads (e.g., transformers, stators and fuel flask transporters), they would have an internal curb radius of 25 m and width of 6.75 m; the slopes of such roads should be limited to 1 in 15.

The main station buildings would be ringed by a main road designed to allow alternative access to all parts of the station in the event of a closure for maintenance or crossing purposes. The workshops and stores would be provided with access from two sides and have suitable areas for parking lorries.

3.7 Station operation considerations

During the development of a station layout careful consideration has to be given to the movement of personnel, vehicles and suppliers after the station has been commissioned. These aspects, together with security considerations can influence the positions of certain buildings. It is intended that all private cars are parked outside of a security fence that encloses the station buildings. A station gatehouse and security office would be located on the main access road inside the security fence. Lay-bys would be provided at the entry to and exit from the security gates for inspection of vehicles and for a suitable weighbridge.

In order to give the shortest possible route for personnel from the car parks to the administration and welfare building, it is necessary to locate these buildings as close as possible to the car parks. This in turn tends to dictate the positions for workshops and stores as explained in Section 3.5 of this chapter. The location of these buildings is also influenced by the objective of minimising the extent to which day-to-day delivery vehicles have to intrude into the station complex.

Figure 1.25 shows a diagrammatic arrangement of the principal personnel and vehicle movement sequences for station operation.

3.8 Cooling water system

Two types of cooling water (CW) systems are mainly used in power stations:

- Direct cooled systems — in which the cooling water is abstracted from the source (which may be river, estuary or sea), pumped directly through the condensers and returned to the source.
- Closed cooling tower systems — in which the condenser cooling water is pumped round a closed system incorporating cooling towers wherein the heat rejected by the condenser is dissipated; an external source of water (river, canal, estuary, etc.) is required to make good the tower evaporation losses and purge losses.

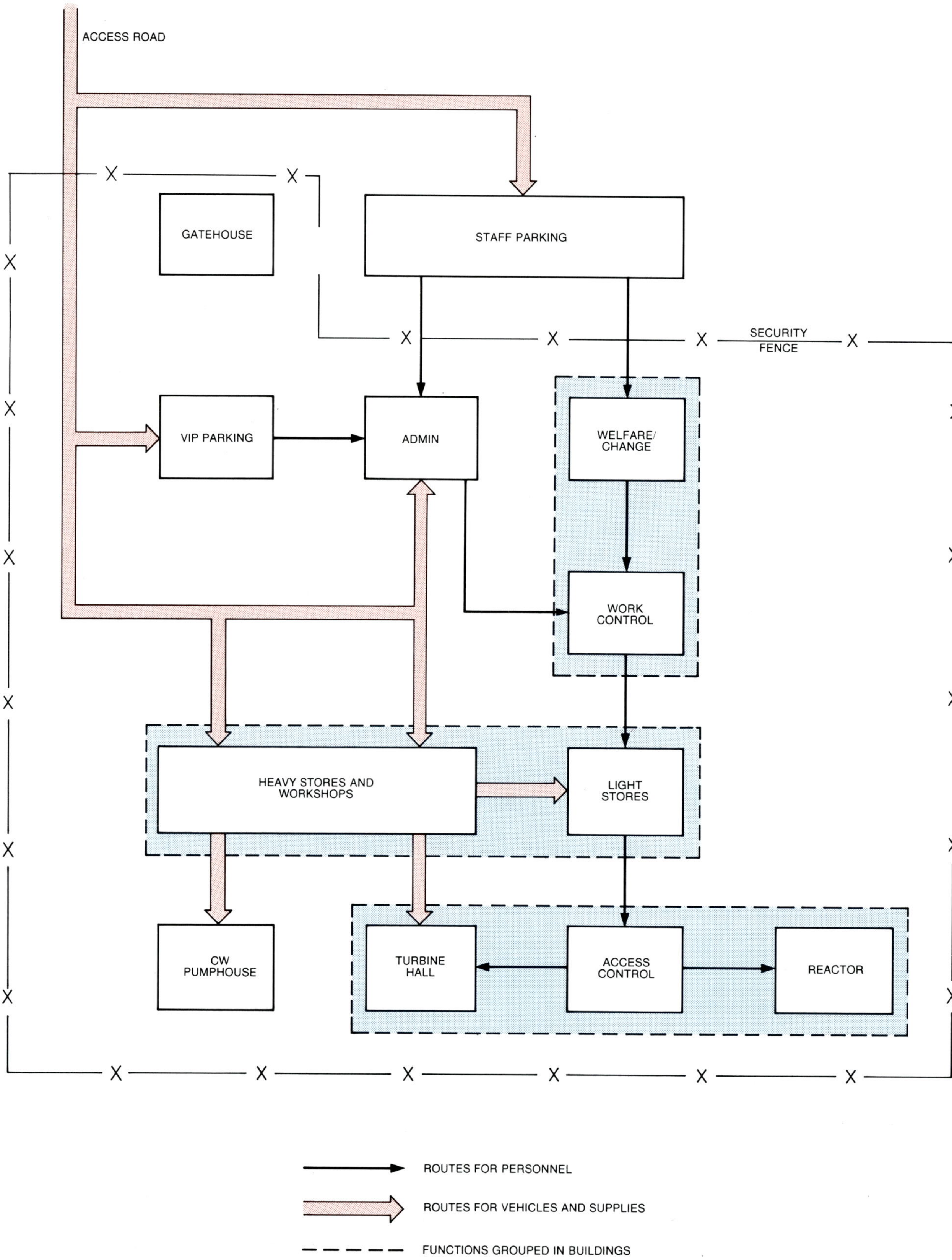

FIG. 1.25 Principal movement sequence for station operation

3.8.1 Direct cooled system

This system is usually adopted when the station is sited on the coast or on a river estuary where there is an adequate supply of water available at all times. A typical arrangement is shown in Fig 1.26. The essential elements of the system consist of an intake structure, cooling water pumphouse, intake culverts, outlet culverts, syphonic weir, seal pit and outfall structure.

It is necessary to establish the correct relative position of the intake and outfall, these points often being decided from a hydrographic survey or hydraulic model tests (see Section 2.3.3 of this chapter). If they are too close together, recirculation will take place and this will reduce plant efficiency and increase running costs. If they are moved further apart the capital cost of the connecting mains may be increased. They should, therefore, be in the position relative to each other that produces a balance between these two costs.

This aspect is very much site related and depends on the geographic arrangement of the coastline and the offshore water movements. It can result in simple intakes and outfalls located at the coastline or a more extensive system resulting in tunnelled intakes and outfalls; very often a combination of these two extremes is used.

The cooling water pumphouse is usually located at the coastline because either a simple channelled intake can be provided or, to minimise the length of costly tunnels, an offshore intake if this is the preferred arrangement. The exact position of the pumphouse can depend on several factors but one important item is constructability and juxtaposition to other buildings. As can be seen from Fig 1.27 the pumphouse can be a very large structure requiring excavations of the order of 60 metres diameter and 20 metres deep. It is therefore important to ensure that within the site layout, sufficient space is allowed between the pumphouse and other buildings to allow satisfactory construction to proceed. It may be that it needs to be located near an existing pumphouse and consideration has to be given to sufficient separation so as not to affect the continued operation of any existing station.

For the connecting mains from the intake to the station and the return to the outfall, reinforced concrete culverts, spun concrete pipes, steel pipes or concrete-lined steel pipes are used depending on the quantity of the water, the subsoil conditions, the avail-

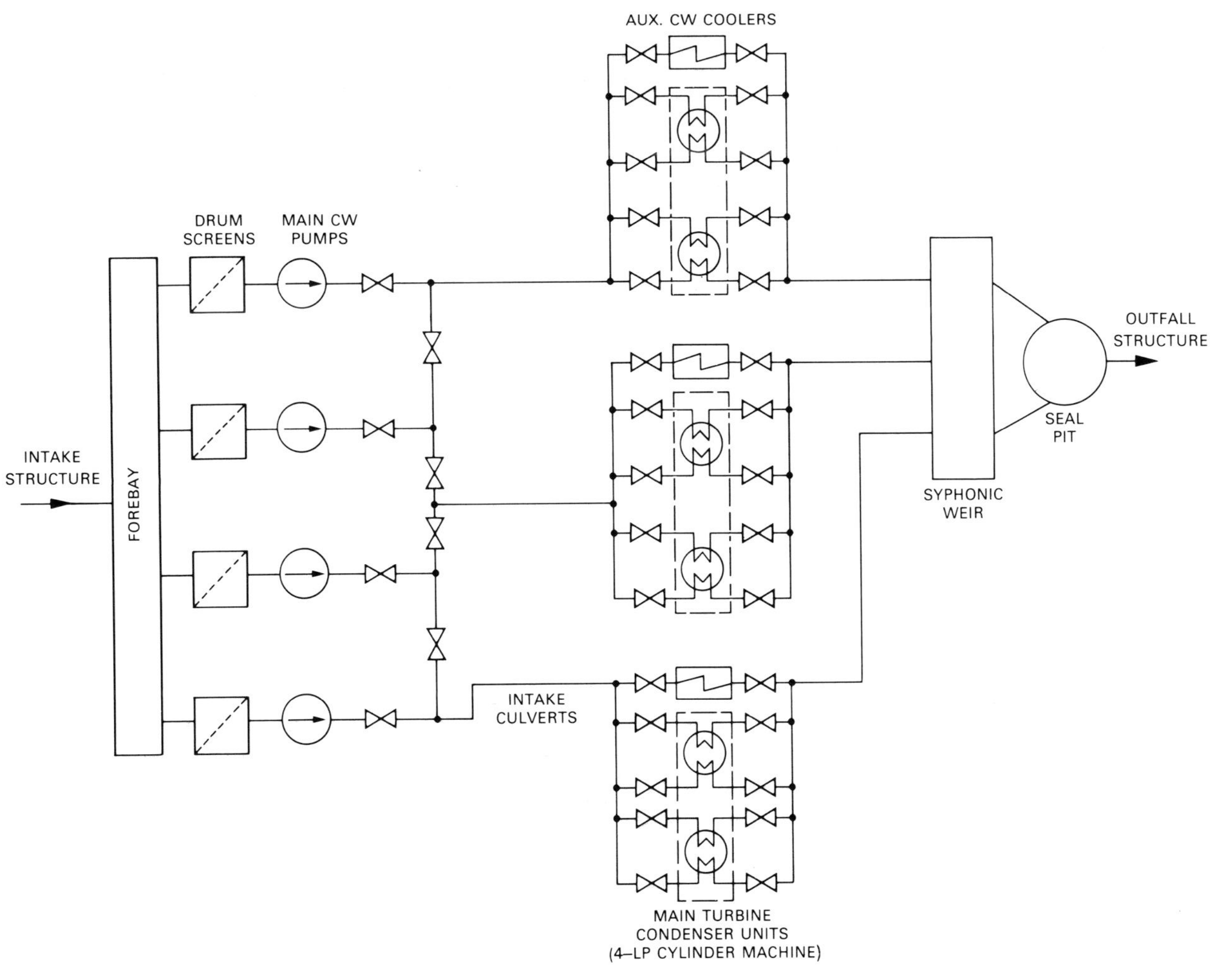

FIG. 1.26 Typical diagrammatic arrangement of a direct cooling water system

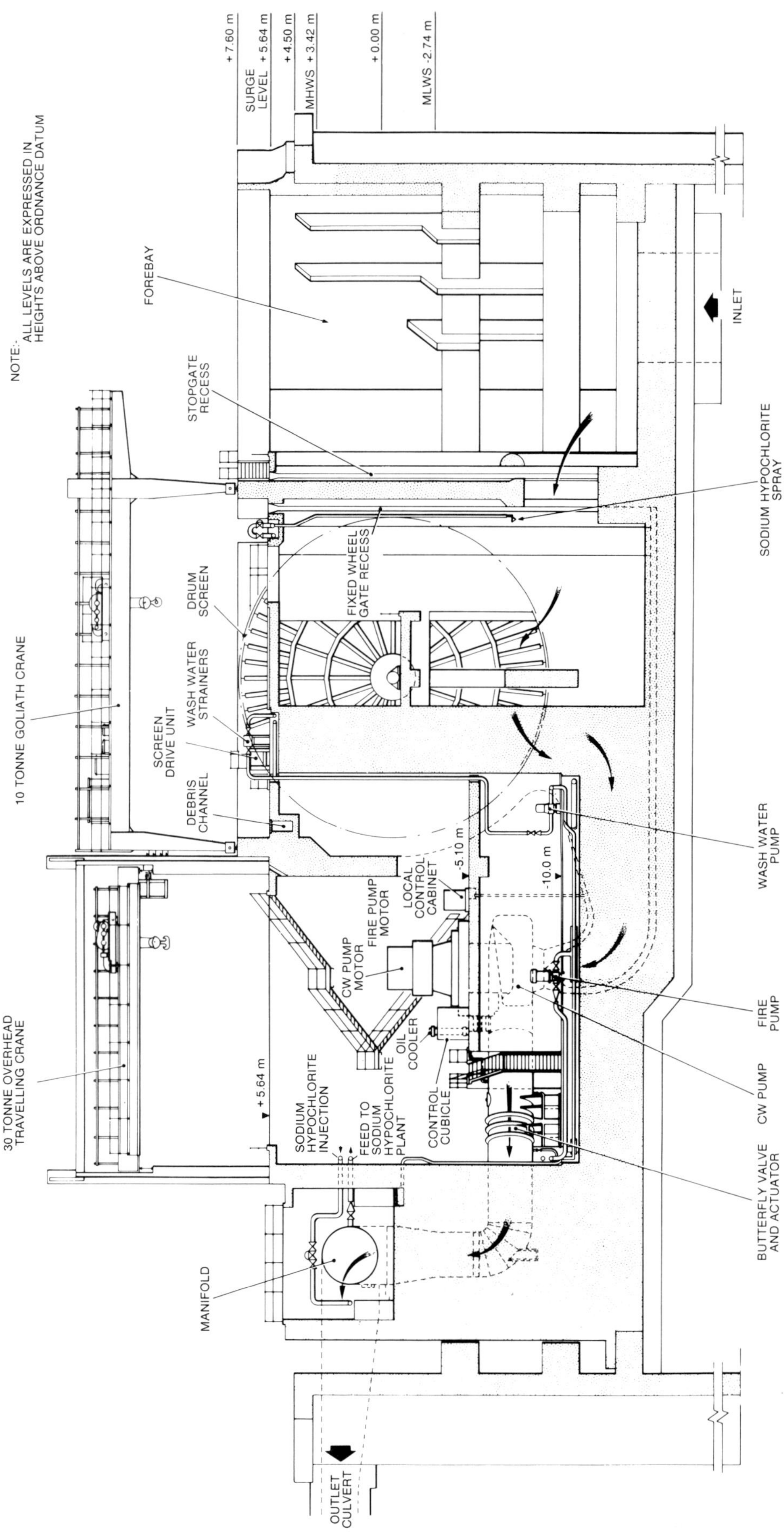

FIG. 1.27 Typical CW intake and pumphouse for a direct cooling water system

ability of materials and the relative costs based on the capital and running costs. The connections from the pumphouse to the turbine hall should then be kept as short and as straight as possible to minimise costs and pressure losses in the system.

Care should be taken in routing the culverts or pipes so as to avoid the crossing of culverts or obstacles that could create air pockets in the system.

The syphonic weir and seal pit should be located as close as possible to the turbine hall to limit the extent of the more expensive pressurised cooling water culverts. The outlet culverts from the seal pit can be designed to a lower pressure. Care should be taken when routing culverts to ensure that satisfactory clearances exist to other services and structures to enable their satisfactory construction.

3.8.2 Closed cooling tower water system

A closed cooling tower water system, or indirect system, is used when the water supply available is inadequate for direct cooling, and the condensers operate on a closed circuit. A typical arrangement is shown in Fig 1.28. The essential elements of the system include a cooling water pumphouse and forebay, intake and outlet culverts, cooling towers and return culverts to the forebay. A make-up and purge system is provided to control the salts concentrated in the system due to continuous recirculation and to replace the amount lost by evaporation from the cooling towers.

The location of the cooling water pumphouse (see Fig 1.29) is again dictated by construction aspects and the need to minimise the pressurised cooling water culvert lengths.

The principles previously explained in Section 3.8.1 relevant to culvert routing equally apply to indirect-cooled systems. The major impact on layout of a closed cooling tower water system is without doubt the cooling towers. At Drax power station (4000 MW), which was the most recent to be built by the CEGB using cooling towers, there are a total of twelve cooling towers, each being approximately 115 m high and 93 m diameter at the pond cill levels. It may be appreciated therefore that the towers, together with the open return culverts to the pumphouse forebay, require a considerable area of land.

Cooling towers are usually grouped and sited so that the CW system as a whole is as compact as possible, maintaining an adequate clearance between adjacent towers, and between the towers and any object which might obstruct the air flow into them. Towers should be so positioned that any spray at the base of the tower does not blow on to public or station roads in frequent use, or onto the coal delivery rail sidings, and it is desirable to reduce the risk of coal dust and ash dust blowing into the tower ponds.

The location of the towers is also influenced by the preferences of the Consultant Architects commissioned by the CEGB for layout studies.

The layout of the make-up and purge system is dictated by the relative locations of the nearest river source and the cooling towers.

3.9 Fuel supplies and storage

3.9.1 Coal plant

The costs of coal deliveries by British Rail (BR) from British Coal are dependent not only on the distances involved, but also on demurrage rates for locos and rolling stock and the efficient use of rail capacity by high-speed permanently-coupled wagons. These costs can be minimised by providing the most rapid and efficient turnround at the power station unloading point. For this reason, the favoured arrangement for coal unloading at any power station site is the merry-go-round system, whereby bottom-opening hopper wagons unload the coal into underground hoppers, with the train running-on to leave the site without stopping. For the train to turn round and return to the loading colliery, a loop is required, with a 250 m minimum radius of track, and having the appropriate standing room for signal delays, means that a considerable area is required for such a loop arrangement. The land within the loop provides a convenient coal stock-out area.

Where insufficient area is available, or where access problems exist, a compromise solution can be adopted with sidings before and after the unloading track hopper and with provision for the loco to run round the train prior to exit from the site.

Figure 1.30 shows typical siding layout schematics.

The track hoppers are situated as close to the boiler house as possible to minimise conveyor lengths, but still providing sufficient distance for the rise from underground hoppers to boiler house bunker tops to be achieved at a suitable inclination angle, allowing for junction towers as required.

Another factor in the coal plant layout derives from the British Coal working arrangements requiring a week's coal burn to be delivered in five working days. Thus, on average, two-sevenths of each day's delivery must be stocked out for reclamation at the weekend. Consequently, stock-out and reclaim on a regular basis must be facilitated, and large travelling bucket wheel machines on rail tracks are often used for stocking out and reclaiming from the appropriate parts of the total fuel stocks. Longer term strategic stocks are held as part of the total stock, but transport to and from these more remote areas of the coal stock area is more economically achieved by bowl scraper mobile equipment.

The total area required for coal store, rail arrangements and handling equipment can be up to 20 hectares for a 2000 MW station.

Similar layout considerations may apply to coastal stations with sea-borne coal deliveries and short-term stocks as a 'buffer' between ship arrivals and longer term strategic stocks.

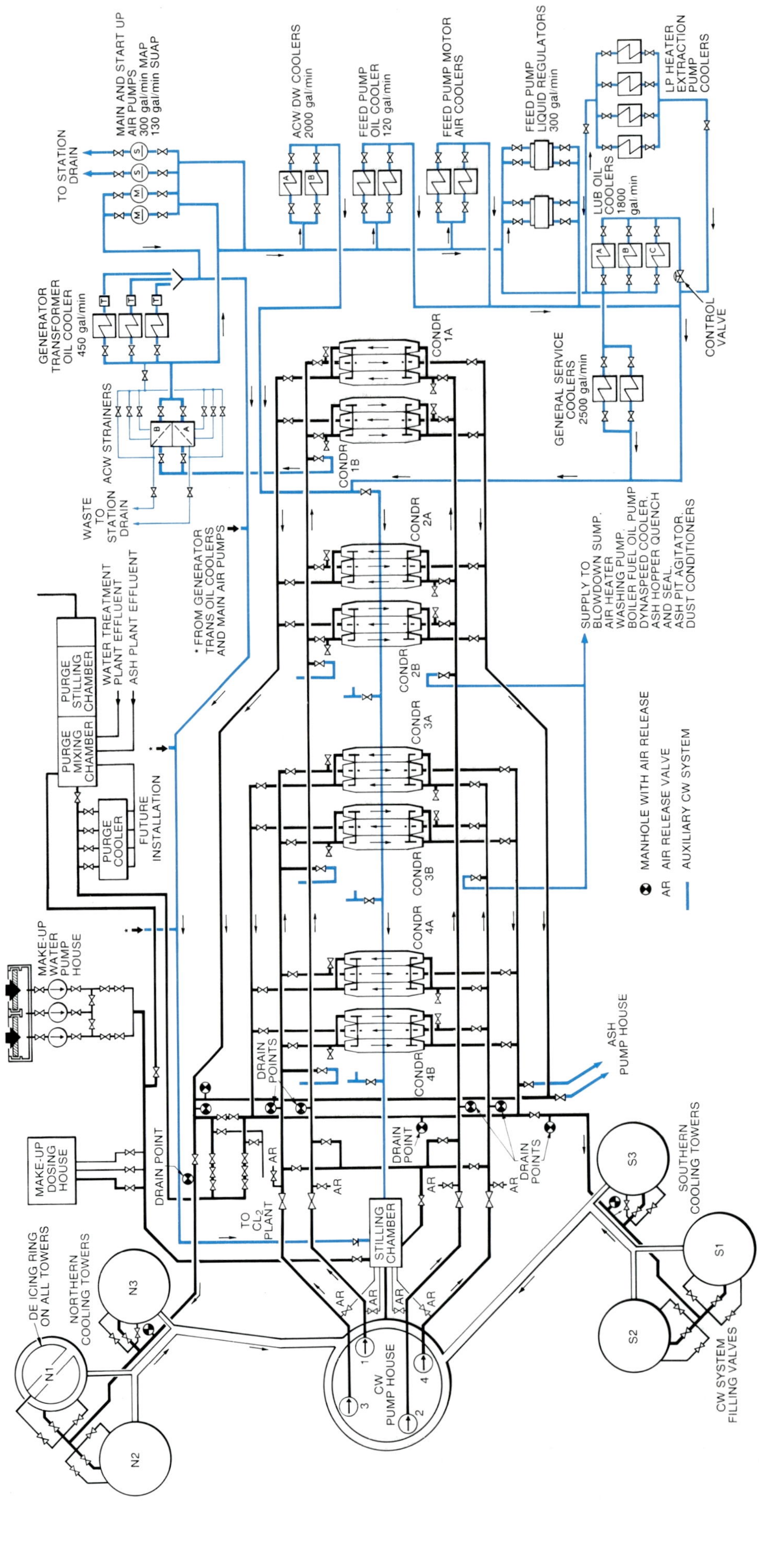

FIG. 1.28 Typical diagrammatic arrangement of a closed cooling tower water system

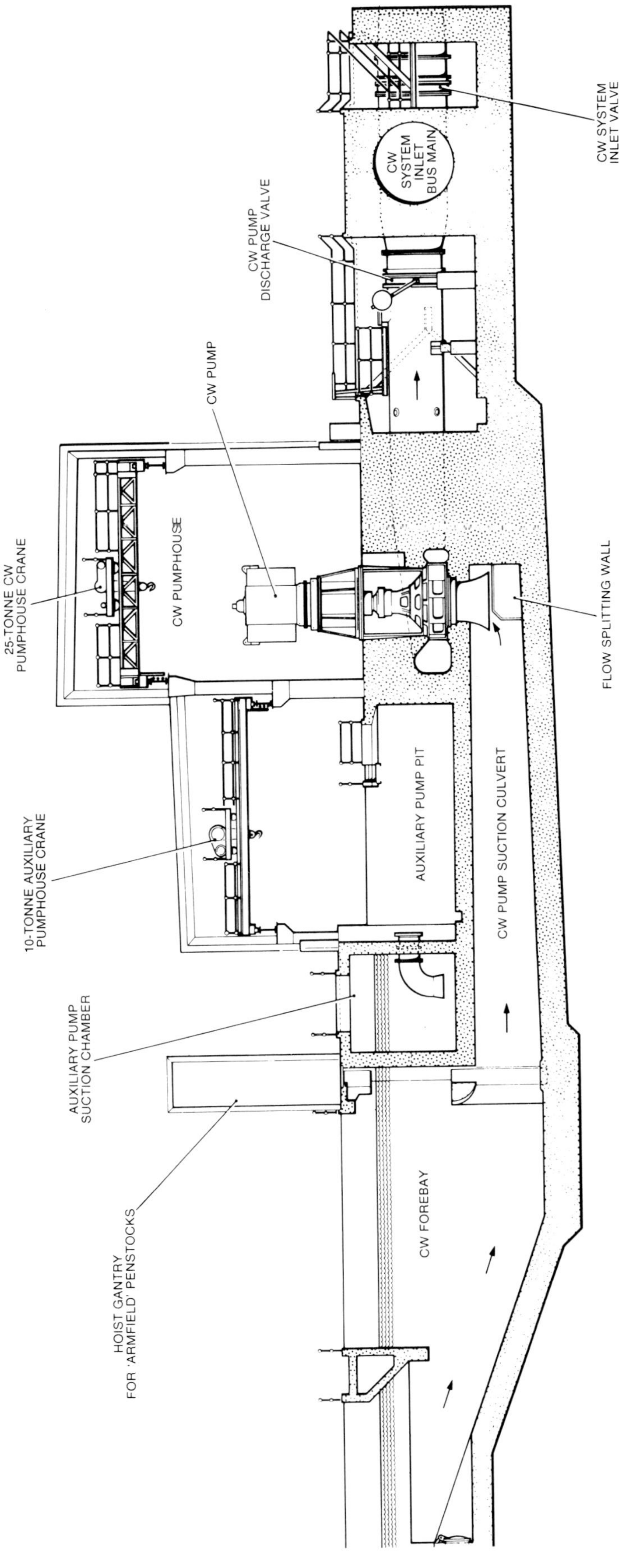

FIG. 1.29 Typical CW forebay and pumphouse for a closed cooling tower water system

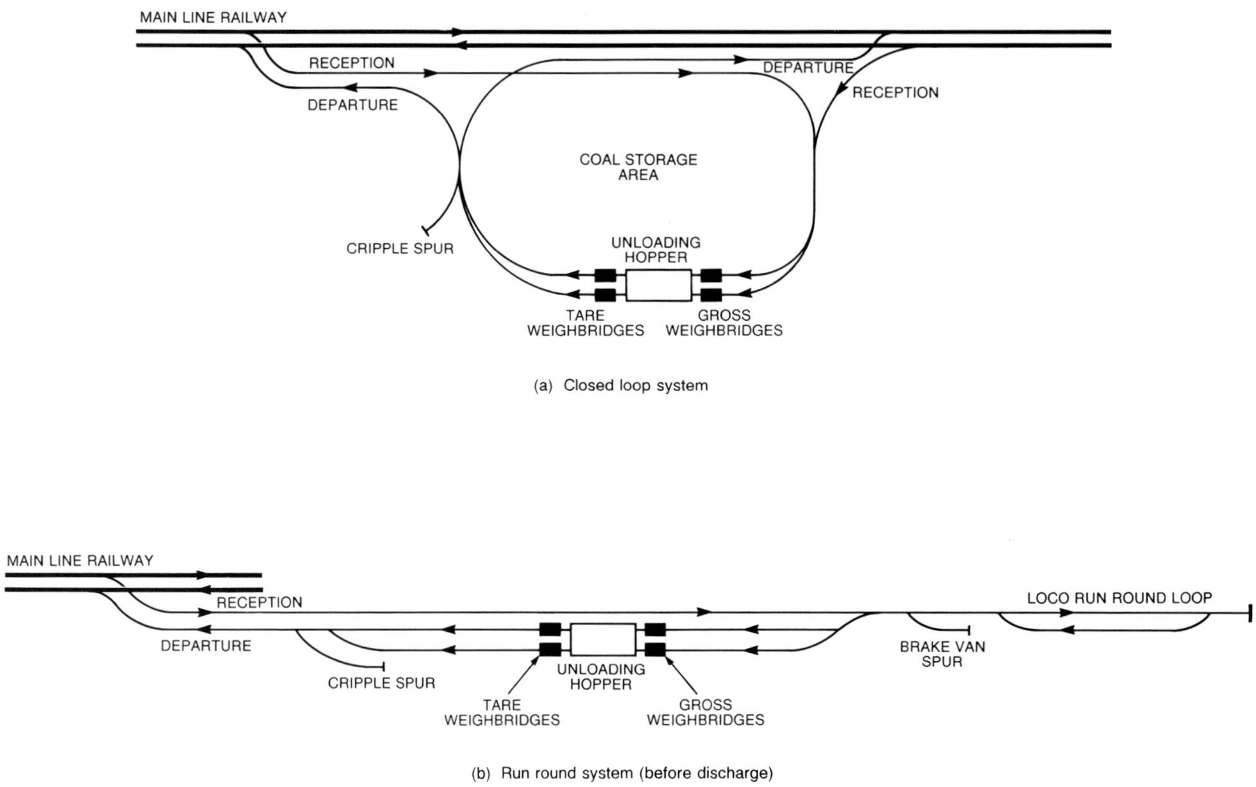

FIG. 1.30 Typical siding layout schematics

3.9.2 Fuel oil plant

To date the CEGB has not located an oil-fired station where its supply of heavy fuel oil would be dependent on road or rail-borne transport. Stations have been sited either close to oil refineries where direct piped fuel is available, or on coasts and estuaries where deliveries from sea-going tankers can be received. Quantities stored depend on a judgement of the security of supplies according to the proximity or otherwise of the source, and factors such as whether import and export to other nearby consumers is required. At least two and possibly up to five, large storage tanks are required. Ideal situations would be close to the main buildings on the 'fuel delivery side', but leaving adequate distances to minimise fire hazard to the station, and from other plant and equipment to the tanks themselves. Also adequately firm ground conditions are required and a suitable area large enough for a bund to contain the contents of one tank in case of a tank fracture.

Figure 1.31 shows the fuel oil delivery and storage arrangements at the 3 × 660 MW Littlebrook *D* power station.

On coal-fired stations, the need for boiler lighting-up oil requires delivery and storage arrangements. The quantities of the lighter grade of oil needed are relatively small and so delivery is normally by road tanker. Storage is in tanks within a bund located as close to the main boiler house as other layout considerations allow.

3.10 Ash and dust disposal

The site layout must provide means of disposal for furnace bottom ash and for the large quantities of pulverised fuel dust produced as waste products. Although purchasers can be found at times for certain quantities of these waste products in the construction industry, in concretes or simply as landfill, long term dumping provisions are required. These can be close to the site or some distance away involving the pumping of dust as slurry, for example, to local natural or artificial lagoons, or transport by rail or sea in a dry condition, or by road in a wet condition. Market opportunities vary over the life of the station; some dumping grounds may become full or otherwise unavailable and disposal economics vary. Consequently the layout is likely to require several disposal options to be kept open in the longer term, whatever the immediate or initial short-term disposal may be (see Fig 1.12).

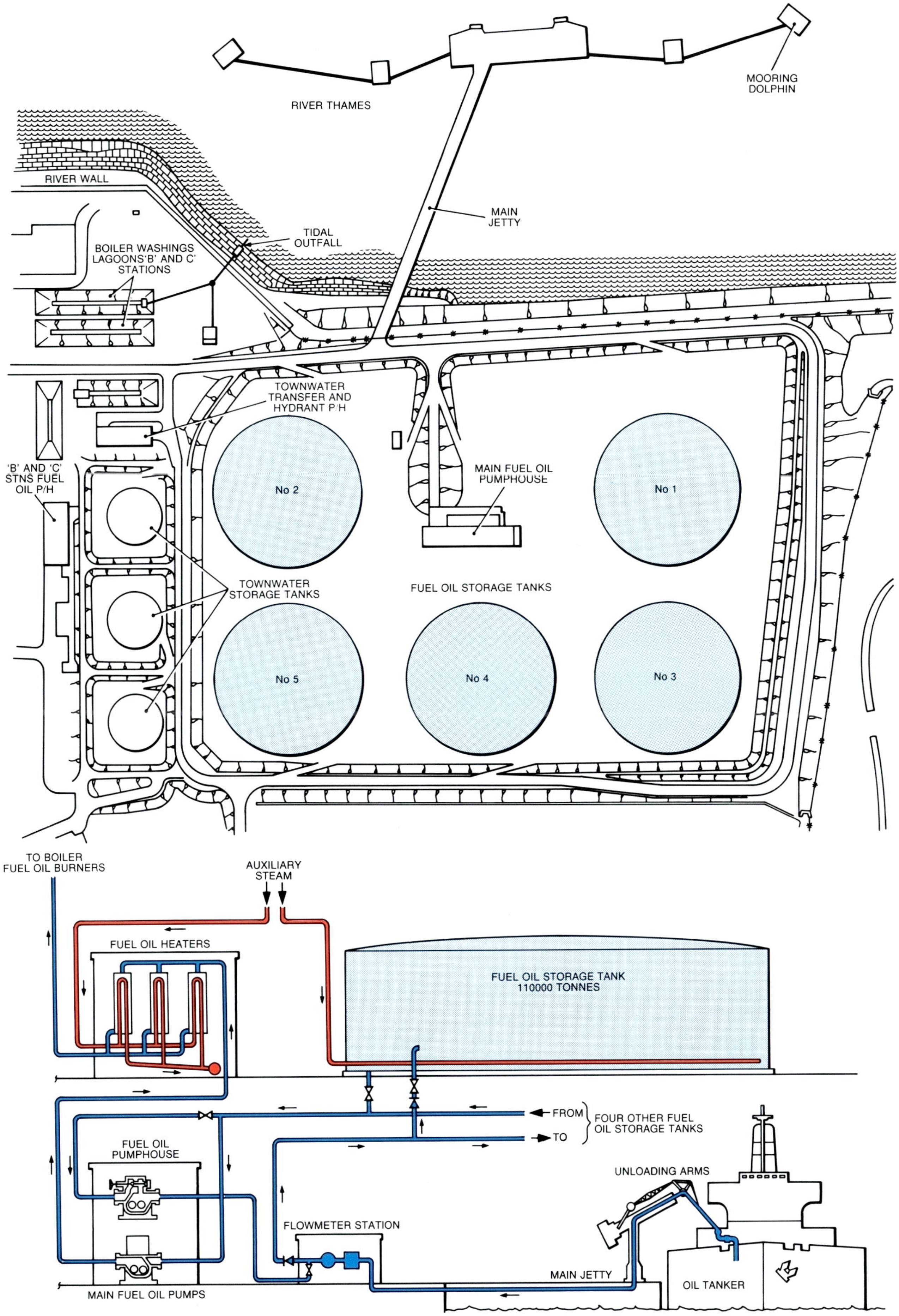

FIG. 1.31 Fuel oil supply and storage — Littlebrook *D* power station

3.11 Flue gas desulphurisation plant materials

Whilst studies are currently being done within the CEGB on possible flue gas desulphurisation plants, it has been identified that the plant based on the limestone-gypsum process will be the most onerous on layout. This plant for a 2000 MW station will require the delivery of some 0.3 million tonnes of limestone per annum and the disposal of some 0.5 million tonnes of gypsum per annum. It is anticipated that the movement of these materials will be by rail and can be incorporated within the rail arrangements provided for coal delivery. However, warehouse storage will be required for strategic stocks of limestone and gypsum producing a further demand on available land.

3.12 Transmission requirements

It is current CEGB policy to use metal-enclosed gas-insulated indoor substations. This type of substation is considerably smaller than the open switchgear compounds and is less onerous from a layout aspect. The plan area of a typical 400 kV metal clad substation for a 2000 MW power station is of the order of 1 hectare. This includes the associated electrical and auxiliary plant buildings and perimeter roads for access. The use of an indoor substation is more acceptable visually and is not affected by potential problems such as seawater spray, cooling tower spray or coal and dust pollution.

It is preferable that the substation be located adjacent to and in front of the turbine hall as this shortens the generator transformer connections. Ideally this will also be on the side of the site from which the transmission lines emanate so that the outgoing feeders can be arranged in an economical manner. It is also preferable that the outgoing circuits be overhead lines for as far as possible, as the use of 400 kV cables is very costly.

3.13 Construction requirements

The size and location of individual contractors' areas depends on the contract strategy adopted for placing the orders for equipment, on the number of contractors involved, and would be based to some extent on information supplied by the contractors. Locations would then, as far as possible, be arranged to coincide with programme requirements. However, it is recognised that certain areas may be required early and would need to be close to the excavations for the main buildings. In the case of a PWR, for instance, the contractors for the civil works, the containment liner and the structural steelwork would be in this category. Where only restricted areas would be available immediately adjacent to the station, the orientation of the station may be important to provide adequate locations for all these areas (see Fig 1.32).

The location of the contractors' offices, mess huts and the car and bus parks should be within a reasonable distance of both the temporary construction areas and the working areas in the main station complex.

The contractors' areas should be on level well-drained land, but if necessary they could be on a number of terraces, providing there are adequate access roads with suitable gradients between the terraces and the station. Typical contractors' working and storage areas for modern nuclear and conventional stations would be of the order of 25–30 hectares. In addition, areas of some 3–4 hectares would be needed for construction car parks. Storage space would also be required for topsoil storage and late excavation/backfill material.

In order to reduce the length of the construction programme, consideration is given to shipping more components to site ready assembled as modules. For instance, steam turbines have previously been assembled and tested at the manufacturer's works, then dismantled into sections for shipment and reassembled on site. It is possible to reduce the amount of dismantling by sending the high pressure and intermediate pressure cylinders still boxed up with their rotors in place. It may also be possible to despatch condensers as assembled modules. Quality assurance is also better controlled under factory conditions and such items as PWR pressure vessels and steam generators can be shipped ready assembled. However, shipping these fully completed plant modules by road causes problems because of their size or weight, or both. This can be solved to a large extent for coastal stations by using a large sea-going barge to deliver these items to a barge berth specially installed as near to the site as possible. Such a berth could also incorporate docking facilities for roll-on/roll-off vessels enabling many other deliveries to be made by sea. This would reduce the volume of construction traffic on the roads near the site. The use of rail access would also be of benefit if it can be provided economically. Additional land would be required for sidings and offloading facilities.

3.14 Amenity considerations

Whilst recognising that production of a reliable supply of electricity at the lowest possible cost is the paramount consideration, it is the CEGB's statutory duty to pay attention to the appearance of new power stations, both in detailed architecture and in its suitability for the environmental amenity.

Very often the architect may suggest a number of arrangements of buildings or cooling towers in order to achieve the correct massing in the landscape and to improve appearance. This work is done in close collaboration with the engineering design staff to ensure that the optimum construction and operational design is still achieved at minimum cost. Landscape architects are also engaged with a view to integrating the station

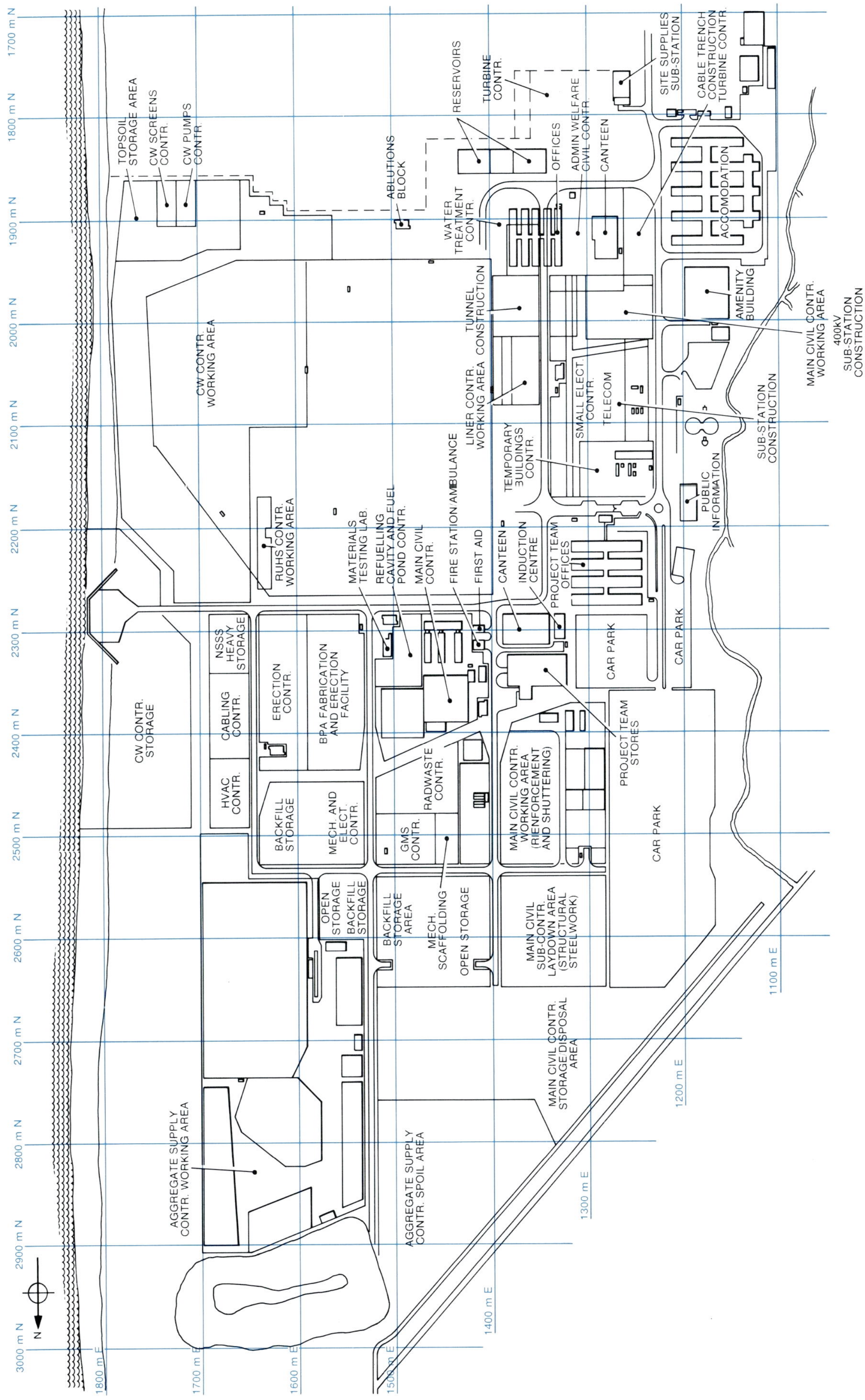

FIG. 1.32 Construction site layout — Sizewell *B*

and transmission equipment in the immediate locale so far as is practicable with the surrounding countryside.

Figure 1.33 shows the successful blending of landscape and power station at the Didcot coal-fired station site.

3.15 Typical site layouts

As stated earlier it is almost always impossible to satisfy every requirement perfectly. Three different solutions to site layout problems are illustrated in Figs 1.34, 1.35 and 1.36 and are described as follows:

Figure 1.34 shows the site layout for a 2000 MW oil-fired station using direct cooling and with a seaborne oil supply. The site area of 21 hectares which was available for the construction of this station was a comparatively small area on which to build a 2000 MW power station. The factors which influenced most of the station layout and plant orientation were:

- The limiting boundaries for river and road access.
- The suitable locations of construction storage and contractors' areas.
- The need to commission gas turbine plant early in the overall construction programme.
- The effect of the extensive cooling water civil works location and access.
- The need to complete the construction by working generally from north west to the access in the south east of this restricted site.

The existence of the transmission routes, together with the knowledge that fuel would be delivered by sea, determined that the boiler house and therefore the chimney should be located near to the river. Consequently, the location and orientation of the boiler drum, turbine hall and generator compounds together with their access routes were established.

The location of construction storage areas and contractors and CEGB site offices influenced the location of reserve feed water tanks and the water treatment plant which were located at the north east corner of the site. The same considerations influenced the location of gas turbines and their associated fuel tanks. The three gas turbine exhaust flues were directed into a single chimney which also included the flue from the auxiliary boiler, thereby influencing the location of the auxiliary boiler house.

The fuel oil heater house was located between the fuel oil storage tanks and the boiler house, with the sootblower air compressor house also in close proximity.

The location of the cooling water intake works in the river dictated the location of the pumphouse on the west side of the site. The outfall shaft was placed at the same end as the pumphouse so that the culvert excavation did not seriously affect access to the boiler house. The chlorination plant was consequently located adjacent to the pumphouse.

The administration block, which also contained the central control room, and the workshops were required to be as close as possible to the turbine hall and therefore located in the area to the south east of the main plant buildings.

Figure 1.35 shows the site layout of a 4000 MW coal-fired station comprising 6 × 660 MW units, utilising a closed cooling tower system and with railborne coal supply. Here, a balance between the engineering and architectural requirements was achieved. The 400 kV switchhouse was placed outdoors and situated parallel to the turbine hall, while the cooling towers were grouped in two sets of six at either end of the station; this is an architectural requirement, which though not detracting much from operational convenience, required an additional pumphouse. However, views of the station from the surrounding country were greatly improved.

A loop system of sidings was adopted for coal delivery. The workshop and stores were located in the turbine hall and the boiler make-up water treatment plant was located central at the front of the turbine hall. The administration block, canteen and welfare services were located adjacent to the access road.

A major factor affecting the layout of this station was that it was built in two phases with three units being initially constructed and then the additional three units being completed later. This meant that construction of the later units had to be phased such that minimum disruption was caused to the operation of the first units.

The provision and layout of ancillary services, e.g., coal handling plant, ash and dust handling plant, cooling water make-up and purge systems, etc., had to take into account the requirement of early operation for three units with the later addition of a further three units. The physical size of the whole station, however, led to the adoption of a split recirculating cooling water system, each half having its own self-contained system.

Figure 1.36 shows the site layout of a 1320 MW AGR station using direct cooling. The station was the second stage of a two-station development on the same site.

The site investigation revealed the existence of a geological fault running approximately north–south and bisecting the useful area of the site. Triassic sandstone exists to the west of the fault and is suitable for the support of power station loads. A complex sequence of Namurian mudstones, sandstones and siltstones, which are not suitable for heavy ground loadings, exists to the east of the fault.

The lines of the sea wall and the geological fault converge towards the south of the site and thus create, to the south of Stage 1, a roughly triangular area on which Stage 2 could be located.

At the time that planning permission was sought for Stage 1, the Stage 2 development was envisaged and shown on the planning application as a mirror image of Stage 1. Although Stage 2 could not, in the event,

FIG. 1.33 Didcot coal-fired station landscape
(see also colour photograph between pp 66 and pp 67)

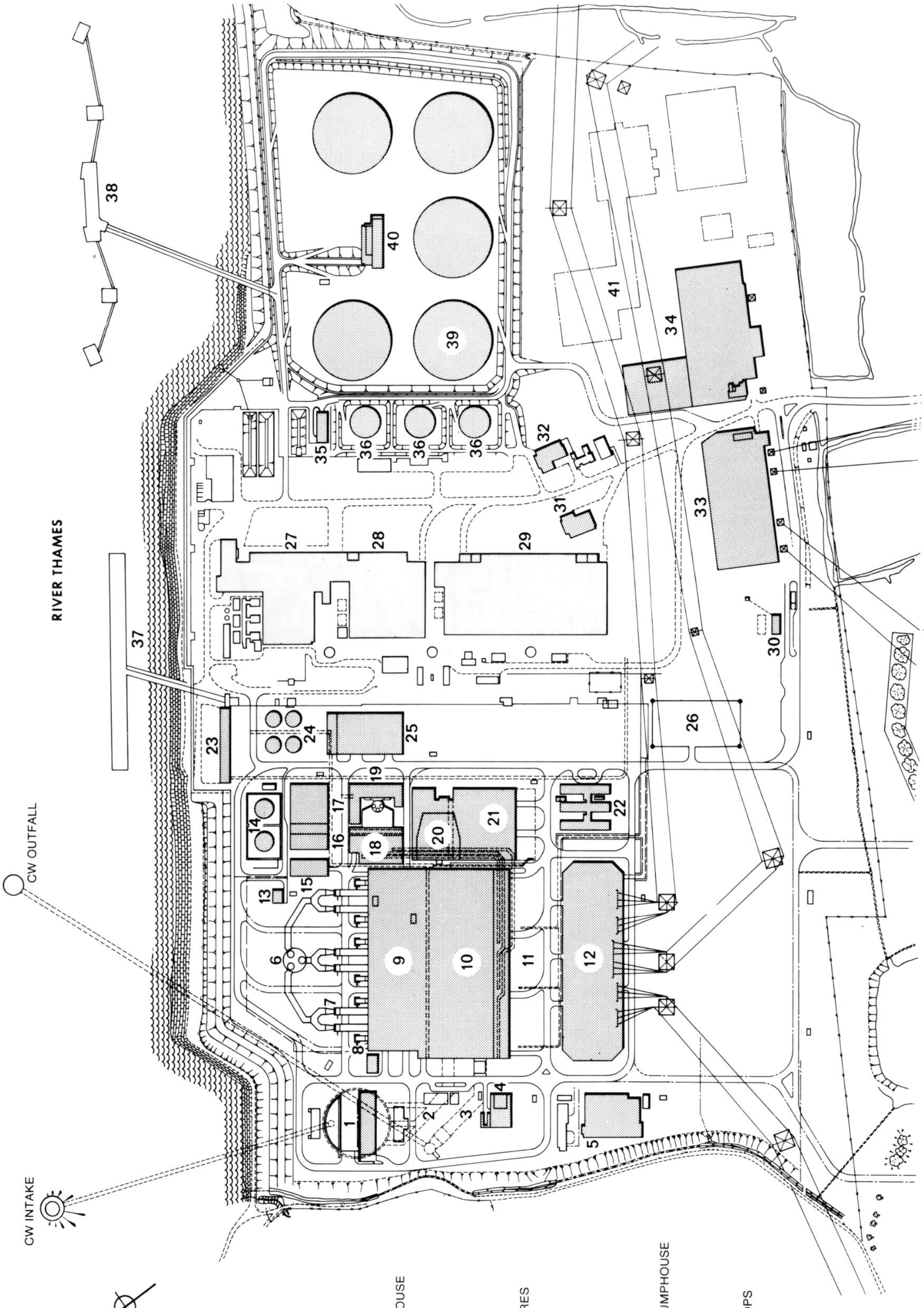

FIG. 1.34 Site layout for a 2000 MW oil-fired station using direct cooling and with sea-borne oil supply

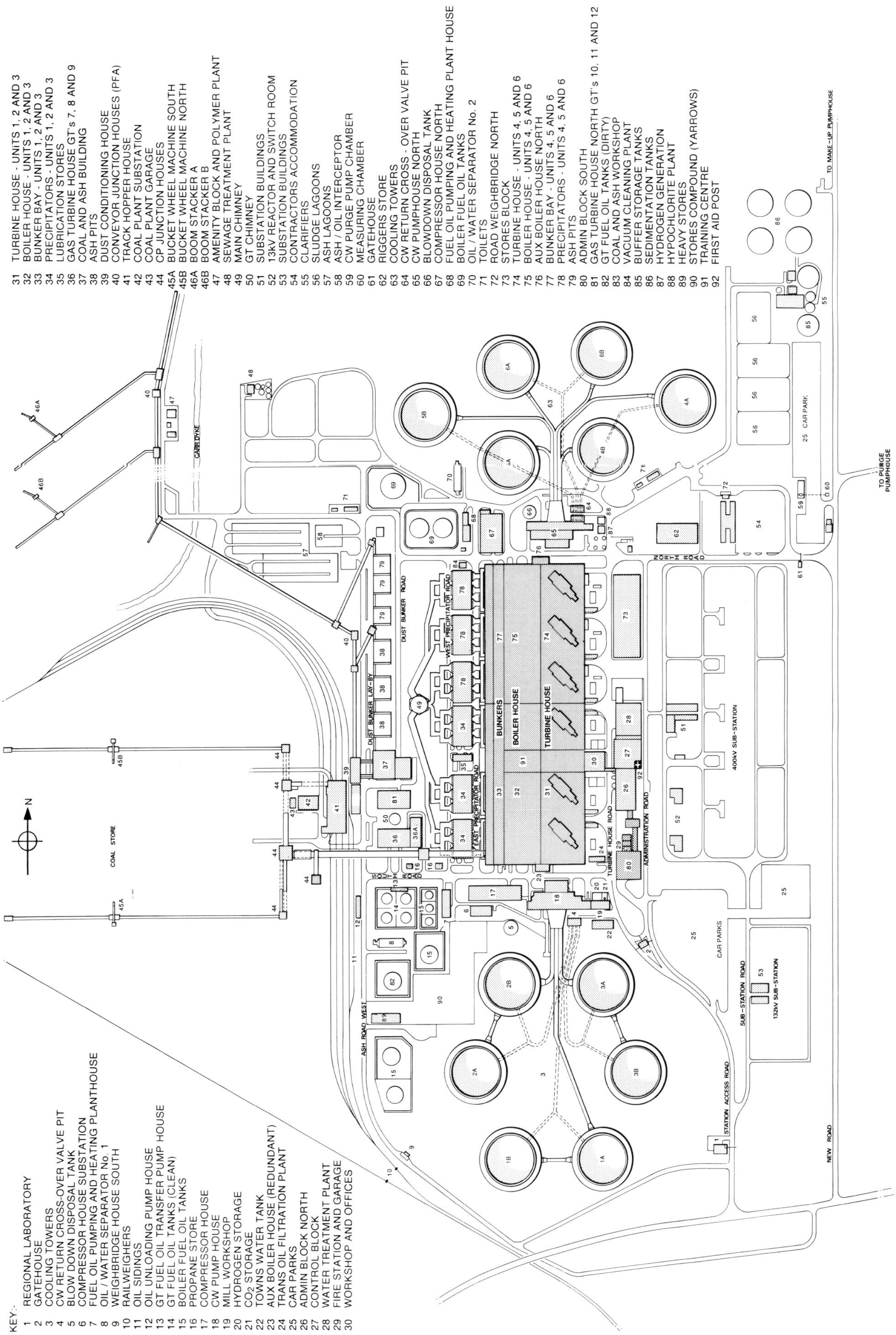

FIG. 1.35 Site layout for a 4000 MW coal-fired station using a closed cooling tower system and with rail-borne coal supply

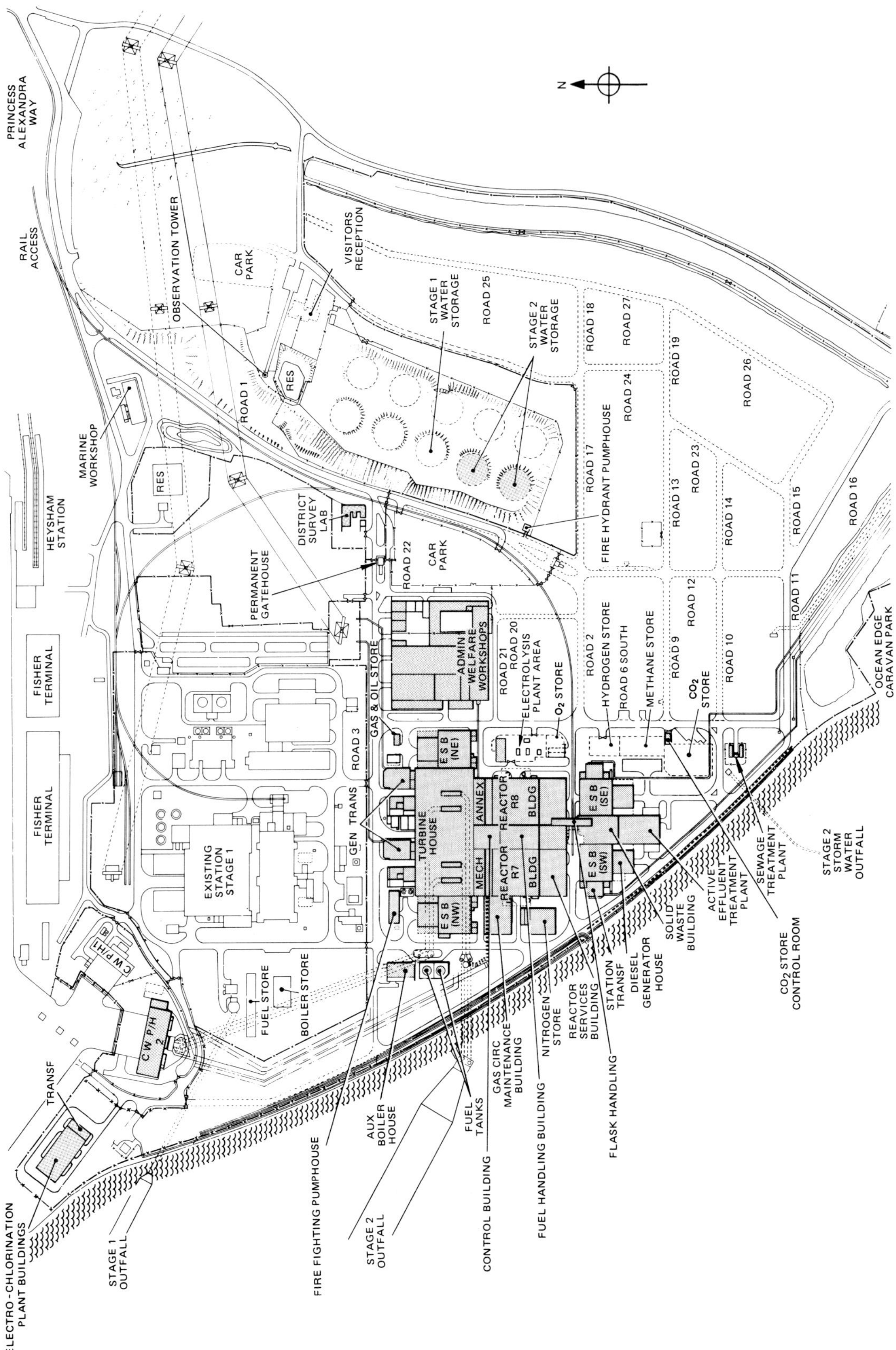

FIG. 1.36 Site layout for a 1320 MW AGR station using direct cooling

conform with that concept, the implied commitment has been recognised to the extent of maintaining parallelism of the station centre-lines in the north–south direction.

The arrangement of the main plant was governed by the need to keep the cooling water culverts and the 400 kV cables as short as possible. The CW pumphouse was located in the harbour alongside the Stage 1 pumphouse and the Stage 2 400 kV transmission lines leave the station area in a north easterly direction. It was therefore economically advantageous to site the turbine hall and generator transformers at the northern end of the available area and the nuclear island to the south of the turbine house.

This arrangement led to the location of the active effluent treatment plant and the solid waste store in a position to the south of the reactor building with access from the nuclear island by means of a bridge.

The essential supplies buildings were strategically positioned around the main building envelope to satisfy the safety requirements.

The block containing the administration, welfare and workshop facilities was located so as to satisfy a number of conditions, namely:

- To fit into a rational pattern of personnel movement.
- To occupy land unsuitable for other uses.
- To permit the movement of station personnel and visitors to and from the block with little or no contact with other station activities.

Sea water for cooling purposes is drawn from the harbour and is discharged in a westerly direction from the turbine hall through culverts and tunnels to an outfall to the bay.

The 400 kV and 132 kV substation is located about 0.6 km east of the power station site and is separated from it by a golf course. Two parallel double-circuit 400 kV overhead lines carry the outputs of Stages 1 and 2 to the substation by a route which leaves from the north east corner of the site, while 132 kV supplies are cabled to the station transformers by a southerly route.

4 Pumped storage

4.1 Introduction

It is not feasible to store electricity directly but the CEGB needs to have a reserve capacity to cope with its system requirements such as:

- Immediate reserve which has a response of only a few seconds to cope with rare major breakdowns.
- Sustained breakdown reserve, available in about five minutes to cover the period until replacement plant can be synchronised.
- Reserve for variations in demand on all timescales to cope with load prediction errors.

The inherent ability for rapid loading of hydro and pumped storage plant and then flexibility in changing from one operating mode to another (e.g., pumping to generating), makes it ideally suited to providing the required reserve capacity.

The geological formation of the UK means that there are relatively few true hydro sites and the majority of those that do exist have already been developed. Pumped storage offers the capability of increasing the hydro capability by pumping water from a lower reservoir to an upper reservoir at times of low demand (e.g., during the night), and then allowing the water to fall back to the lower reservoir to drive turbines and hence generators during periods of peak demand or in order to meet a reserve requirement. The hydraulic machines can be separate as at Ffestiniog where there is a separate pump and turbine on the same shaft, or combined into a reversible pump/turbine as at Dinorwig. The latter reflects current world practice.

In siting a pumped storage power station there are a number of important features as follows.

4.2 Suitable topology

The basic requirements for a pumped storage scheme comprise the following, which are shown in Fig 1.37 and relate specifically to the Dinorwig project:

- Upper reservoir.
- Low pressure tunnel.
- Surge shaft and pond.
- High pressure shaft and tunnel.
- High pressure penstock tunnels.
- Machine hall.
- Tailrace tunnel.
- Lower reservoir.

The most conventional arrangement of pumped storage scheme utilises an existing lake as the lower reservoir and a convenient mountain topology which would allow a high level feature such as a cirque or corrie to be transformed into a reservoir by the addition of a dam. Economics and environmental issues will dictate whether the hydraulic system and conduits are above or below ground. There are two other arrangements which have been considered. Underground reservoir pumped storage uses a nominal sea level reservoir as the upper reservoir and some arrangement of tunnels or caverns as the lower reservoir. Salt caverns, disused coal or mineral mines and purpose-built tunnel systems have been considered for this type of duty. The other possibility is the use of the sea as one of the reservoirs. In this case there is a significant potential difficulty with contamination of fresh water sources with salt water.

Pumped storage has only been shown to be economic when the available head between the two reservoirs is

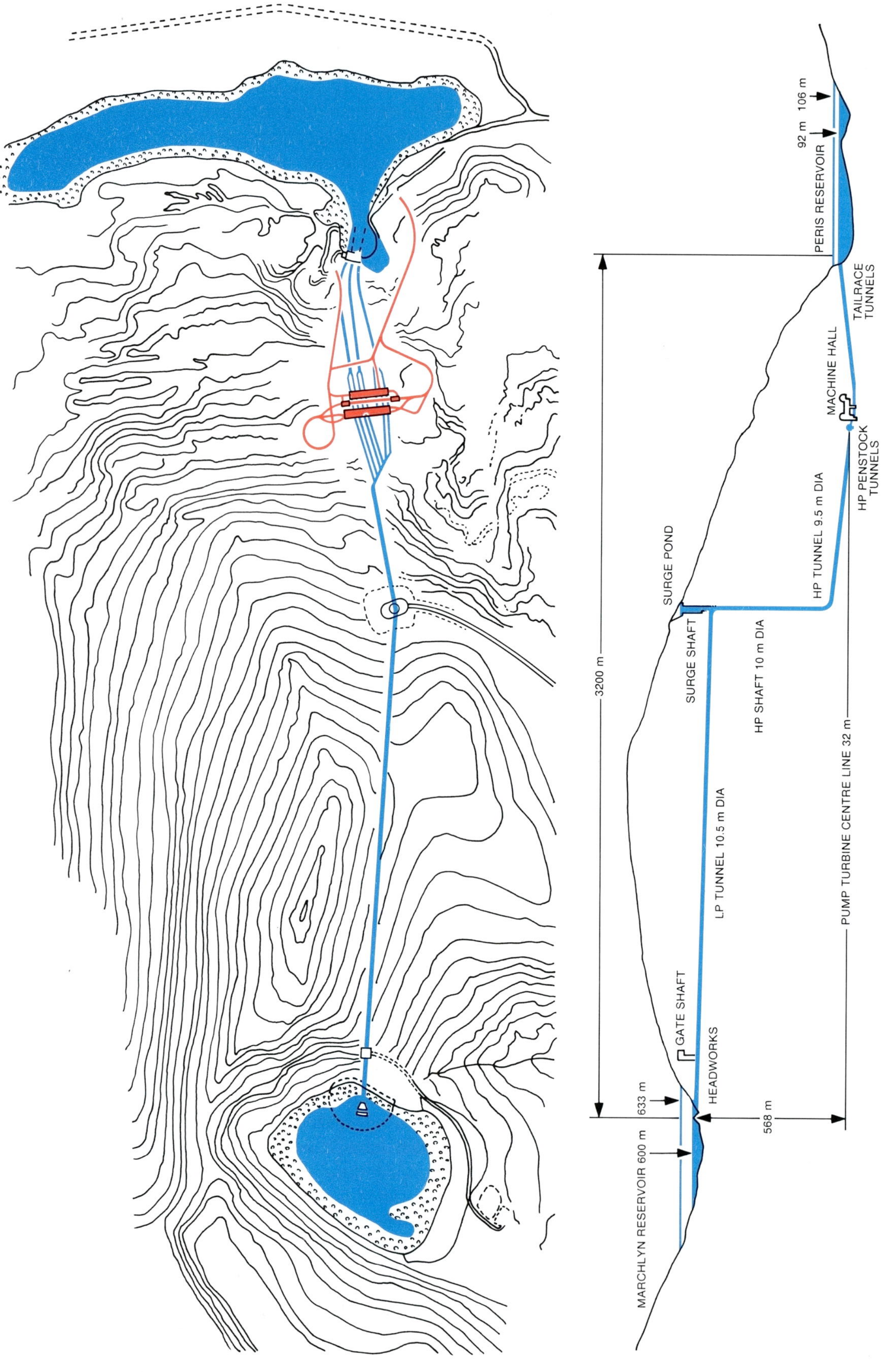

FIG. 1.37 Topographical section of Dinorwig pumped storage power station

relatively high (about 300 m) and in order that the hydraulic losses can be contained within reasonable bounds, the two reservoirs should be horizontally as close together as possible. The position of the power station is obtained by projecting a 10% gradient access tunnel line from the surface down to the level required for pump runner submergence and then checking that there is sufficient rock cover of suitable quality to withstand the operating pressures. This will locate the minimum length of the low pressure end of the system, and an economic evaluation must then be done to estimate the cost changes arising from further extension of the low pressure and access tunnels, in order to shorten the more expensive high pressure tunnels.

This requirement for high vertical displacement and small horizontal separation is only satisfied, in England and Wales, in the mountainous districts to the west of the country and in the Pennines, although a lower head scheme has been examined in North Devon. The studies in the early 1970s which led to the selection of the Dinorwig site, investigated three sites in detail. All of them were in the Snowdonia area of North Wales and were either wholly or partly inside the National Park.

4.3 Ground conditions

The requirements here are for ground that is either impervious to water leakage or which can be made impervious, e.g., by the injection of a chemical or concrete grout. All of the sites mentioned have slate as the underlying rock. These would be classified as hard rock sites allowing the construction of significant underground galleries without bracing and which can be made sufficiently watertight. The reservoir works would be developed using rockfill embankment dams.

In the early planning stages, assumptions must be made about the suitability of the ground and rock based on a very limited amount of data. Some data can be obtained by strategically placed boreholes and supported by at least one exploratory tunnel down into the power house location. Excavation costs can be estimated and a balance must be made between minimising the size of underground chambers and the difficulties which this strategy creates in terms of constructability and access for maintenance during station operation.

4.4 Site capacity

Pumped storage schemes are designed according to the following criteria:

- Pumping capacity and the time it takes to refill the upper reservoir, assuming all the pumps are in operation at full output.
- The return amount of energy which can be generated as a percentage of the pumping energy (charge energy factor).
- The requirements for quick start-up in the generating mode.

The first two criteria determine the size of pumping plant and the capacity of the upper reservoir, whilst the third has important effects on the hydraulic layout of the system.

As an example, the Dinorwig power station is designed for a pumping period of 6 hours each night, the six reversible pump-turbines are each nominally rated at 300 MW and the charge energy factor is 0.78. This allows the station to meet a generating criterion of full minimum head output for 5.4 hours per day. In order to meet this storage requirement at Dinorwig, the working volume of the reservoirs is 7 million cubic metres.

4.5 System and transmission requirements

Pumped storage can be economic in a number of scenarios, for example, to meet a reserve capacity and to limit the two-shifting operation of large units.

The requirement for reserve provides the biggest economic benefit because it allows expensive running of large plant at part load to be limited. In general, for a large modern pumped storage scheme, this reserve requirement would be met by the first station. Subsequent pumped storage schemes would be more difficult to justify until such time as the nuclear contribution (which is difficult to load cycle) exceeds the night-time trough demand.

Ideally, for the scheme to provide the best reserve capacity, it should be located close to the major load centres so as not to be at risk from grid disconnections. Unfortunately, the geology of the UK does not allow this and the Dinorwig station, which provides the system reserve, is located in a relatively isolated region. This places significant pressure on the security of the grid connections and the potential faults which occur during split grid operation. Particular care is therefore required in designing the electrical systems to protect the plant against dynamic oscillations and pole slipping.

Large generating stations in the UK would normally be connected to the grid by at least three circuits. In the case of Dinorwig, this was reduced to two single circuits by using underground cables for much of the routes. This avoided the problems of double circuit faults caused by storm conditions. The solution of using underground cables also solved the amenity problems of taking overhead lines through a National Park.

4.6 Hydraulic system requirements

The design of the hydraulic system and the layout features which need to be accommodated are determined to a large extent by the system requirements.

For a plant to meet the system reserve capability, the requirement is for a quick response.

At Dinorwig the requirement is for the plant to generate 1320 MW in 6 to 10 seconds. The plant is also designed to meet the other spinning reserve requirements and provide a frequency regulating duty. This imposes a requirement for up to 40 mode changes per day and 400 000 pressure cycles over the station life, so that fatigue is one of the design criteria for the high pressure parts of the system.

Fast start-up and mode changing is best achieved with as short a hydraulic system as possible to limit the pressure surge effects. The hydraulic machinery can be arranged for fast load pick-up in several ways:

- Running the system in hydraulic short-circuit with some of the units pumping, while the other units are generating. Load pick-up is achieved by tripping the pumping plant and rapidly bringing the turbines to full load.
- Running the generators synchronised with the system but generating no load. The operation of the inlet guide vanes is then the critical factor in achieving the required loading rate.
- Spinning in air is similar to the foregoing item except that the machines are motored from the system, with the pump-turbine dewatered by means of compressed air to reduce losses.

Studies may be required in the latter case to confirm the way in which the air is purged from the pump-turbines during the loading process. The operation of the main inlet valve is critical to this method of fast load pick-up which is the one adopted at Dinorwig.

In order to simplify construction, the power station complex should be located as close as possible to the lower reservoir. It must also have sufficient submergence at minimum water levels to avoid cavitation at the pump-turbine inlet. The power system conduits must have a very smooth profile in order to minimise operating friction losses and the design and construction process is simplified by maintaining the conduit operating pressures within a small number of fixed envelopes. The HP penstocks should be as short as possible as they represent the most highly stressed section of the hydraulic system and provision must be made for a surge shaft as close upstream of the power complex as possible to alleviate the upsurge during loss of generation. The size of the various hydraulic structures is optimised following a detailed surge study of all the normal transient events, as well as the credible sequences of events which can intensify the surge effects. These might include two successive station trips from full load pumping with the second station trip timed to give the worst level variations in the surge chamber.

At Dinorwig, a number of alternative tunnel schemes were examined including: a single high pressure shaft with three tailrace tunnels, twin high pressure shafts and tailraces; three high pressure shafts and tailraces.

The first scheme was the most economical, and after civil and reliability engineering design studies had shown it would give the required availability and taking account of tunnel inspection times, the arrangement shown in Fig 1.37 was adopted. Maximum station water demand is 420 m^3/s. The power/time criterion was the most important factor in tunnel sizing because of the need to accelerate the 2 km water column from standstill to full flow in six seconds. The system velocities were then checked and fixed by balancing the cost of various tunnel sizes and their energy losses, within the limits of previous experience, to ensure a satisfactory tunnel lining integrity and acceptable pressure surge levels. Figure 1.38 shows the optimised surge shaft design and gives the nominal design conduit velocities. Figure 1.39 shows the flow and water hammer pressures following complete trip of six turbines from full load.

4.7 Heavy load access

Because of its size and weight the generator motor is often built in-situ so that the largest loads brought to

System velocities corresponding to the extreme maximum station generating flow of 420 cubic metres per second are:

10.5 m dia LP tunnel	4.8 m/s
9.5 m dia HP tunnel	5.9 m/s
3.3 m dia penstock tunnel	8.2 m/s
2.5 m dia main inlet valve	14.3 m/s
3.75 m dia draft tube valve	6.4 m/s
8.25 m dia tailrace tunnel	2.6 m/s

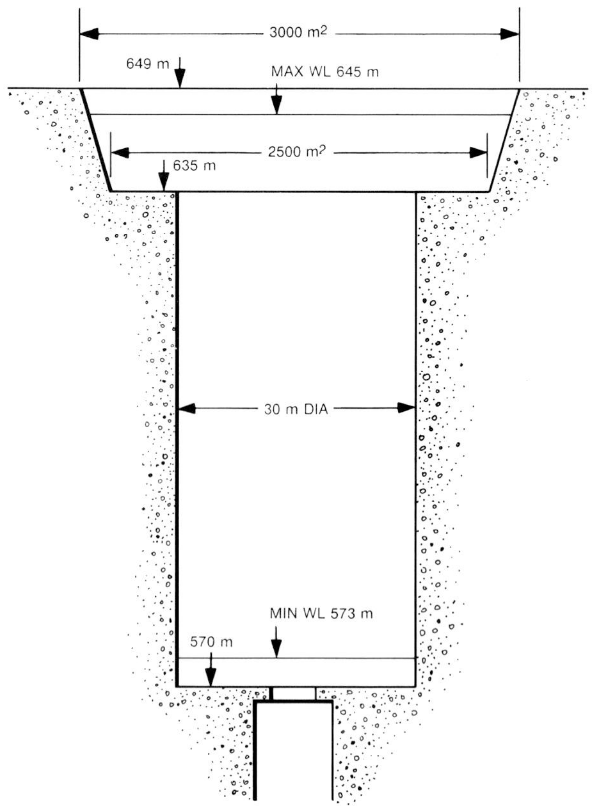

FIG. 1.38 Dinorwig conduit parameters

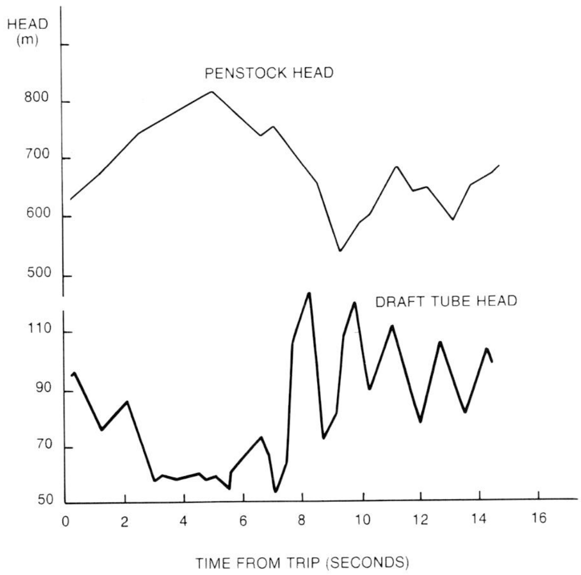

FIG. 1.39 Hydraulic transient response of 6-unit trip at Dinorwig.

site are the three-phase transformers. These may weigh up to 300 t. The shape of this load on its transport will determine the size of the access tunnels. The access tunnels can have a maximum gradient of 10%.

4.8 Availability of construction labour

The construction of a pumped storage scheme in mountainous regions away from centres of population provides problems of accommodation, education, transport and training and local resistance to the use of travelling men on construction works. Many of the construction activities can use local labour, possibly involving retraining for the job, but this strategy also has difficulties at the end of construction where the numbers employed in the finished power station may be very small. This can act as a demotivating influence on the workforce as the project nears completion.

4.9 Environmental impact

The environmental impact of a pumped storage power station raises many issues:

- Visual impact of station and electrical connections.
- Fisheries.
- Land taken for construction of reservoirs.
- Improvement of access routes and traffic movements.
- Import and export of construction materials.
- Restoration and repair of the natural landscape.
- Disposal of spoil from the construction works.
- Hazard studies such as flooding by overtopping of either of the reservoirs.

At Dinorwig, the environmental problems were particularly difficult because of the location of the site in the Snowdonia National Park. The decision to place the majority of the works underground solved most of the problems and careful attention to detail has limited the long term visual impact of the station by strategies such as: removing construction roads, landscaping to hide access roads, and collection and storage of topsoil prior to construction followed by replacement and seeding after construction with locally-collected heather and grass species (see Fig 1.19).

5 Gas turbines

5.1 Introduction

Due to the nature and overall simplicity of gas turbine plant, great flexibility may be exercised when considering its suitability for siting at specific geographical locations. Compared to other forms of power generation plant, gas turbine stations require relatively small land areas and a limited number of site resources. The major considerations for siting gas turbine plant are the transportation of fuel to the site and the integration of the electrical output into the transmission network. There is no requirement for a steam cycle which reduces both the overall capital cost of the station and the water requirements for the site. The size and loads of individual plant items installed are such that compared to a major power station development, access requirements to a gas turbine station site are comparatively modest. Consequently, gas turbine stations may be sited within large urban developments or near to major load centres with relative ease. For example, Fig 1.40 shows the location of the Watford gas turbine station.

5.2 The role of gas turbines

Gas turbines have been installed on the CEGB network for two main reasons; auxiliary power generation and peak load generation (*or peak lopping*).

5.2.1 Auxiliary power generation

The employment of gas turbines by the CEGB dates from 1961 when it was recognised that disconnection of a section of the grid could result in a severe lowering of grid voltage and frequency. In such circumstances, station outputs are progressively reduced through the lowered output of their frequency-conscious auxiliaries, and ultimately cascade tripping of stations occurs. Once disconnected from the grid, the loss of the auxiliaries means that the station cannot build up to power again, even though it is otherwise fully operational.

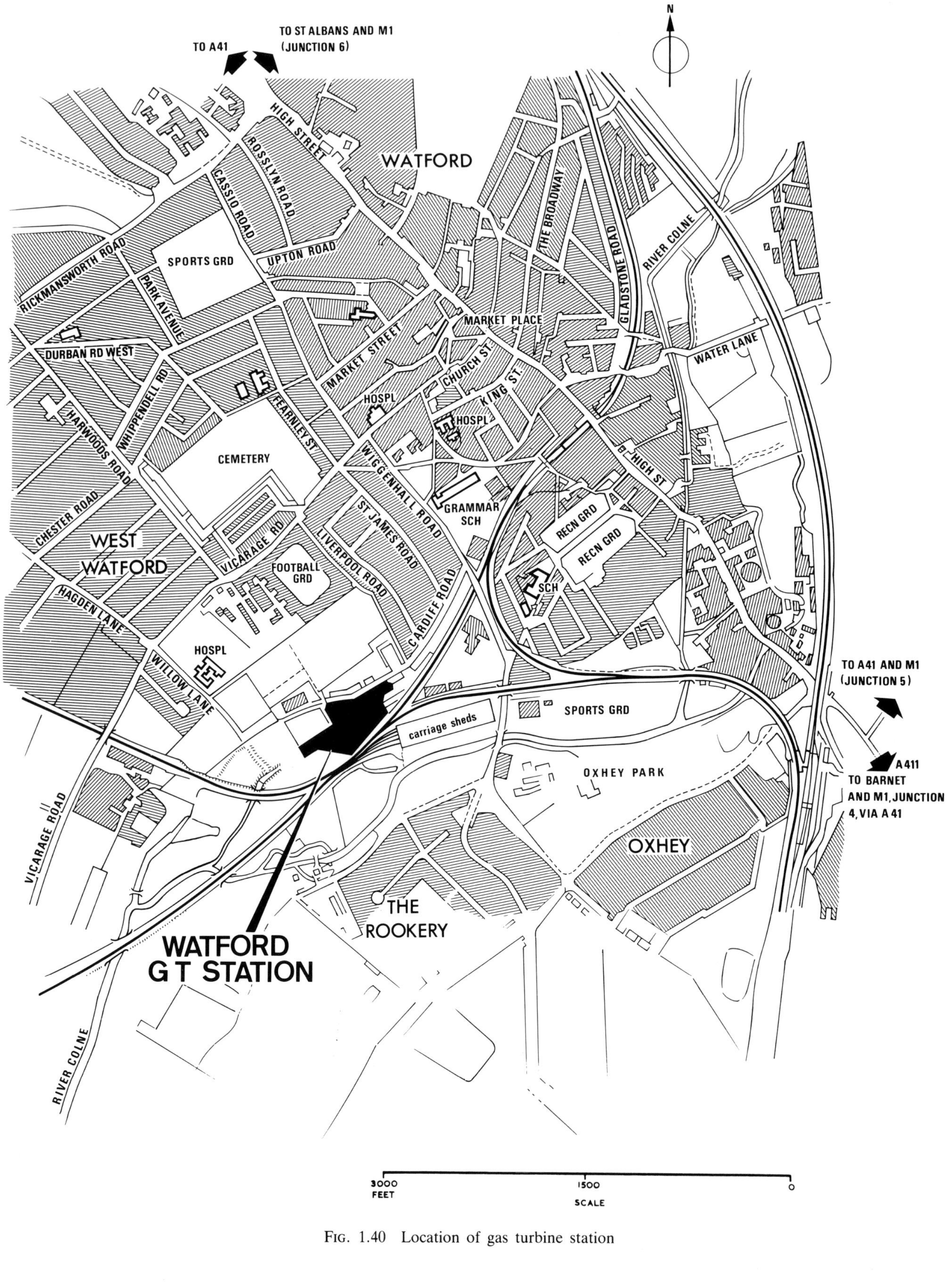

FIG. 1.40 Location of gas turbine station

It was therefore decided to arrange for the auxiliaries of the 500 MW unit stations, then being built, to be fed under lowered frequency conditions by an independent supply from a fixed gas turbine station, located within the site and adjacent to the main power station buildings. Gas turbines, in which the high pressure exhaust from jet engines is ducted to mechanically separate power turbines, were chosen for this purpose since no other prime mover was available which was of the right size and able to reach full load within several minutes of a start. The power turbines were directly coupled to AC generators whose size was fixed by the auxiliary needs of the stations in the range 17.5 MW to 29 MW.

Such gas turbines, based on aero engine technology, are able to provide the capability for starting main generating units from cold when disconnected from the grid (called a *black start*) as well as adding to its output for meeting peak load demands. On nuclear power stations these units may be used to supply power to emergency reactor cooling plant and other safety-related equipment in the event of a reactor trip.

5.2.2 Peak load generation

With the launching of the emergency sets installed at several large power stations, the CEGB was aware of a need for turbines of large capacity which could be placed on sites of old, inefficient steam stations. The result was the development of gas turbine power stations designed to meet peak load demands containing sets with outputs between 55 MW and 70 MW capacity which could easily be incorporated into the 132 kV network.

During times of high demand, these gas turbines are used in a *system support* role to back up the generation available from the major power stations. Their relatively short start times enable generation to be added to the power network quickly. This is important in controlling and stabilising the system frequency and in meeting sharp increases in consumer demand if there is little or no extra high merit plant capacity available.

In total, the CEGB has six gas turbine power stations currently in operation to meet such peak load requirements, each equipped with two generating sets. The total capacity of these stations is approximately 1300 MW, comprising units of up to 70 MW with four gas generators combined together in various ways to produce the required output (see Fig 1.41).

Although gas turbines have the advantages of low overall capital cost and relative ease of siting close to load centres (with consequent transmission economy and suitability for remote control operation), the low thermal efficiency and high fuel costs mean that this type of unit is only economic for annual load factors of up to approximately 5%. Hence, gas turbine stations are confined to peak lopping operations and to cope with system emergencies.

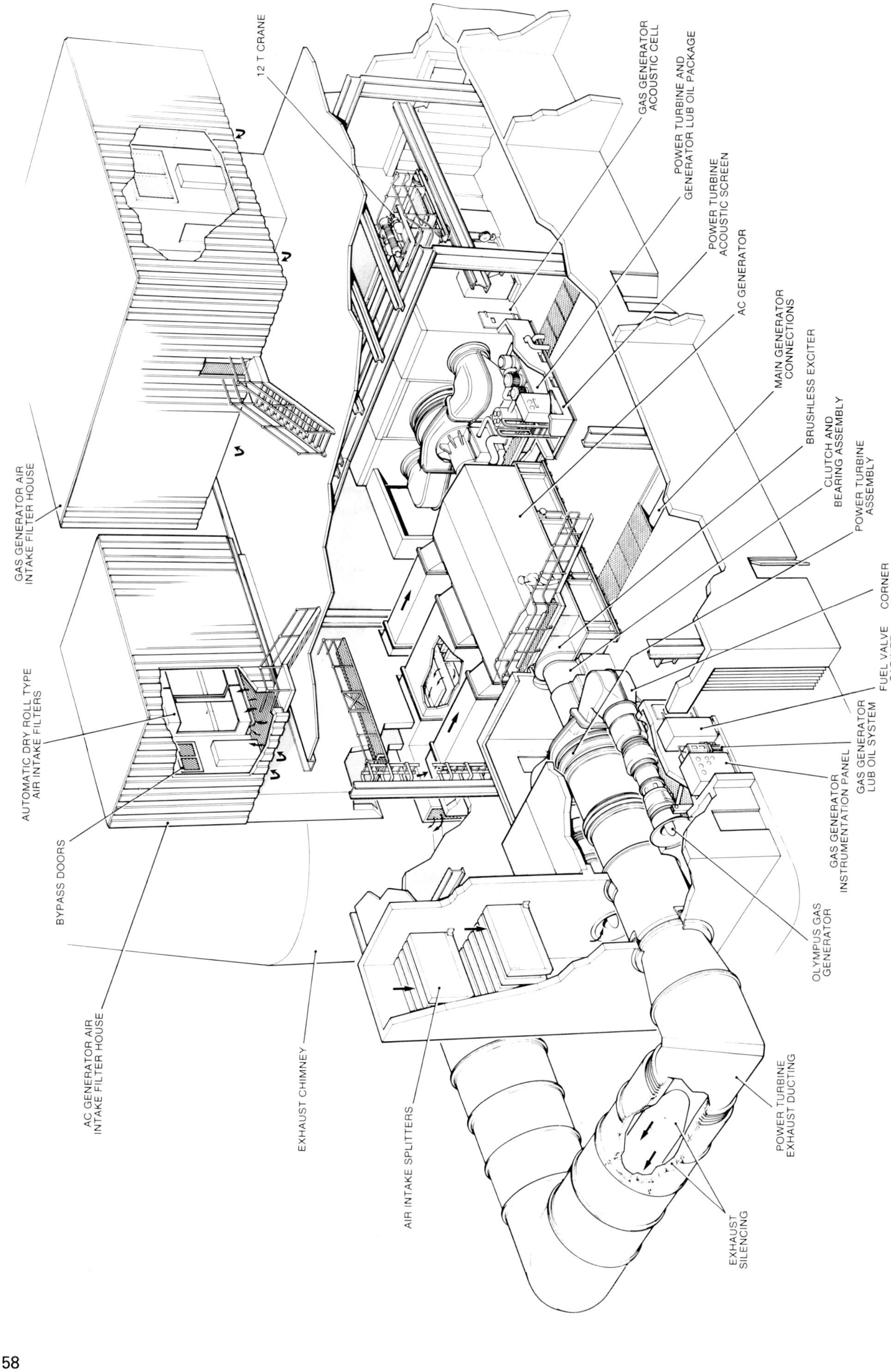

FIG. 1.41 Typical gas turbine station

CHAPTER 2

Station design and layout

1 Introduction

Power stations are complex arrangements of individual plant items, equipment and mechanical and electrical engineering systems. The term 'station' in its widest sense can be taken to include all the plant equipment, engineering systems and buildings which are normally accommodated within the confines of the site boundary, but it is often convenient to consider the design process as being sub-divided into two areas. Firstly, the main station buildings which contain the major plant items and systems such as the steam raising process and turbine-generators, and secondly, the auxiliary supporting systems and services such as the coal handling plant, ash handling plant, cooling water pumps, etc., which are often located around the site outside the main buildings. Whereas the design of the main building is, in the main, independent of site-related factors above foundation level, the design and layout of the major auxiliary systems is often influenced to a significant extent by site-specific features.

The content of this chapter follows this philosophy where, following a review of the major factors influencing the design process and the types of power stations operated by the CEGB, details of the layout considerations, which influence the design of the main plant areas, are given. The following sections of this chapter describe the features which have a major influence on auxiliary equipment and systems.

2 Power stations used on the CEGB system

The CEGB operates a number of different power station types to ensure a secure and cost effective supply of electricity, as is required by its statutory duties. Presently these may be classified under the following headings:

- Coal-fired.
- Oil-fired.
- Dual-fired (combination of coal and oil or natural gas).
- Nuclear.
- Hydro and pumped storage.
- Gas turbines.

The first four categories mentioned comprise the majority of stations operated on the CEGB system and utilise a similar concept, which is a steam raising source supplying steam to a turbine-generator to produce electrical power.

2.1 Fossil-fired stations

Coal, oil and dual-fired (either coal or oil) stations have many similar design features, with the main difference being the type of fuel used to generate steam in the boiler. Coal-fired stations require extensive fuel storage and handling facilities, ash collection and disposal facilities, and larger boilers than oil-fired ones due to the generally lower calorific value of the fuel. They are therefore inherently more expensive from a capital cost viewpoint than oil-fired stations. By implication, dual-fired stations are even more expensive due to the need for duplicate fuel storage and handling facilities.

Fossil-fired stations can be constructed employing large individual unit sizes; currently up to 660 MW is operational in the UK utilising both subcritical and supercritical steam cycles. Fossil-fired stations provide the bulk of generation on the CEGB system, currently accounting for more than 80% of the total system capacity. Figures 2.1, 2.2 and 2.3 illustrate typical stations on the CEGB system, showing coal-fired, oil-fired and dual-fired stations respectively.

FIG. 2.1 Drax 6 × 660 MW coal-fired station
(see also colour photograph between pp 66 and pp 67)

FIG. 2.2 Littlebrook D 3 × 660 MW oil-fired station
(see also colour photograph between pp 66 and pp 67)

FIG. 2.3 Kingsnorth 4 × 500 MW dual-fired station
(see also colour photograph between pp 66 and pp 67)

2.2 Nuclear stations

The CEGB currently operates two basic types of nuclear stations, magnox and advanced gas cooled reactor (AGR), both of which utilise gas cooled reactor technology.

In both cases steam is raised in boilers which are heated by the reactor coolant carbon dioxide, but because a magnesium alloy is utilised as the fuel cladding for the early magnox stations, the steam temperatures are limited. Similarly, the advanced gas cooled reactor stations use carbon dioxide as a coolant, but by utilising a more advanced design of fuel element, higher steam temperatures comparable with the most up to date fossil-fired stations can be reached.

Nuclear power stations currently contribute base load generation amounting to over 16% of the total system capacity.

Figures 2.4 and 2.5 show Oldbury magnox and Heysham *2* AGR nuclear stations respectively.

FIG. 2.4 Oldbury 2 × 313 MW magnox station
(see also colour photograph between pp 66 and pp 67)

FIG. 2.5 Heysham 2 2 × 660 MW AGR station
(see also colour photograph between pp 66 and pp 67)

2.3 Hydro-electric and pumped storage stations

The water resources of the UK do not allow for large scale run-of-river hydro-electric stations, but a few small scale schemes have been developed where they have been shown to be economic and environmentally acceptable.

Water storage schemes have also been used which allow some flexibility of operation in relation to system needs, but again these have only been generally viable on a modest scale.

Figures 2.6 and 2.7 show the pumped storage station at Ffestiniog and the hydro station at Kielder, respectively.

FIG. 2.6 Ffestiniog 4 × 90 MW pumped storage station (see also colour photograph between pp 66 and pp 67)

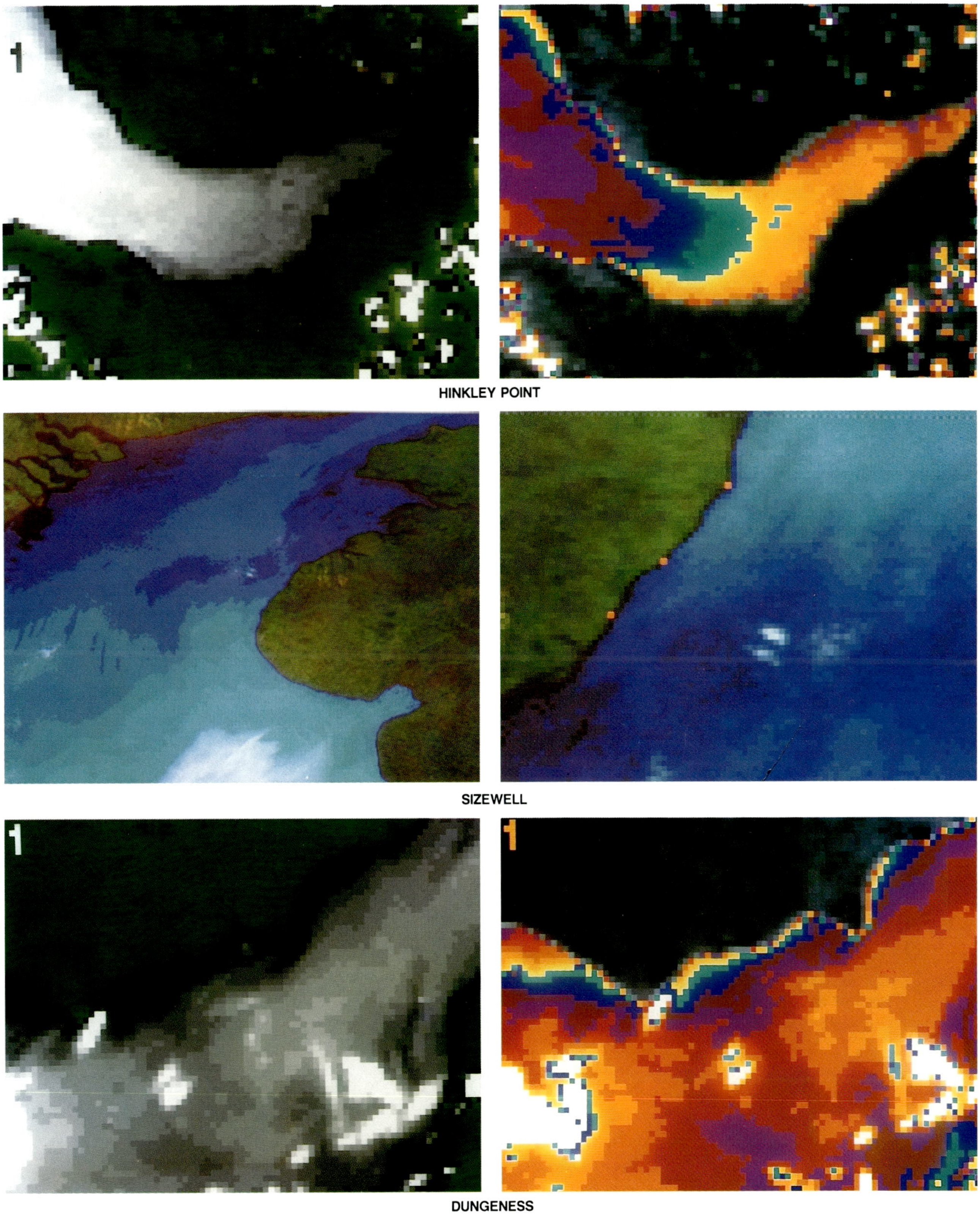

FIG. 1.9 Satellite imagery of thermal discharges from power stations

(a) AIR CUSHION TRANSPORTER

(b) ROLL-ON / ROLL-OFF VESSEL

FIG. 1.11 Transport of heavy or abnormal loads

Fig. 1.12 Ash disposal site

FIG. 1.15 Cooling towers at Drax power station

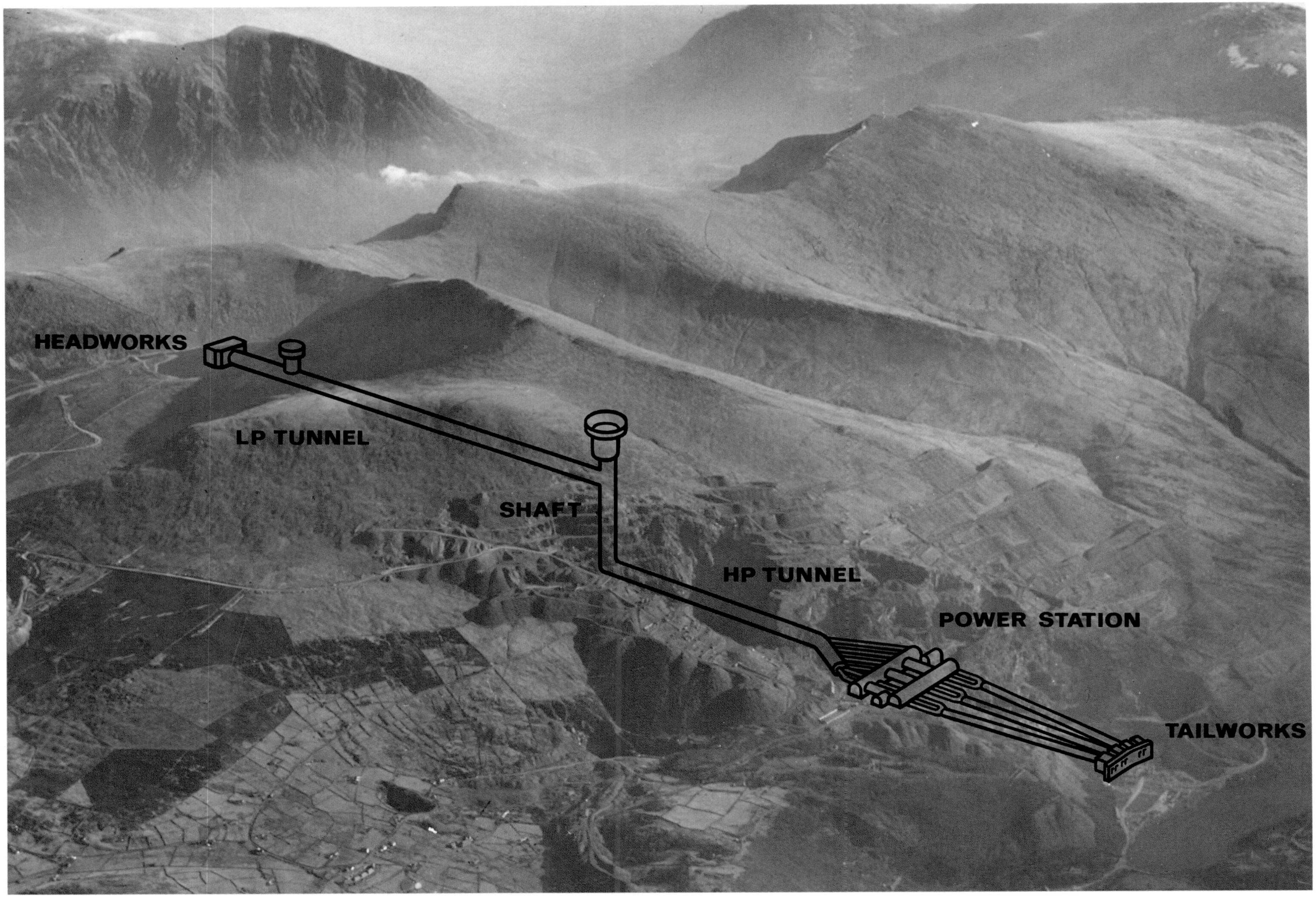

FIG. 1.19 Hidden power: Dinorwig pumped storage power station

FIG. 1.20 View of Drax power station nearing completion

FIG. 1.33 Didcot coal-fired station landscape

FIG. 2.1 Drax 6 × 660 MW coal-fired station

FIG. 2.2 Littlebrook D 3 × 660 MW oil-fired station

FIG. 2.3 Kingsnorth 4 × 500 MW dual-fired station

FIG. 2.4 Oldbury 2 × 313 MW magnox station

FIG. 2.5 Heysham 2 2 × 660 MW AGR station

FIG. 2.6 Ffestiniog 4 × 90 MW pumped storage station

FIG. 2.7 Kielder 1 × 5.5 MW hydro station

Fig. 2.8 Cowes 2 × 70 MW gas-turbine station

FIG. 2.9 Leicester 2 × 70 MW gas-turbine station

FIG. 2.14 Wind turbines at Carmarthen Bay site

FIG. 2.26 Turbine island concept

FIG. 2.67 View of pumphouse and intake screens — Littlebrook D

FIG. 2.80 Bucket wheel stocking-out reclaiming machine

FIG. 2.7 Kielder 1 × 5.5 MW hydro station
(see also colour photograph between pp 66 and pp 67)

Pumped storage power stations can offer an economic and flexible generation capacity where sites with suitable rainfall, geological strata and topography allow the construction of two reservoirs at different levels. These stations have a specific role on the CEGB system.

At off-peak periods cheap power can be provided from highly efficient base load fossil-fired and nuclear plant to pump water, through reversible turbines, up from the lower reservoir into the higher level storage reservoir. At times of peak or emergency demand the water is allowed to flow under gravity (as in a natural hydro-electric system), and its hydraulic head is used to generate power. By this means, improved use is made of the best thermal power stations in the system and power is provided at peak periods at a lower cost than that otherwise provided by less efficient power stations. The fuel which is saved, often more than offsets the additional fuel consumed in providing energy for the pumping. The re-use of water allows a much greater generating capacity without increasing the civil engineering costs, which make up the bulk of the capital cost of pumped storage schemes.

In addition to providing a peak supply, the plant can also be used for standby duty since it can be ready for operation at very short notice to cover unexpected load increases, or breakdown elsewhere on the system.

Considerable flexibility is possible in planning storage schemes. The variables of pumped capacity, pumping periods and reservoir capacity can be considered together to give the most economical scheme when integrated with the system loading and the other plant in the system.

Because of its less complicated design, pumped storage can be built at a lower cost than conventional thermal power stations, and it needs only a small staff to operate it.

2.4 Gas turbine stations

Gas turbine stations are a further method of providing generation capacity. Two types of gas turbines are available for this purpose, the aero-engine-derived units employing modified aircraft engines as the gas generators, and a heavy duty or industrial type which is purpose-built for power generation purposes. Both types employ a similar conversion process where high temperature gas at low pressure is expanded in a power turbine which is directly coupled to an electrical generator.

The two types of plant have different performance characteristics. The aero-derived units range in capacity to about 70 MW by utilising up to four separate jet engines to provide the necessary gas flow rates. A particularly useful feature of this type of gas turbine is its ability to reach full load within 1 and 2 minutes of starting. Figure 2.8 shows the CEGB gas turbine station at Cowes, Isle of Wight.

The heavy duty gas turbines have longer starting times, typically about 15 to 20 minutes from cold, but have individual capacities ranging up to some 190 MW at the present time. Two such heavy duty units are installed at the CEGB's Leicester Power Station, see Fig 2.9.

Gas turbine plant utilises a simple generation cycle which, apart from fuel delivery and transmission connections, requires little in the way of site facilities. Such simplicity of cycle leads to the plant being comparatively cheap in capital cost terms and fairly cheap to construct.

To achieve peak performance, clean premium fuels such as natural gas or distillate oil is required, but because of the relatively high cost of these fuels, the overall economics of operation tend to limit their application on the CEGB system to peak lopping or emergency duties. Typical annual load factors of less then 5% are achieved when used in the system support role. Less expensive oil fuels can be used with the heavy duty machines provided pre-treatment is undertaken to remove some impurities. However, operational experience tends to be limited and that which is

FIG. 2.8 Cowes 2 × 70 MW gas-turbine station
(see also colour photograph between pp 66 and pp pp 67)

FIG. 2.9 Leicester 2 × 70 MW gas-turbine station
(see also colour photograph between pp 66 and pp 67)

available suggests some penalty on performance and increased maintenance attention is incurred.

Gas turbine units can be installed in common buildings as conventional fossil-fired units, or can be supplied in individual packaged units ready for installation on a prepared base.

3 Future development options

The CEGB, as a major operating utility, regularly reviews the technical and economic status of the various generating systems which could be available for future use.

The outline reviews discussed in the following sections are obviously related to the current state of development of the CEGB system, and apart from future projects which are actually committed for construction, they must be regarded as somewhat speculative. Hopefully, however, it will illustrate to the reader that the CEGB retains and fosters an active interest in future generation options and expends considerable resources in this direction.

3.1 Fossil-fired plant

The CEGB considers that coal-fired stations will continue to provide a substantial contribution to UK energy needs for the foreseeable future. A 'reference design' for future 2 × 900 MW coal-fired stations, utilising a subcritical steam cycle, was completed in 1989 and, subject to the satisfactory conclusion of siting studies and consent applications, the CEGB hopes to commence construction of the first station in the near future. Outline details of the station layout are shown in Fig 2.10.

A major implication on the layout and operation of new fossil-fired stations is the requirement for flue gas desulphurisation (FGD) equipment to be fitted. This plant, installed to meet increasing environmental concerns, requires additional buildings for the process plant and land for the handling and storage of the feedstocks and by-products of the flue gas treatment. Studies into coal gasification and supercritical steam cycles are continuing for possible future adoption.

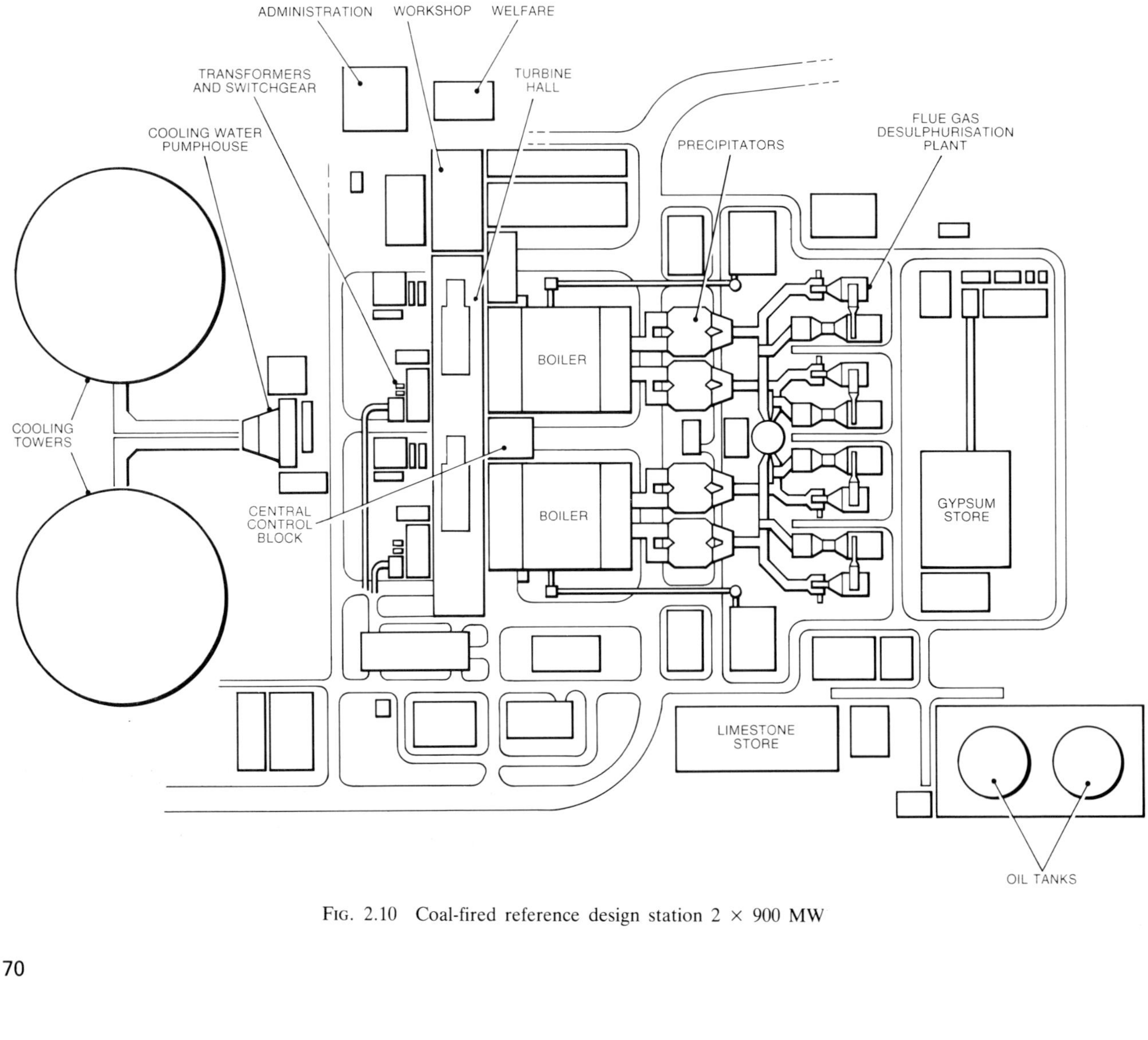

Fig. 2.10 Coal-fired reference design station 2 × 900 MW

3.2 Nuclear stations

The CEGB has adopted the pressurised water reactor (PWR) as the basis for its new nuclear station which is being built at Sizewell *B*; construction works commenced in late 1987. It is hoped this 1200 MW 'reference design' will form the basis for a small programme of stations utilising similar technology which will be located at other suitable sites within the UK, subject to consents being given. Figures 2.11 and 2.12 show respectively a station plan and section through a typical PWR station.

3.3 Combined cycle gas turbines

These plants are an extension of gas turbine station developments where waste heat boilers are installed to recover heat from the gas turbine exhaust. This heat is utilised to produce steam which is employed to drive a steam turbine-generator, which can be either unitised with the individual gas turbines or ranged across a number of units.

The technology of such plants is well-established and their main benefit is the higher overall cycle efficiency which can be achieved. However, a premium fuel is normally required, and careful consideration of the relative merits of increased capital cost, long term fuel prices and reliability is required before such a plant can be installed. The CEGB has considered the station design options which could be adopted for this type of plant and Fig 2.13 illustrates a possible general plant configuration.

3.4 Wind power

Wind turbine-generators promise to play a prominent part in the field of alternative sources of energy. A 250 kW horizontal axis and a 100 kW vertical axis wind turbine generator have been installed as demonstration units on the CEGB's Carmarthen Bay site, and a 1 MW machine is to be constructed at the Richborough site in the near future. There are also several other units with capacities up to 3 MW operational in the UK.

Figure 2.14 shows the wind turbines at the Carmarthen Bay site.

The CEGB is supportive of the efforts of manufacturers to develop this source of energy and besides co-operating with development proposals as they arise, is actively looking at the potential for wind turbine arrays for providing a contribution to its system capacity.

3.5 Tidal power

The potential for harnessing the tidal power of the River Severn and the River Mersey estuaries is being investigated. If found to be practicable and economic, such schemes could provide about 6% of the CEGB system needs, but probably not before the year 2000 at the earliest.

3.6 Geothermal energy

The CEGB will be providing substantial support for investigations into hot dry rock technology which will be carried out in the UK over the next few years. This involves tapping into hot dry rocks of 200°C and circulating water down to 6 km depths through fissures enlarged by hydraulic fracturing in order to capture heat. The best sites appear to exist in National Parks at exposed granite quarries in the south-west and north-west of England. Reservoir behaviour, however, is proving currently more complex than anticipated.

3.7 Combined heat and power

Combined heat and power schemes have been used extensively by overseas utilities but not on a significant scale to date in the UK. The CEGB, however, remains interested in pursuing such schemes as and when suitable opportunities arise.

4 Station design concepts

4.1 Basic considerations

Major power stations on the CEGB system have for many years been planned from the onset to be completed with a given number of units of the same rating and to a similar layout. Where additional capacity is found to be necessary at a particular site in the future, then this has been accomplished by constructing a second *B* or third *C* station and so on until the total site capability has been developed.

The advantages of this process are that internally to an individual station's development, the units are in all major aspects identical, and therefore the detailed station design is to a great measure reduced to designing the first unit only. There are also operational advantages in having a standard arrangement for each unit in a station and a reduction in the quantity of spare parts which it is necessary to retain in stock. This concept of identical layout for each unit has become current practice for stations containing two or more units.

In terms of overall site layout, the construction of additional generating capacity by station rather than individual unit extensions has some practical advantages. The design and construction of any new capacity can proceed independently of any existing generation facilities on the site and the designer is not required to make provision for some unknown future requirement. This allows the station design to be optimised with the proviso that it does not preclude future development of the remaining site area.

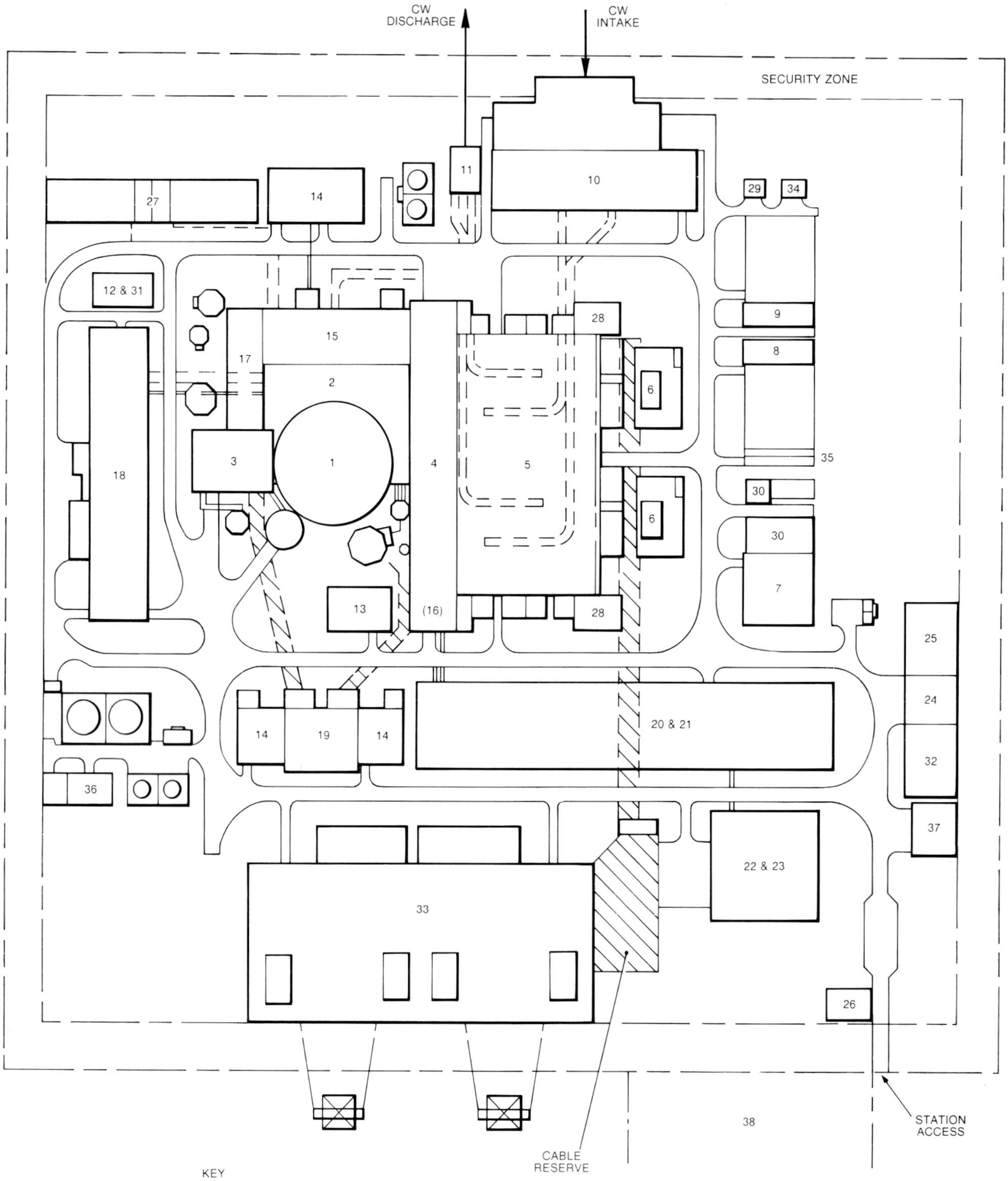

KEY

1 REACTOR BUILDING
2 AUXILIARY BUILDING
3 FUEL BUILDING
4 TURBINE HOUSE ANNEXE
5 TURBINE HOUSE
6 GENERATOR TRANSFORMER
7 WATER TREATMENT PLANT
8 HYDROGEN PRODUCTION PLANT
9 HYPOCHLORITE PRODUCTION PLANT
10 CIRCULATING WATER PUMPHOUSE
11 SEAL PIT
12 NITROGEN STORE
13 AUXILIARY BOILER HOUSE
14 DIESEL HOUSE
15 CONTROL BUILDING
16 ACCESS CONTROL ANNEXE
17 DECONTAMINATION SHOP
18 RADWASTE BUILDING
19 AUXILIARY SHUTDOWN BUILDING
20 WORKSHOPS
21 MAIN STORES BUILDING
22 WELFARE BUILDING
23 ADMINISTRATION BUILDING
24 FIRE FIGHTING PUMPHOUSE
25 TOWN WATER RESERVOIR
26 GATE HOUSE
27 RESERVE ULTIMATE HEAT SINK
28 STATION TRANSFORMER
29 CO_2 STORE
30 BULK CHEMICAL STORE
31 OXYGEN STORE
32 EMERGENCY FEED WATER MAKE-UP RESERVOIR
33 400 kV SUBSTATION
34 GAS BOTTLE STORE
35 BULK HYDROGEN (CYLINDER) STORE
36 LUB OIL / GREASE / SOLVENT STORE
37 GARAGES / FIRE STATION
38 CAR PARK

FIG. 2.11 PWR station layout

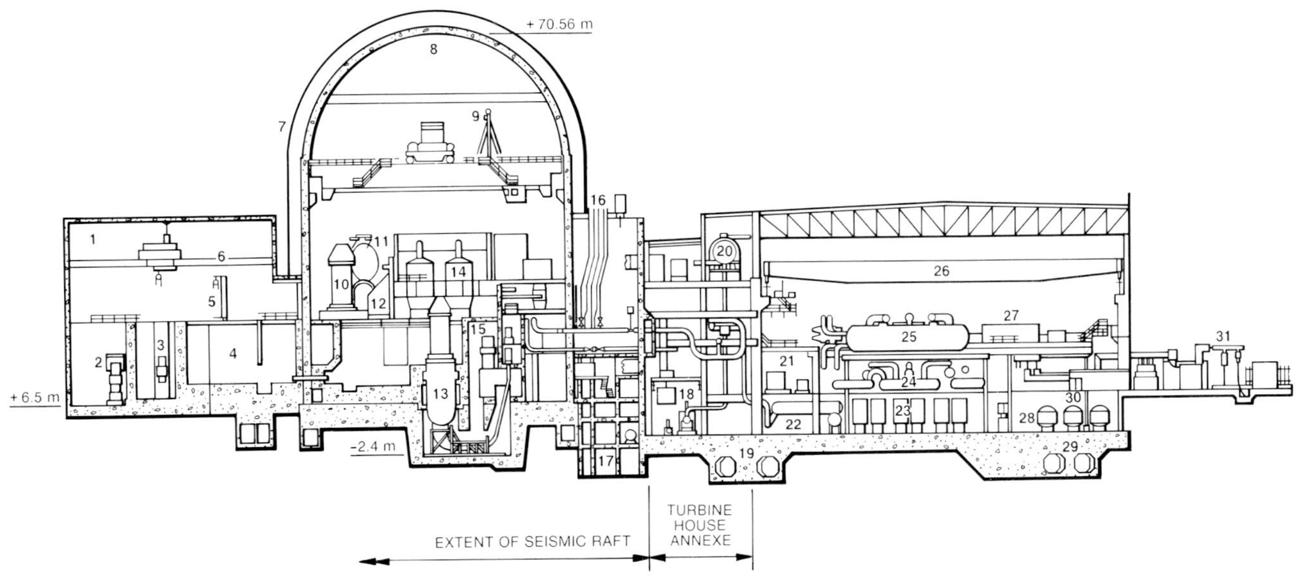

KEY

1 FUEL BUILDING
2 FLASK RECEPTION
3 FLASK DECONTAMINATION
4 FUEL TRANSFER CANAL
5 FUEL HANDLING MACHINE
6 FLASK HANDLING CRANE
7 SECONDARY CONTAINMENT
8 REACTOR BUILDING
9 POLAR CRANE
10 HEAD PACKAGE MAINTENANCE
11 PLANT ACCESS HATCH
12 REFUELLING MACHINE
13 REACTOR PRESSURE VESSEL
14 STEAM GENERATORS (4)
15 REACTOR COOLANT PUMPS
16 SAFETY VALVE VENTS
17 CABLE TUNNELS
18 FEED PUMPS (6)
19 CW OUTLET
20 DEAERATOR
21 LUB OIL PLANT
22 HP HEATERS
23 CONDENSERS
24 LP HEATERS
25 MOISTURE SEPARATOR REHEATER
26 TURBINE HOUSE CRANE (2)
27 TURBINE GENERATOR
28 EXTRACTION PUMPS
29 CW INLET
30 POLISHING PLANT
31 GENERATOR TRANSFORMER

FIG. 2.12 Section through PWR station

Station developments allow technological changes to be accommodated in discrete steps and with a reasonable number of installed units, which allows an effective and efficient supporting infrastructure to be provided. It also permits an operationally-efficient staffing regime to be established avoiding the complexities of different technologies and operating procedures on the same station.

Ultimately, decommissioning and demolition of time-expired stations can be undertaken in a discrete package with the cleared area again being available for redevelopment as and when required, independently of any other generation facilities on the site. Figure 2.15 shows how this philosophy is currently being applied to the CEGB's Hams Hall site.

4.2 Design objectives

The objectives of any aspect of power station design, given a specific fuel type and choice of steam cycle, are to achieve the lowest capital cost and ease of construction, together with simplicity and efficiency in the operation and maintenance of the station over its projected life. These objectives are not easily achieved. Realistic station designs are evolved over long periods of time and result from the input of experience and the continued re-evaluation of the variables which influence the design process. The experience and the rules and regulations which contribute to this process may vary from utility to utility and therefore a differing emphasis may be placed on the contributing elements which, ultimately, can influence the final result. However, whichever station design is finally adopted the process will require consideration of a range of factors which influence the primary objectives and which are outlined as follows; the order is not intended to denote priority as these will be utility specific:

Efficient operation

- Reliability of operation.
- Safety in operation.
- Simplicity of operation.

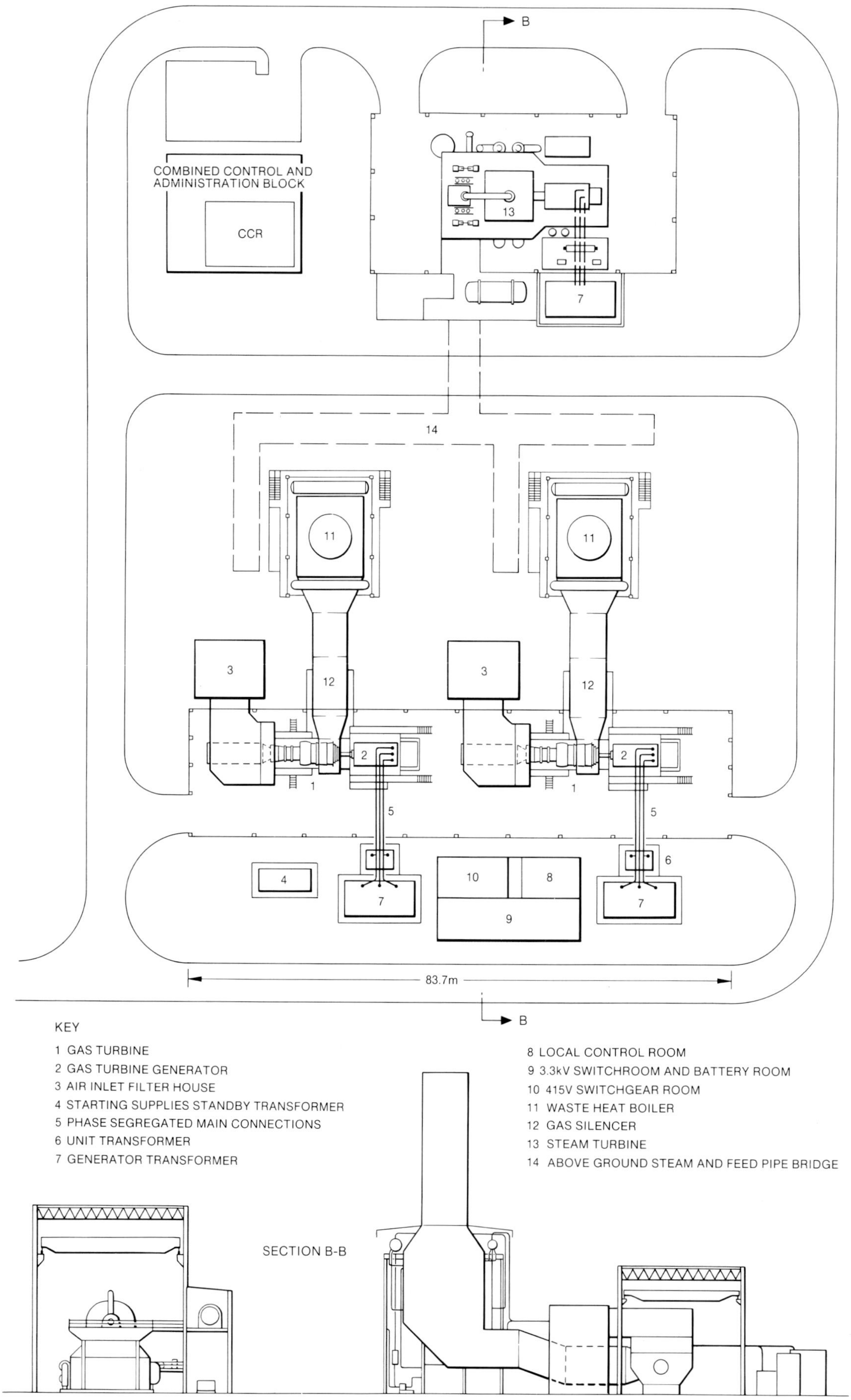

FIG. 2.13 Combined cycle gas turbine station layout

FIG. 2.14 Wind turbines at Carmarthen Bay site
(see also colour photograph between pp 66 and pp 67)

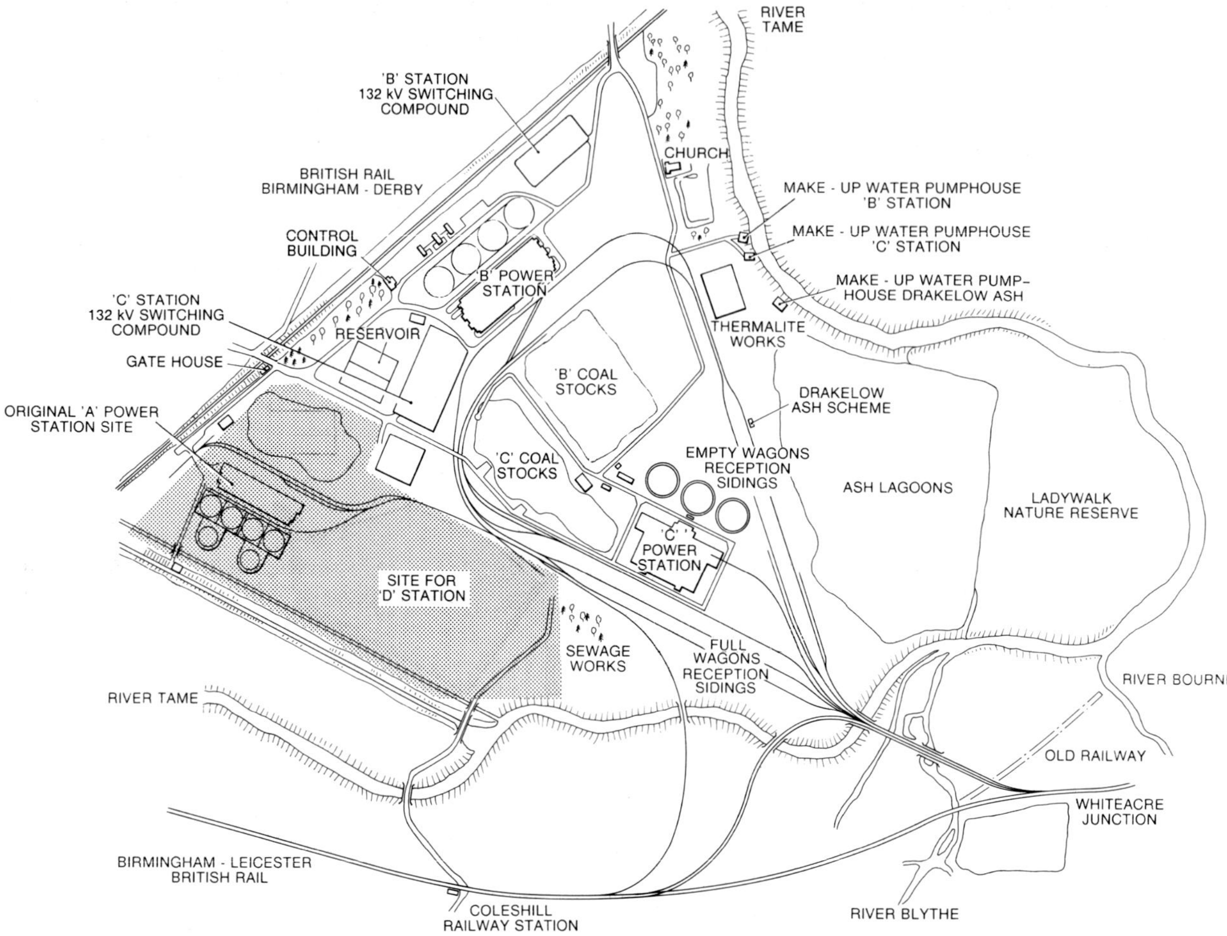

FIG. 2.15 Redevelopment of existing power station site for new station

- Good working conditions.
- Ease of maintenance.
- Service experience.

Minimum overall cost

- Low capital cost.
- Low operating cost.
- Construction efficiency.
- Plant life.
- Service experience.

Within each of the major design objectives consideration of the factors listed will also need to take into account the following:

- Statutory legislation.
- Environmental protection measures.
- Local resources and expertise available.
- Siting restraints.

The role of the designer is to evaluate these factors and arrive at the optimum design for a given proposal. Although in control of some parameters and not others, the designer is required to undertake the engineering and cost optimisation studies within the limits of imposed constraints to meet the project objectives.

The following sections of this chapter deal with basic station and plant layout from a CEGB viewpoint. Other volumes discuss in more detail the specific technical considerations relating to the design of plant for CEGB power stations.

5 Plant operation

The ability of station staff to efficiently manage the plant operation depends not only on the plant and equipment specifications but also on the layout. The development of main lines of communication and maintenance access throughout the plant are a fundamental layout activity and need to be established at an early stage in the conceptual design of the station.

Personnel movement around the station can be established by assessing the relationship between the main permanently-manned areas of the station and the major plant items, and between the manned areas themselves. Main personnel assembly points are likely to be:

- Administrative offices.

- Staff welfare and amenities centre.
- Workshops, work planning and stores centre.
- Main control room.
- Subsidiary control rooms such as for coal and ash handling plant.
- Laboratories.

All of these centres must service each of the installed generating units equally and provide for efficient and rapid personnel movements between centres where an important interrelationship can be established.

Reviews of personnel movements within operational stations of the CEGB have enabled these factors to be evaluated and station layouts developed which promote economic and efficient operational practices. Essentially, these studies have shown the benefits of establishing a station management centre where all major activities are grouped together, apart from those directly involved with plant monitoring and control such as the main and subsidiary control rooms.

These centralised facilities are positioned as far as possible to meet the needs of the site overall, and Chapter 1 Section 3 reviews their location in relation to the principal movement sequences for both station personnel and vehicles.

The position of the station central control room (CCR) is of high importance because it is the centre of plant monitoring and control. It needs to be located in an area where it is central to all unit operations, yet accessible during, but secure against, any major incident however unlikely.

The CCR may be at the centre of the station at operating floor level, typically on the boiler house side in the mechanical annex or between boilers, or on the side of the turbine house remote from the boilers. This latter position has been adopted on many multi-unit stations of the CEGB as illustrated in Figs 2.16 and

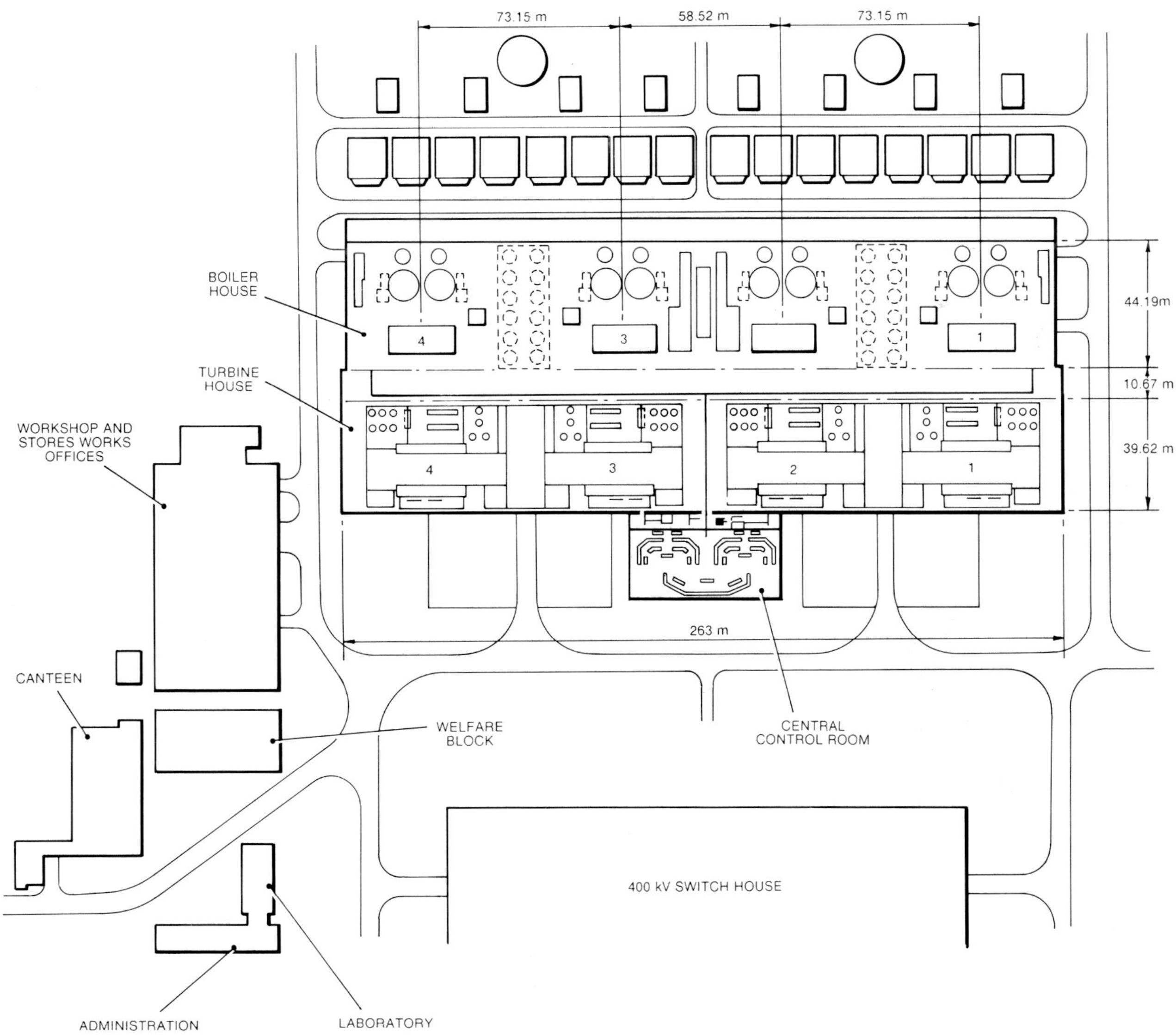

Fig. 2.16 Typical main buildings plant layout 4 × 500 MW longitudinal sets

2.17. Consideration is being given to adopting a central control room location adjacent to the mechanical annexe for the future 2 × 900 MW stations.

Vertical communication must also be given adequate consideration and stairways and lifts need to be correctly integrated with horizontal access facilities. For larger items of equipment and infrequent major plant movements, hoisting wells should be provided at strategic locations throughout the station, again being co-ordinated with horizontal access routes.

6 Station layout

6.1 General

Station layout is concerned with the logical and economic use of space and the relationship of one piece of plant with another.

An efficient layout of plant and systems minimises losses and therefore running costs. Ideally, plant items should be located as close as practical, but the designer has to ensure that adequate access for operation and maintenance is provided. The best overall design therefore is one which strikes the correct balance between lowest cost and the best arrangement from both constructional and operational points of view.

Given that the overall station design is a compromise of the various factors, it is still the aim on the grounds of economy to keep the overall main building dimensions to a minimum, and it is left to the ingenuity of the designer and the utility experience to determine the correct balance to accept between the competing variables.

Over the years, the basic layout of stations in the CEGB has developed into a general standard pattern, irrespective of the number of generating units installed. However, the overall plant arrangement within the

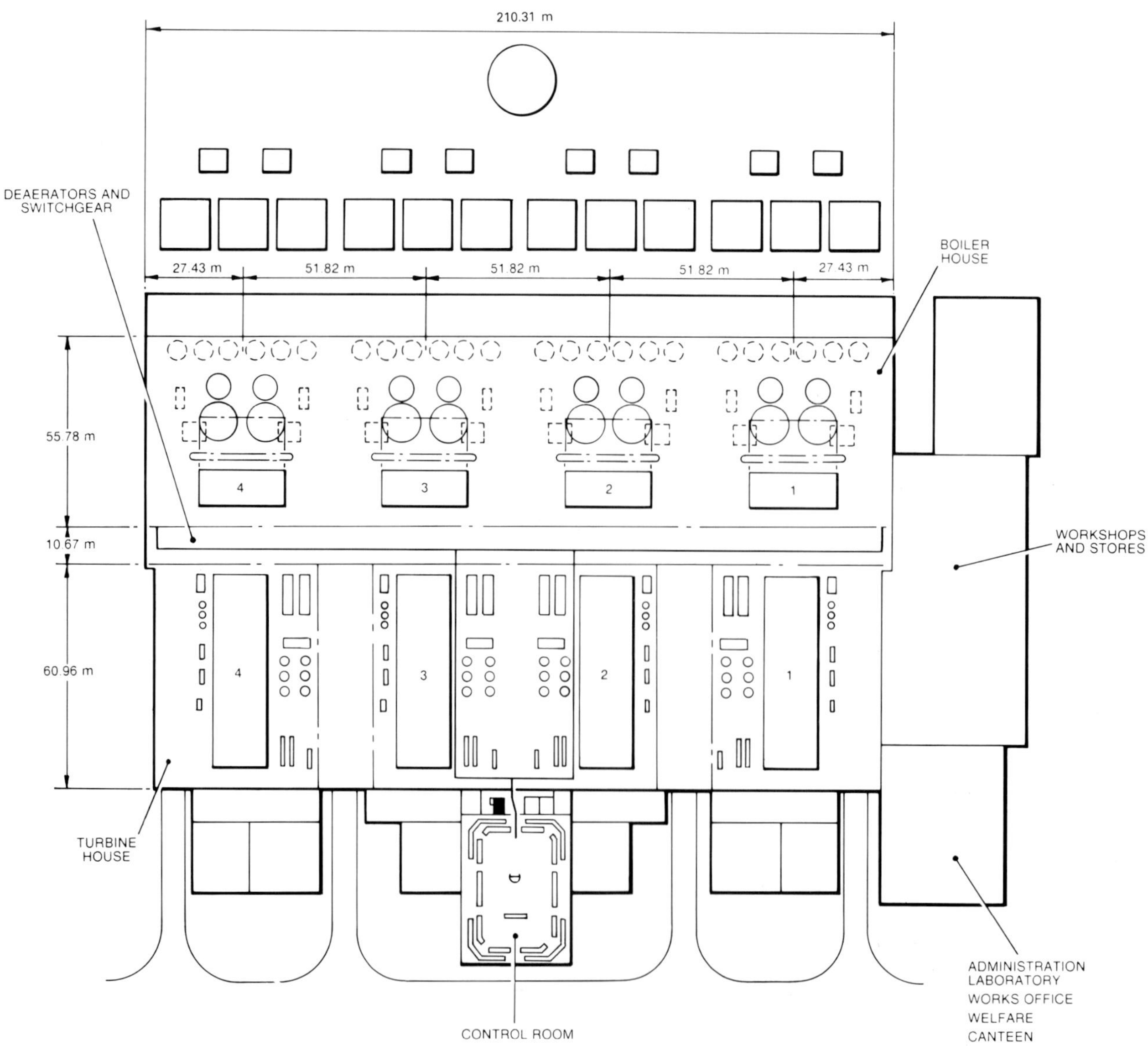

FIG. 2.17 Typical main buildings plant layout 4 × 500 MW transverse sets

building envelope has varied depending upon the design and manufacture of the main plant items, particularly boiler and turbine.

This basic pattern consists of an integrated building structure with boiler and turbine houses arranged in parallel but usually separated by an annexe containing mechanical and electrical auxiliary plant items and systems. To the rear of the boiler house are located the precipitators and chimney, whilst on the *open* side of the turbine hall are usually located the transformers and some auxiliary switchgear. Figures 2.18 to 2.21 inclusive show a number of station arrangements illustrating these principles.

6.2 Main plant orientation

The single most important decision which influences overall station layout is perhaps the choice of the relationship between the turbine-generator and the boiler, since both the boiler centreline spacing and dimensions of the turbine hall can be significantly influenced by this decision. It is not a simple decision. Traditional practices, the choice of plant supplier, plant specification, the relationship between civil and electrical and mechanical engineering costs, the scope of supply of particular contractors, site restraints and perhaps even the engineers' likes and dislikes may influence the final outcome. The overall number of permutations which can be developed considering all the potentially variable factors is large, but in practice the decision often comes down to the choice of a few options which experience within a utility has shown to provide economic and practical solutions. These are illustrated in Fig 2.18 to 2.21.

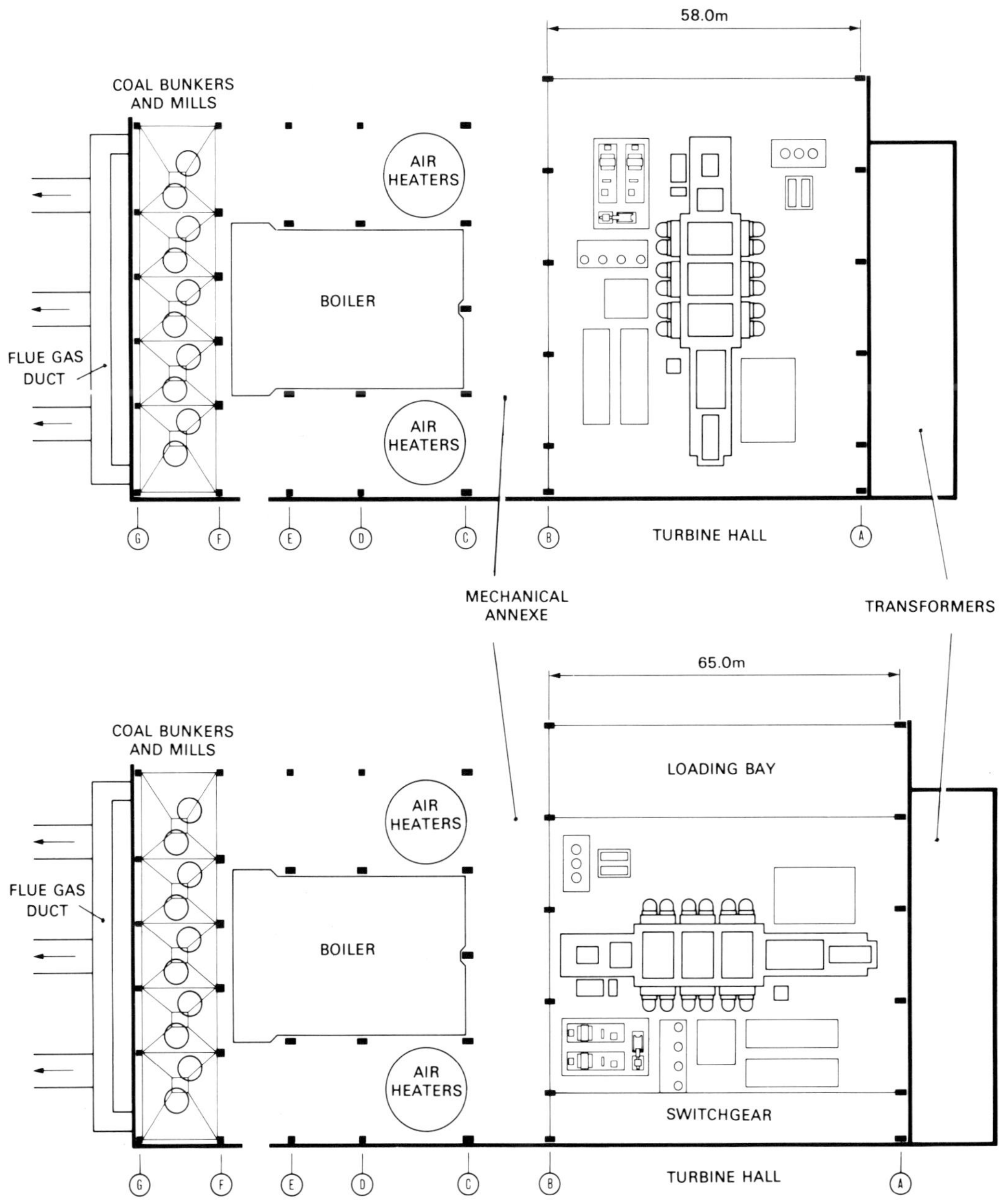

FIG. 2.18 Longitudinal and transverse arrangements of 6-flow LP turbine-generators

6.2.1 Turbine-generator plant

With fossil-fired plant, the initial determination is between transverse, longitudinal or angled layout of the turbine-generator. With 4-flow low pressure (LP) turbines, the overall dimensions of the turbine with the condensing, feedheating and general turbine auxiliary plant alongside produces a plan area approximately square. With a 6-flow LP turbine, a similar arrangement of auxiliary equipment produces a rectangular area.

These areas must be covered by the main overhead cranes or have individual facilities, so the plan area of the turbine hall follows from the alignment of the approximate number of units installed. Provision must also be made for adequate access and loading bays and for the provision of laydown areas for machine parts during overhaul.

The 660 MW unit plan in Fig 2.19 shows that with 4-flow LP turbines, the turbine hall dimensions are little affected whether a longitudinal or transverse arrangement is adopted. Generator rotor withdrawal space may marginally increase the width needed for a transverse arrangement and so requires a slightly larger crane span, but the advantage of the shortest possible symmetrical main steam and reheat pipework routes offsets this feature.

An angled 660 MW 6-flow LP turbine-generator can produce a compact turbine hall arrangement as shown in Fig 2.20. Complications can arise in utilising the

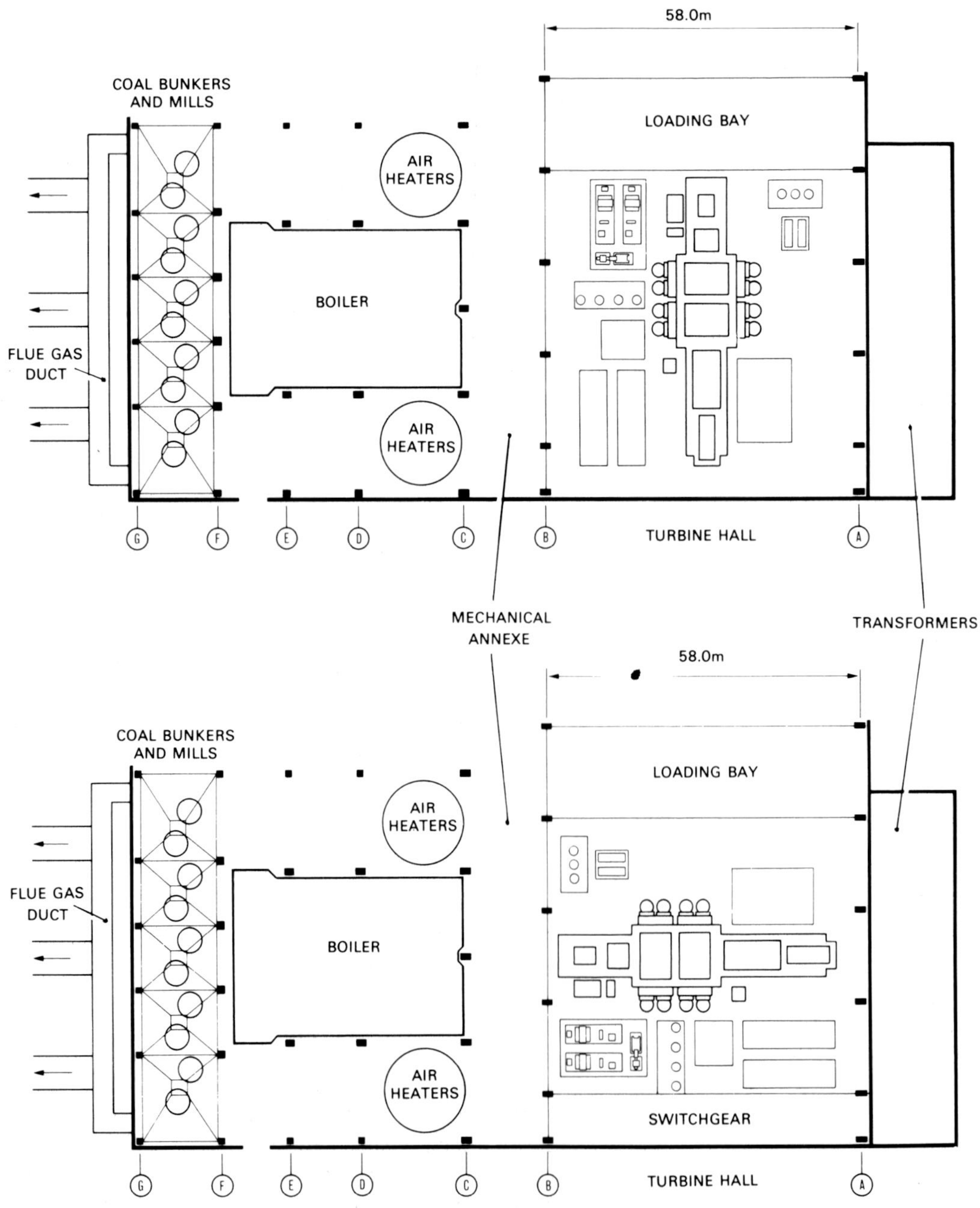

FIG. 2.19 Longitudinal and transverse arrangements of 4-flow LP turbine-generators

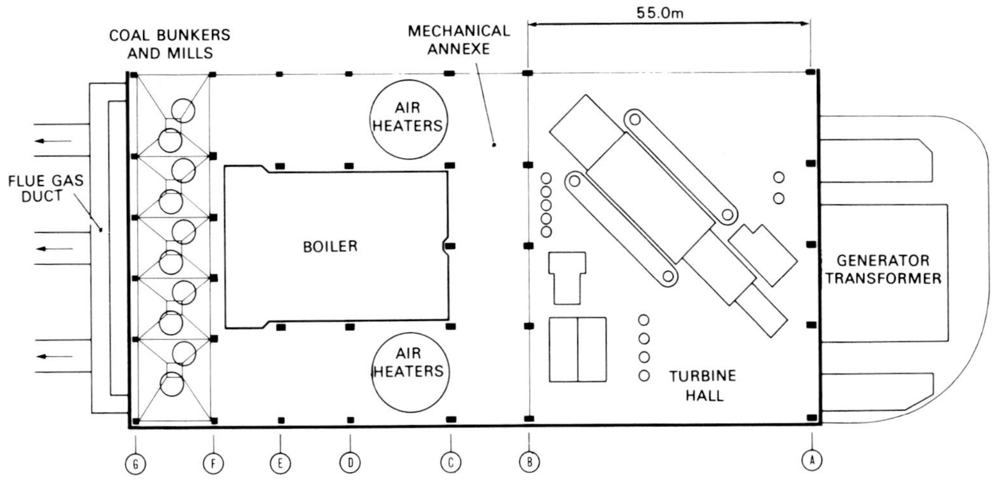

FIG. 2.20 Diagonal layout arrangement of 6-flow LP turbine generators

remaining triangular areas to give a good layout of auxiliary plant and in the control of the crane motions.

The 6-flow LP turbine arranged longitudinally produces a narrower turbine hall with minimum span main cranes. Against this is the disadvantage of longer, asymmetrical main steam and reheat pipework and generally an increase in boiler centreline spacing.

With a 660 MW 6-flow LP turbine arranged transversely, a wider turbine hall is required necessitating a larger span crane, but again this is offset by gains in main steampipe layout. The long crane span can be split by providing two cranes on separate parallel longitudinal rails, but the complication of the supporting columns splitting the turbine hall and the consequent associated blind spots, not approachable by the cranes, must be considered.

The same disadvantage applies if, in a transverse arrangement of turbine-generator sets, separate cranes are provided longitudinally over each set. Erection and maintenance requirements are likely to demand a minimum of four independent cranes.

6.2.2 Boiler plant

The minimum boiler spacing or centreline distance is normally controlled by the layout of auxiliary plant items such as long lance sootblowers, mills, bunkers, fans, ducting and airheaters which often extend beyond the net furnace width. These minimum dimensions can only usually be achieved when utilising a transverse turbine-generator layout and boiler spacing is often increased when using longitudinal turbine arrangements.

It is also necessary to consider which way the boilers will face relative to the turbine hall; whether the *firing wall* (or the furnace in the case of corner-fired boilers) is to be on the *turbine side* or the remote side, as illustrated in Fig 2.21.

If the firing wall is to be on the turbine side, then front-firing favours the positioning of coal bunkers, feeders and pulverised fuel (PF) mills between the boiler and turbine for economic pulverised fuel pipe routing. A possible disadvantage with this arrangement is the introduction of noise and potential dust problems into the centre of the station and lengthening of main steam and reheat pipes through housing to cross the bunker bay. This arrangement however minimises flue duct lengths to the precipitators.

Turning the complete boiler plant through 180° to locate the coal bunkers and mills at the rear of the boiler house, reduces some of the potential problems of the turbine side arrangement, but the gas flows are now remote from the precipitators. Additional ducting is required to carry flue gas around the boiler and the boiler width overall is then increased.

With oil-fired boilers, a longitudinal arrangement of 6-flow LP turbines requires the boilers to be opened out from minimum centres. With a 4-flow LP turbine the transverse arrangement with boilers on minimum centres gives a compact overall station layout.

An alternative arrangement of boiler plant is possible where the coal bunkers, feeders and pulverised fuel mills are located on either side of the boiler. Such an arrangement extends considerably the spacing or centreline distances of the boilers and therefore is most suited to a station design utilising longitudinal turbine arrangements.

The CEGB has only adopted this layout practice when corner-fired boilers have been installed and the furnace is located towards the turbine hall.

The boiler is the largest single item of plant in the power station and in conjunction with its auxiliary plant, such as pulverising mills, fans, airheaters, sootblowers, etc., presents a major layout exercise to ensure the often conflicting requirements of minimum capital cost and adequate access for erection and maintenance are correctly evaluated.

Many utilities prepare *reference designs* for overall power station designs, but it is usually in the boiler area where such a philosophy cannot always be applied due

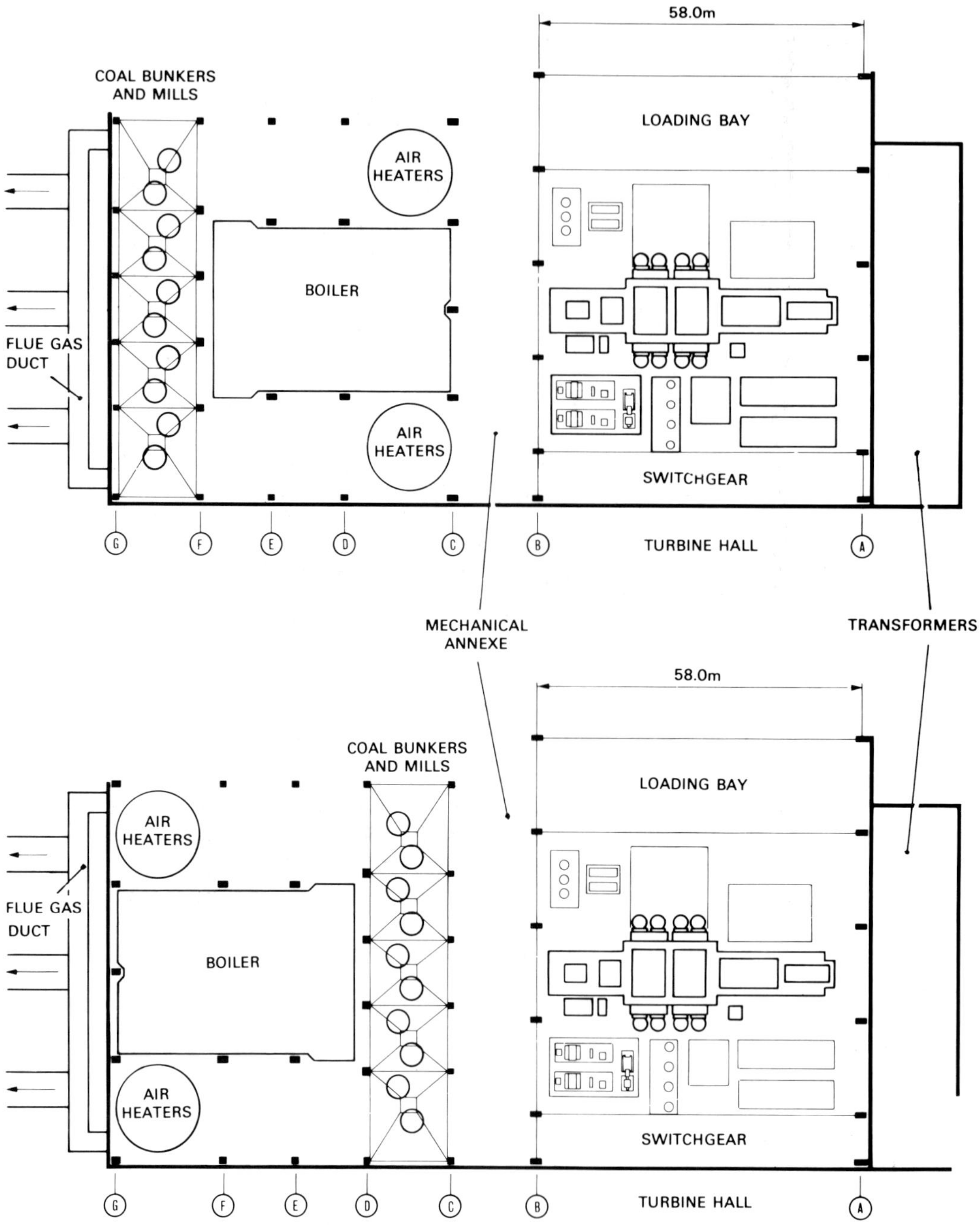

FIG. 2.21 Layouts showing alternative boiler orientations

to the influence that the choice of fuel and its properties exert on the plant design.

Figures 2.22, 2.23 and 2.24 show how such factors influence both the boiler design and also the choice and layout of auxiliary plant. Such variations can obviously have a major impact on overall station design directly by the addition or omission of equipment, or by the relationship of weight and size of the boiler to the supporting structure and enclosure envelope. Secondary areas such as the storage volume of coal bunkers and the type and capacity rating of auxiliary plant such as coal pulverising mills, fans, etc., are also determined by these factors and can influence some of the principal overall dimensions of the station layout, particularly the minimum boiler centreline spacing.

6.3 Layout conventions

In reviewing the basic layout concepts outlined, the reader should have identified a number of layout practices which are consistent between all the options discussed. These practices have evolved over the years within the CEGB and have been demonstrated to provide economic and practical designs in a UK situation. Such practices are regularly reviewed in relation

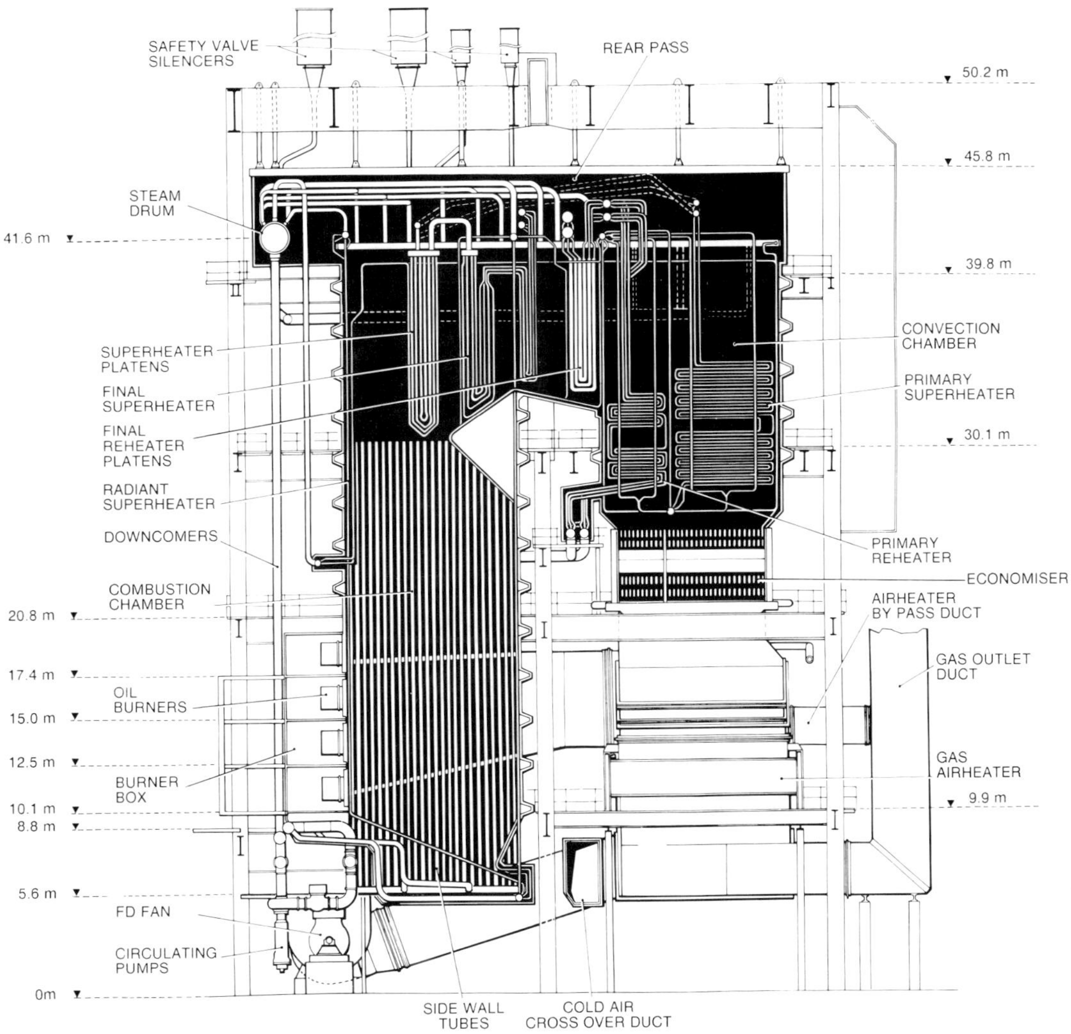

FIG. 2.22 500 MW oil-fired boiler and auxiliary plant

to new plant designs and operational procedures and amended accordingly, but represent the knowledge gained by years of experience within the CEGB on station design, construction and operation of large fossil-fired units.

6.3.1 The unit principle

The unit principle is the association of a single turbine-generator and boiler, together with its immediate auxiliary services, to form a complete, virtually self-contained generating unit.

Many power station engineers will be familiar with 'range' stations where all boilers steam into a common receiver and from which individual turbine-generators take their supplies. Feedwater is similarly returned to a common receiver from which individual boilers draw their supplies. This arrangement gives the maximum flexibility in relation to overall boiler and turbine availability, but CEGB experience is that the reliability of large modern units makes such a provision uneconomic. It has therefore been CEGB practice for many years to build completely integrated units where the boiler and turbine are matched in capacity.

Many services however remain as station-based systems, e.g., cooling water system, town water services, fuel handling and storage systems, etc., but this does not invalidate the unit concept.

6.3.2 Mirror imaging

'Mirror imaging' is the complete or partial handling of plant and systems between pairs of boiler/turbine units to give a symmetrical layout about their common centreline. Such a concept is shown in Fig 2.25.

A *mirrored* design gives an attractive layout but CEGB experience suggests it has many practical disadvantages. It requires the production of two sets of engineering and design details, encourages manufacturing and construction errors and it potentially precipitates operator error.

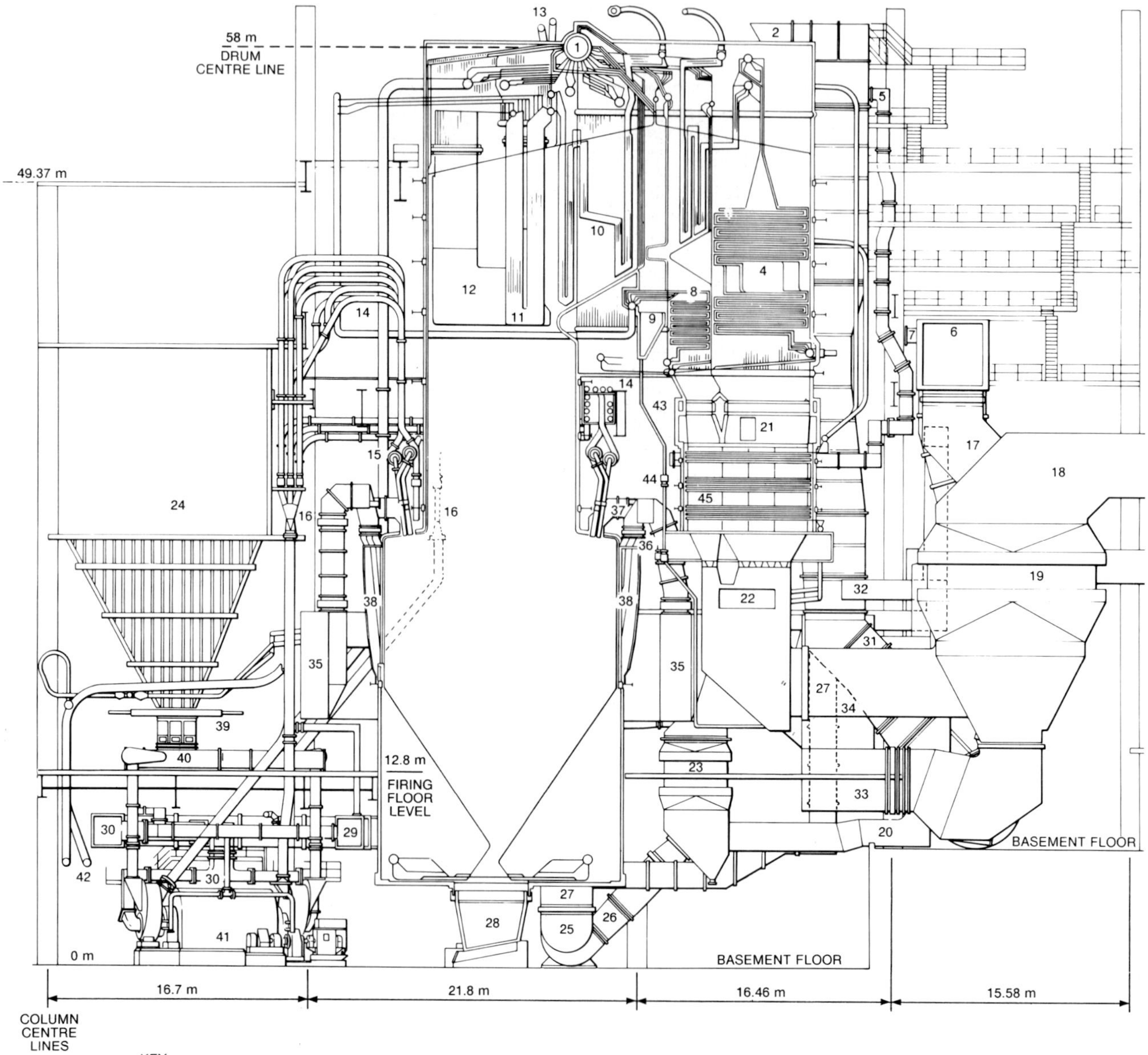

KEY

1 STEAM DRUM
2 FD FAN SUCTION DUCT
3 SECONDARY REHEATER ELEMENTS
4 PRIMARY REHEATER ELEMENTS
5 REAR WALL BURNER AREA VENTILATION DUCT
6 FD DISCHARGE AND CROSS OVER DUCT
7 SOOTBLOWER AIR CONNECTION
8 PRIMARY SUPERHEATER ELEMENTS
9 FURNACE GAS EXIT HOPPER
10 SECONDARY SUPERHEATER ELEMENTS
11 PLATEN SUPERHEATER ELEMENTS
12 RADIANT SUPERHEATER
13 DESUPERHEATERS
14 PULVERISED FUEL PIPES
15 HORIZONTAL CYCLONES
16 THREE - WAY DISTRIBUTORS
17 MAIN AIRHEATER AIR INLET DUCT
18 MAIN AIRHEATER GAS OUTLET DUCT
19 MAIN AIRHEATER
20 MILL AIRHEATER GAS OUTLET DUCT
21 HOT GAS TAP
22 MILL AIRHEATER GAS INLET DUCT
23 MILL AIRHEATER
24 COAL BUNKER
25 PA FAN
26 PA FAN DISCHARGE DUCT
27 PA FAN SUCTION DUCT
28 FURNACE HOPPER
29 COLD AIR BUS MAIN
30 HOT AIR TO MILLS DUCT
31 MAIN AIRHEATER AIR BY-PASS DUCT
32 HOT AIR RECIRCULATION
33 MAIN AIRHEATER AIR OUTLET DUCT
34 GAS TO MAIN AIRHEATER
35 SECONDARY AIR BUS MAIN
36 SECONDARY AIR TO PLENUM CHAMBER
37 TERTIARY AIR DUCTS
38 PLENUM CHAMBER
39 BUNKER OUTLET VALVE
40 COAL FEEDERS
41 BALL MILL
42 MAIN STEAM PIPES
43 DUST PIPES
44 DUST VALVES
45 ECONOMISER

FIG. 2.23 500 MW coal-fired boiler designed for downshot firing and low volatile fuel

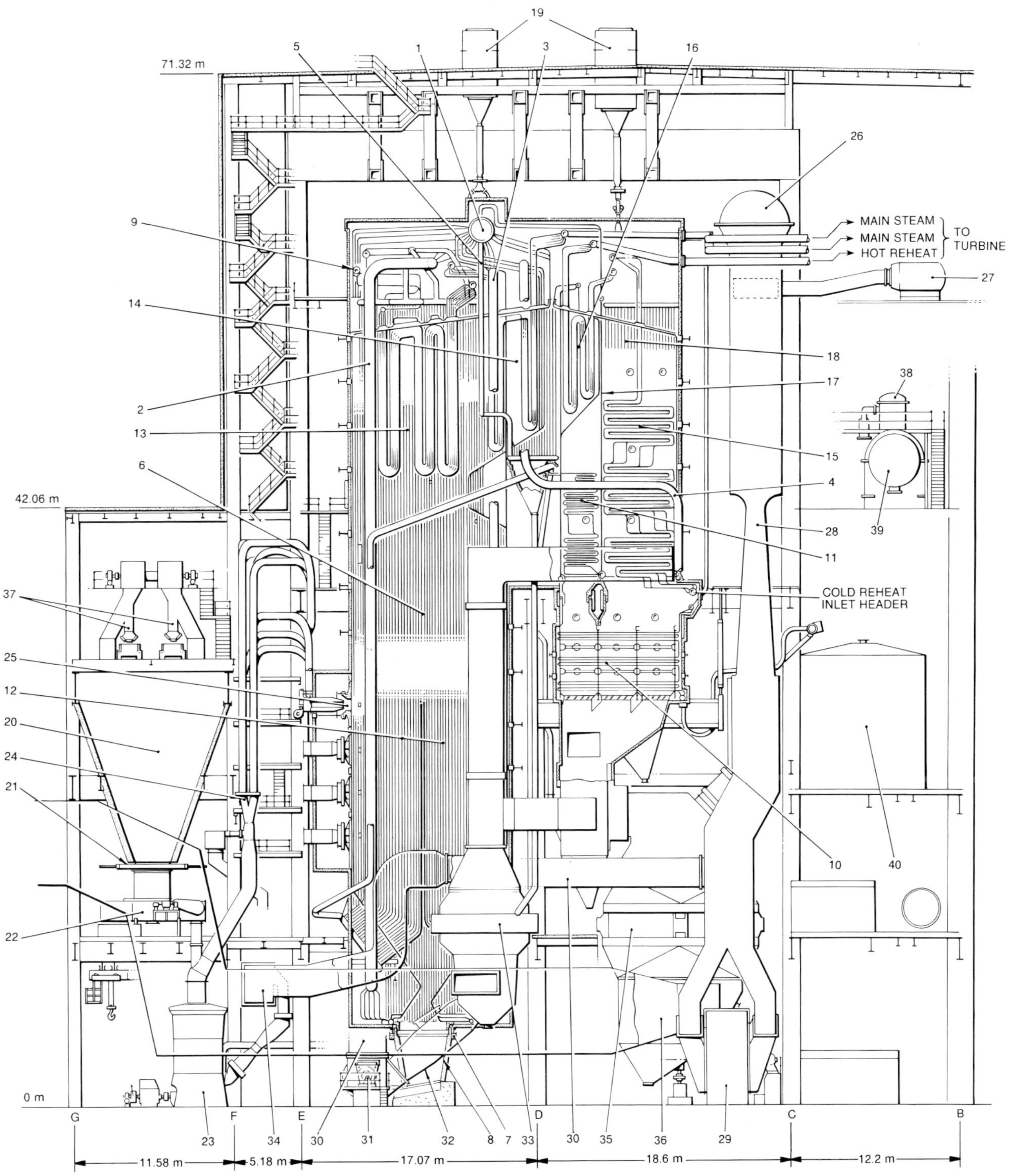

KEY

1 STEAM DRUM
2 DOWNCOMER A
3 DOWNCOMER B
4 DOWNCOMER C
5 DOWNCOMER D
6 FURNACE SIDEWALL
7 WATER SEAL
8 ASH HOPPER
9 DESUPERHEATER
10 ECONOMISER
11 PRIMARY SUPERHEATER
12 STEAM DIVISION WALL
13 PLATEN SUPERHEATER
14 SECONDARY SUPERHEATER
15 PRIMARY REHEATER
16 SECONDARY REHEATER
17 STEAM PARTITION WALL
18 REAR ENCLOSURE DIVISION WALL
19 SAFETY VALVE SILENCERS
20 COAL BUNKER
21 BUNKER OUTLET VALVE
22 COAL FEEDER
23 PULVERISING MILL
24 FOUR - WAY DISTRIBUTOR
25 BURNER
26 FORCED DRAUGHT FAN AIR INLET
27 COMBUSTORS
28 VENTURI
29 FORCED DRAUGHT FAN
30 PRIMARY AIR FAN SUCTION DUCT
31 PRIMARY AIR FAN
32 PRIMARY AIR FAN DISCHARGE DUCT
33 MILL AIRHEATER
34 HOT AIR BUS MAIN
35 MAIN AIRHEATER
36 GAS OUTLET TO PRECIPITATORS
37 COAL CONVEYORS
38 DEAERATOR
39 DEAERATOR STORAGE TANK
40 RESERVE FEED WATER TANKS

FIG. 2.24 500 MW coal-fired boiler designed for frontwall firing and UK bituminous fuel

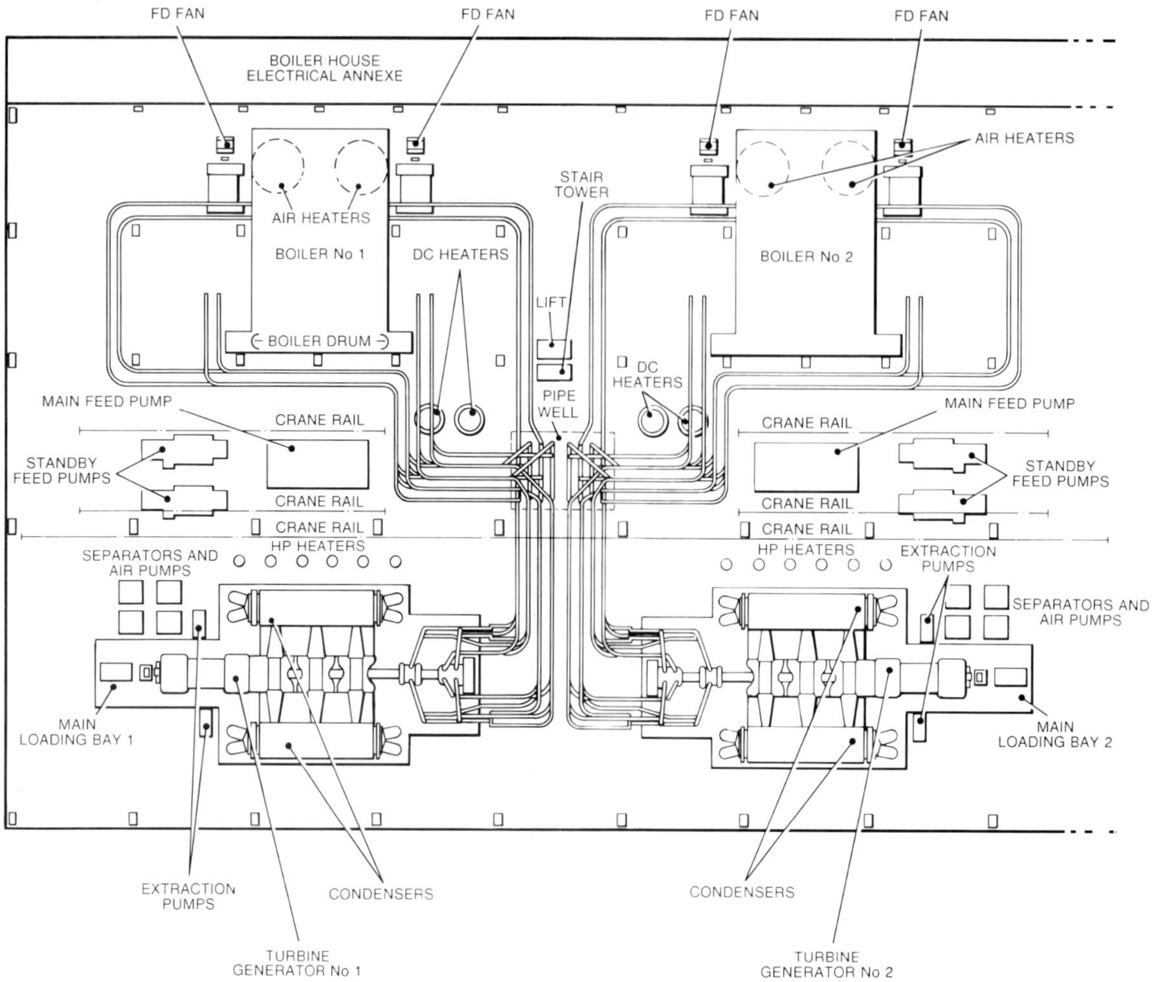

Fig. 2.25 Main plant layout showing mirror image symmetry

Current station layout practice within the CEGB does not favour mirror imaging plant layout but is based on replication of complete boiler/turbine packages.

6.3.3 Turbine island concept

It is conventional practice for turbine-generators to be supported by foundation blocks which are elevated above the basement or ground floor level of the power station. The height at which the turbine-generator is located is termed the *operating floor* and the designer has to decide whether to install a complete floor throughout the turbine house at this level or just to elevate the turbine-generator and its immediate walkways and local laydown areas for small turbine components. The latter concept is shown in Fig 2.26.

Elevation to operating floor level of the turbine-generator alone is termed an *island* layout and has been adopted by the CEGB as its preferred option. This preference is based on the need to provide clear and unhindered access to main plant items. The layout allows direct crane access to all parts of the turbine hall, good maintenance access and efficient lighting and ventilation to all areas. Defined laydown areas for heavy plant are provided at basement level as are clearways for personnel and equipment movement.

A continuous operating floor level throughout the turbine hall requires that the supporting structure be designed to accommodate plant loads imposed when the unit is dismantled for maintenance. Additionally, separate provisions for plant maintenance, ventilation and lighting are required for the basement area or any intermediate floor levels.

The benefits claimed for the continuous operating floor are the opportunities for equipment location in intermediate floor levels, better facilities for turbine maintenance, ample laydown space local to the turbine and the generally pleasing and uncluttered appearance of the turbine hall.

Overall, the CEGB considers that the island concept provides the most economic and practical plant layout, and gives the best access and facilities for plant maintenance.

The appearance of the turbine hall is a question of preference to be exercised by an individual utility, but the CEGB approach is that the station should be seen

Fig. 2.26 Turbine island concept
(see also colour photograph between pp 66 and pp 67)

to be functionally and operationally efficient in the production of electricity.

6.3.4 Mechanical annexe

Many CEGB stations have an annexe between the turbine hall and the boiler house. This annexe is used mainly to accommodate some elements of the feed-heating system, particularly the de-aerator, which on CEGB stations has traditionally been located at high level to ensure an adequate suction head for the boiler feed pumps. The annexe is also used to accommodate various water storage tanks associated either with the feed make-up system or auxiliary cooling water systems.

Although traditionally termed a *mechanical annexe*, the central location within the station makes it suitable for the location of several other facilities, and it is common practice to locate switchrooms in the annexe at various floor levels housing both unit and station switchboards.

6.3.5 Boiler house enclosure

CEGB practice is to adopt totally-enclosed boiler houses for stations within the UK. A small number of units were built some years ago utilising external boilers with only a roof canopy, as is common practice with many overseas utilities, but service experience has clearly demonstrated that such a practice is ill-advisable to UK weather conditions.

7 Turbine-generator systems

7.1 Feedheating plant

The feedheating plant forms an integral part of the generating process by raising the temperature and pressure of the condensate returning from the turbine to the boiler and, if necessary, accepting make-up supplies from the reserve feedwater system.

The plant is essentially a number of pumps and heaters arranged in series and which are linked by a pipework system. The location of each of the components in the system is required to follow logical and defined sequence, and it is important in terms of the overall system economics and hydraulic performance, for each element to be correctly located in relation to the other and to the turbine in particular.

The feedheating system diagram, Fig 2.27, shows that the system has numerous interfaces and pipework connections to the turbine, and to minimise the overall cost and system losses, the CEGB practice is to locate the main elements of the system in logical groups around the turbine. The de-aerator and its associated storage tank however need to be positioned to provide water with a suitable suction head for the boiler feed pumps. Traditional CEGB practice has been to adopt a high level location within the mechanical annexe, but a low level position either in the annexe or turbine hall can be used provided acceptable pump suction conditions can be maintained.

The layout of the feedheating plant for the CEGB's Littlebrook *D* 3 × 660 MW oil-fired station is shown in Fig 2.28 and Fig 2.29 shows the equivalent for a proposed 2 × 350 MW station.

7.2 Condenser and auxiliary plant

The CEGB has experience of pannier and integral condenser arrangements but recent stations have adopted the more conventional underslung arrangement where the condenser itself is located directly beneath the turbine LP cylinders.

The condenser is essentially an integral part of the turbine but the layout engineer needs to consider the construction and maintenance activities together with the disposition of auxiliary equipment. The major space and access requirement is for tube withdrawal and insertion. Construction practice may allow delivery of full-tubed condensers, and due consideration of this fact needs to be made in the station design, but provision still requires to be made for possible tube replacement during the life of the station. With a transverse turbine arrangement this is often possible by utilising loading bays or laydown areas located between units, but for longitudinal turbine layouts, removal sections of the building cladding or a separate local enclosure may be required to give adequate access, as indicated on Fig 2.30.

An acceptable cooling water pipework or culvert routing to the condenser inlet and outlet waterboxes needs to be established. The design and routing in the immediate vicinity of the condenser is influenced by the need to ensure a good water distribution between condenser flow paths and a uniform flow through the condenser inlet and outlet isolating valves which on CEGB stations are typically of the butterfly type. Elsewhere within the turbine hall it is generally necessary to avoid the foundations of the main building columns, turbine block, heavy auxiliary plant items and areas of high superimposed floor loadings such as laydown areas.

The dimensions of the cooling water conduits are established by optimisation studies with the level below the station basement being determined by reference to the hydraulic gradient and acceptable civil construction techniques. Since the foundation system for the station is very *site specific*, the overall integration of the conduits into the civil engineering design needs to be resolved on a site-by-site basis.

The main auxiliary plant associated with the condensers is air extraction equipment for the steamside and, depending on the system hydraulic conditions, for the waterside also. Steamside air extraction plant usually consists of a number of vacuum maintaining

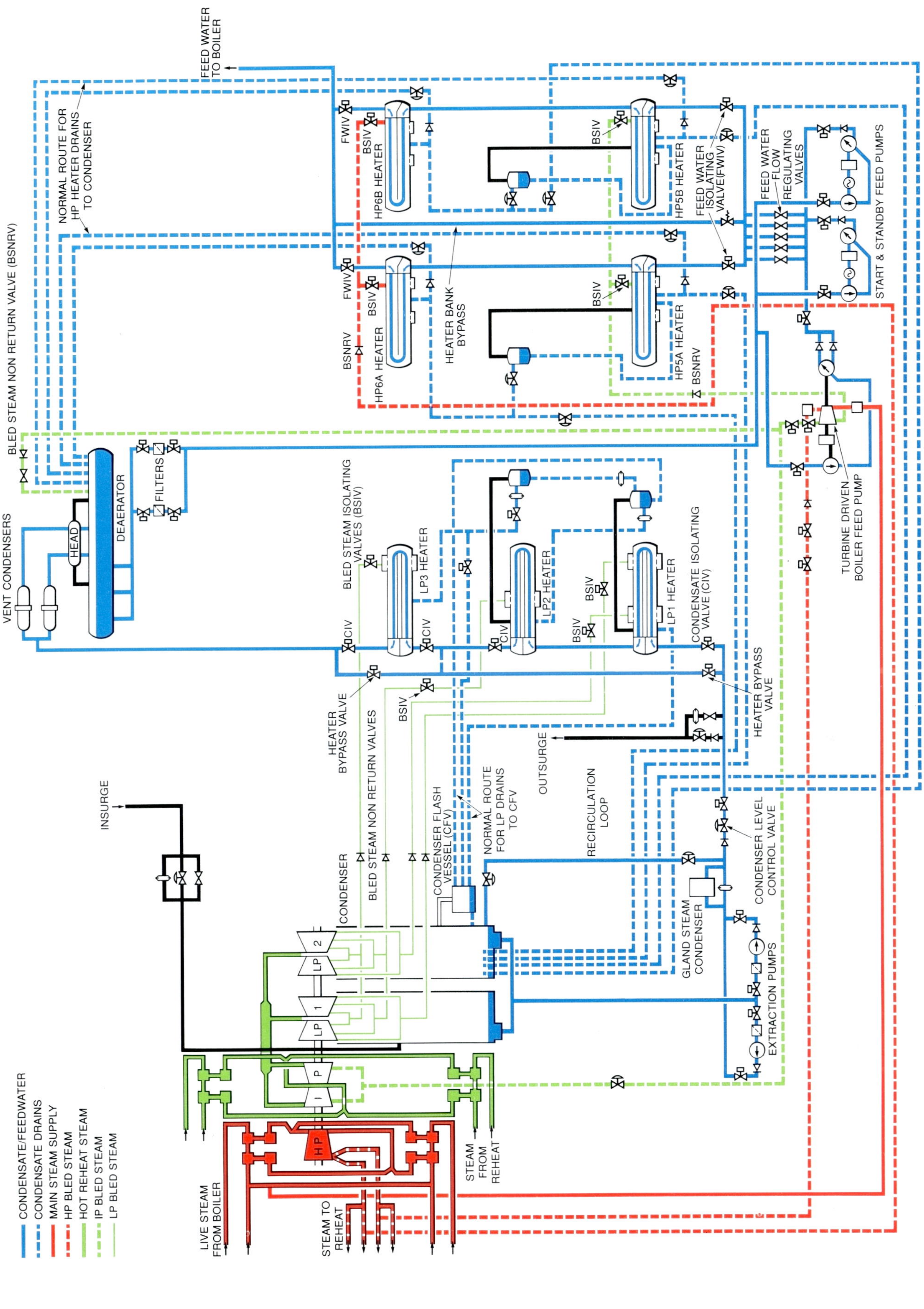

FIG. 2.27 Feedheating system

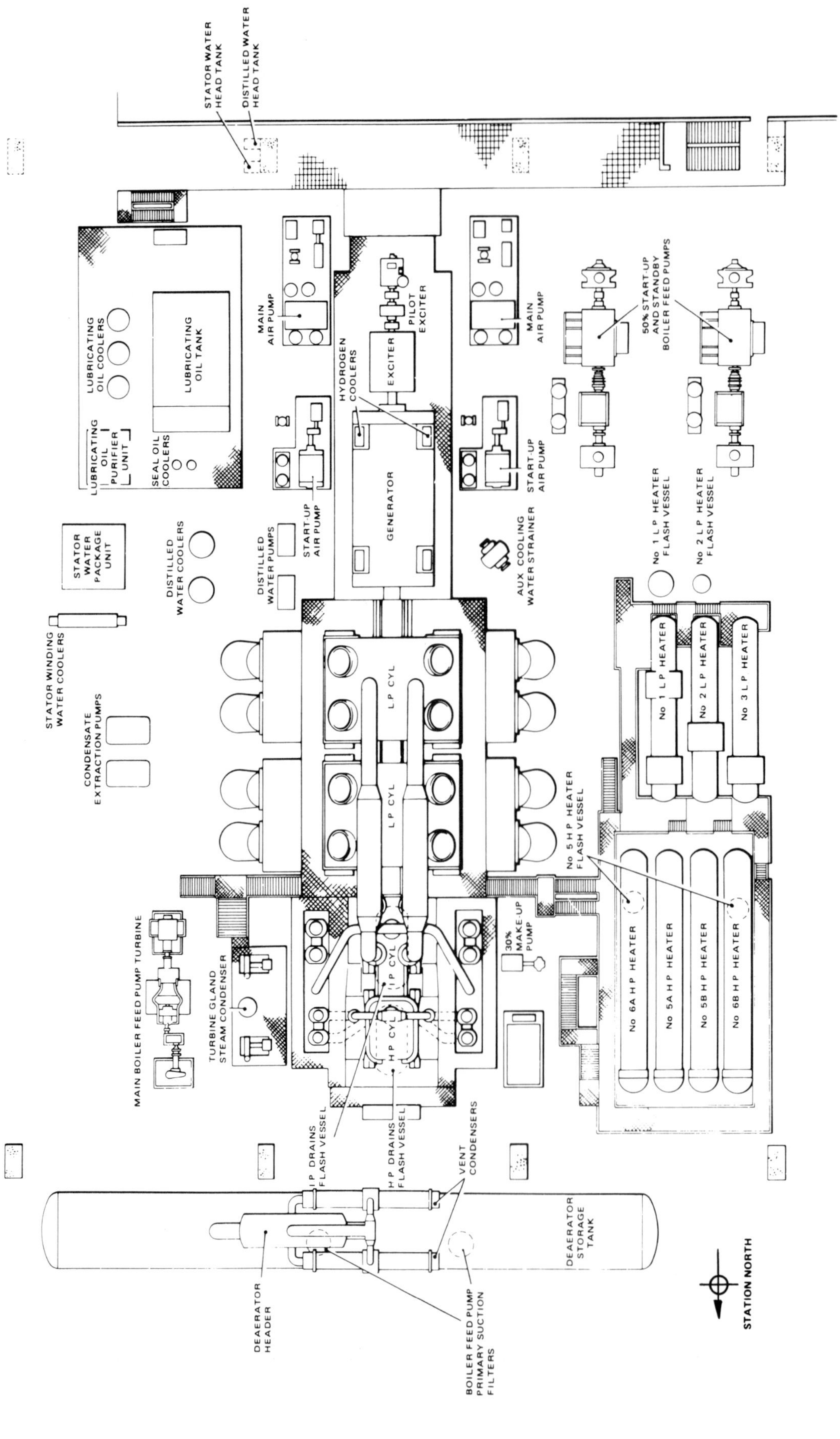

FIG. 2.28 Feedheating plant layout — 3 × 660 MW station

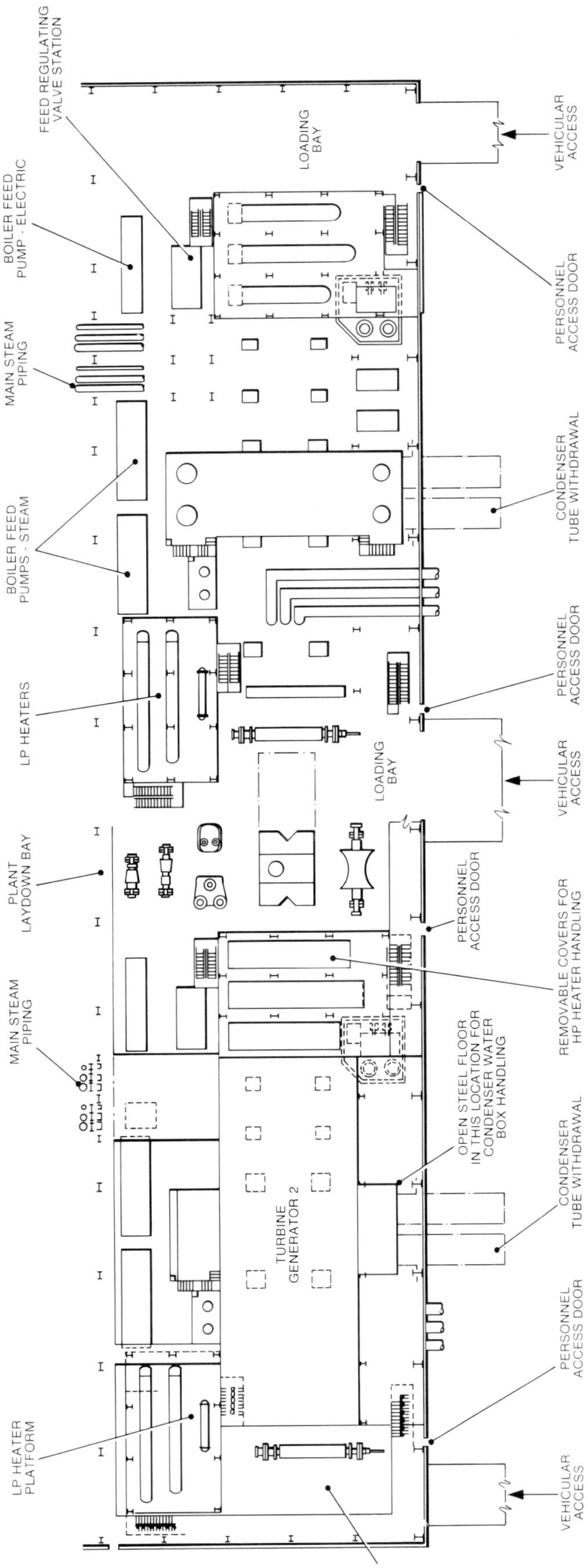

FIG. 2.29 Feedheating plant layout — 2 × 350 MW station

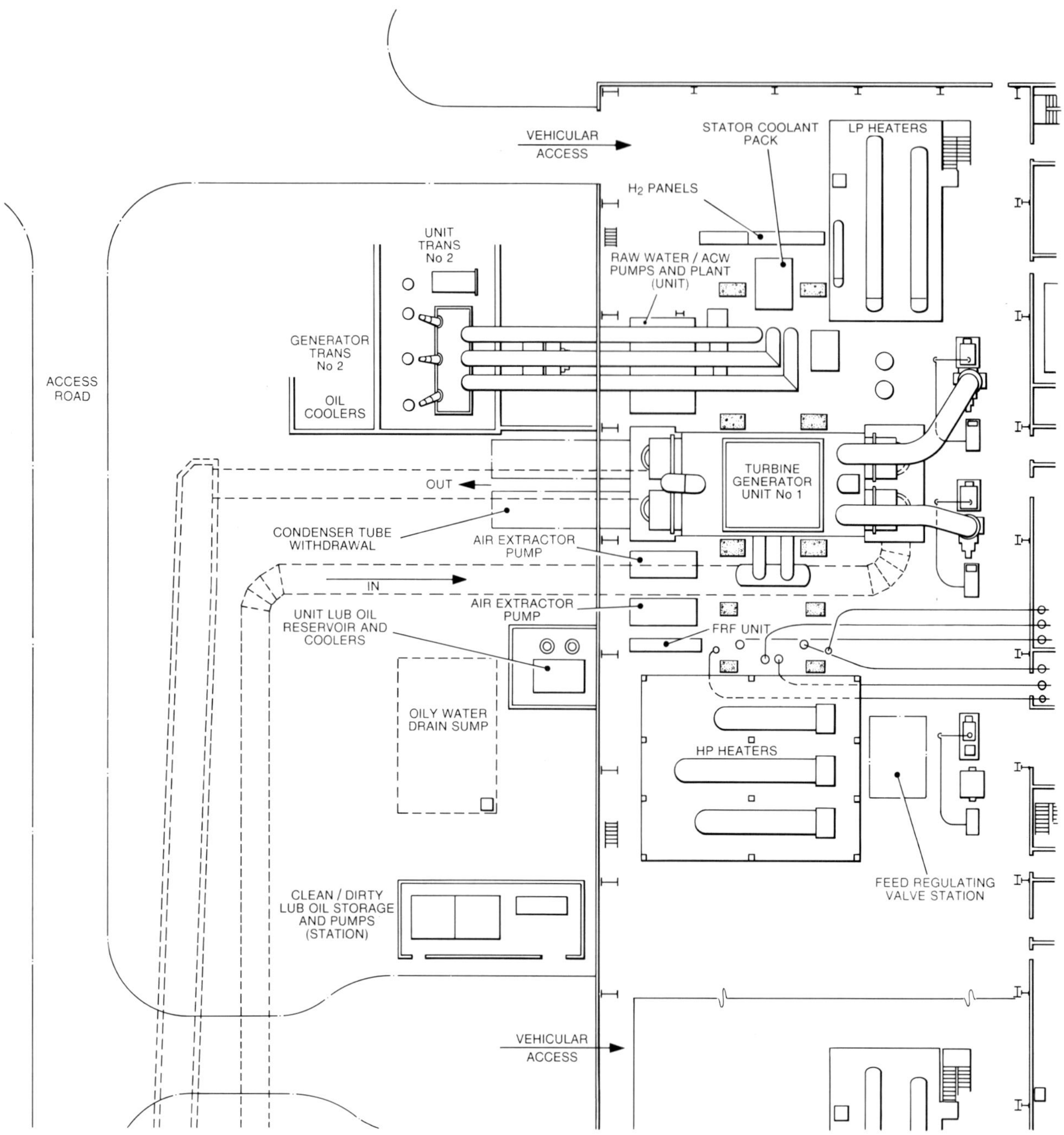

FIG. 2.30 Plant laydown areas for a 2-unit station

units and quick-start pumps arranged on a unit basis. The preferred location for this plant is in close proximity to the condenser.

Waterside air extraction equipment is sometimes fitted to the outlet waterboxes on direct cooled stations, when system hydraulic studies have indicated that external assistance may be required in maintaining top condenser tubes full of water. Large air extraction pumps are normally provided for a system priming function, but because of their generally infrequent usage, are often provided on a station basis with individual turbine condensers connected by a busmain. Smaller capacity service pumps are usually installed on a unit basis to provide continuous air extraction during unit operation. The service pumps should be located adjacent to the individual unit condenser outlet water boxes, but the main priming air pumps can be conveniently located to suit pipework and electrical connections.

7.3 Erection and maintenance

The relationship between the boiler and turbine is of basic importance in establishing the overall station layout including the spacing of the individual units, but the total enclosed floor area and overall volume of the turbine hall needs to be determined by considering the plant layout in relation to the erection, operation and routine maintenance requirements.

The majority of modern turbine-generator plant is assembled from modular packages and their adoption necessitates consideration being given at the design stage to the means of access and the handling arrangements to be used for installation and maintenance. The designer is required to assess whether the loads to be handled are 'one off' installation packages which are dismantled for normal maintenance, or components requiring routine handling facilities during such periods.

Current design practice on CEGB stations is to utilise specialist lifting equipment for the 'one off' installation lifts and install permanent cranage facilities for those components requiring routine maintenance attention. Typically the generator stator would be installed utilising specialist jacking equipment, but permanent lifting facilities would typically be provided to lift the generator rotor, steam turbine top covers and steam turbine rotors.

On multi-unit stations, where up to six large units have been installed, the opportunity was taken to provide a permanent lifting capability for the generator stator by operating two overhead cranes in tandem with a load spreader beam. The cranes would normally be provided with a low speed selector on the main hoist gearbox to give this facility.

The need to size the overhead cranage facilities to handle all major plant components requiring attention during an overhaul period, and the adoption of the *island* layout concept, means that the majority of remaining equipment and auxiliary systems within the turbine hall can be adequately serviced by the overhead cranes without recourse to local handling equipment. To ensure that adequate handling for such plant is provided, a 'hook approach and plant laydown diagram' is produced to illustrate the area of coverage and the accommodation provided for the various plant items and components dismantled during the maintenance periods. The extent of laydown space provided will depend on both the number of installed turbine-generator units and the maintenance programmes, but for a 2-unit station it would be CEGB practice to provide sufficient laydown area for one complete unit, as shown in Fig 2.30.

When nominating suitable plant laydown areas, consideration needs to be given to access and personnel safety issues, and to avoid potentially hazardous situations being created. Irrespective of whether a unit is being overhauled, clear and safe access is required for normal operational procedures, and it is necessary to make provision for dedicated access clearways through the turbine hall. These restricted zones should be clearly marked to preserve the integrity of the areas and operator familiarisation.

The provision of such laydown areas, access clearways and the layout of the plant itself determines the overall plan area needed for the turbine hall, bearing in mind on both the transverse and longitudinal layouts the need for generator rotor withdrawal. The overall turbine hall height is determined by the lifting height required for dismantling and movement to laydown areas of the longest vertical plant item, together with the necessary clearance room for the crane facilities themselves.

8 Boiler systems

8.1 Pulverised fuel system

Coal pulverising mills in CEGB stations are usually located on individual foundation blocks which are isolated from the main station foundations by the use of an insulating material which absorbs and reduces transmission of the vibrations generated by this equipment. The minimum spacing of the mills is determined by the overall size of this foundation system and the need to provide maintenance access to the equipment.

Pulverising mills are usually a high-level maintenance item and such frequent attention, involving heavy lifts, necessitates the provision of a suitable component handling system. Normal CEGB practice is to provide a dedicated facility and a continuous access clearway through the mill bay for equipment movement. The mills need to be located in close proximity to the boiler firing face to minimise the length of pulverised fuel (PF) pipework. PF pipework should be arranged so that for any one coal mill, the burners are uniformly distributed to ensure that for whichever combination of coal mills is in operation, the heat input to the furnace is evenly distributed.

The relationship of the mill bay with respect to the firing arrangement of the boiler has been discussed in Section 6.2.2 of this chapter.

8.2 Draught system

Figure 2.31 shows how the basic boiler layout options influence the layout of the air and gas ducts.

A rear bunker and mill bay, results in the airheaters being centrally located within the station and requires the routing of large flue gas ducts along each side of the boiler, which must pass through or around the bunker/mill bay to reach the precipitators. This often results in an increase in the boiler centreline spacing, all other factors being equal. The central bunker and mill bay eliminates this disadvantage and allows short and direct gas duct routes to the precipitator inlets.

The adoption of enclosed boiler houses has allowed advantage to be taken of recovering heat losses which invariably occur from the boiler plant. Reference to Fig 2.22 shows that the forced draught fan suction ducts are routed from the upper areas of the boiler house in

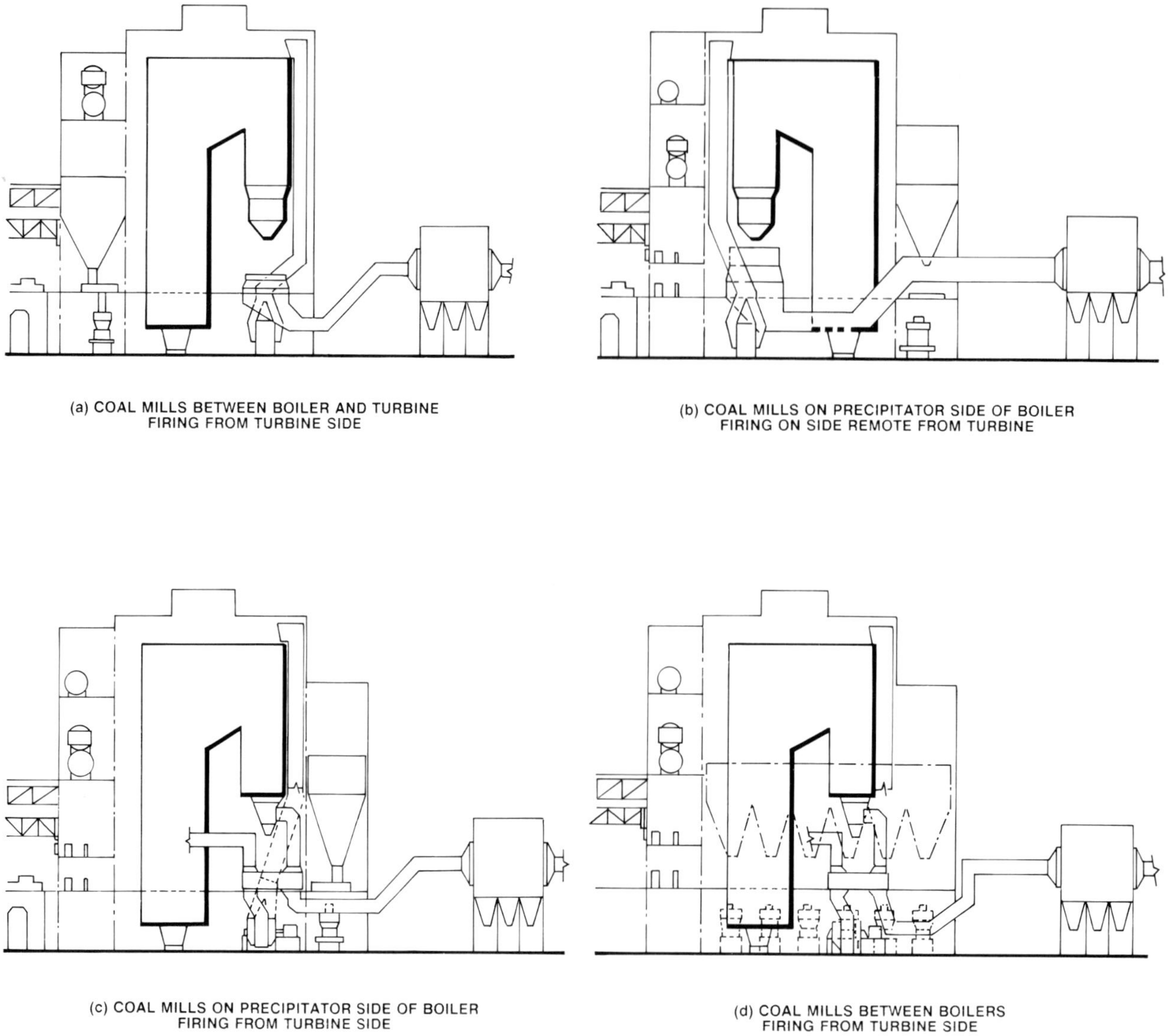

FIG. 2.31 Air and gas ducts layout

order to recover the hot air which rises to the top levels of the building by convective action. The arrangement of the ducting is comparatively simple, but a route reservation down each side of the boiler which allows straight runs without offsets is necessary.

The gas and air ducting which connects the boiler unit to the airheaters, fans and precipitators occupies a considerable space within the boiler house. The duct cross-sections must be large to reduce gas and air flow resistance to an economic minimum and be designed to avoid sharp bends and changes in section which, besides inducing pressure loss and turbulence, can, in the case of flue gas ducting, precipitate fall-out of pulverised fuel ash particles.

Typical CEGB practice on larger fossil-fired units is to locate the airheater as close as possible to the economiser gas outlet flues at about operating floor level, with the flue gas ducts being run below this floor level to the precipitator inlets. The combustion air ducting from the forced draught fans is also normally run below operating floor level to the airheater inlets.

8.3 Oil firing system

The majority of conventional stations have an oil firing capability, although the extent varies from a fairly minimal requirement for boiler lighting-up on some coal-fired stations through to a full load capability on totally oil-fired stations.

A fully oil-fired boiler is attractive from a layout viewpoint in that the boiler house design is simplified by the omission of the coal pulverising mills, bunkers, ash hoppers, etc., and the boiler orientation can be chosen to provide the best arrangement of air and gas ducting.

Equipment provision for oil firing is reasonably simple. A station-based ring main is usually adopted which continuously circulates heated oil from the main storage tanks. Individual boilers take oil from this ring main via additional pumping and heating units which control the oil condition to suit the boiler burner characteristics. The oil passes from these unitised pumping and heating units to another ring main which circulates the boiler firing face and allows individual burners to tap off their oil supply through a control valve. The ring main spills back to the pumping and heating units.

On coal-fired boilers, a supply of oil is used for lighting-up and initial operation of the boiler until a sufficient load is established to maintain a coal mill in service. The design requirements for such a lighting-up system are similar to full oil firing, but are generally sized to carry only a small percentage of full boiler load. On some coal-fired stations provision for a larger oil firing capability is made. Such 'overburn' facilities can be used to support unit output in the unlikely event of an excessive number of coal mills being out of service at the same time.

8.4 Boiler fittings

The boiler has many externally-mounted items of equipment to which pipework and electrical services are required together with personnel and maintenance access provisions. The most important are:

- Main steam stop valves and integral pipework.
- Reheater inlet and outlet connections.
- Feedwater inlet connections to economiser.
- Blowdown and drain valves.
- Sootblowers.
- Burners.
- Boiler safety valves and silencers.
- Steam and water sampling equipment.
- Chemical injection equipment.
- Air release valves.
- Drum level gauges and alarms — local and remote.
- Control and instrumentation sensors and transmitters.
- Various access and observation doors.
- Ash and dust removal points.
- Gas ducting dampers.

The frequency and extent of access for operation and maintenance activities will vary for each of these activities, and it is necessary to review the full scope of activities required at the design stage to ensure that adequate and safe provisions are provided. This is primarily the task of the layout engineers.

Many operational activities on the boiler will concentrate on the burner area, including on-load maintenance and adequate attention to the provision of artificial lighting to workshop standards, fire protection and means of escape for personnel.

The layout of platforms and galleries at the firing floors should allow for easy withdrawal of oil burners and for filters to work comfortably and safely. Provision for accommodating oil spillage should be provided with adequate drainage and collection arrangements to ensure the minimum fire risk.

Generally the equipment identified here is located over the whole boiler area and demands a comprehensive series of galleries around the boiler casing at suitable levels, all the way from the ground up to drum level (see Fig 2.32).

8.5 Dust extraction plant

Electrostatic precipitators are now routinely used for the removal of pulverised fuel ash from the boiler flue gases. They need to be located between the boiler and the stack in a position which allows the gas duct routing to be as short and straight as possible. The most suitable location is therefore immediately to the rear of the boiler house.

It is common practice to divide the flue gases into streams and utilise a number of precipitator units in parallel with typically three, but sometimes four, passes being adopted for each boiler unit. The overall size of the precipitator plant is influenced by the fuel characteristics and in some instances the overall width of the plant may affect the minimum boiler spacing.

8.6 Flue gas desulphurisation plant

Flue gas desulphurisation (FGD) equipment is likely to be installed at all major fossil-fired stations in the future, and in fact retrofit installations are also being carried out at a number of existing stations.

A number of treatment systems are already commercially available with several more in the development stage, and the CEGB has selected both the limestone-gypsum and a regenerative system for its initial installations of this plant. In terms of layout, both systems have similar basic requirements in that the plant is divided into unit and common equipment.

Figures 2.33 and 2.34 illustrate the studies which have been carried out for the two systems at the Drax site for units 4, 5 and 6, the most recently installed plant.

In each case the unit plant, consisting mainly of absorbers, fans, reheaters and ducting is located in the area immediately behind the unit precipitators between the induced fan discharge and the dust bunker access road.

The common plant area containing either the limestone and gypsum treatment or the regeneration equipment is located to the north-east of the station area and

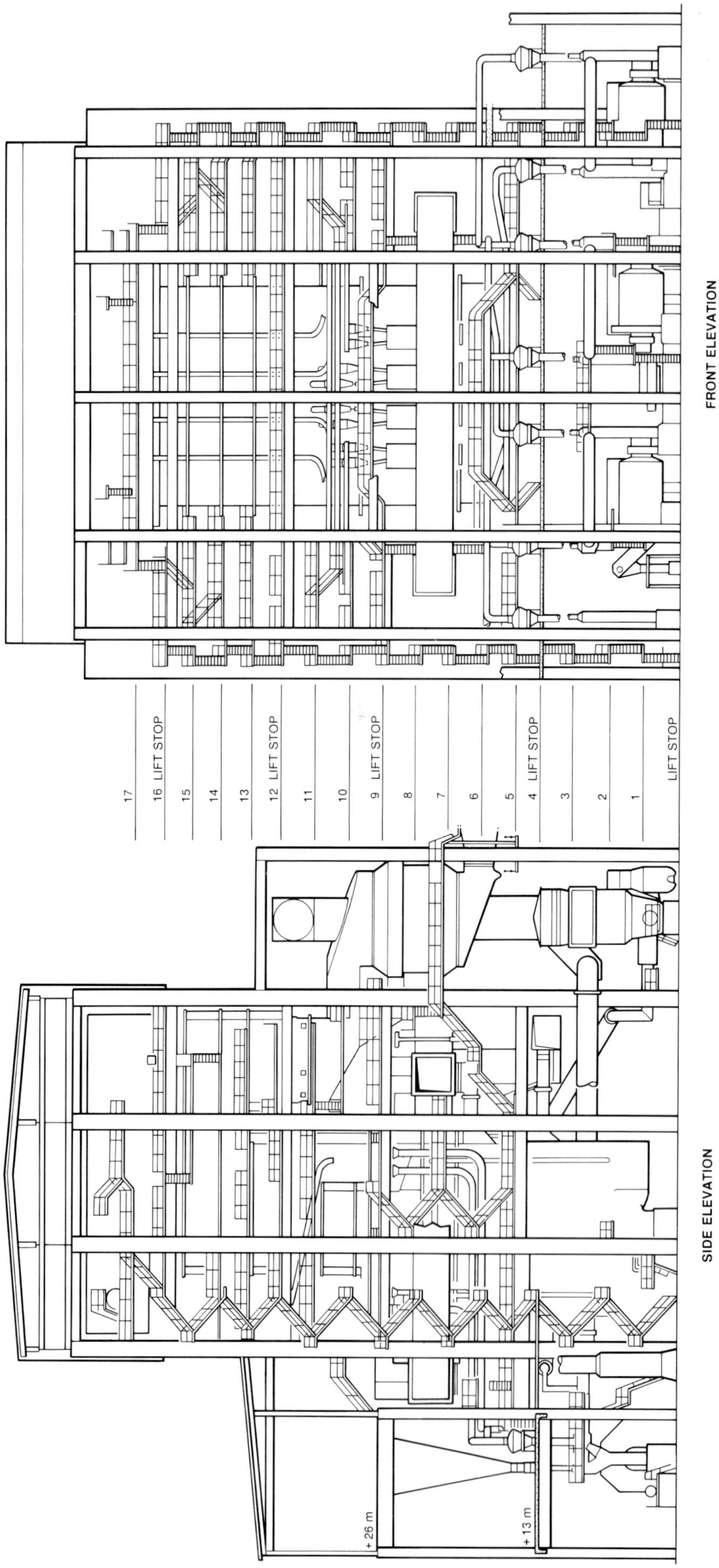

FIG. 2.32 Boiler levels and galleries

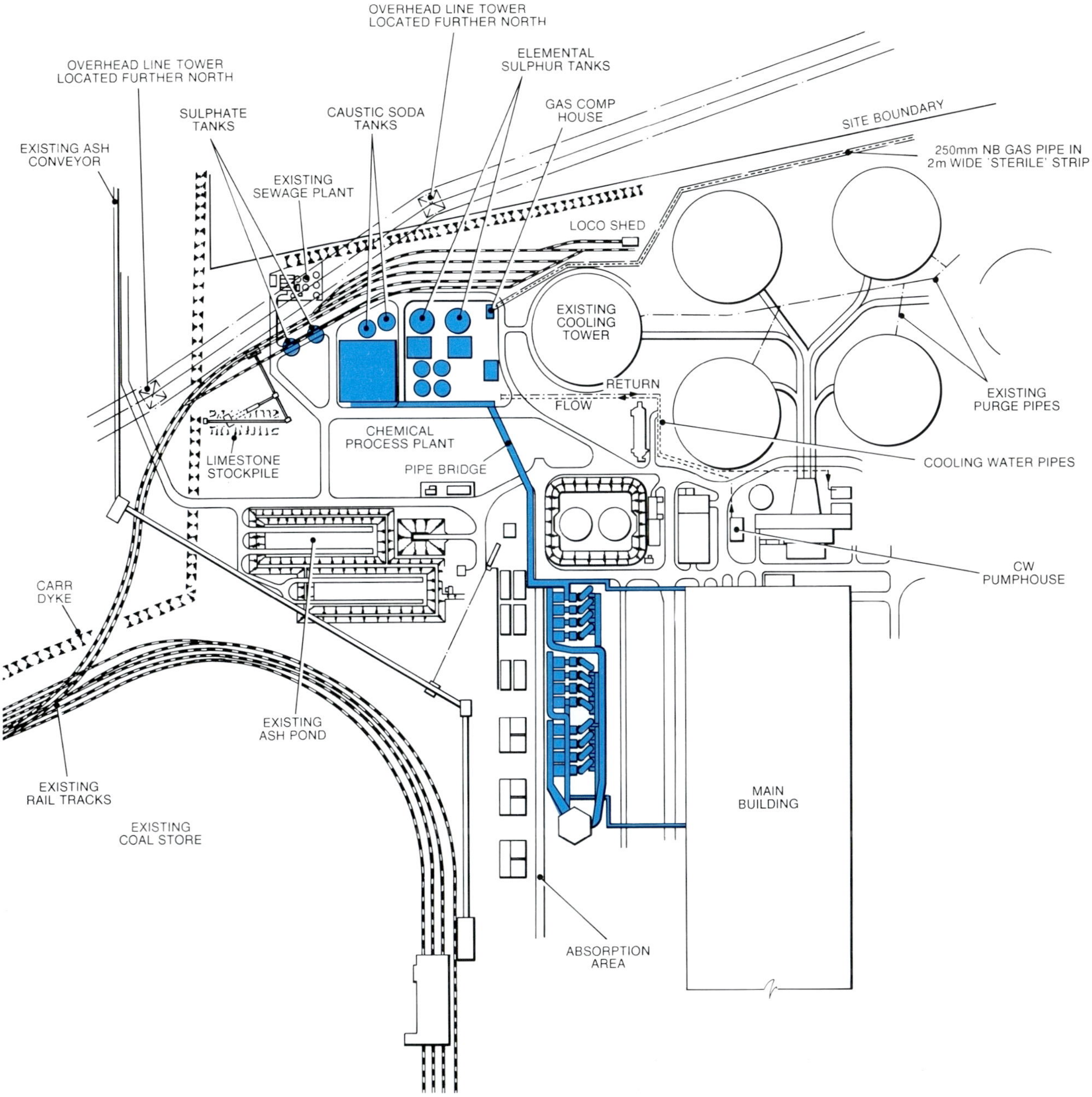

FIG. 2.33 Site plan showing location of the regenerative FGD plant

is connected to the unit plant by means of a pipe bridge and cable trench.

The plant is serviced by many external facilities such as rail sidings, water services, electrical services, etc.

The location of FGD plant on a greenfield site would need to be considered as part of the overall site layout exercise.

See Volume B, Chapter 4 for detailed descriptions of FGD processes.

9 Main steam pipework

The main concern in developing the layout for main steam pipework is to provide for flexibility and to minimise loadings at the terminal points of the boiler and turbine. In practice, because of the high cost of this pipework, a minimum total piping run is preferred and often the layout adopted will be a compromise based on economic as well as technical factors.

The thermal expansion and contraction of the pipework needs to be accommodated by the inherent flexibility designed into the system, and by the use of cold pull-ups and constraints which are applied to the pipework where needed. *Cold pull-up* is the term used to describe the prestressing of a pipework system in the cold condition, such that it is in a neutral state after expanding to the hot operating condition.

Flexibility is a major concern during the layout process, and it is advisable for continuous reviews of

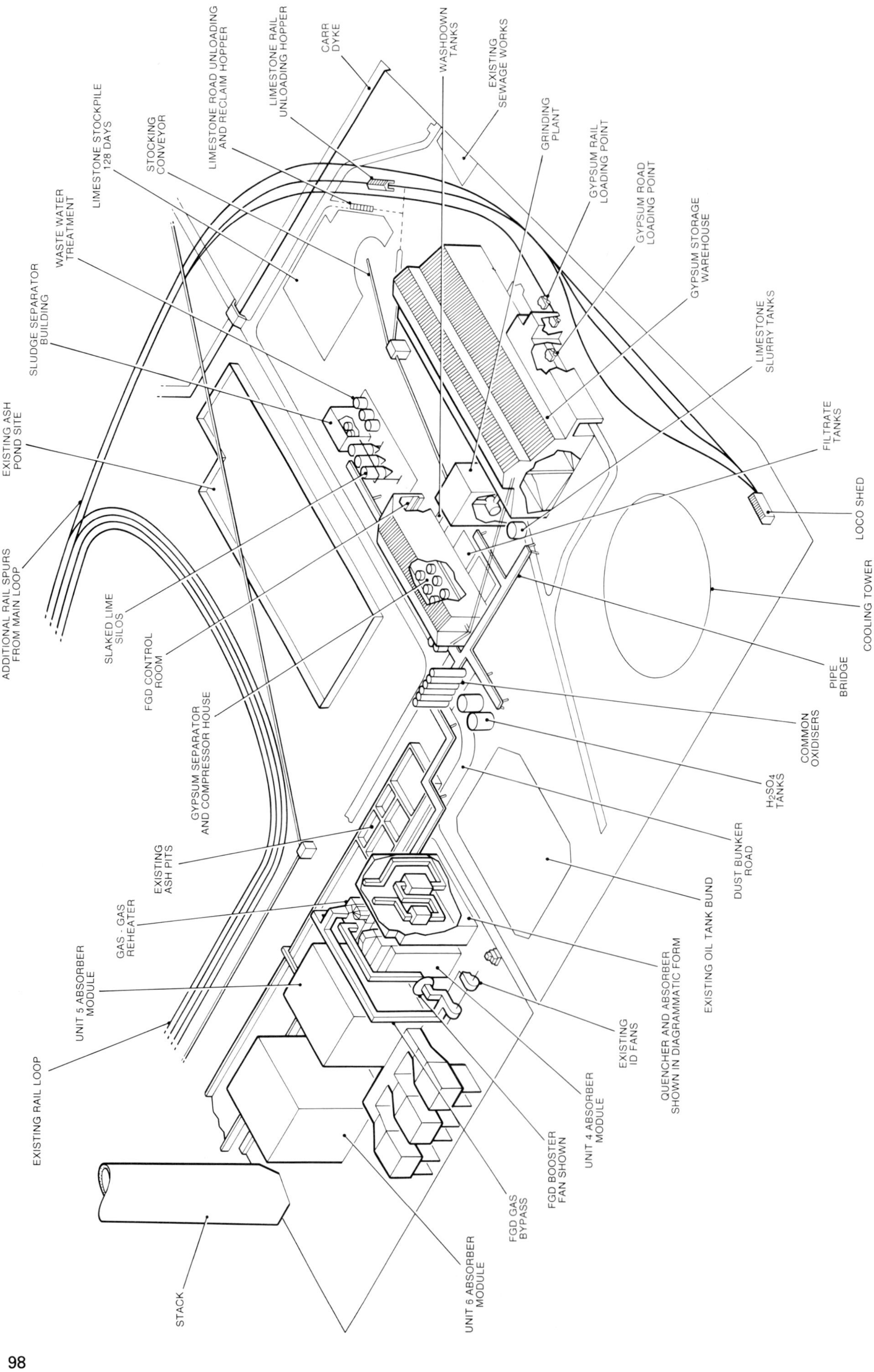

FIG. 2.34 Perspective arrangement of the limestone-gypsum FGD plant

pipework design to be undertaken as station layout development proceeds.

The thrusts and turning moments at the terminal points can be difficult to calculate manually, but modern computing methods based on mathematical analysis allows rapid reviews and layout to move forward with the knowledge that design changes can be readily assessed.

The position of the turbine relative to the boiler can have significant influence on both the overall flexibility and cost of the pipework system. Transverse turbine layouts located on the boiler centreline generally allow for short symmetrical pipe routes to be adopted as shown in Fig 2.35.

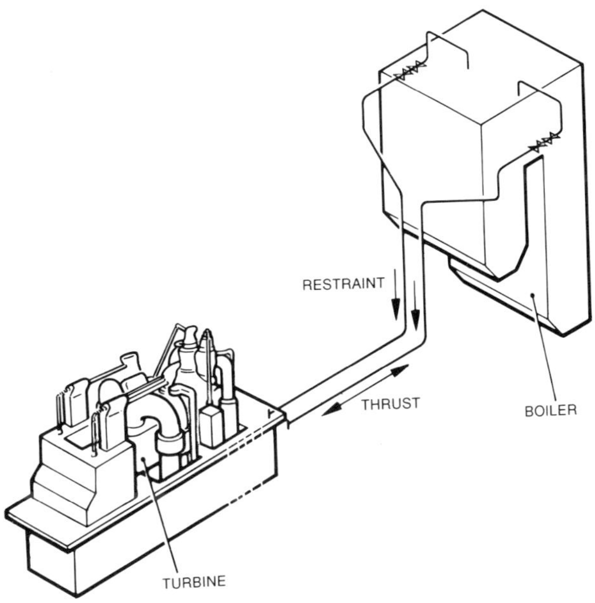

FIG. 2.35 Main steam pipework arrangement for transverse turbine

The longitudinal turbine arrangement often results in a longer asymmetrical piping layout as shown in Fig 2.36, unless the *ideal* layout can be adopted by off-setting the turbine-generator unit from the boiler centreline. In an overall station layout context, whilst off-setting of the turbine hall relative to the boiler house may be viable in a technical sense, the overall effect on the station's visual appearance may need to be considered. Additionally, if an asymmetrical piping layout is adopted than a review of steam pressure and temperature variations should be undertaken to ensure that the steam conditions at the turbine inlets are within acceptable tolerance levels.

To permit the monitoring and recording of pipework expansions and creep, a measurement system has been introduced on more recent stations within the CEGB. These measurements are taken following completion of construction at regular intervals in the life of the station. Suitable access facilities should therefore be provided in the layout for this activity.

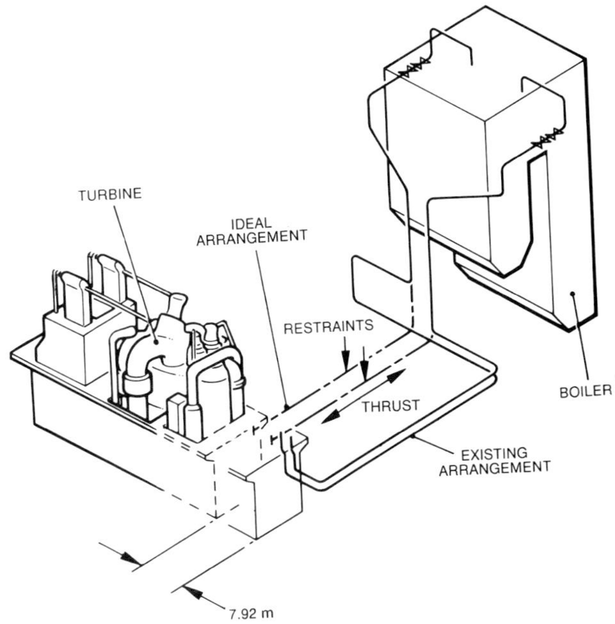

FIG. 2.36 Main steam pipework arrangement for longitudinal turbine

10 Low pressure pipework and valves

Low pressure pipework and valves are usually associated with the following services systems within the power station:

- Town water.
- River water.
- Treated (demineralised) water.
- Chlorine solution dosing.
- Reserve feedwater.
- Blowdown water recovery.
- Fire hydrant system.
- Fixed water spray protection.
- Ash sluicing supplies.
- Nitrogen supplies.
- Potable water supplies (fit for drinking).
- Compressed air supplies.

Many of these services lend themselves to trunk main routing throughout the length of the station with branch supplies out to each unit. Ample space for the routing of the pipework must be reserved, for example in the heater bay between the sets and the boilers, or below the operating floor walkways.

In the early stages of design it is essential that diagrams are prepared for all low pressure services and a pattern established of valve and pipework requirements. The diagrams are then constantly developed as the requirements of the main plant items services become known. Low pressure pipework system design and layout must be responsive to the overall main plant layout. As with all pipework design, the route and pressure drop must be optimised to give best overall efficiency.

Pipework should not be run in trenches or small ducts out of sight. It should be visible and accessible for maintenance together with its associated valves. There must be correct drainage falls with air release valves fitted where necessary. Pipework should be well-supported and secured. Pipe joints are usually the weakest links in a pipework system and demand the highest standards of materials and workmanship. The number of joints should be minimised by the use of welding.

11 Water storage tanks

Water storage represents a large space requirement and could, for instance, be in excess of 9090 m^3 of town water to cover station requirements for 24 hours. Reserve feedwater should be available for at least 24 hours maximum continuous rating (MCR) supply for each boiler.

Storage facilities for town water are normally located externally to the main building, perhaps adjacent to the water treatment plant, but certainly with consideration being given to other uses and needs on this particular commodity, i.e., specific reserve for fire fighting purpose, FGD requirements, etc.

The design of the storage facility is dependent on site space availability and various forms can be considered, e.g., low profile pressed steel tanks, concrete reservoirs or vertical cylindrical tanks. The vertical cylindrical type of tank is most common for reserve feedwater storage use, the shape naturally allowing optimisation of base diameter to height ratio to cover such aspects as available space and ground load bearing capability.

These large quantities of water are not now stored at high level in the boiler house or turbine hall, as it is uneconomical in terms of steelwork to support the load. In addition, a disaster situation would arise if a tank weld or interconnecting pipe connection failed. Figure 2.37 shows a typical town water and raw water system for an oil-fired station.

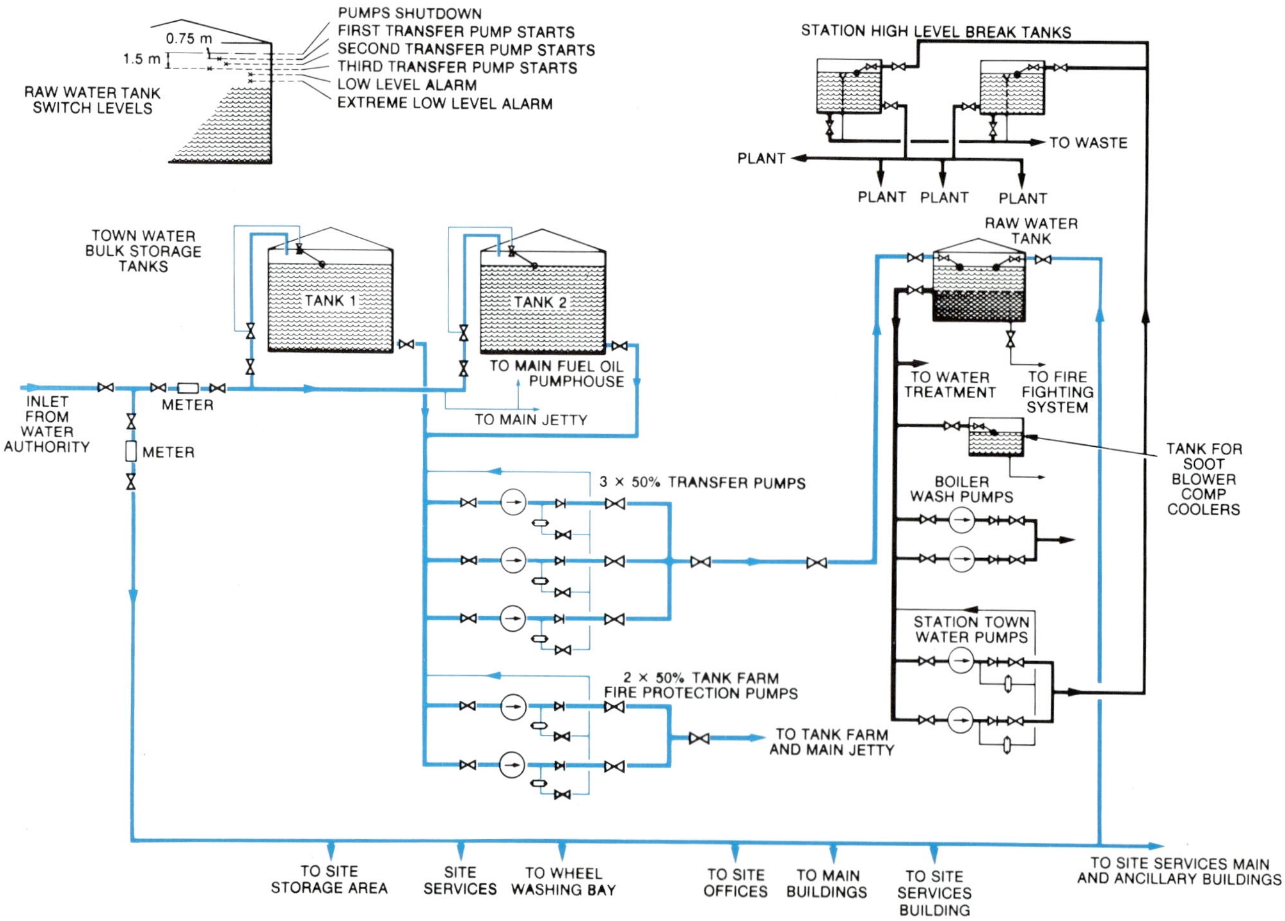

FIG. 2.37 Typical town water and raw water system

12 Cranes

Cranage is required within the turbine hall to cater for the erection of plant during the station constructional phase and maintenance during the subsequent station operational phase for the turbine-generator units and their respective auxiliary plant. The cranage is needed to cover the maximum possible floor area with due regard to the disposition of the plant, laydown areas and loading bays.

There have been three different configurations of steam generating units to turbine layouts within the current programme of 660 MW sets, i.e., longitudinal, transverse and diagonal. The selection of the earlier stations' turbine hall layout was constrained by the then available design of turbine hall cranes which had limits of span and capacity. Ultimately the design of cranes permitted spans and capacities large enough to accommodate the full length of the turbine, i.e., transverse designs. The span for a 660 MW set is about 65 metres. The transverse design was generally adopted because optimisations of turbine hall span, high pressure and reheat pipework runs, the type of low pressure and high pressure feedheaters when used, the laydown requirements and particularly in the case of nuclear stations the consequences of turbine disintegration, showed the most economic design. However, with larger units, it is possible that economics may change the disposition of the turbines from the preferred transverse design.

Figure 2.38 shows an arrangement of turbine hall cranes.

The cranage hook approaches at the end of and at the sides of the building will then determine the travel and span required for the cranage. The height of the crane is determined by the clearances that are required for erection and dismantling of plant, although it should be noted that the heaviest load is not necessarily, and probably unlikely to be, that which determines the cranage rail height. From the foregoing information it is then possible to define the major building dimensions with the exception of the turbine roof height which is dependent on the cranage lifting philosophy.

There are two basic lifting philosophies that can be adopted, firstly to lift the heaviest indivisible load by cranage, or secondly to use a synchronous jacking technique to lift the heaviest loads. There are a number of different ways of providing the former, viz., by a single crane or by two cranes lifting in tandem. The method that is generally chosen is lifting in tandem because of the lighter wheel loads spread over a longer track length which results in smaller column sizes. The depth of bridge beam is smaller, thus decreasing the height of the turbine hall roof. More importantly perhaps is the improved flexibility of operation of cranes during the construction and maintenance periods.

The latter, the synchronous jacking technique, permits a much lower capacity crane to be used and has the benefits of even lighter wheel loads and small bridge beam depths, and can be accommodated into the overall station programme without increase in programme time. This technique however has some detrimental effects in that floor areas have to be strengthened locally, it is cumbersome in operation during the construction of the station and interferes with other contractors working in the vicinity.

On an economic basis there is little difference in the cost of the two options when account is taken of the relative crane, building and specialist equipment costs. However, there must be a cost advantage to the improved flexibility offered with a two crane system lifting the heaviest indivisible load which is difficult to quantify over the life of the station, but which will be of major benefit if the heavy indivisible loads, such as the generator stator, need to be replaced within the station life.

The majority of modern stations thus contain two electric overhead travelling cranes in the turbine hall which will lift on their main hooks, either the heaviest indivisible load in tandem or the heaviest maintenance load individually. Where a station has more than four units a third crane can be added. The turbine hall cranes also usually have an auxiliary hook which is capable of lifting smaller loads at much faster speeds than the main hook, thus making the crane a versatile lifting machine.

13 Fire protection

13.1 Introduction

The primary objective of any fire protection philosophy must be to minimise the probability and the consequences of any postulated fire. In spite of steps taken to reduce the probability of a fire, it is possible that fires may occur from time to time. The fire protection philosophy for a modern power station should therefore be arranged to achieve a balance in:

- Preventing fires from starting.
- Limiting the consequences of a fire.
- Reducing the severity of any fires that do start.

It is not anticipated that any one of the above items can be totally effective, but the strengthening of one area can compensate for weaknesses in other areas.

13.2 Prevention of fires

The prevention of fire involves systematic consideration of the use, nature and physical arrangement of combustible materials, the potential of plant to become ignition sources, and the relative location of combustible material with respect to ignition sources. A careful check is therefore kept on the use of combustible

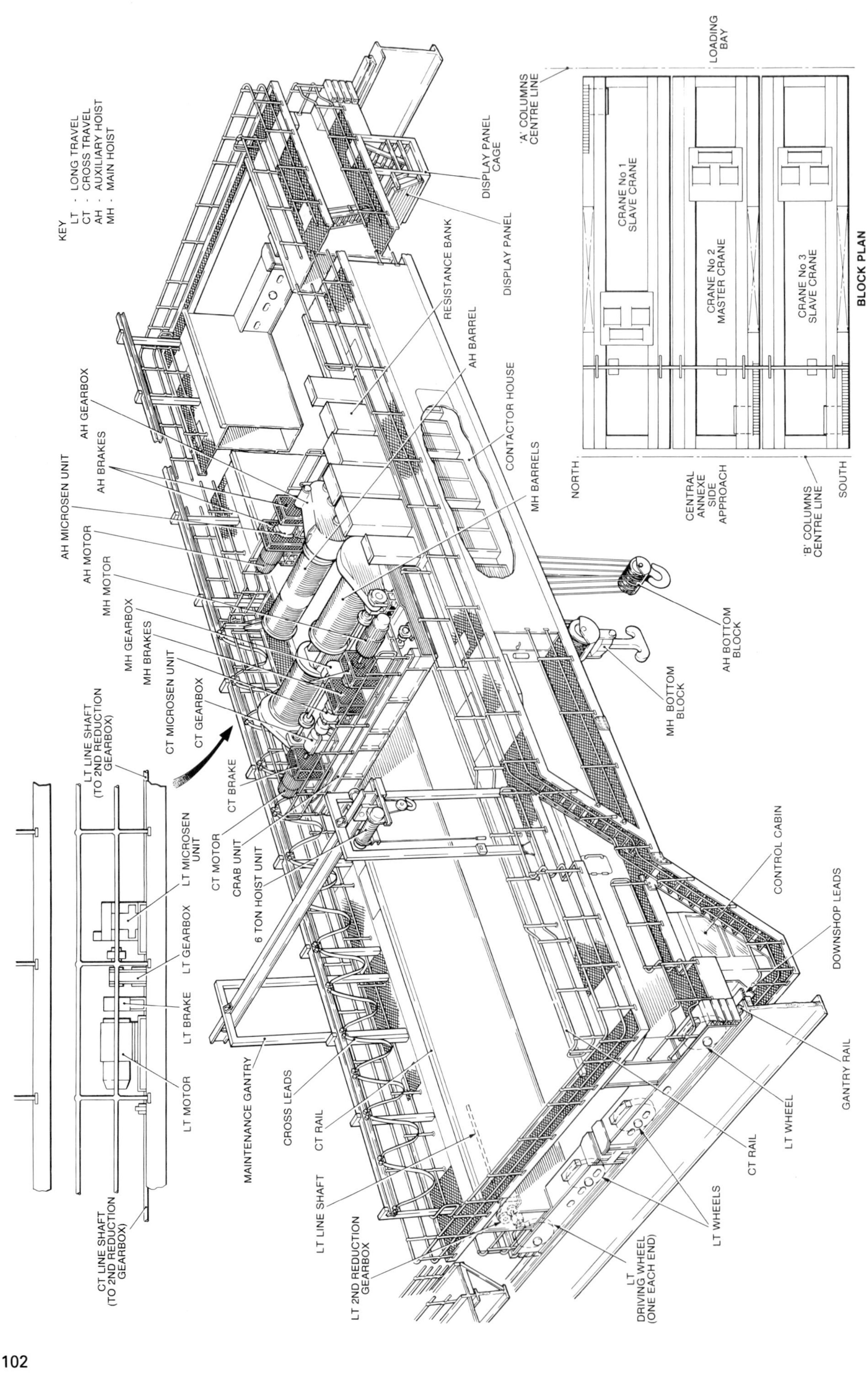

FIG. 2.38 Arrangement of turbine hall cranes

materials throughout the design stage of a station, and materials which have beneficial characteristics such as limited ignition and reduced fire propagation are specified. In addition, plant is designed, selected and laid out to limit the risk of ignition and the spread of fire from one area to another. Maintenance activities such as welding and cutting are tightly controlled as is the movement and storage of large quantities of combustible material such as fuel oil or cleaning fluid.

13.3 Limiting the consequences of a fire

It is obviously desirable, and in the case of nuclear stations it is often essential, to limit the amount of equipment affected by any potential fire so that station availability can be preserved or plant can be safely shut down. In addition, there is often a requirement to provide a safe-means-of-escape route for personnel in the event of fire. The primary method of achieving this is by the sub-division of the station into separate areas so that the extent of any fire is limited to that area. This sub-division is normally achieved by separation of buildings or plant by an intervening space and is of particular importance at the site layout stage. Further physical segregation of plant into fire compartments and fire zones takes place within the buildings in order to reduce the average combustible loading, to segregate sources of ignition from combustible material, to segregate redundant plant or to create escape routes. Very often natural fire barriers are created within the building due to the structural requirements of the building.

There may however be a requirement to further segregate the building into smaller compartments, and in this case additional fire barriers are erected. These fire barriers would be rated and certified as having been tested against a recognised standard such as BS476 Part 22 [1].

The need for segregation imposes a significant burden on space requirements during the layout stage, especially in the design of cable routes. It is therefore important that they are identified and established at an early stage in the development of the project.

13.4 Reducing the severity of fires

The severity of a fire can be reduced if it is detected early enough and measures are taken to extinguish the fire or to prevent it from spreading to adjacent areas. Water is the most common form of extinguishant used on a power station although foam systems are used on some fuel storage areas, and gases such as Halon or carbon dioxide are used in electrical equipment areas. The water extinguishing systems take two forms, an external hydrant system serving each building, and a permanently installed water spray system covering selected plant. Each system is separated from the other and has its own water supply and dedicated pumping plant.

In the case of the hydrant system, water is pumped around the site via underground pipes fed from an inexhaustible supply such as the CW forebay. In general no building is greater than 75 m from any one hydrant and most buildings are served by several. Hydrants conform to basic Fire Brigade patterns and are used as a source of water for fire tenders or dry rising mains in the buildings. Figure 2.39 shows an external hydrant system layout.

The fixed fire protection system employs nozzles which spray extinguishing water over the envelope of the risk; the following areas are usually covered:

- Turbine-generator.
- Transformers greater than 100 kVA.
- Boiler fronts.
- Cable tunnels and cable flats.
- Standby diesel generators.
- Coal conveyors.
- Gas circulators.
- Oil or gas storage vessels.

Figure 2.40 shows a typical spray water system for a 3-unit oil-fired station. The water is pumped from a central pumping station which includes standby pumping plant, a dedicated water supply with enough water stored to supply the largest risk, and a pressurised water tank to allow for instantaneous response.

Detection systems are selected from a range of commercially available equipment to suit the characteristics of the working environment and the expected fire. With the exception of the turbine, systems are arranged to be automatic in operation and great care is taken to ensure that spurious operation is avoided. Typical areas of coverage are as follows:

- Turbine plant — frangible bulb heat detection and rate-of-temperature-rise detection.
- Cableways — line type heat detecting cable.
- Switchgear — smoke detectors.
- Computer suites — smoke detectors.
- Boilers — thermocouples and frangible bulbs.
- Fuel oil tanks — low expansion foam injection.

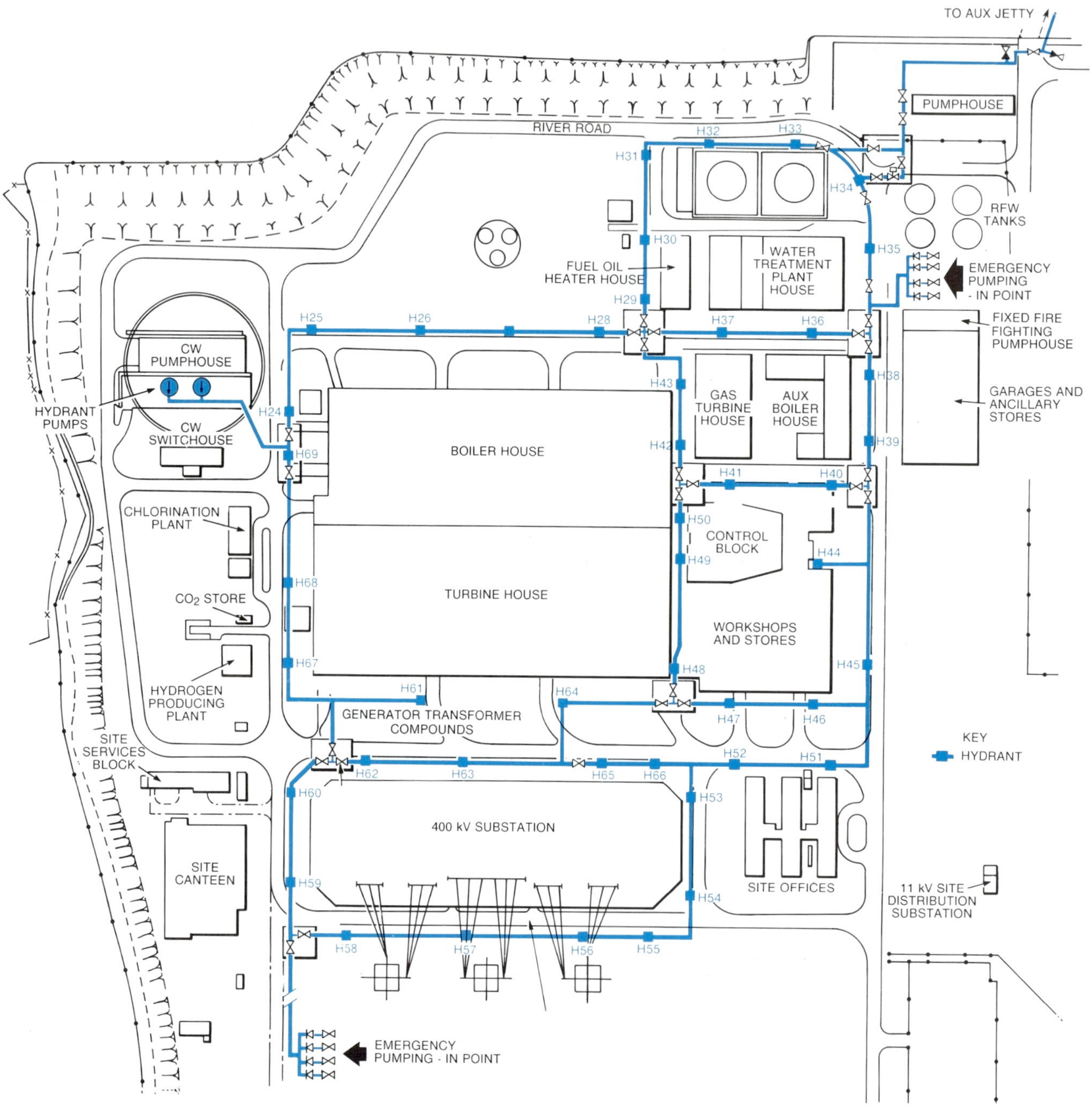

FIG. 2.39 External hydrant system

14 Electrical plant layout

14.1 Introduction

The main objectives when establishing the layout of electrical plant for a power station are to:

- Achieve minimum construction costs.
- Minimise energy flow losses.
- Provide access for operation.
- Provide adequate space and facilities for maintenance.

The transmission grid lines, grid substations, generator transformers and station transformers, together with the boiler or reactor, turbine-generator, fuel supply, waste removal and cooling water supplies are all considered for their individual requirements before arriving at a compromise station layout.

Design of the electrical auxiliaries system is determined by the size and number of generating units and

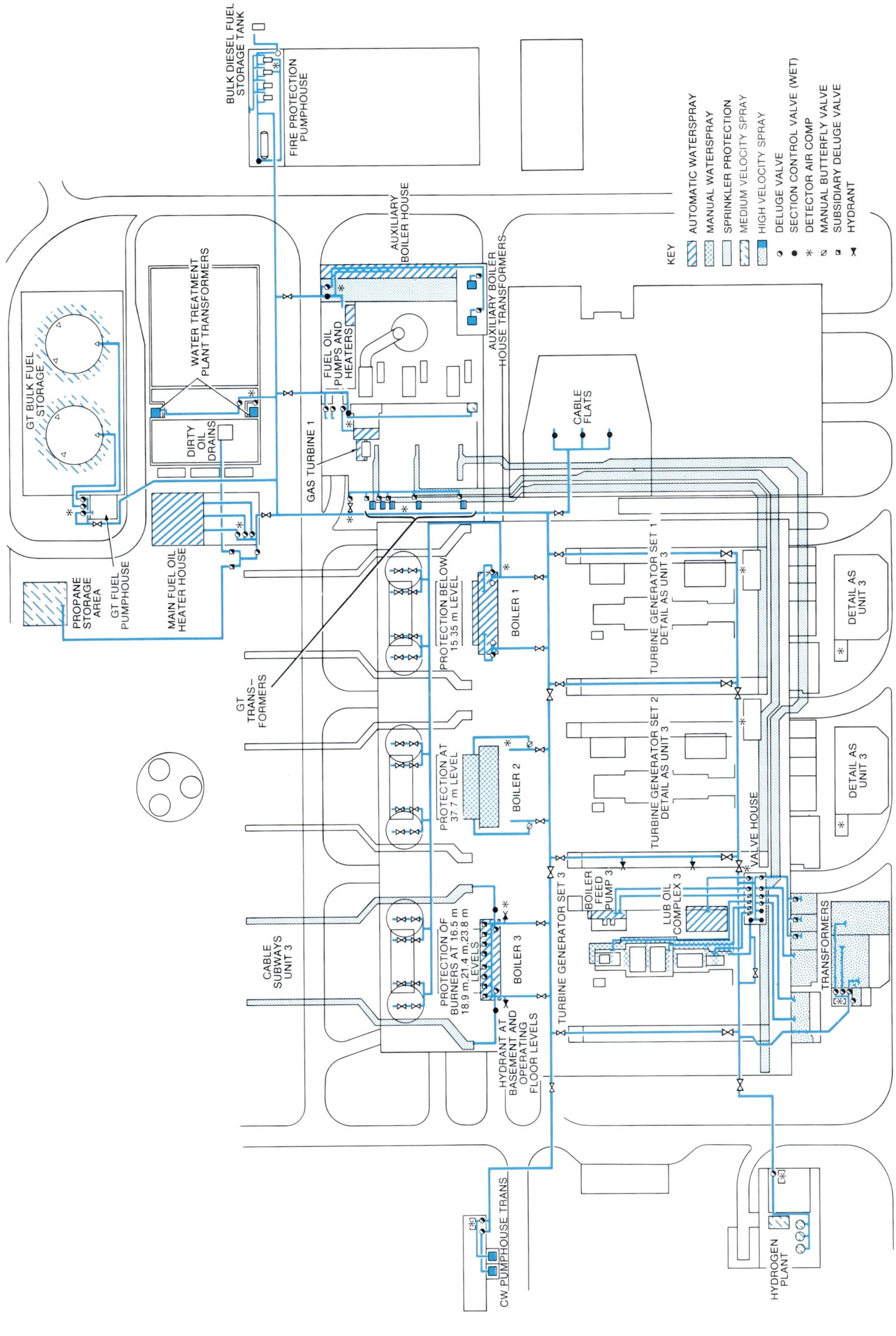

FIG. 2.40 Spray water system

the required electrical drives. An outline of the station electrical system, in the form of a power system diagram must be produced before the layout of electrical plant can proceed. This diagram will indicate the number and size of transformers, switchgear voltage levels and allocation of main drives to switchboards. On a typical CEGB station this will result in a 3-voltage level system operating at 11 kV, 3.3 kV and 415 V. Having designed the electrical auxiliaries system, the following major equipment must be located:

- Transformers — generator, station, unit and auxiliary.
- Auxiliary switchgear — 11 kV, 3.3 kV and 415 V.
- Generator main connections.
- Power and control cables — allocate space for major routes.
- Control room equipment — including control and instrumentation (C and I) rooms housing computer and alarm equipment.

The layout of the power station electrical equipment must take account of the following design considerations:

(a) All switchgear and electrical control equipment should be housed in clean and dry conditions.

(b) The electrical switchgear and distribution equipment should be located as near as possible to the centre of the electrical loads.

(c) The location of the central control room (CCR) should be positioned to suit the best man-movement arrangement, together with the best position for cabling facilities.

(d) Cable and electrical plant fire risk precautions should be taken in the form of fire protection systems and physical segregation of plant and cables. The principal considerations are that any incident on one unit should not affect any of the other units and, especially on nuclear power stations, sufficient alternative systems and plant, designed within the necessary segregation provisions, should be provided to allow for the safe shutdown of the plant.

(e) Oil-filled transformers should be located externally to the main buildings.

Where cable routes extend outside of buildings, and therefore run under the ground, close co-ordination with other disciplines of the design team is essential to avoid clashes with other services such as surface and storm drainage, mechanical services pipework, etc. This is most important with regard to high voltage (HV) cable connections between the generator transformers, station transformers and grid substations, which may require to be routed in concrete troughs alongside roads and rail tracks, thus sterilising large areas of ground.

It is of prime importance that all electrical plant layout design shall conform to the requirements laid down in relevant design codes and standards.

14.2 Auxiliary switchgear

Dedicated switchrooms are often used for the auxiliary switchgear; this gives unit-to-unit segregation and segregation within a unit if this is required for reactor essential cooling. The switchgear is protected from any possible damage from external sources such as leaking pipes or vessels and any serious incident in the main plant areas. The switchrooms provide a degree of security and the correct environment for operation and maintenance of the switchgear.

There are several points to be considered regarding the location of the switchrooms and annexes. The layout of the main plant dictates the locations of ancillary plant, and, in particular, the location and orientation of the turbine-generator determines the location of the generator transformer and unit transformer which in turn affects the location of the unit board. An economic and practical compromise has to be made on the positioning of the unit board to try and eliminate long and difficult cable runs from transformer to switchgear and then from switchgear to high voltage drives and feeds.

These considerations also apply to the station transformer and switchgear, except that the main factor determining the locations of these items is the entry of the grid feed to the station transformer.

The ideal location for, say, the 11 kV unit boards is a centrally located position, i.e., between boiler house and turbine hall. This gives reasonably short and easy cable routes to boiler and turbine plant. An important parameter in the overall station cost is the specific cost per unit volume of the structure. A central position created specifically for electrical plant increases this cost considerably and possibly unnecessarily.

In a two-unit station if the central position is not selected, the auxiliary switchgear would be located in external switchgear annexes positioned as near as possible to suit the plant associated with switchgear. This allows an economic cable layout and unit segregation is achieved with the minimum of main cables crossing the station.

Switchgear annexes external to the main buildings provide relatively easy access routes for initial installation and provide a logical and ready identification of various switchboards for operation and maintenance. Another consideration is the need for availability of the 11 kV station board and its lower voltage feeds for early commissioning and proving of plant. External annexes are conducive to this need for early completion. With the external annexe concept most of the auxiliary switchgear would be located in the annexes, i.e., 11 kV, 3.3 kV and 415 V with items such as coal plant switchgear located near to its plant area.

If a central location is selected, this would typically house the 11 kV unit boards and feed the boiler and turbine auxiliaries, the 3.3 kV and 415 V switchgear would be in annexes on the wings of the main building.

Other factors to be considered in the location of auxiliary switchgear are:

(a) Access routes must be practicable — for initial installation and future possible removal, overhaul and replacement. All switchrooms must have at least one escape door ideally positioned diagonally opposite the main access doors; in large switchrooms more than one door is required.

(b) Existing facilities, such as cranes, may be utilised and if necessary hatchways may be provided in the roof of switchrooms.

(c) Adequate provision must be made for cable routes bearing in mind fire risks and separation or segregation considerations.

(d) The location must be protected against water ingress from overhead water tanks, pipework, etc.

The unit transformers, station transformers and 11 kV/3.3 kV unit and station auxiliary transformers are of the oil-cooled outdoor type and are located remote from the switchgear. The 3.3 kV/415 V transformers are now of the air-cooled type and are integral with the 415 V switchboards.

14.2.1 11 kV and 3.3 kV switchgear

Once the locations of the switchgear have been decided, the actual layout of the switchrooms must be considered.

The switchrooms must be large enough to accommodate any make of switchgear with sufficient space for possible future extensions to the switchboards. Adequate rear access is required to allow the cabling to be installed and terminated and to allow inspection and removal of rear mounted equipment. Sufficient space at the front of the switchboard is required for complete withdrawal of the circuit-breaker truck. The floor must be designed to include fixings, floor levelling and truck runners as specified by the switchgear manufacturer and also to take into account the effect of the truck wheels when a circuit-breaker is removed and manoeuvred.

Space should be allowed for maintenance activities in the switchroom. Wall space is required for control and alarm junction boxes and miscellaneous equipment such as lighting and small power distribution boards and transformers, telephone junction boxes and fire detection and protection panels.

The cabling is laid in dedicated cable races below the switchgear and passes through holes or slots in the floor onto cable supporting steelwork in the race. The positioning of the switchgear and the holes or slots in the floor must take into account the civil design of the building. Care must be taken to avoid positioning the main cable entry over floor beams or any other major obstruction (see Figs 2.41 and 2.42).

The clear height under beams and services of the 11 kV and 3.3 kV switchrooms must be adequate to allow arc-shute removal, using the manufacturer's lifting frame. The arc-shutes are removed for inspection and maintenance of the contacts and arc-shutes. A runway beam is required, positioned over the rear of the switchgear for installation and future removal of the voltage transformers.

Access doors of sufficient height and width are provided to allow an assembled circuit-breaker panel to be manhandled into the switchroom.

The switchroom lighting and small power installations should allow for safe operation and routine maintenance activities and testing. Specialist maintenance jobs will be carried out in the station workshops. The switchrooms must be conditioned to prevent condensation and in areas of high ambient temperatures sufficient ventilation provided to prevent any overheating problems.

14.2.2 415 V switchgear

The considerations applied to the high voltage switchgear are also applied to the lower voltage switchgear. However, 415 V switchgear locations are much more diverse, e.g., water treatment plant, chlorination plant, etc., and to avoid any possible corrosion of the switchgear, dedicated switchrooms within these plant areas are necessary.

The 3.3 kV/415 V transformers are of the natural air-cooled type and are integral with the switchgear; this obviates the necessity of large low voltage (LV) cables from transformer to switchgear. Adequate space for installation, testing and maintenance of the transformers must be allowed for, bearing in mind that the transformers are bulkier than their associated switchgear. The floor has to be suitably designed for the weight of the transformer.

14.3 Turbine-generator auxiliaries

Whilst the greater part of the turbine hall layout is the result of co-operation between the mechanical plant engineers and the turbine manufacturer, the electrical layout engineer must take a very detailed interest at an early stage to ensure that provision is made for personnel and cable access to all electrical items, and suitable locations are agreed for electrical equipment cubicles away from hot and potentially wet locations.

Some equipment imposes its own special restrictions; for example, the hydraulic control fluid used to operate the main steam valves must not be allowed to come into contact with the PVC insulation which is used on most cables. Chemical reaction causes the PVC to decompose.

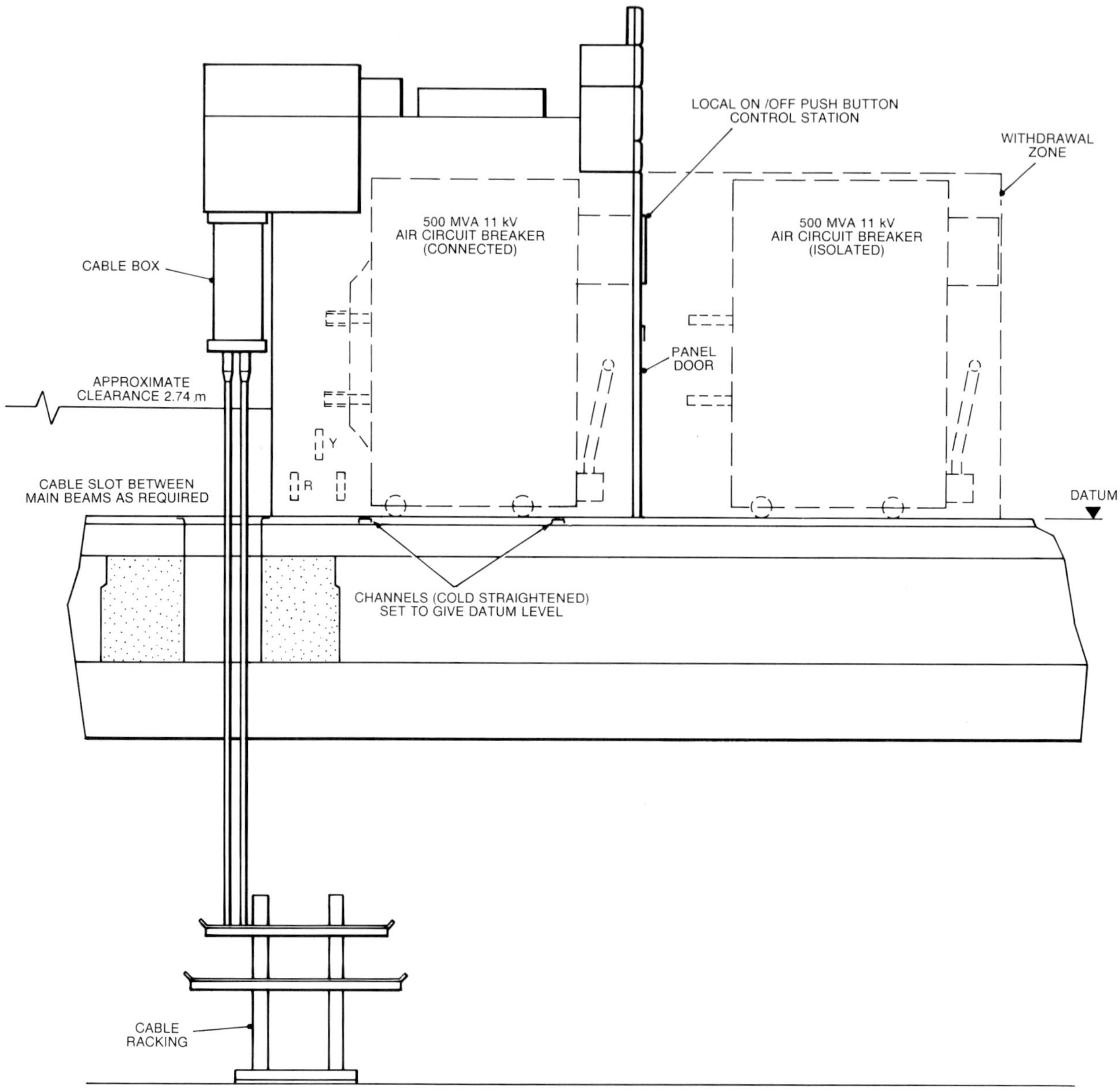

FIG. 2.41 Section through typical 11 kV switchroom and cable race

From the examples of station layout shown in this chapter it will be noted that the relative positions of many auxiliaries vary considerably from one station to another; the amount of equipment being dependent on the operating philosophy of the station. A base-load unit may possibly have one quick-start air pump and three maintaining air pumps, whilst a unit intended for flexible operation could have two quick-start and two maintaining air pumps.

The principal turbine-generator auxiliary plant which must be considered when preparing the electrical equipment layout for the turbine hall is as follows:

- Auxiliary oil.
- Seal oil.
- Jacking oil (normally one pump at each main bearing).
- Barring gear.
- Stator winding water.
- Distilled water.
- Hydrogen cooling.
- Condenser air extraction.
- Auxiliary cooling water.
- General service water.
- Hydraulic control fluid.
- Forced-air cooling compressors.

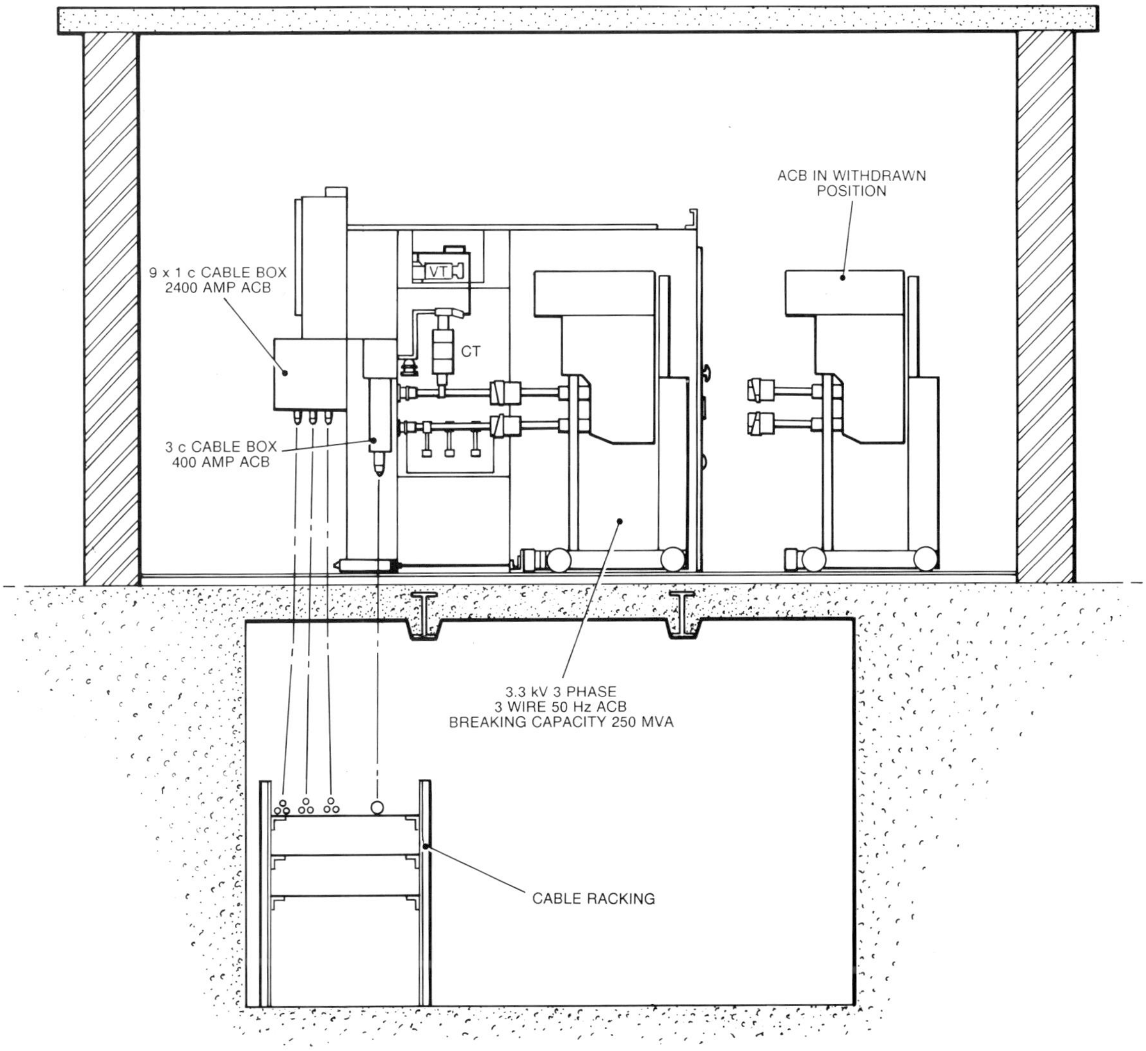

FIG. 2.42 Section through typical 3.3 kV switchroom and cable race

- Exhaust spray.
- Gland sealing steam.
- Feedheating.
- Electrical distribution cubicles.
- Instrumentation cubicles.
- Local equipment housings (possibly 20 per unit).

Most pumped systems have duplicate 100% duty units with instrumentation to measure pressure and temperature and give an automatic changeover should the running pump fail. Also, separate cable routes are provided for each pump to meet the requirements for segregation — see Section 14.6.1 of this chapter.

In addition to the auxiliary equipment, the turbine-generator will itself require extensive cable steelwork to cater for the many instrumentation and control devices along the length of the set. Most of these will be marshalled by the turbine manufacturer who will fit flexible metal harnesses, containing heat or oil resistant wiring as necessary, between the actual devices. The devices may be located in a hostile environment and then junction boxes positioned in less hostile locations.

A large unit can have approximately 150 power operated valves, each of which need provision for power and control cables.

14.3.1 Excitation equipment

All the latest generator excitation systems make use of an AC exciter with diode rectification and thyristor control. The diodes may be static and mounted in the automatic voltage regulator (AVR) and excitation cubicle, in which case the cubicle must be mounted close to the generator to allow for the solid busbars necessitated by the heavy excitation current (5170 A for a 660 MW unit). Final connections are made to the pilot and main exciters by sliprings and brushgear.

Alternatively the diodes may be mounted between the exciter and the generator and rotate with the shaft, control and switching taking place between the main and pilot exciters. The lighter current permits the use of cable connections and allows greater flexibility in positioning the equipment cubicle. With this arrangement however, diode monitoring and maintenance are more complicated.

14.4 Main connections

The standard 660 MW generator produces over 19 000 A at 23.5 kV, the earlier 500 MW generators produce 14 500 A at 23.5 kV and the proposed 900 MW will produce 24 750 A at 26 kV. This power has to be transmitted to the LV terminals of the generator transformer by the main connections.

The magnetic and phase-to-phase fault problems associated with transmitting such large currents led to the development of the phase isolated busbar (PIB) and this is now standard for all main connections.

In a PIB system, each line connection consists of a current carrying high conductivity aluminium conductor supported by post insulators inside a high conductivity aluminium sheath of fully-welded construction which is electrically continuous over its complete length, see Fig 2.43. The sheaths have insulated sections inserted at both the generator and the generator transformer ends, and at the ends of the tee-offs to the voltage transformers and unit transformers. Bonding the sheaths together at their ends ensures that the current induced in the sheath is in anti-phase to that in the main conductor. The sheaths are insulated from adjacent metalwork, bonded together at the generator end and earthed to the station earth bar. Hence the magnetic field from the conductor and that from the sheath cancel each other out both in normal working and under short-circuit conditions. There is little heating of surrounding or supporting steelwork nor is there electrodynamic strain between different phase connections.

Despite the large reduction in magnetic effects achieved by phase isolation, in order to further minimise the effects of any induced currents, there must be

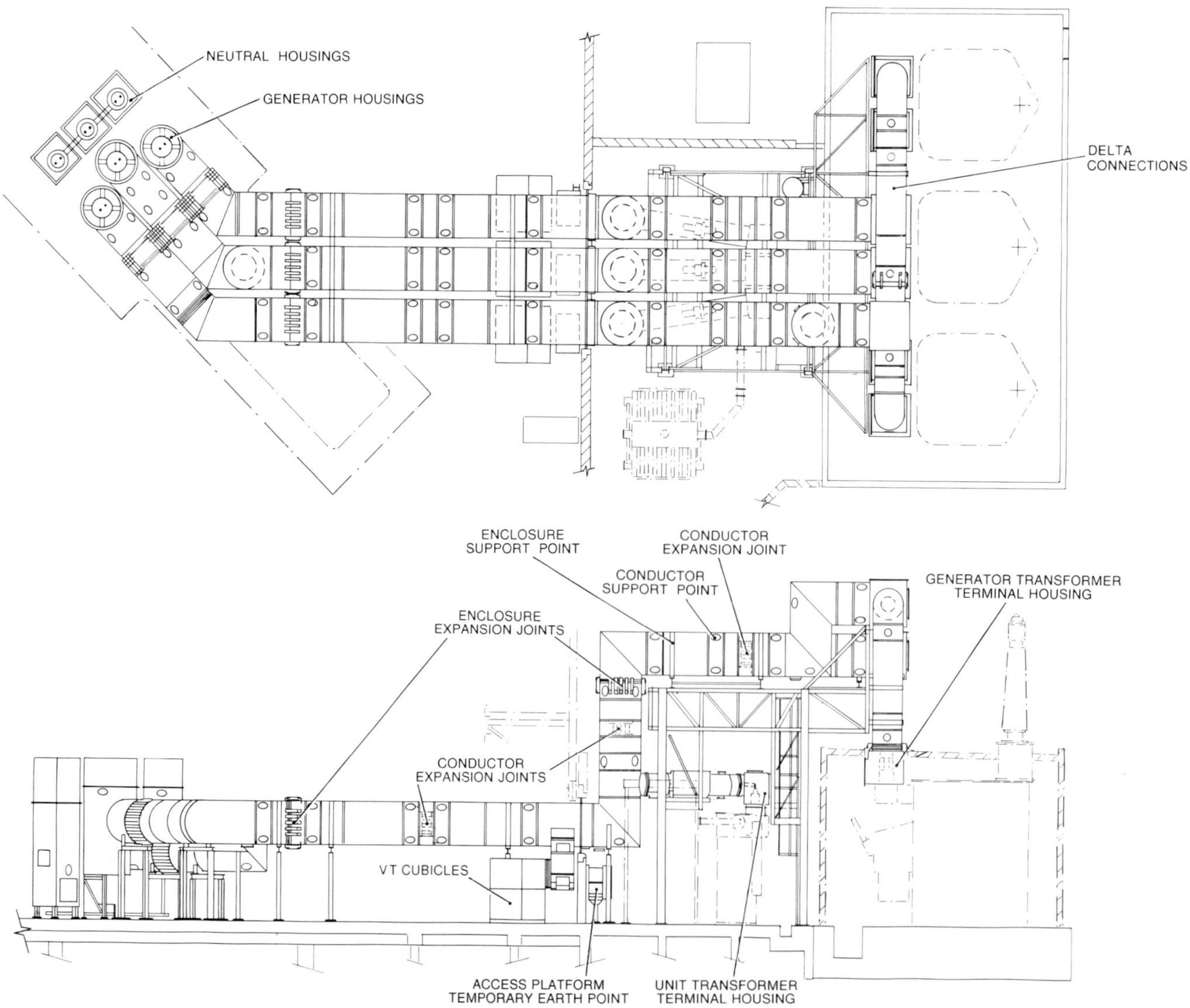

FIG. 2.43 Phase isolated busbar main connections

a gap of at least 300 mm between the main connections and any steelwork or pipework running parallel to them. When the run of steelwork is at right angles to the main connections the gap can be reduced to 150 mm. Closed loops of metalwork must be avoided, also the main connections must be insulated from all supporting steelwork.

Natural air cooling is practicable up to about 20 kA but a significant increase in current rating for the same conductor size is possible by forced air cooling.

Phase isolated busbars arrive on site in make-up sections which may be 10 m or even more in length. They are then supported in position and welded together. Being continuously welded they are naturally vermin and drip proof over their length so they only need sealing at their ends. Special provision is made at the generator to ensure that hydrogen, if it should leak from the generator, cannot collect in the main connections, similarly provision is made at transformer connections to prevent the ingress of bushing oil. In order to prevent dust entering, and to guard against condensation, which can cause a reduction in the dielectric strength of the insulators and corrosion, the connections are pressurised to approximately 12.5 mbar with dry conditioned air.

The main conductors are joined to the generator and other items of plant by means of flexible connections. This prevents the transmission of vibration and consequent work-hardening of the aluminium. Their route should be as straight as possible because bends can create hot spots. Expansion joints are required in long straight runs; however, these are usually designed so that they do not particularly increase the sheath diameter.

The line and neutral connections are, in the majority of cases, brought out from the underside of the stator with the line connections close to the exciter end of the generator and the neutral connections nearer to its centre line. The simplest arrangement of connections is when the unit is arranged transversely and they are then run straight out to the generator transformer passing between the generator supporting steelwork, through the turbine hall wall and over the unit transformer (see Fig 2.42). The voltage transformers are then also positioned between the generator supporting steelwork legs with their connections teed-off from the main connections passing overhead.

When the unit is arranged longitudinally this simple arrangement cannot be used. The longitudinal arrangement creates additional bends and also sterilises part of the turbine basement area. Bringing the connections out of the top of the generator alleviates many of these problems but then creates one of its own because they are now susceptible to damage from loads being moved on the overhead cranes.

The LV side of the generator transformer is delta connected. The generator transformer now consists of three single phase units arranged to form a three-phase bank. This means that the delta must be formed externally. On early designs this was done using a delta box across the three transformers, but more recently the delta box has been made in the main connections just prior to their terminating point on the generator transformer (see Fig 2.44).

Also associated with main connections are the voltage transformers (VT), the neutral connections, the excitation busbar connections and the mounting of the current transformers.

The VT connections are teed-off the main connections. The simplest position for the VTs is immediately below the main connections in the turbine basement.

The excitation connections interconnect the exciter, the rectifier cubicle and the field switch cubicle. The two cubicles are positioned side-by-side in the turbine basement, level with and slightly to the side of the exciter. Three-phase connections from the exciter are taken from above the main busbars and at right angles to them, they then turn through 90° and drop vertically onto the rectifier. The busbars connections between the rectifier and the field switch run in the canopy. The connections from the field switch back to the generator brushgear take a similar route to those from the exciter (see Fig 2.45).

The neutral connections and the star bar are supplied with the main generator, but the neutral earthing transformer and resistor form part of the main connections. The neutral earthing transformer and its associated resistance banks are supplied mounted on a frame as one unit which must be positioned as close to the star bar as is possible. This may well require it to be supported at some considerable height above the turbine hall basement. On the neutral connections immediately below the stator and above the star bar are mounted current transformers. There may be as many as seven of these on each phase. They are supported from the neutral housing which screens the whole of the neutral connections. In addition to the current transformers on the neutral, there are others mounted immediately before the unit and generator transformers.

14.5 Transformers

14.5.1 Generator transformers

The generator transformer is the largest transformer on a power station and connects the generator output to the grid. There is a generator transformer for each generating unit and it is rated according to the size of that unit. Like any other oil-filled transformer, the generator transformer should be located within an outdoor compound, protected by a high pressure water-spray fire protection system and surrounded by a bund wall capable of containing not only all of the oil contained in a single phase transformer, but also the discharge from the fire protection system over the whole bank of transformers on that raft following rainfall prior to the incident. For the generator transformer this could mean up to 320 000 litres, of oil and water

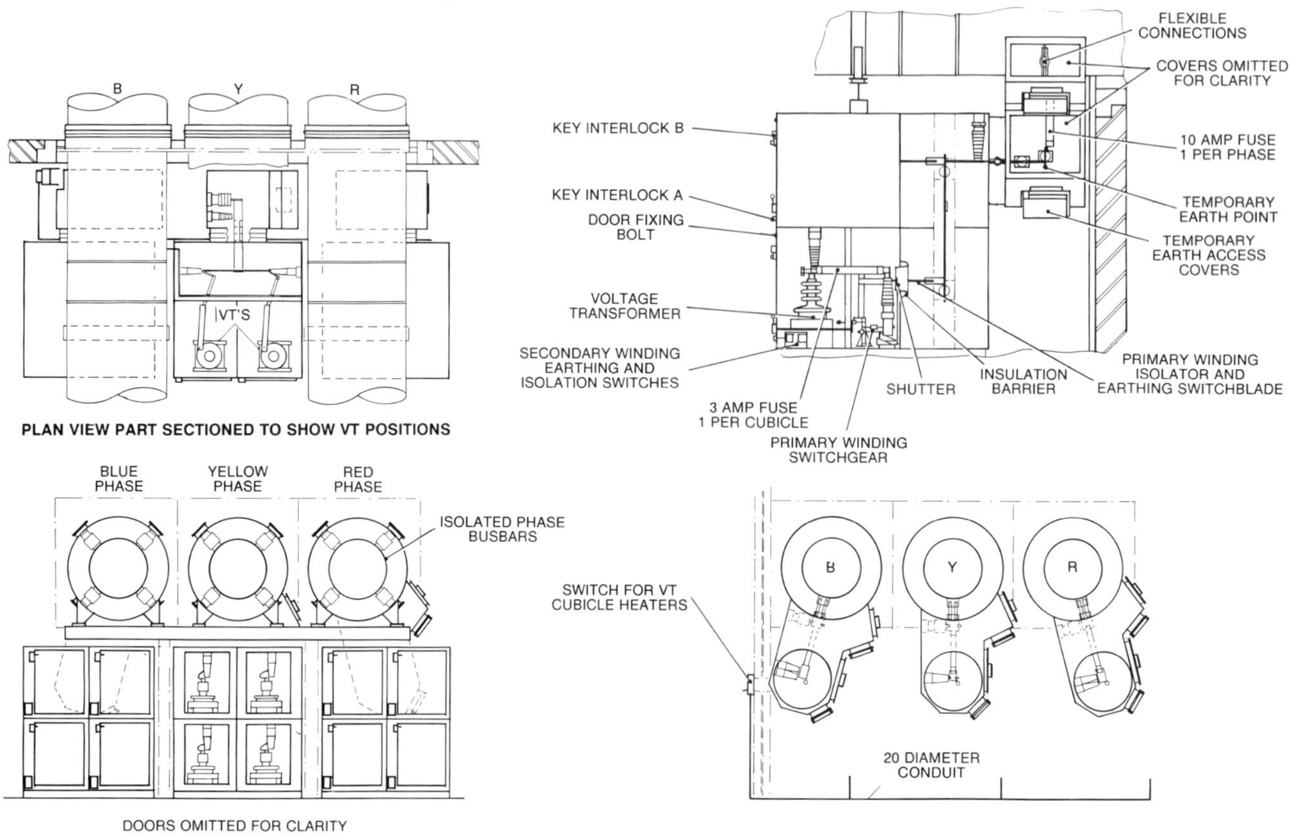

FIG. 2.44 Arrangement of voltage transformer cubicles and tee-off connections

mixture. Also, as any other oil-filled transformer, the raft must drain into a special system equipped with an oil separator and interceptor, and there must be no possibility of puddles forming or being left after a spillage.

However, certain other considerations determine the location of the generator transformer (see Fig 2.46):

- It must be as close to the generator as possible so as to keep the main connections as short as possible.
- The generator transformer is one of the heaviest loads delivered to site so its location will seriously affect the position of the site perimeter road along which it must be delivered and from which it must be manoeuvred into position.
- Its cooler bank needs an area of approximately nine times its own plan area to be free of major obstructions over 1.3 m high, although this area may include fences and roadways.

All single phase units of a similar rating are currently designed to be interchangeable. This means that all their interface dimensions shall be the same and there shall always be the same distance between the individual phases forming the three-phase unit. This dimension is 4.9 m for the 800 MVA transformer and 5.1 m for the 1145 MVA transformer.

Whilst the oil connections to any transformer tank are identical, the location of the conservator and cooler bank within the compound can be adjusted to suit any particular site.

On early designs of single-phase generator transformers, the LV delta was formed in an oil-filled delta box which spanned the three tanks, but this has now been superseded by making an air-cooled delta in the main connections just before they connect onto the transformer. Mounted on the main connections just prior to the delta are the protection current transformers. These transformers and the delta need to be supported in position and if the route of removal of the transformer unit is underneath this support then adequate clearance must be provided.

On the more recently-built power stations, the HV connections from the generator transformer have been made through SF_6 (sodium hexafluoride) insulated isolators and earth switches into 400 kV cables which then run in concrete troughs out to the grid substation. This gives a much more compact arrangement than the air-insulated equipment and overhead connections previously used. It also improves access and site safety because there are no longer 400 kV overhead wires crossing the site perimeter road.

Each transformer has a cooling system comprising two oil pumps and four cooler fans. All of these are

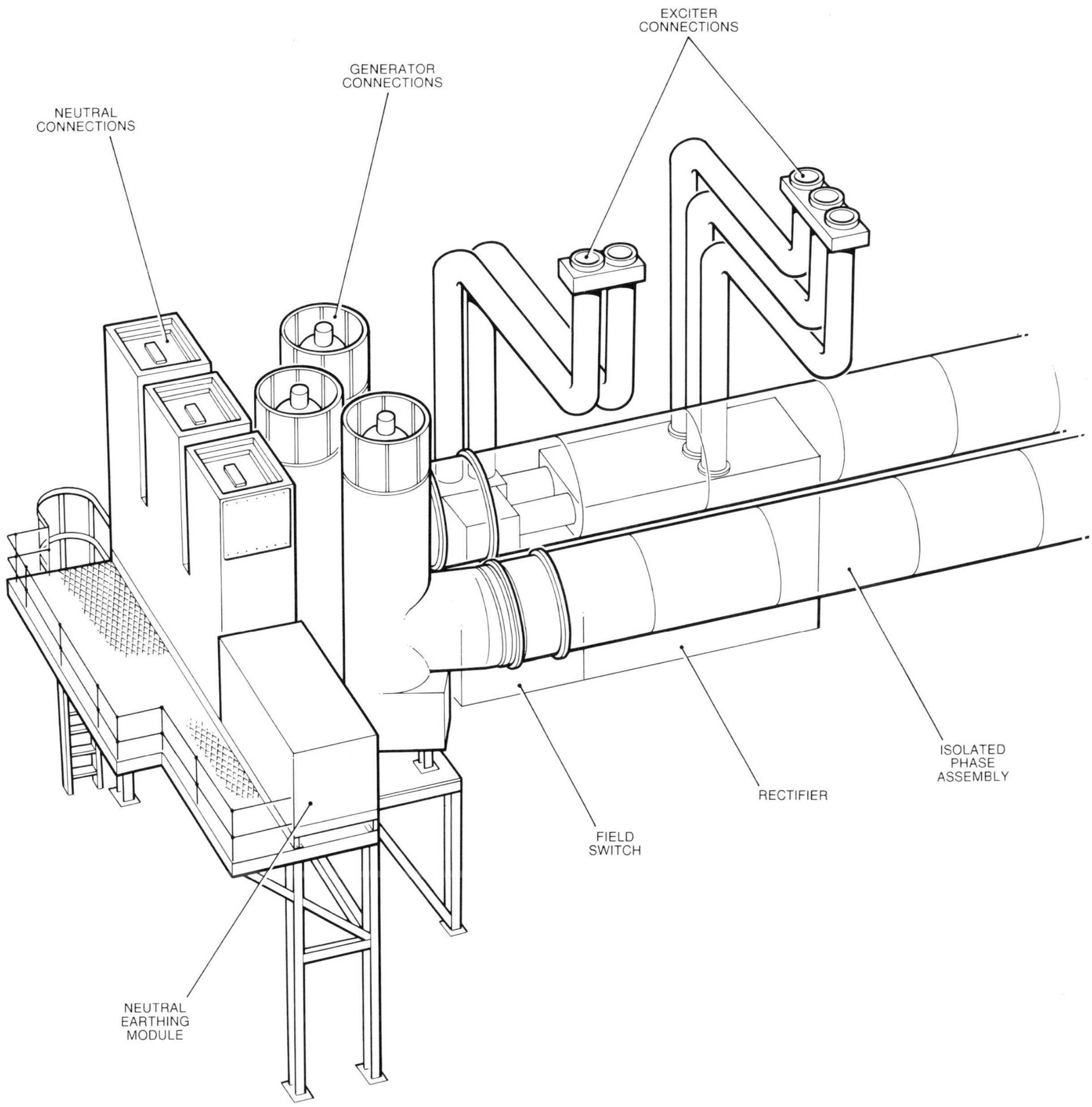

FIG. 2.45 Layout of exciter connections and main connections below the generator

supplied and controlled from the transformer marshalling kiosk which is located just outside the bunded area. The marshalling kiosk also controls the on-load tap changer and marshalls all local cables for alarms, etc., originating on the transformer.

14.5.2 Station transformers

The second largest transformers on a power station are the station transformers. These step down the grid voltage to that of the highest plant auxiliaries which on modern power stations is always 11 kV. They are required for commissioning the first plant on a new station and supply loads not specifically associated with the generating unit such as lighting, cranes, CW pumps, etc.

The location of the station transformer within the power station is not as critical as that of the generator transformer, but it should be as close as possible to the buildings and the 11 kV switchboards that it is supplying. Its compound must fulfil all the general conditions applying to oil-filled transformers with respect to fire protection, drainage and bund walls (see Fig 2.46).

The HV connections to the station transformer may be overhead or underground depending on the site. The LV connections may be either 11 kV cables or phase isolated busbars depending on the relative locations of the station transformer and its associated 11 kV station switchgear. Phase isolated busbars are only favoured when the transformer and switchgear are close together because they are inflexible and need straight runs with a minimum of bends. When cables

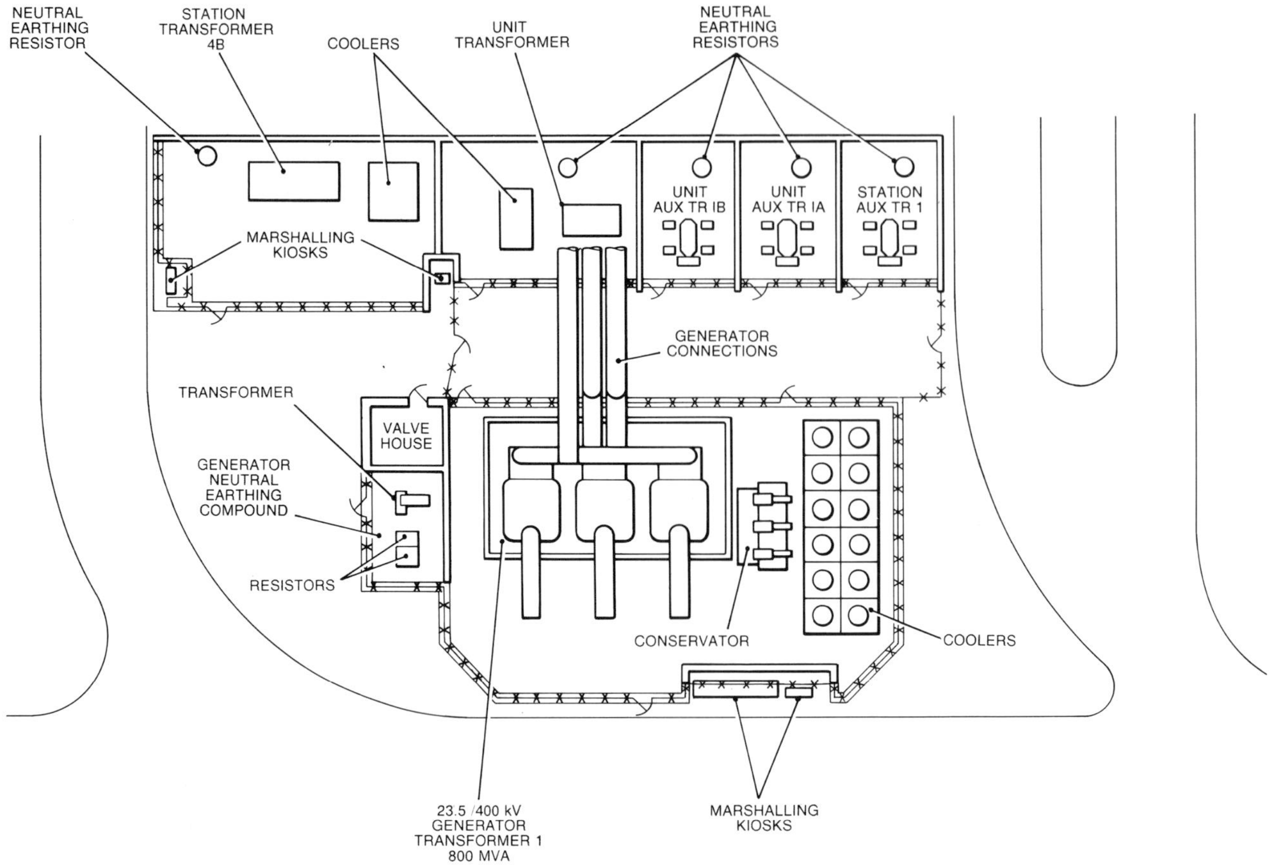

FIG. 2.46 Transformer compound and electrical annexe

are used, they usually leave the transformer compound via underground ducts or tunnels. These must be adequately sealed to ensure that there is no possibility of oil or water escaping from the compound into the cableways.

14.5.3 Unit transformers

The HV side of the unit transformer is teed straight off the main connections, so this transformer should be positioned very close to the generator transformer (see Fig 2.46). Its compound is equipped with high pressure water spray fire protection, bund walls, and drainage into an oil/water separator. Where two oil-filled transformers occupy adjoining compounds, they are separated by a fire barrier wall.

The 11 kV connections from the unit transformer are by cables, two, three or four per phase depending on the rating. These leave the compound via ducts into a trench or tunnel. The ducts must be tightly sealed against the ingress of transformer oil. The star point of the LV is taken to earth via a liquid neutral earthing resistor (LNER) which limits the earth fault current. This LNER is also accommodated within the transformer compound.

The unit transformer is a heavy load so adequate haulage facilities must be provided to enable it to be placed in position.

The maximum permitted noise emission from the unit transformer is 30 dB at 400 m, and a noise enclosure is not normally required.

14.5.4 Auxiliary transformers

Auxiliary transformers on power stations are many and vary greatly in size, rating, insulation, etc. On CEGB stations, the largest are the 11 kV/3.3 kV auxiliary transformers which can be 12.5 MVA, 10 MVA or 8 MVA. They are oil-filled and hence require all the safeguards of any oil-filled transformer, i.e., outdoor location, high pressure waterspray fire protection, fire barrier walls, bund walls and drainage into an oil/water separator. Where possible they are positioned to suit the 3.3 kV switchgear.

Transformers of 3.3 kV/415 V are usually rated at 2 MVA, 1.6 MVA or 1 MVA as these values match the standard current ratings of 415 V switchgear. These transformers may be oil-filled, as are the majority of those on existing power stations. However, the requirement that oil-filled transformers be located out of doors tends to create long runs of 415 V cables with a consequent increase in either cable size or volt drop. This has led to the consideration of alternatives. Of these, the type most favoured by the CEGB is the dry type air-cooled class C insulated (AN). These, as their description implies, contain no fluid at all and have the advantage of being considered totally fireproof and they can be made integral with the 415 V switchgear. They do, however, have the disadvantages of higher cost, heavier weight, and lower reliability than oil-filled transformers.

When dry type transformers are fitted in switchgear there are certain layout requirements to be considered. Firstly, they are relatively heavy and bulky so the switchgear increases in both weight and size necessitating bigger switchgear rooms with stronger floors. Also they dissipate heat so this must be considered when designing the heating and ventilation system. Being integral with the switchgear, the LV connections are busbars into the switchgear and the HV connections are cables.

14.6 Cables

The cabling on a power station performs the essential function of providing electrical interconnection between the many items of electrical and control equipment. During station erection and commissioning, the completion of the cabling systems is dependent on the timely installation of plant items. It is evident therefore that station cabling is a very important consideration at the overall design and planning stages.

Layout of the main cable routes is to a large extent dictated by the location of plant, transformers, switchgear and the central control room. However, segregation of unit electrical services also helps to establish the layout, particularly for nuclear stations where segregation between quandrants has to be considered in more detail (see Fig 2.47).

It is a basic requirement that cabling for one particular unit be segregated from the cabling to other units, and cable tunnels are an ideal method of obtaining this segregation on the major cable routes. Also, because cable tunnels are located at basement level they give the added advantage of being completed early in the civil programme, hence delays in the cabling installation can be avoided (see Fig 2.48).

The importance of providing adequate accommodation for cables cannot be over-emphasised, and a typical cable tunnel with a capacity of sixteen arms has on many occasions proved insufficient. Apart from installation difficulties, congestion of large quantities of cables creates many problems such as overheating, overloading cable supports, loss of separation parameters and high levels of combustible PVC insulation which increase the fire hazard. Hence at the design stage consideration must be given to the expected number and size of cables running in the network of tunnels.

When the basic cable tunnel design is established, arrangements should be made for ventilation, drainage, inserts for the cable supports and cable drum access facilities for cable pulling. From the safety point of view, personnel access points and emergency exits have to be located in conjunction with fire barriers and doors which divide the tunnels into sections and so restrict the spread of fire and dense smoke. Further, all major cable tunnels have a fixed waterspray fire protection system which is automatically initiated by linear heat detection cables running above and below each cable rack.

14.6.1 Segregation

Cables are vital for the control and operational activities that take place in a modern power station. The failure of a data or power cable due to a small fire can have catastrophic effects on such activities. Therefore, where possible, it is important to design the cable system layout to limit the effects of such a situation and one such method is to have segregated cable routes.

For conventional stations, the basic principle for major cableways is that the cables for each unit shall be kept segregated, whilst on the minor routes segregation is achieved by routing the cables in different directions. Segregation is required to limit generation loss by preventing the spread of fire and damage to other units, hence not more than one unit should be lost. It is however possible to keep a fire-damaged unit on load by transferring to standby feeds which have been taken by a different route to the main feeder; this segregation within the unit is generally referred to as the *A* and *B* routes. Segregation will depend on the system design and may affect cabling to such items as unit transformers, station transformers, cooling water pumps, boiler feed pumps, gas turbines, etc. However, segregated routes must be taken where duplicate DC supplies for switchgear tripping are provided, also where *main* and *emergency* supplies are provided, e.g., turbine lubricating oil pump. Where cables are installed direct in the ground, a distance of 1 m between segregated groups is considered adequate. For cables running parallel, in cable tunnels, etc., a 600 mm separation distance is necessary between control and single core power cables, this is to avoid inducing interference currents in the control cores, particularly under fault conditions. A separation distance of 300 mm between multi-core power and control cables, and also between single core and multi-core power cables is acceptable. On plant where control and power cables run side-by-side for a short distance, this length is limited to a maximum of 5 m for total run of cable. In cableways it is considered good practice to install the power cables on the uppermost racks to reduce unnecessary heating and hence thermal ageing of the control cables.

For nuclear stations, however, additional segregation is necessary for the safety of personnel, plant and

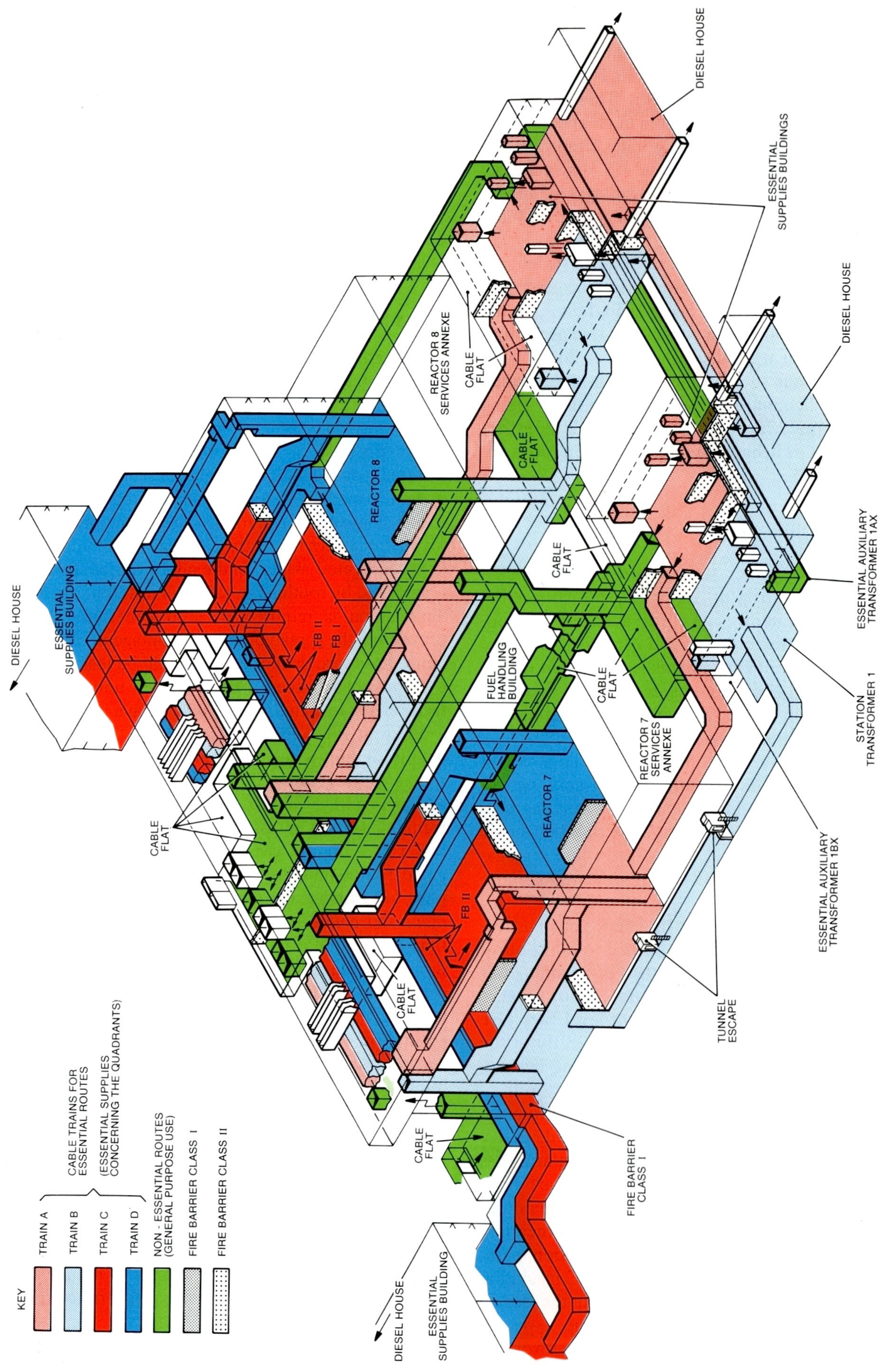

FIG. 2.47 Typical segregated cableway complex

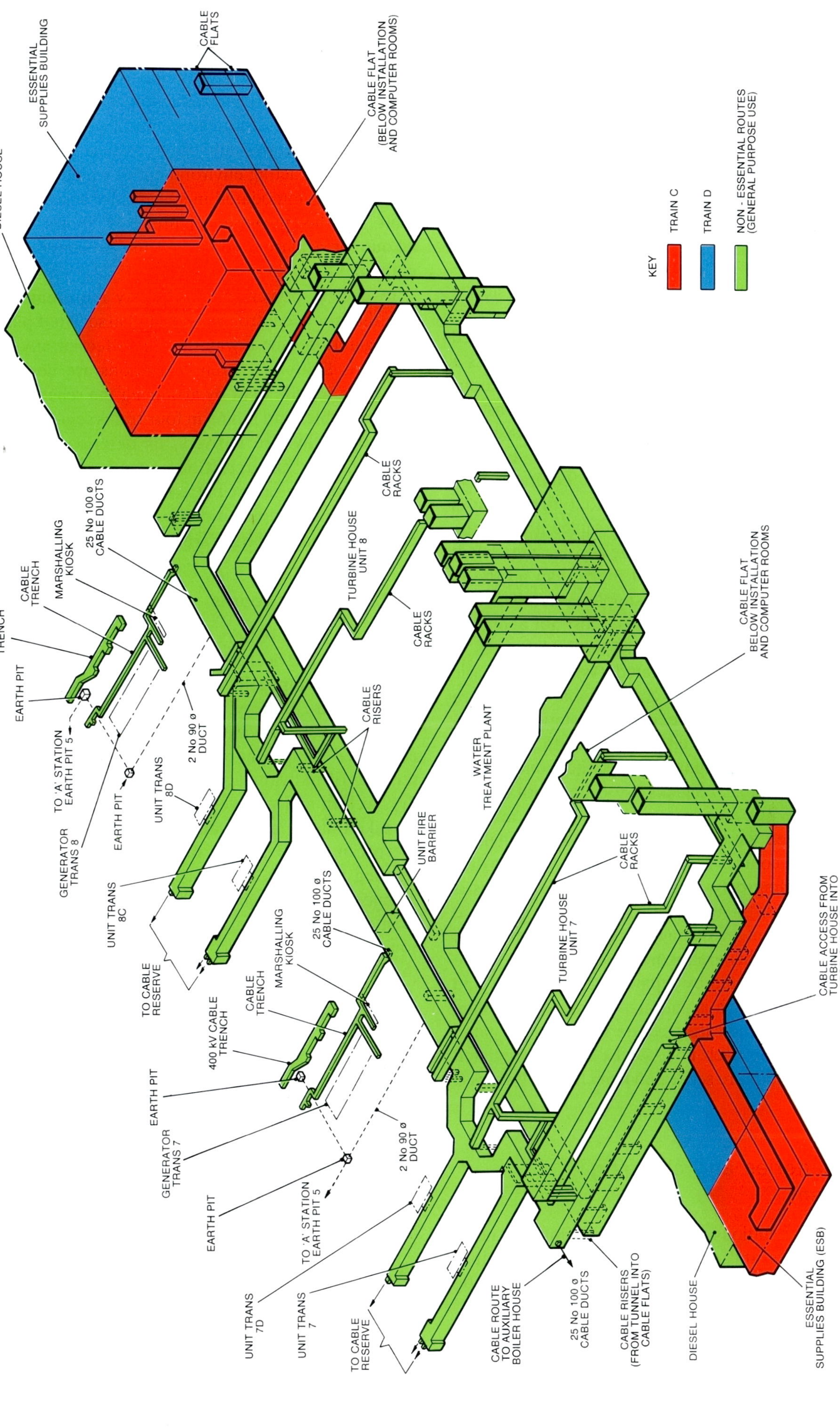

FIG. 2.48 Typical layout of cableways for turbine hall

general public. To fulfil this safety criteria means that no major incident, whether it be caused by fire, flood, earthquake, turbine disintegration, hot gas release or any other occurrence, should prevent the reactor being tripped when required or impede any reactor cooling operations. This is achieved by having two separate reactor cooling systems (*X* and *Y*) which are fed from four electrically-independent sections of the essential supplies system (trains *A* to *D*), each of these trains having its own power and control cabling system. In addition to these separate systems, segregation of cabling on a nuclear station is divided into two classes. For segregation class 1 cables installed direct in the ground, cable groups should be at least 4 m apart, whilst with segregation class 2 cables the recommended minimum spacing is 1 m. For both segregation class 1 and 2 cables laid in concrete trenches, the groups should be spaced 6 m or more to protect against oil spillage or mechanical damage. In addition, segregation class 1 cables, plant and equipment of different groups must be separated by a 4-hour rated fire barrier, whilst for segregation class 2, a 1-hour fire rating is acceptable. All this segregation of cables around the reactor must be so designed in the layout stage, such that for any major incident, not more than all the *X* supplies (or all the *Y* system supplies) are lost. An alternative choice is that the design must not allow the loss of more than half the *X* and *Y* supplies.

Apart from these safety-related systems, the basic segregation requirements for conventional stations are equally applicable to nuclear power stations. Also, whilst the power supplies ensure safe shutdown of the reactor, equally as important are the control and instrumentation cables. Signals associated with reactor safety trip are run in separate cables which are run directly between equipments and are not marshalled. In the central control room area, control and instrumentation cabling between essential supplies trains is designed to class 2 segregation, this being provided to limit the extent of damage only; consideration of reactor safety in this area is not necessary due to the reactor being protected by separate safety circuits and post trip cooling which is automatic. Similarly, cabling to the emergency indication centre is again segregated to class 2 between each unit, whilst cables associated with trains *A* and *B* are separated from those of trains *C* and *D*.

14.7 Batteries and charging equipment

Modern power stations and substations require a number of batteries and chargers of different capacities and voltages for a variety of uses. Ever increasing emergency and standing loads imposed on the DC systems have made it necessary for separate batteries to be provided for each boiler/turbine unit as well as common station services, with adequate arrangements for interconnection. In this way faults are confined to the particular unit and are therefore more readily identified. In addition, this allows the battery equipment to be sited closer to its electrical loads, leading to shorter cable routes and reduction of the effects of volt drop, facilitating a more economical cable installation.

Because batteries and their associated equipment demand an exacting environment in which to function, they are accommodated in specifically designed rooms built for this purpose. Figure 2.49 shows a typical layout for a battery room and its associated charging equipment.

In nuclear power stations, safety requirements dictate that some control systems and associated plant are segregated in such a way that the effect of any system failure is limited. Currently, stations are designed to meet this requirement by arranging the reactor control systems cables and piping in four trains segregated from each other by fire barriers, as described in Section 14.6.1 of this chapter. Consequently, nuclear power stations require larger batteries and chargers than the conventional stations to cope with the essential duties, and hence require larger areas to house the equipment.

14.7.1 Battery rooms

The siting of the battery rooms should be central to the associated electrical loads. This alleviates the need to install uneconomically large cables to compensate for voltage drop. Battery rooms in CEGB power stations are exclusively provided for housing batteries and any associated equipment required for inspection and maintenance of individual cells. The environment must be dry, well-lit and well-ventilated at all times. Care must be taken at the design and siting stage to ensure that there can be no ingress of moisture from fixed fire-fighting apparatus in rooms above the battery room.

Since battery capacity and performance is affected by temperature, a stable ambient temperature of 20°C is sought within the battery room. It is at this temperature that the batteries are rated, although a temperature differential of 5°C to 40°C can be tolerated. The effects of any cyclic solar heating can be minimised by siting the battery room on the north side of any building, conversely temperature drop is arrested by electric heaters of the totally-enclosed type.

Battery cells may be arranged in a single or double tier configuration, and in single rows for positioning against walls, or in double rows where access is available to both sides. Each cell must be readily accessible for maintenance and inspection (see Fig 2.49).

14.7.2 Charging equipment and switchgear

Independent charging equipment is provided for each battery. If any one charger should fail there is the facility for interconnection with other system chargers of similar output voltage.

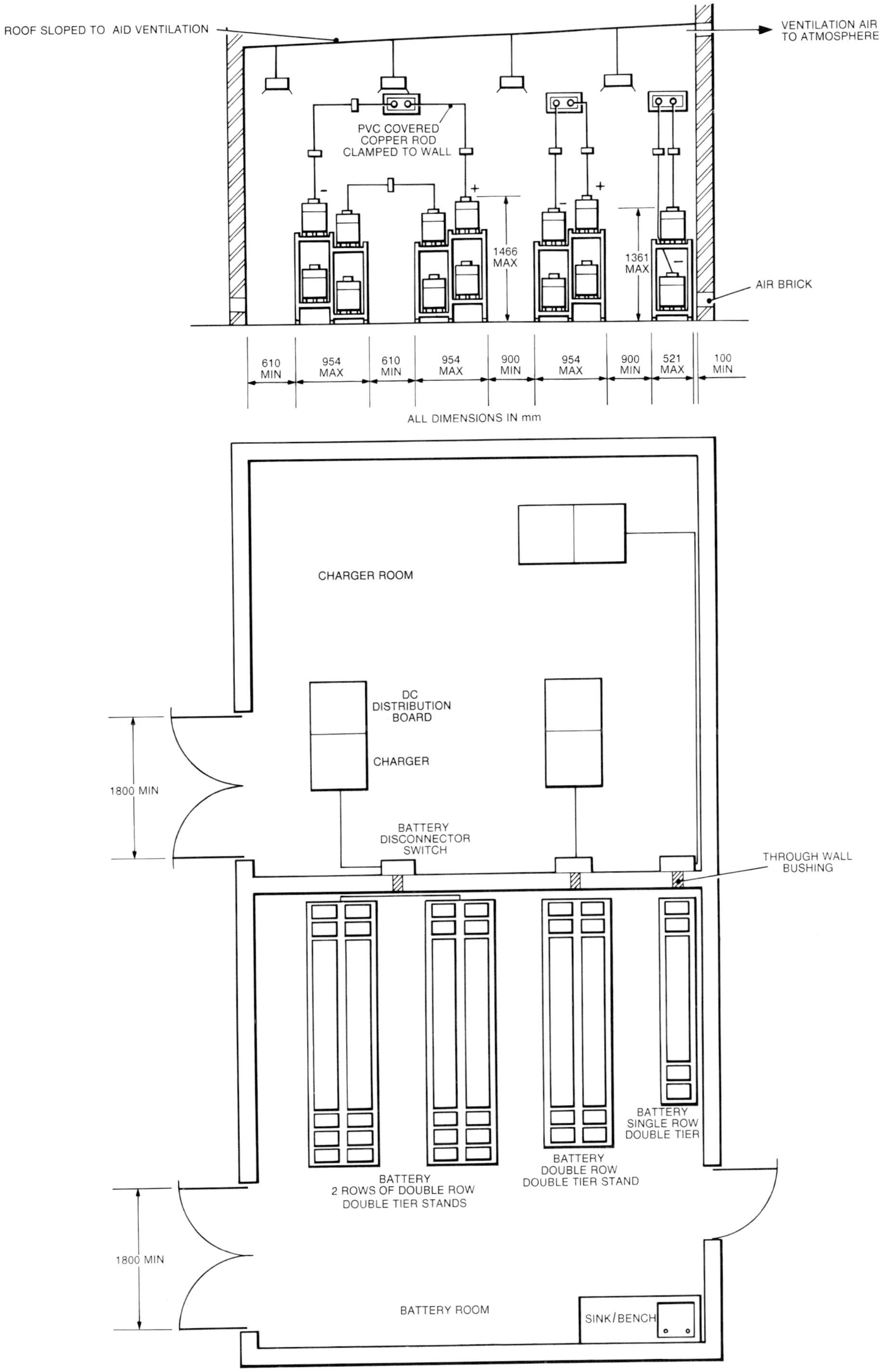

FIG. 2.49 Layout of battery room and charging equipment

Figure 2.48 shows that the chargers and switchgear are housed in a separate room adjacent to the battery room. This keeps them isolated from the hazardous battery byproducts, i.e., the hydrogen gas and airborne acid molecules, and maintenance is achieved without the need to enter the battery area.

14.8 Control rooms

The practice has been developed in recent power station designs to provide a central control room (CCR) for the station which includes all the facilities for controlling each of the generating units together with the controls for both main and auxiliary electrical equipment.

A considerable amount of equipment is involved in controlling the various items of plant in a power station and this centralised location covers the provision of computers, automatic control systems, control panels, remote control equipment, telecommunications equipment and alarms. It is essential to provide this equipment with a clean and controlled environment and this is achieved by installing a centralised heating and ventilating system.

The location of the CCR is determined by the type of power station, site area limitations, cost, personnel movement, security control of personnel access, confines of mechanical plant and electrical connection requirements. It is essential to keep the length of cable runs to the practical minimum. For a 2-unit station the CCR would be located between the units, on a 4-unit station the CCR would be between units 2 and 3 whilst on a 3-unit station the practice has been to locate the CCR at one end of the turbine hall. The CCR is generally located in a multi-level control building in order to house all the equipment involved and provide cable access to this equipment. There are many possible solutions to the location of the CCR and on any one project a number of factors will influence the result. Figures 2.50, 2.51 and 2.52 show CCR locations on coal, oil and nuclear stations respectively.

The CCR contains control desks and panels necessary for the operation of the power station. A number of these are common to all types of power station and others which only occur on specific types. All CCRs contain a supervisor's desk, an operator's desk and a panel for each unit, a station services desk/panel, electrical auxiliaries supply panel and transmission control panels. Desks and panels are floor-mounted but the panels may be free-standing or inset within a false wall cladding with only the display surfaces visible from the supervisor's and unit operator's desks.

On fossil-fuelled power stations, additional desks and panels may be provided for monitoring burner flame performance and sootblower control, whilst on nuclear power stations a post-trip mimic panel would be provided. On a hydro-electric power station, reservoir and river levels and flow may be controlled or displayed on a separate desk or panel. Figures 2.53, 2.54 and 2.55 show typical CCR layouts.

14.8.1 Desks and panels

The supervisor's desk provides facilities for communication to grid control, unit desks, plant areas and local emergency services. Provision is also made for VDUs to display generator unit information and transmission system details.

The controls and instruments are arranged on a modular basis on the unit control desk to facilitate maintenance and changes. The controls allow for cold start and hot restart of the unit as well as normal running and are grouped in plant areas on a functional basis. Figure 2.56 shows the basic outline design of a unit control desk and Fig 2.57 details a small group of modular controls on the desk.

The unit panel provides controls, indications and recorders which require less-frequent operator attention or adjustment than those on the unit control desk. This panel may be either modular or non-modular or a combination of both.

The station services desk uses modular construction where possible and accommodates controls and indications not associated with a particular unit and typically includes the CW system, fuel storage, water treatment, auxiliary boilers and fire protection and may also include a mimic display of the CW system. Figure 2.58 shows the CW system mimic of a station services desk/panel.

The electrical and auxiliary electrical panel displays, in mimic form, the electrical system from the 415 V system up to the generator main connections output to the grid substation. It includes any standby diesel or gas turbine generating units and is normally located so that each unit operator can see the electrical system of the unit he controls as well as any station-based system. The layout design attempts to give the supervisor a clear view of the majority of the mimic area, particularly that of the station-based section.

The transmission panel displays in mimic form the electrical system starting from the generator main connections through the local substation up to the input point of the grid, i.e., from 26 kV up to 132 kV or 400 kV.

14.8.2 Cable access and terminations

With the major part of the control of the station concentrated in the CCR, and the introduction of computer control for start-up and shutdown of the generating units, a large number of cable terminations are necessary near to the CCR. These are normally provided below the CCR and take the form of large cable termination marshalling frames or cubicles. On nuclear power stations the introduction of 4-train cable segregation adds constraints to the civil design in order to form physically separate cableways to the CCR area.

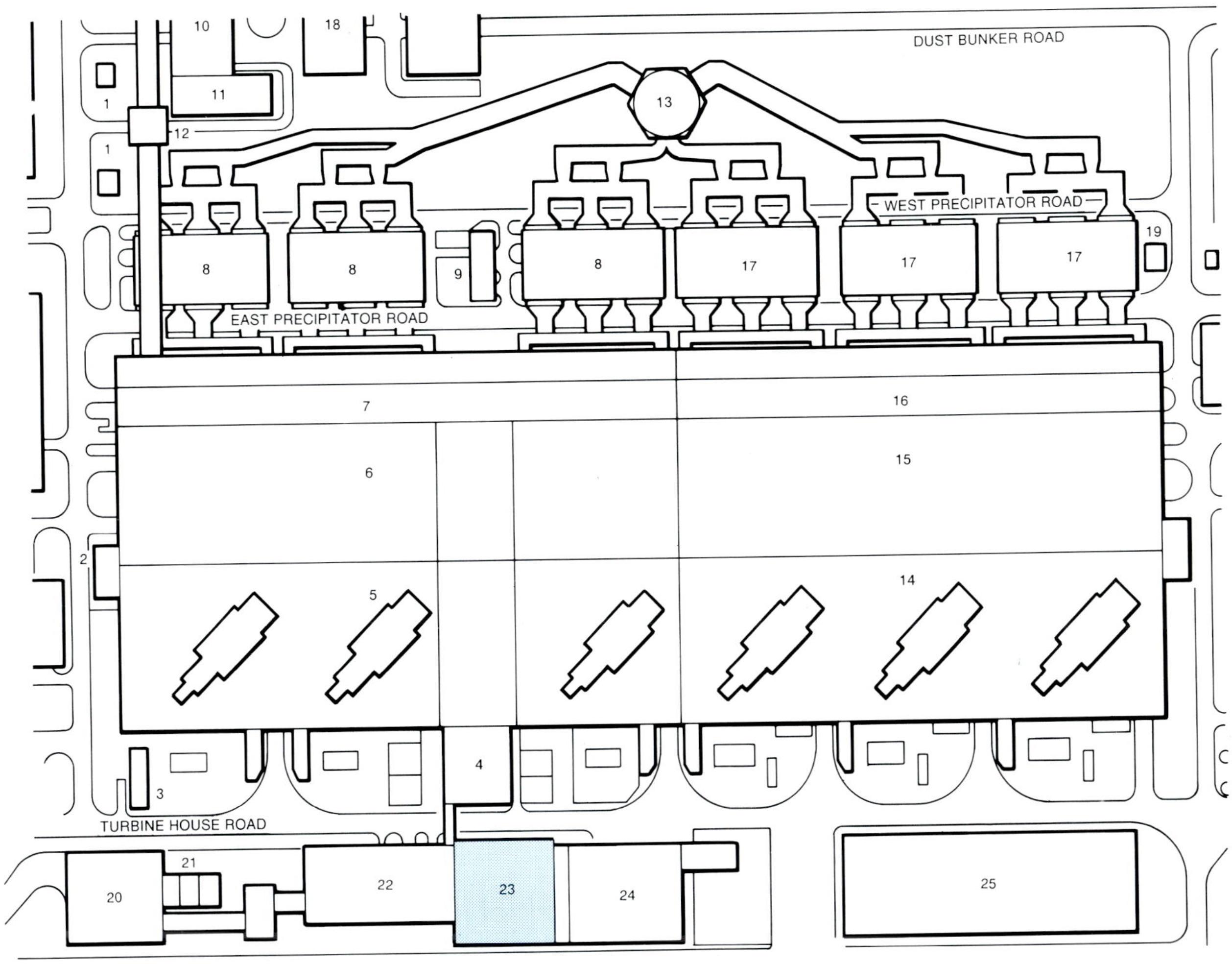

1 PROPANE STORE
2 AUX BOILER HOUSE
3 TRANS OIL FILTRATION PLANT
4 WORKSHOP AND OFFICES
5 TURBINE HOUSE - UNITS 1, 2 AND 3
6 BOILER HOUSE - UNITS 1, 2 AND 3
7 BUNKER BAY - UNITS 1, 2 AND 3
8 PRECIPITATORS - UNITS 1, 2 AND 3
9 LUB OIL STORES
10 GAS TURBINE HOUSE
11 GT CONTROL BUILDING
12 CP JUNCTION HOUSES
13 MAIN CHIMNEY
14 TURBINE HOUSE - UNITS 4, 5 AND 6
15 BOILER HOUSE - UNITS 4, 5 AND 6
16 BUNKER BAY - UNITS 4, 5 AND 6
17 PRECIPITATORS - UNITS 4, 5 AND 6
18 GAS TURBINE HOUSE NORTH
19 VACUUM CLEANING PLANT
20 ADMIN OFFICE SOUTH
21 FIRE STATION AND GARAGE
22 ADMIN BLOCK
23 CONTROL BLOCK
24 WATER TREATMENT PLANT
25 STORES BLOCK

FIG. 2.50 CCR location on coal-fired station

Physical separation for cableways is normally continued up to the underside of the control surfaces of the unit control desk and may affect the positions of the desks and panels within the control room itself.

14.8.3 Control room design

The design of the CCR equipment and layout commences at an early stage of the power station design period and continues through to the mid-term of the design period. The process brings together knowledge and experience of a large number of the engineering disciplines and is a fully integrated process.

The first stage is to define the civil *envelope* structure within which the CCR will be located together with a list of panels required for the particular type of power station. The size of each desk and panel is based on the number of controls and displays to be incorporated. These are defined by the engineering disciplines, e.g., electrical, C and I, transmission, etc., requiring service on each desk or panel and by the layout and grouping of the controls and displays. The shape and angles of display and control areas is also affected by ergonomics, i.e., the ability of operators to see and operate the controls. These factors are taken from standard

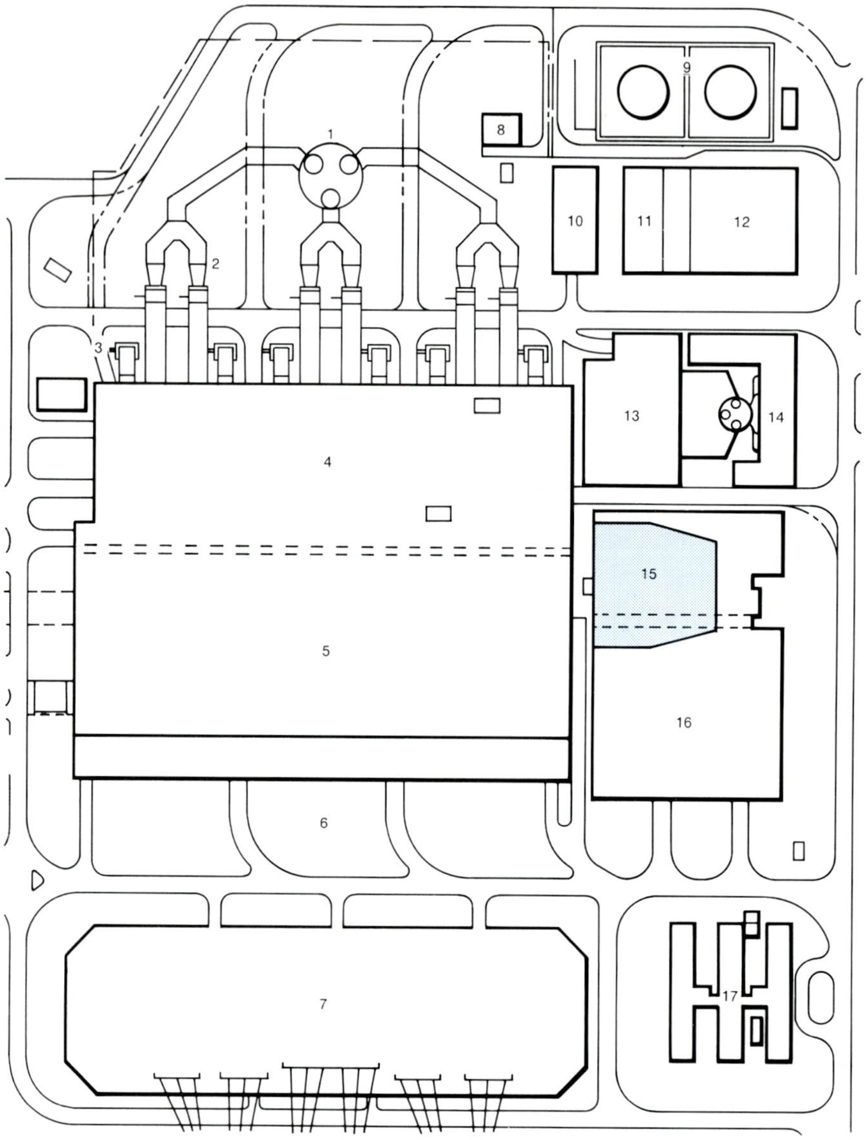

KEY
1 MAIN CHIMNEY
2 ID FANS
3 FD FANS
4 BOILER HOUSE
5 TURBINE HOUSE
6 GENERATOR TRANSFORMERS
7 400 kV SUBSTATION
8 PROPANE STORE
9 GT FUEL OIL TANKS
10 FUEL OIL HEATER HOUSE
11 SOOTBLOWER COMPRESSOR HOUSE
12 WATER TREATMENT PLANT
13 GAS TURBINE HOUSE
14 AUXILIARY BOILER HOUSE
15 CONTROL ROOM
16 WORKSHOPS AND STORES
17 SITE OFFICES

FIG. 2.51 CCR location on oil-fired station

tables of seated or standing eye heights and reach of a representative population cross-section.

The shape of the panels or desks also takes into account the position of cable access to them and the number of cables anticipated to be required at each location. These cable requirements reflect back to the civil structure design which has to be assessed and modified to match these requirements.

As the design progresses, consideration has to be given to the routes through the building which will be used to bring the desks or panels into the CCR. This may require that certain doorways have to be larger than originally anticipated in the civil design, or the equipment manufacturered in sections small enough to pass through doorways which, due to seismic considerations, cannot be enlarged.

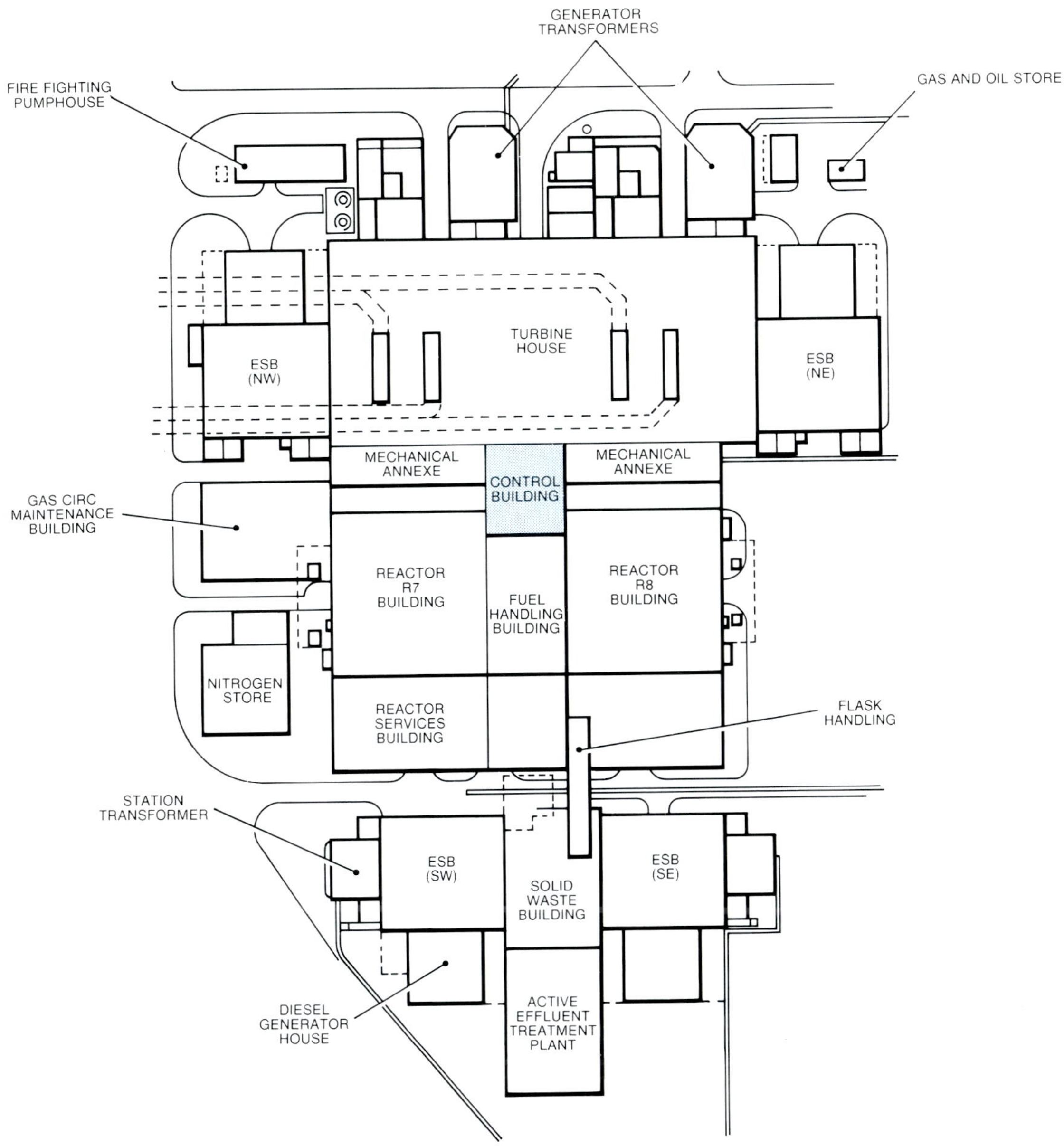

FIG. 2.52 CCR location on AGR nuclear station

Where vertical display or mimic panels are intended to be inset within a false wall cladding, the design of the panels must be integrated with the design of the interior finish of the CCR. Sufficient space must be allowed behind the false wall cladding for maintenance of, and cable access to, the inset panels.

As the size and location of each desk and panel becomes firmly established so the civil design is also fixed. At this stage the movement of personnel within the CCR, and the associated facilities, is assessed from both security and operational viewpoints. This requires discussions with national security advisers to protect against attempts to disrupt normal operations of the power station. At the same time the public relations considerations may indicate a requirement for a visitors viewing gallery to be incorporated in the design. On nuclear power stations the viewing gallery may also be used by the Inspectorate of Nuclear Installations to assess operators' reactions to simulated incidents.

During the period of determining the CCR layout, information will also be passed to consultants on lighting, heating and ventilating and fire protection requirements. The lighting must enable the operator to see the controls and displays in his charge but must not produce undue glare or reflections. Provision must be made for redundant lighting so that loss of a single power supply does not put an area in darkness, e.g., half the power supplies to the lighting over one unit operator's desk is from another unit's power system. The heating and ventilating system should provide a moist air supply over the greatest concentration of VDUs so that the operators do not suffer from a *dry*

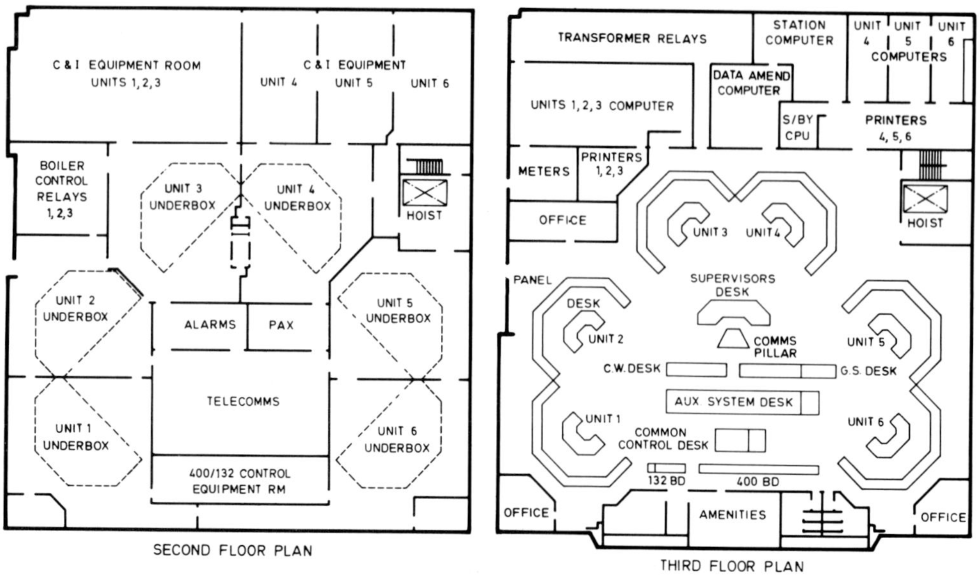

FIG. 2.53 CCR layout on coal-fired station

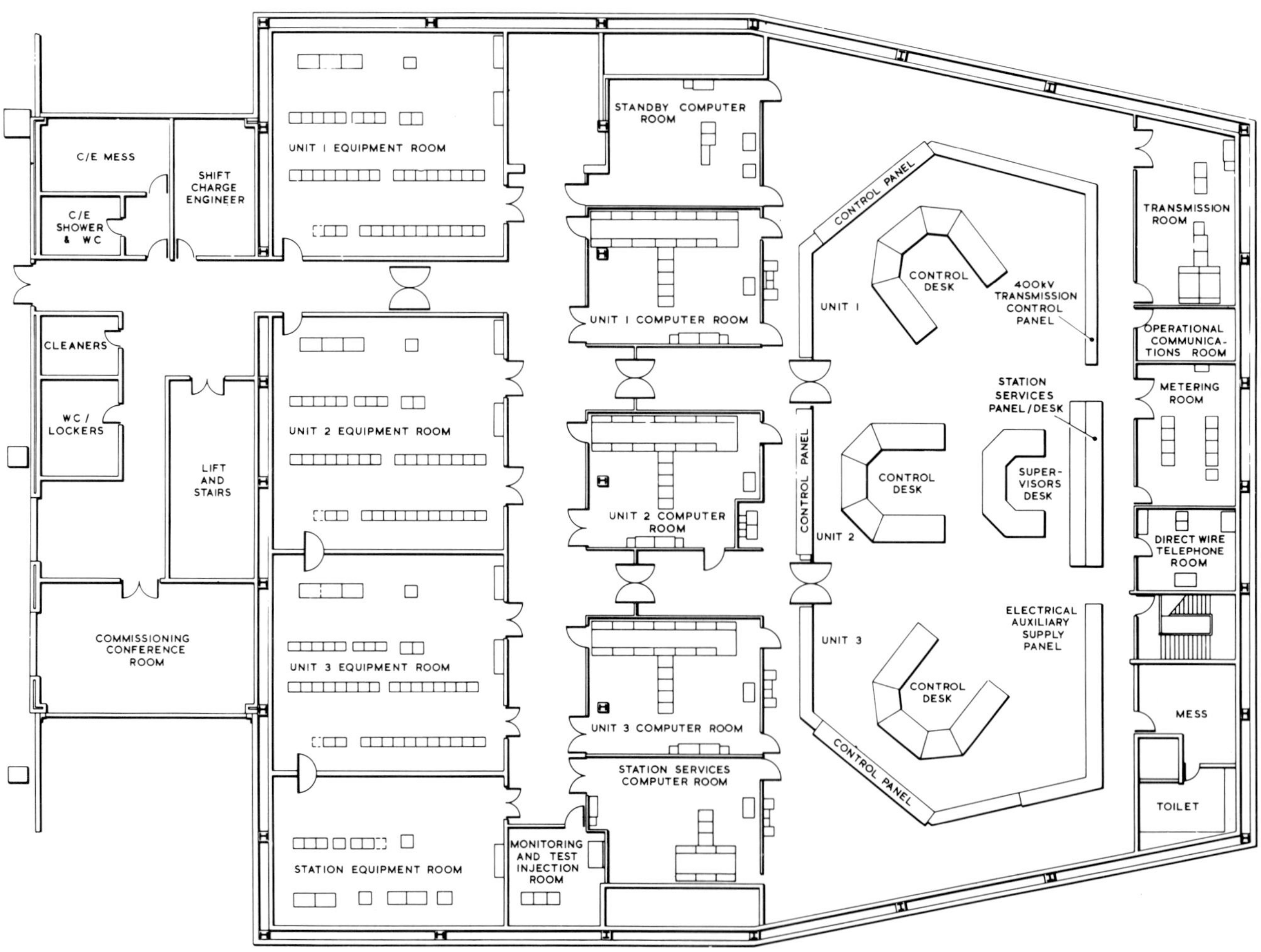

FIG. 2.54 CCR layout on oil-fired station

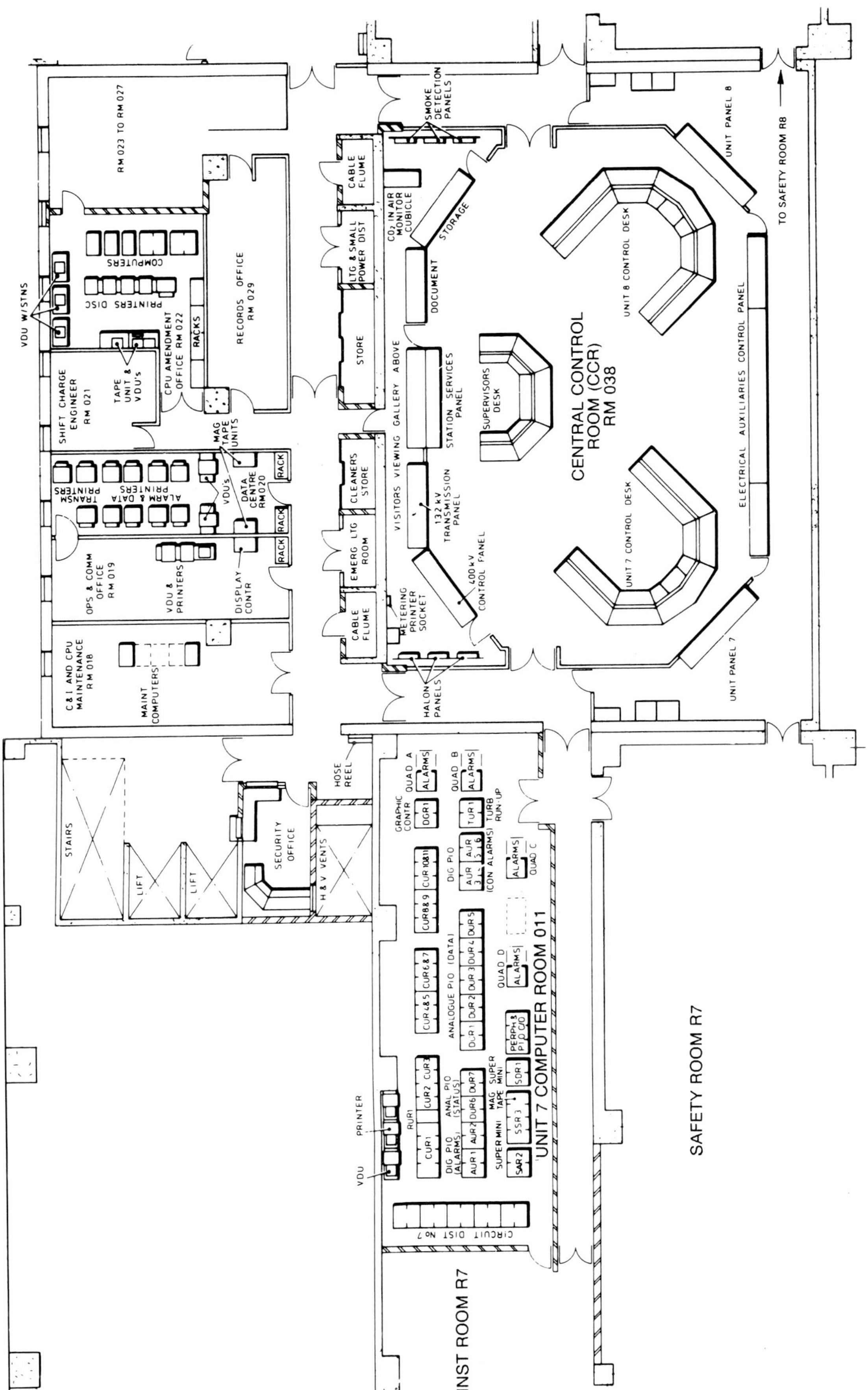

FIG. 2.55 CCR layout on AGR nuclear station

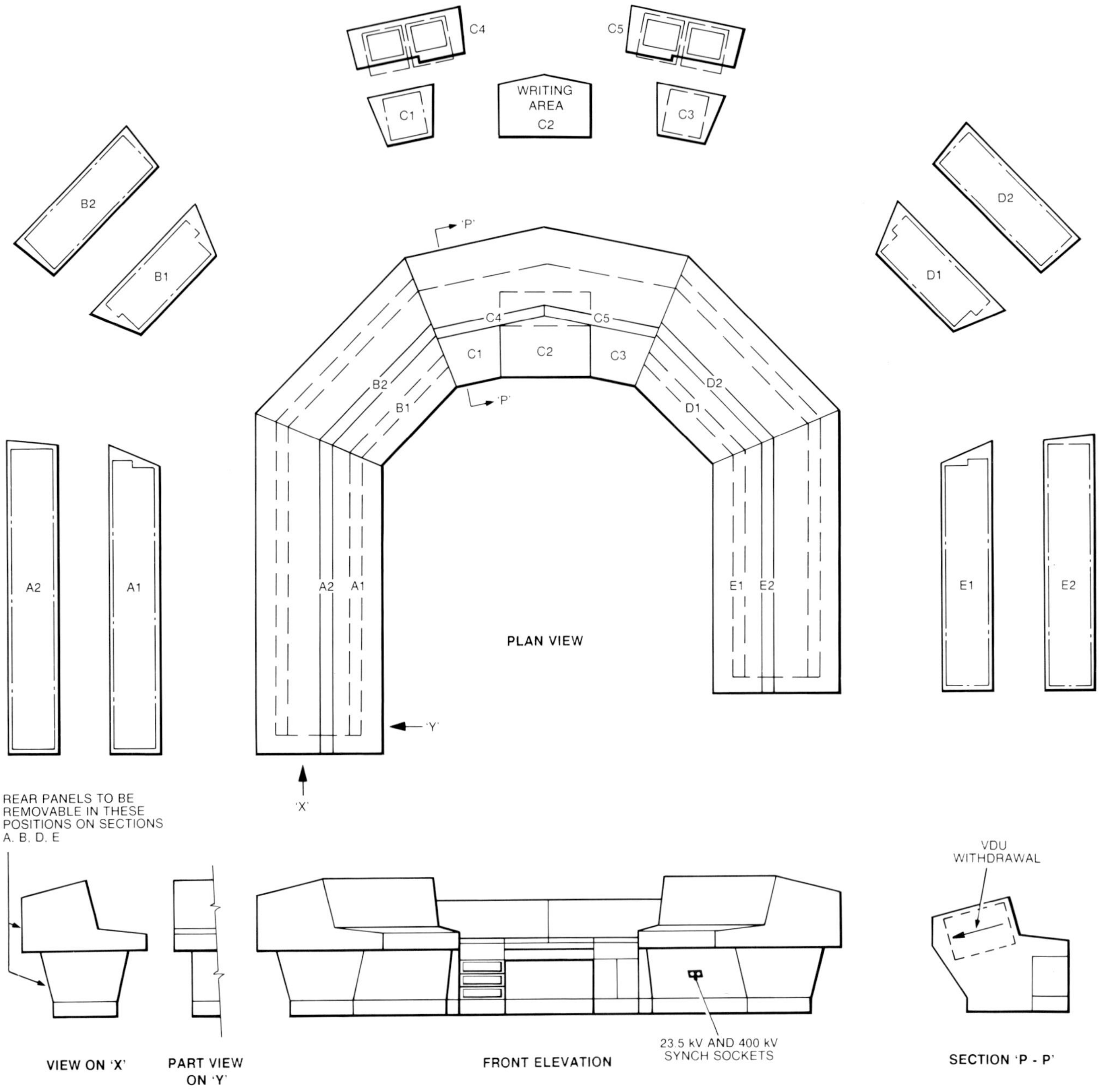

FIG. 2.56 Unit control desk basic layout

eyeball effect. Where a false or *computer type* suspended floor design is incorporated in the CCR a Halon gas system is normally used as a fire protection system in conjunction with heat and/or smoke detectors located in the floor void. Halon gas removes oxygen from the atmosphere and its release must be delayed in order to allow personnel time to evacuate the CCR area.

Associated with the CCR are various offices and facilities, the design and layout of which proceed concurrently with the CCR design. These include the shift charge engineers office, C and I instrument and computer maintenance workshops, operators messroom, toilets and showers. With the increased use of computers to control and record activities within the power station, provision is made to output hard copy records of log routines, etc. As the line printers for these are generally noisy they are normally housed in a separate room to avoid disturbance to personnel.

The movement patterns of maintenance equipment, between the CCR and related equipment rooms, must also be considered when the layout of corridors and fire door positions are established.

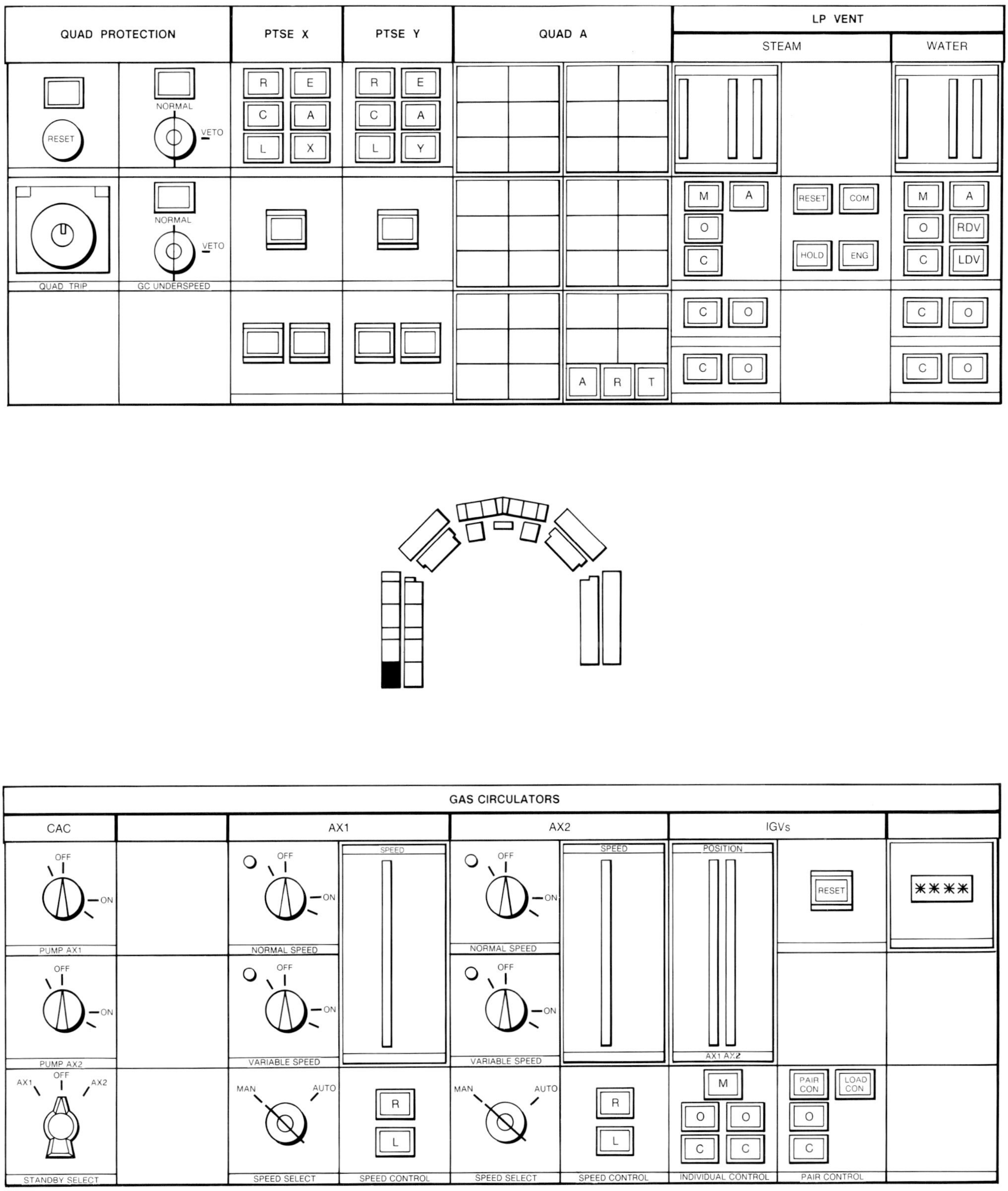

FIG. 2.57 Unit control panel modular controls

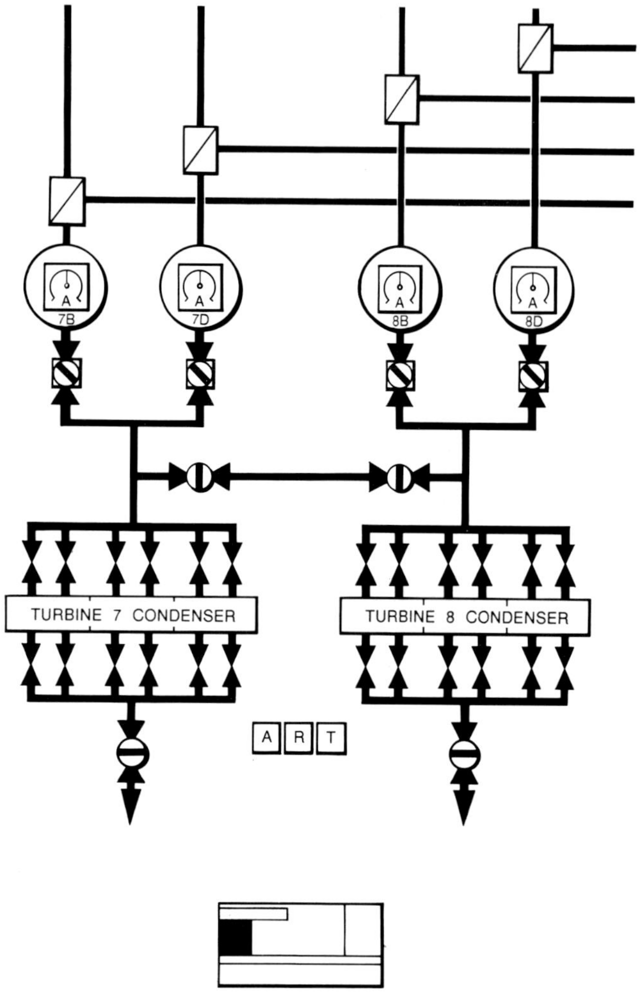

FIG. 2.58 CW system mimic panel

15 Heating, ventilation and air conditioning

15.1 Introduction

The objectives of a heating, ventilation and air conditioning (HVAC) plant on a power station can be considered to include one or more of the following:

- To comply with Statutory Regulations such as the Health and Safety at Work Act and the Nuclear Installations Act.
- To comply with CEGB Policies and Safety Rules.
- To maintain specified environmental conditions for plant and personnel.
- To remove toxic and hazardous materials present in the atmosphere.
- To remove heat or moisture generated by equipment.
- To assist in the containment of radioactive material and to minimise discharges to the environment.
- To assist in controlling the spread of smoke and fire in order to support the escape of personnel and the protection of plant.

To ensure that the identified design objectives are met it is essential that the HVAC requirements are considered as early as possible. HVAC equipment such as fans, ducts and filters require a lot of space and plant rooms and duct routes must be identified at an early stage in the layout of the station.

During the design phase each area of the station is studied in turn. Maximum and minimum temperature limits will be determined from equipment specifications and personnel requirements. Humidity levels will be determined and the need for specialist ventilation to dissipate toxic gases or control smoke and hot gases will be also determined. From these detailed requirements for each area, and with a knowledge of the building fabric, the solar gain to the buildings and the maximum and minimum external design conditions, estimates can be made of the heating, cooling and ventilation loads for each area. With this information a decision can be made concerning the type of system to be adopted for each area and preliminary plant sizes can be estimated enabling plant rooms and duct routes to be drawn up. As the design of the main plant progresses and more information regarding plant heating and cooling loads becomes available, then the preliminary estimates will need to be checked. As soon as a reasonable degree of confidence in the cooling and heating loads is available then final sizing and ordering of equipment can take place.

15.2 Ventilation of nuclear stations

HVAC systems on nuclear stations also serve to supplement physical containment in order to prevent the escape of radioactive particles. Air patterns are arranged so that air is extracted from the areas of higher activity and supplied to the cleaner areas. The arrangement is designed to ensure that air always flows from the cleaner areas at sufficient velocity to prevent the back-diffusion of radioactive particles. Elaborate studies are made and sufficient equipment is installed to ensure that these air flows are maintained under all conceivable operating conditions. In addition to preventing the spread of contamination within the building, the ventilation system has two further important roles. By providing sufficient air changes within the ventilated space, the system also helps to minimise the build-up of activity within the space by continuously purging it with clean air; and by fitting high efficiency particulate filters to the extract system the discharge of activity to atmosphere is kept to within allowable limits. The filtration system has a collection efficiency of greater than 99.9%, and before being discharged to atmosphere the air is sampled to check for levels of activity in case of accidents such as a burst filter. These filtration systems are regularly checked to ensure that the collection efficiency does not deteriorate.

The foregoing requirements inevitably place a high burden of reliability on the ventilation system. This leads generally to the requirement for standby fans and filters, which in turn results in complex control systems. The necessity for standby plant, high air volumes and large banks of filters, inevitably leads to a high requirement for space. It is therefore important that these requirements are identified at an early stage in the design. For a detailed treatment on the design of active ventilation systems, reference should be made to the Atomic Energy Code of Practice 1054 — Ventilation of Radioactive Areas.

Figure 2.59 is a simplified schematic of an AGR nuclear station central control room heating, ventilation and air conditioning system.

15.3 Smoke and fire control

In order to prevent the spread of fire and smoke between fire compartments and to assist in search and rescue operations, smoke or fire venting facilities are usually incorporated into the design. For large single-storey buildings such as a turbine hall, automatically-initiated fire vents are installed in the roof. These ventilators operate at a predetermined temperature, or on a signal from the fire detection system, and open-up allowing fire and hot gases to escape to atmosphere.

To prevent fire and hot gases spreading between compartments which share the same ventilation system, the HVAC systems are fitted with fire dampers at penetrations through the fire compartment walls. These fire dampers possess the same degree of fire resistance as the partitions they penetrate and are automatically initiated on detection of heat or from the fire detection system. In addition, it is common practice to include a facility to enable the ventilation system to be used as a vehicle for extracting smoke following a fire. This is usually accomplished by fitting a fireman's switch to the control system, which enables the system to be switched between *normal operation, smoke extract* and *off*. These switches are usually located in an easily accessible entrance lobby.

15.4 General layout of HVAC plant

15.4.1 Turbine hall and boiler house

Traditionally the main buildings of a fossil-fired station have been naturally ventilated. Advantage has been taken of the tall boiler house, the high internal heat losses and the forced draft (FD) fans to draw air in through inlet louvres located at low level around the turbine hall and boiler house. The air rises through the building and is drawn into the FD fan intake with the balance being rejected through roof vents. The advantages of this system are that no energy is expended in driving the system, and heat losses from the boiler and turbines are collected and passed back into the boilers; collection efficiencies greater than 60% are not uncommon. Disadvantages, however, include problems such as high boiler house temperatures, dust and condensation in the turbine hall and smoke logging in the turbine hall during a fire situation. These problems are usually associated with too little inlet area at low level in the turbine hall. Increasing the area, whilst solving these problems, leads to cold weather problems, for this reason consideration is now being given to separating the buildings with a partition wall and ventilating them separately along the lines of a nuclear station turbine hall.

15.4.2 Coal bunkers

This area is usually mechanically ventilated by means of an extract fan and fresh air inlet louvres. Air patterns are arranged to suppress dust levels, and sufficient air changes are provided to disperse any carbon monoxide which may be given off by a slow burning bunker fire. Consideration is being given on the latest stations to fitting filters on the extract to reduce the dust discharged to atmosphere. This puts a heavy burden on space requirements in the area.

15.4.3 Electrical equipment annexes

With the exception of computer suites and control rooms, which usually have their own packaged air conditioning units, most electrical annexes housing cableways, battery rooms, switchgear rooms, etc., are mechanically ventilated for cooling with duct-mounted heater batteries. Air is usually ducted from a central fan room and the temperature of the room is regulated by a room-mounted thermostat controlling a heater battery located at the duct entry to the room. Occasionally several rooms are served by one heater battery, in which case an average of room temperatures control the temperature. On occasion when equipment is sensitive to high humidity, humidistats are installed, which on sensing a high humidity, e.g., greater than 70%, override the heating system, increasing the room temperature, thereby dropping the humidity. The air volume allocated to each area is determined from an estimation of the maximum heat load and the maximum outside air temperature. During winter conditions 90% recirculation of extract air is employed to conserve heating costs.

Figure 2.60 shows a heating and ventilation system for an essential electrical supplies building.

15.4.4 Auxiliary buildings

Wherever possible, auxiliary buildings such as the cooling water (CW) pumphouse and compressor houses are naturally ventilated. Such buildings are

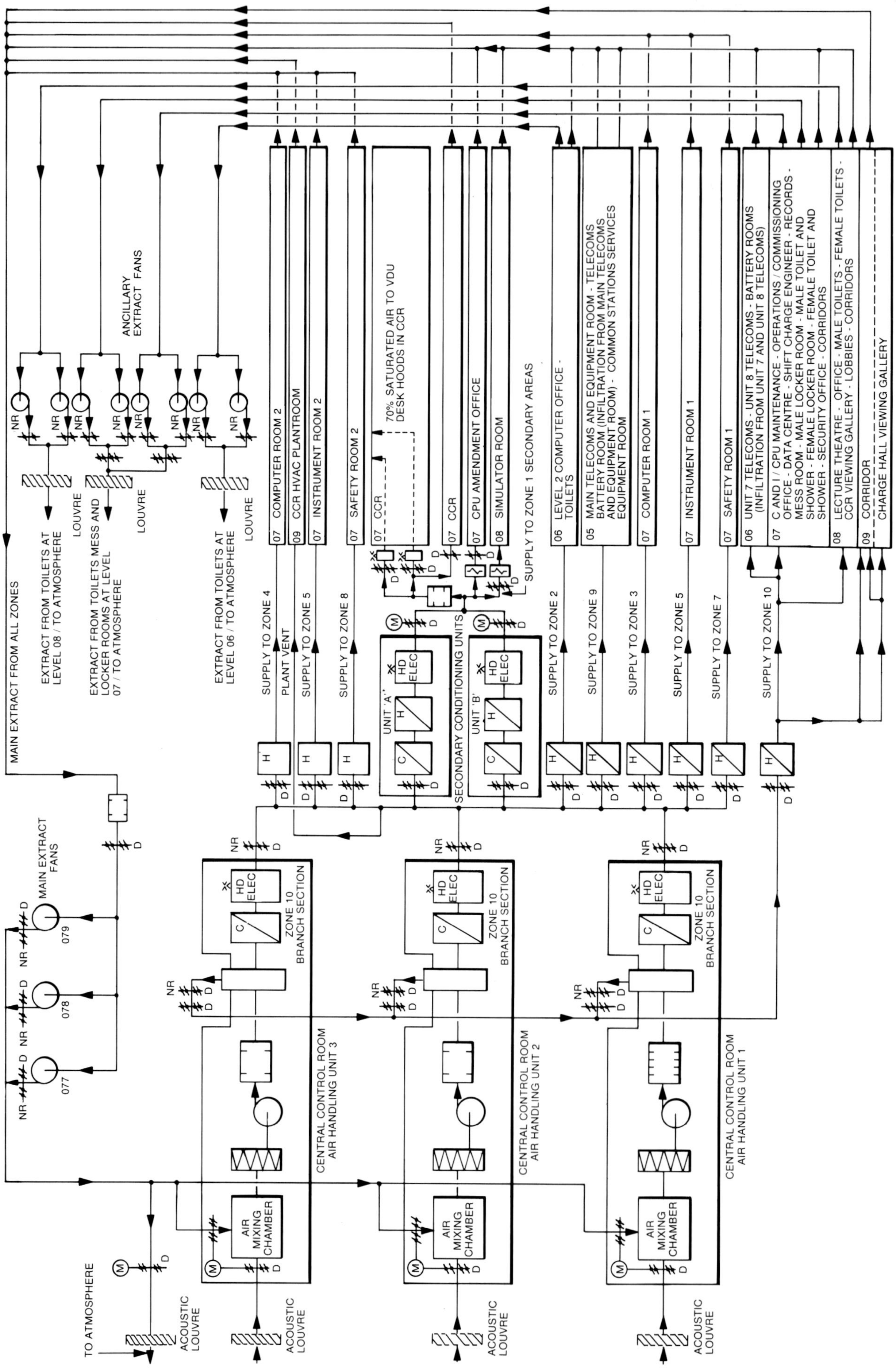

FIG. 2.59 HVAC system for nuclear station central control room

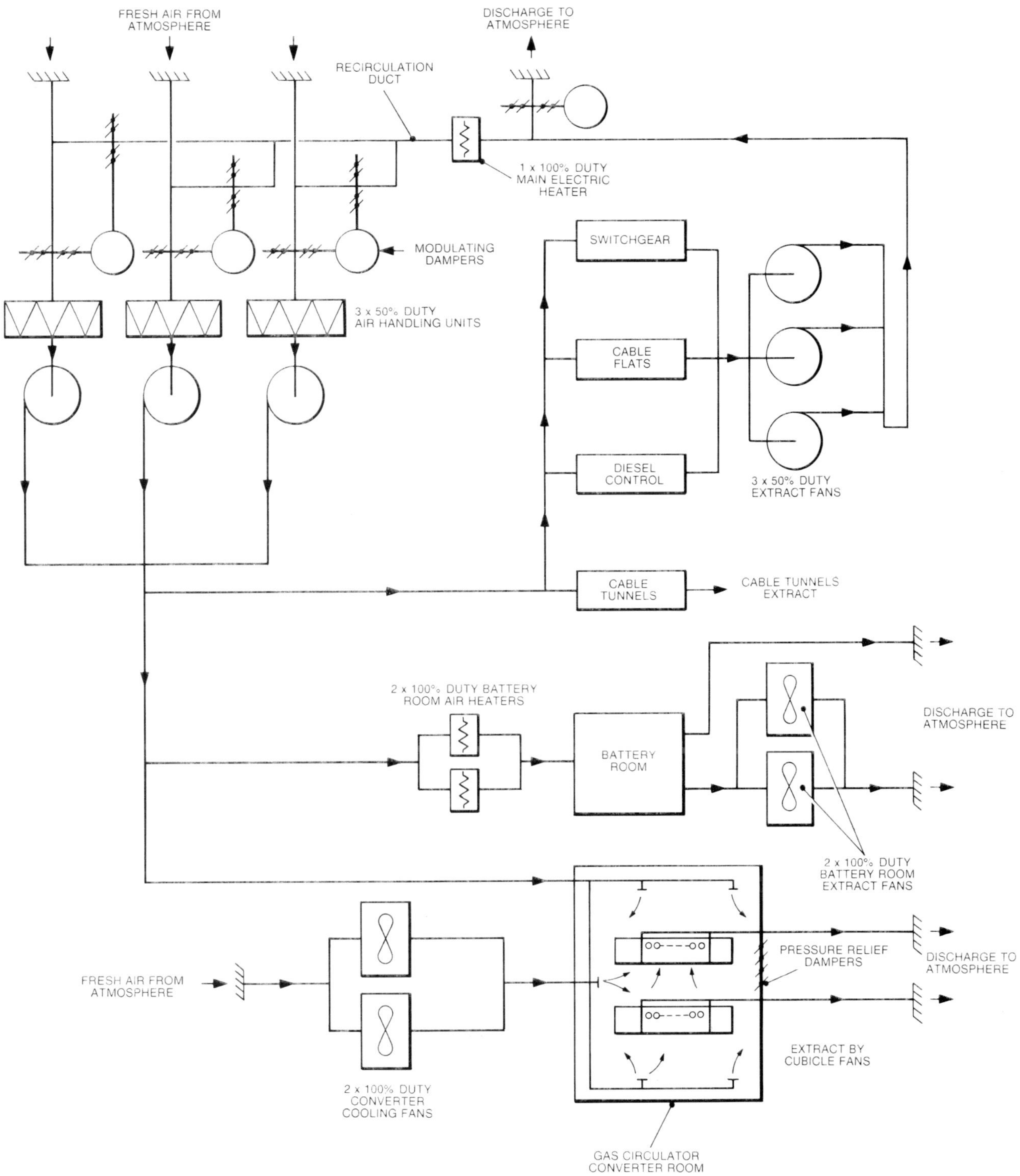

FIG. 2.60 Heating and ventilation system for essential electrical supplies system

rarely heated, reliance being placed on the fact that plant is working during cold spells. Buildings such as the water treatment plant, oil pumphouse, ash/slurry pumphouse, fire fighting pumphouses, etc., which have low heat loads and which are sensitive to cold weather, usually employ thermostatically controlled roof extract fans and fresh air inlet louvres. On occasion heating is also provided, usually in the form of radiant panels or fan-coil units. In the case of the administration block and amenities buildings, these usually employ a mechanically ventilated system with heating coils similar to that used in the electrical annexes.

16 Air services

Air services comprise the following systems:

- *General service air* — this is normally a station system, sized to cater for maximum maintenance demand on one boiler/turbine unit.

  The compressor plant and receivers can normally be accommodated in the boiler basement or adjacent area, or perhaps contained within a compressor house as the situation dictates.

- *Control and instrumentation air* — this is ideally a unit system, giving security to unit output. Plant sizing will depend on the philosophy for pneumatic application, but compressors, filtration and drying plant and receivers will be accommodated on each unit in a convenient location.

- *Turbine forced-air cooling* — this is a station system with a large quantity demand, resulting in perhaps large and heavy compressor plant. As a consequence the compressors and receivers will be located at basement or ground level in perhaps a central position and provided with their own compressor house.

- *Sootblowing air* — where this is used on fossil-fired stations it will be a large quantity demand on a station basis. In order to obtain an economic system of air supply it is possible to utilise the turbine forced-air cooling compressors for sootblowing purposes.

- *Breathing air* (nuclear stations) — this consists of compressor plant and equipment for conditioning the air, together with air receivers and is arranged on a station basis. Being of a strategic nature, the system is provided with a back-up storage facility of breathing air cylinders which function on loss of compressors or low pressure indication. The breathing air compressor plant and cylinder storage will be situated in some convenient location within the building complex at ground level.

Pipework systems for all the foregoing air services take the form of distribution networks to points of application throughout the station.

Figure 2.61 shows the basic component arrangement of a typical compressed air system, and Fig 2.62 shows a compressed air set which provides control air to a water treatment plant.

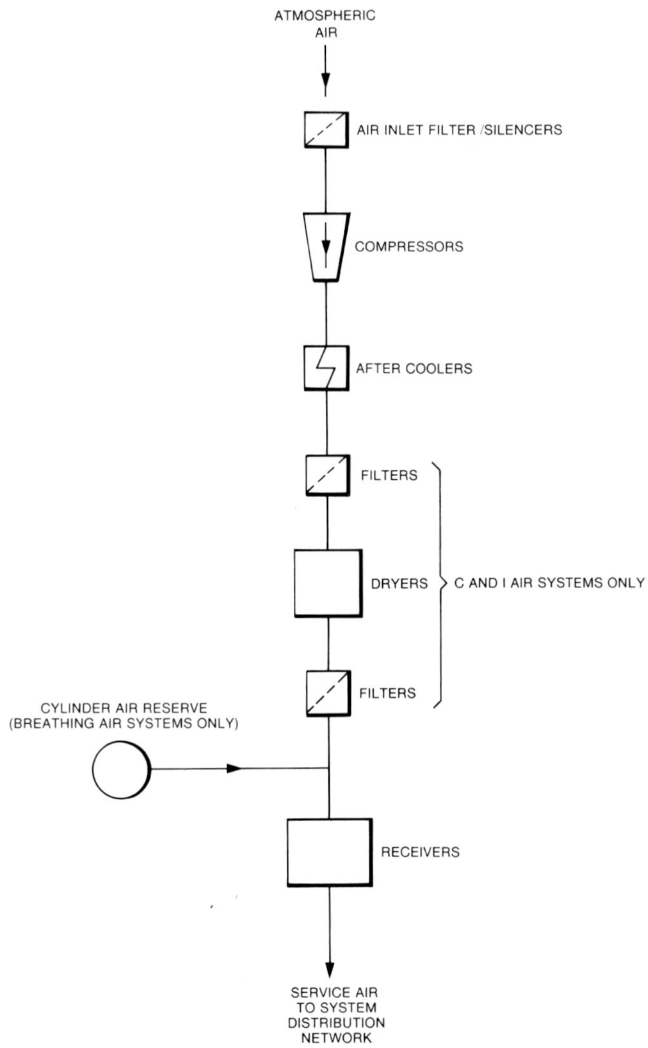

FIG. 2.61 Compressed air basic system

17 Water treatment plant

Individual designs and unit layout of water treatment plant should be left to specialists to optimise on local requirements. However, the overall positioning of the plant is considered in the general station layouts.

Water treatment falls into four possible categories:

- Boiler make-up water treatment, with perhaps condensate polishing.
- Main cooling water system treatment.
- Alternative water supply treatment (if required).
- Flue gas desulphurisation waste water treatment.

Boiler make-up water treatment plant is preferably located in a central position if condensate polishing regeneration facilities are included, but in any event requires easy access to town water supplies, effluent discharge points and regeneration chemical storage facilities. The boiler make-up water effluent neutralisation facility is generally sited as sumps under the treatment plant to allow open discharge of the waste regenerant.

The principal problems associated with the main cooling water system treatment are reduction of large pipework distribution routes and disposal of sludges removed. As the treatment forms a hydraulic break, geographical siting will attempt to suit, if possible, end point supply by gravity flow rather than by pumping. For sludge disposal, access is available into the cooling tower purge for discharge back to source, or suitable

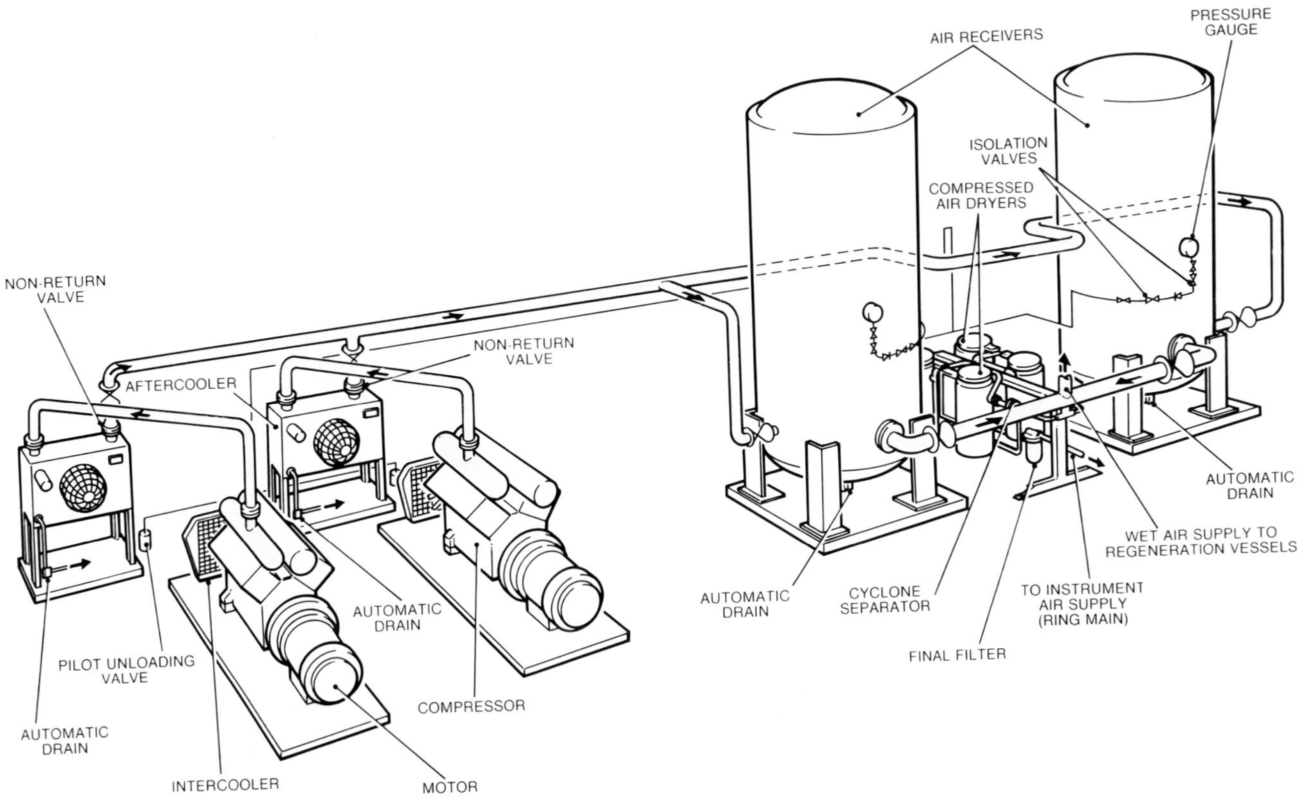

FIG. 2.62 Control air compressor set

equipment is provided enabling removal of the sludge as caked solid.

Any alternative water supply treatment plant is located with good access to town water storage tanks and preferably adjacent to the boiler make-up water treatment plant, thus enabling a sharing of chemical storage and supervision facilities.

The flue gas desulphurisation waste water treatment plant is located within the common plant area with adequate access for chemical intake and caked sludge discharges.

Chemical storage facilities are quite extensive and need good road access for supply vehicles. Storage tanks for hazardous fluids must be provided with all necessary bunding, etc., to contain spillage. The facility will reflect the needs of the storage of differing commodities, amongst which will be alum, lime, polyelectrolyte, sodium hydroxide, sulphuric acid and ferric chloride.

Figure 2.63 shows the basic stages of a typical water treatment plant.

18 Cooling water plant

18.1 General design considerations

The cooling water (CW) plant provides the heat sink for the condensers, turbine and sometimes boiler auxiliaries, and on nuclear stations for the reactor auxiliaries. The plant is divided between all of the engineering disciplines and includes pumps, valves, screens, pipework, strainers, gates, debris filters, heat exchangers, head tanks, intakes, outfalls, surge shafts and weir chambers.

The plant design is realised following a series of studies into the hydraulic, operational, economic and civil engineering requirements of the system. The key parameters in the design can be summarised as:

- Pump head and flow.
- Culvert/tunnel water velocities.
- Variation in pump suction water levels (e.g., tides).
- Screen mesh size.
- Influent water quality (suspended solids, chlorides, weed, fish, etc.).
- System pressure rating.
- Ambient temperatures.

The process of design begins with outline studies and data-gathering to collect the key parameters listed. The system flow is set by an optimisation of the condenser surface area, turbine-generator heat rate, capital cost of construction and through-life pumping costs. Initially, the condenser designer is provided with two figures (usually based on past experience) for the cost of a metre of pump head, and the cost of a m^3/s of flow. The head figure assumes a datum flowrate and indicates the

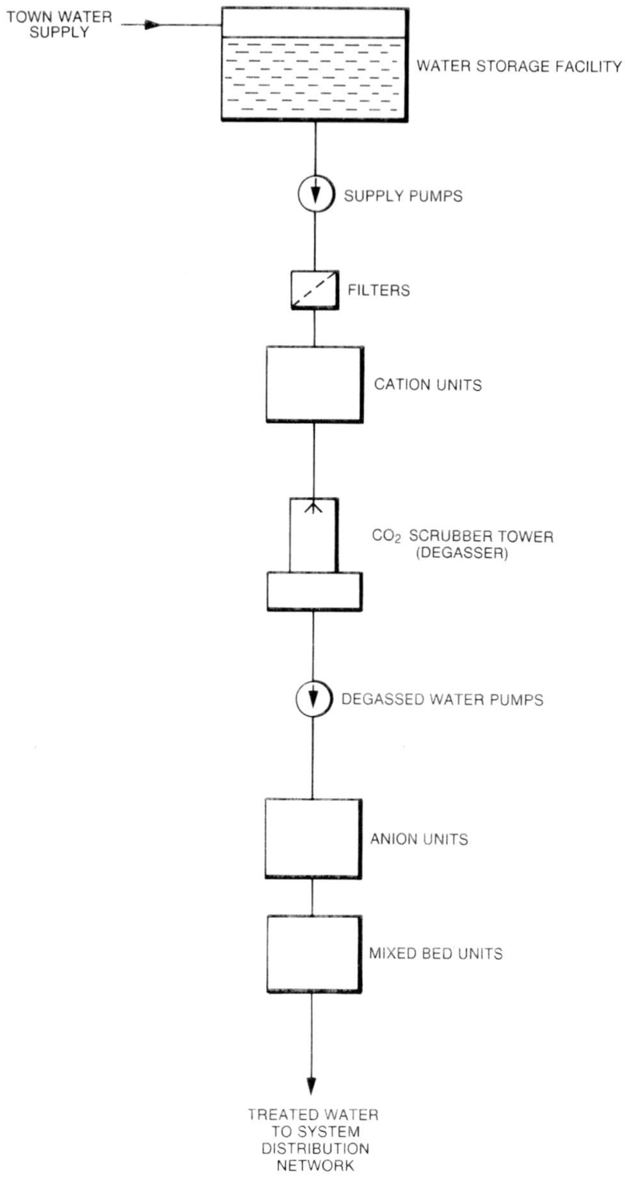

FIG. 2.63 Water treatment plant system

cost changes to the system resistance, e.g., by having fewer condenser tubes or by making the pipes smaller. The flow figure is found by dividing the estimate of capital cost for the system by the nominal flowrate. As the design proceeds these figures are progressively refined when more detailed data are available.

Another crucial factor is the level at which the turbine hall basement is set. An optimisation is carried out which examines a series of levels and evaluates the cost of pumping, heat rate savings, HP pipework and civil works. Calculations at this stage will be to first-order accuracy and the costs or savings must be weighed against the uncertainties of programme extension, unknown ground conditions, and knock-on design changes. Figure 2.64 shows the results of the optimisation for Littlebrook *D* power station (3 × 660 MW oil-fired).

Hydraulic gradient calculations are done with the objective of minimising the pumping power, and at the same time taking into account uncertainties in the calculation of hydraulic losses for complicated system shapes. The British Hydro-Mechanics Research Association (BHRA) have produced a design code for hydraulic calculations [2]. Typical hydraulic gradients for direct and indirect cooled stations are shown in Fig 2.65. The crucial features on the direct cooled station are the siphon height, siphon seal and that part of the system which is sub-atmospheric. Current practice is to adopt a siphon height of approximately 9 m for the lowest barometric head in the area of the site, minimum operating flow, low tide and maximum fluid temperatures. The latter requires discussion with the condenser designer, since temperatures are frequently experienced in the upper condenser tube rows in the order of 6°C to 7°C above the datum design temperature. The system in Fig 2.65 (a) shows a station positioned at high level owing to difficult ground conditions. In order to minimise pumping costs, the maximum condenser siphon is maintained by the use of a siphonic seal design subjected to modelling tests. This feature also ensures that the culvert downstream of the condenser discharge remains full during unit shutdown (particularly useful for two-shift operation). During the basement optimisation and hydraulic gradient studies various culvert sizes, materials and operating velocities are reviewed to make a choice which aims at lowest through-life cost, but also takes account of the potential disruption to other construction activities which can be caused by the installation of CW culverts, for example, prevention of access and excessive dewatering.

Water level variations are calculated by reference to tide tables and meteorological data so that the margins above highest astronomical tide (HAT) and below lowest astronomical tide (LAT) required for storm surges and wind effects can be allowed for. On tower cooled stations, the calculation relates to the allowable variation in tower pond level to allow for interruption of the make-up supplies.

Screen mesh size is determined by the smallest passage through which the flow will have to pass. Current practice is to make the screen apertures between a third and a quarter the diameter of this size. Where auxiliary coolers have smaller tubes than the main condenser, or use plate-type coolers it is usual to have a secondary enclosed type screen dedicated to this duty.

The water quality affects the choice of materials, the coatings which might be required (where the chlorides content exceeds 600 PPM) and the likelihood of the system being effected by chemical or biological fouling. In locations where there is significant sediment in the water special measures must be taken to avoid taking it into the system, or to ensure that water velocities are high enough to keep it is suspension as it passes through the system. Weed and fish inundations can be a parti-

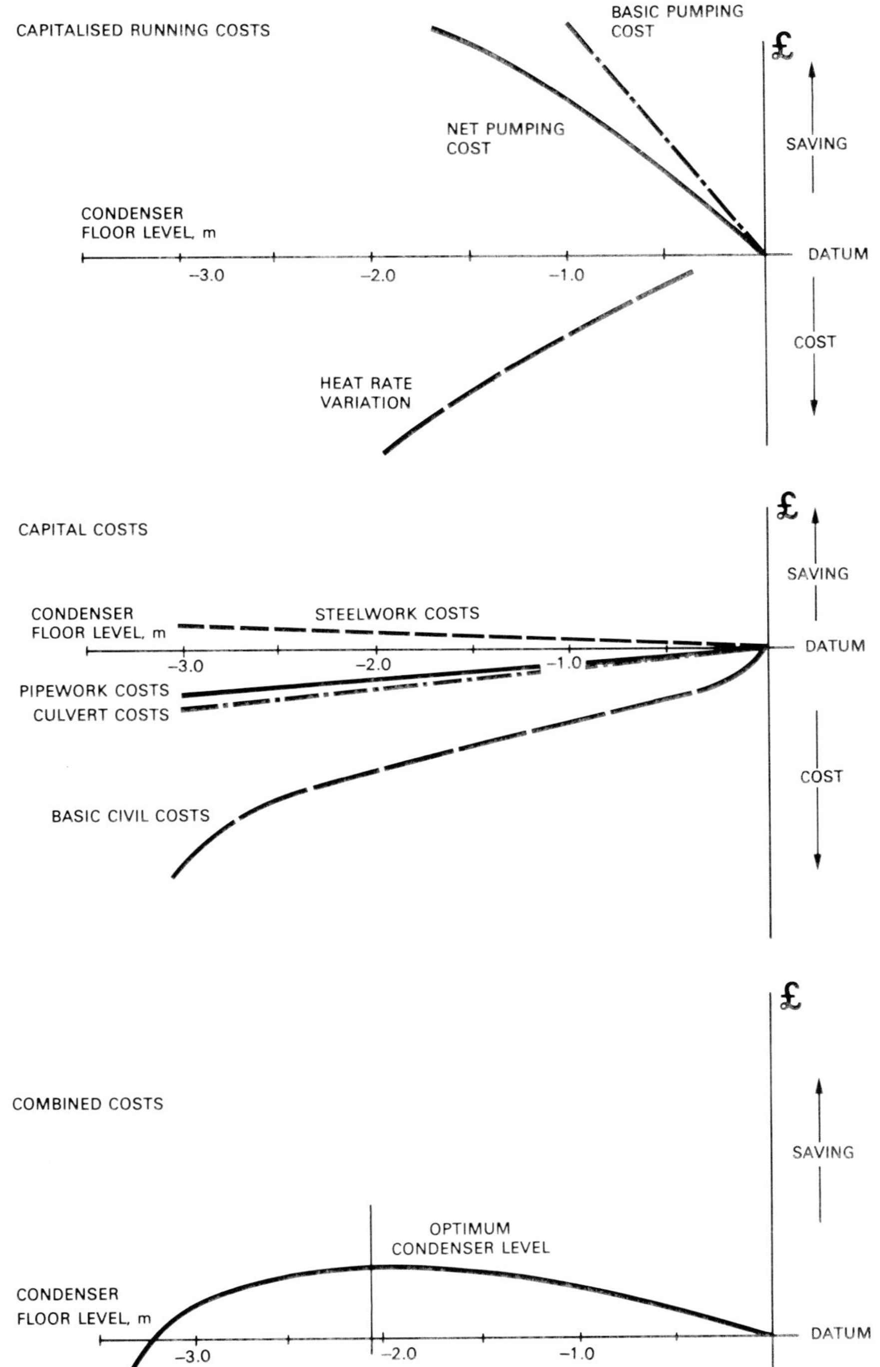

FIG. 2.64 CW system optimisation results

cular problem and they impact heavily on the detailed design of the screening plant. Unfortunately, a screening system which is optimised for the removal of weed is not very efficient at removing fish.

The system pressure rating is set by consideration of the highest steady pressure that the system can see. This usually relates to the pump no-flow head and high water level in the pumphouse forebay. Every effort is made to keep the system transient pressures within this value but often a surge allowance is also required. The transient pressures are estimated using special-purpose computer software to evaluate all of the feasible operating regimes and faults which can occur. Control of the pressure surge levels is obtained by defining the closure rate of the pump discharge valves, fitting of air injection and release valves on the condenser waterboxes, and by the careful sizing and positioning of the outlet surge tank.

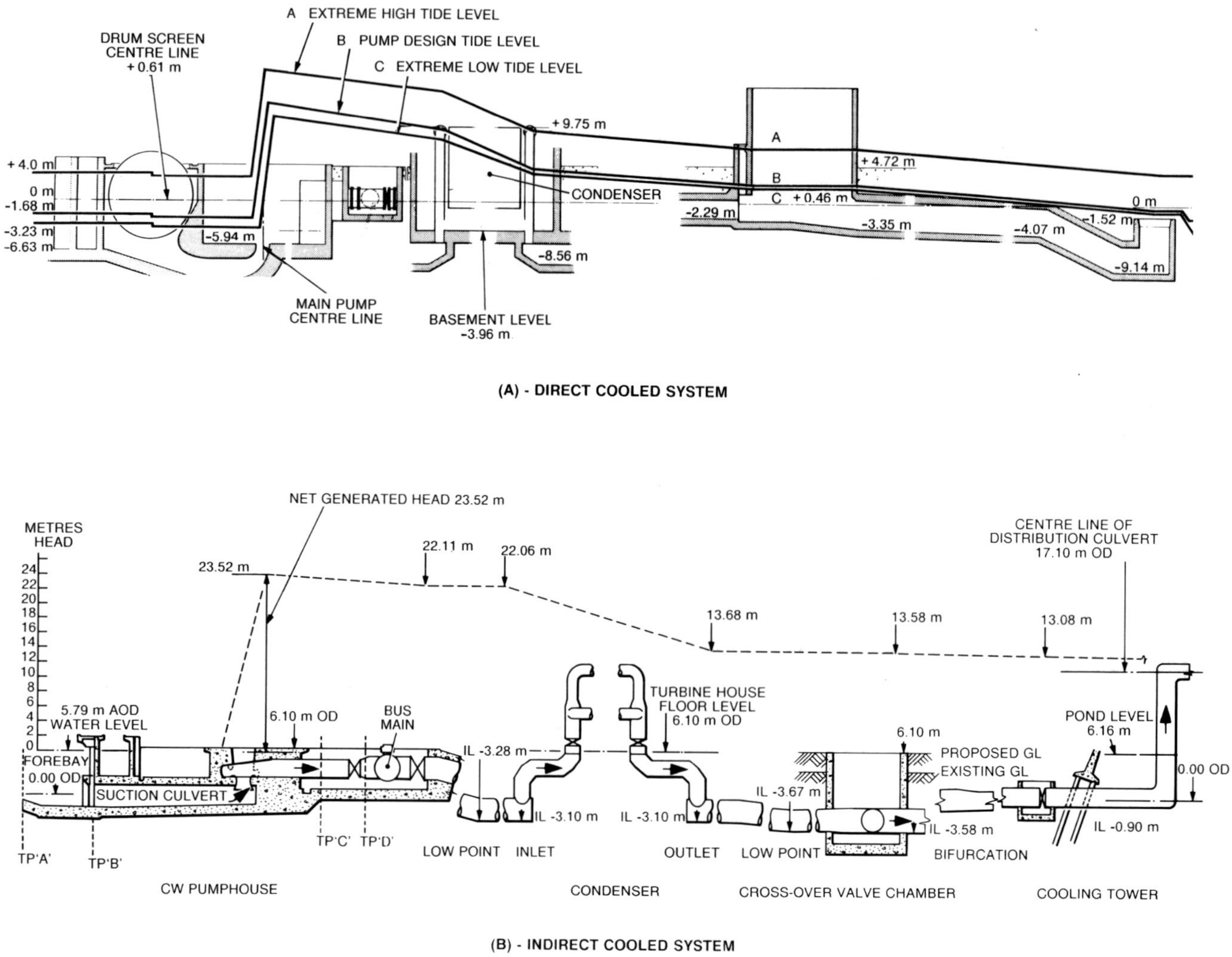

NOTES:
1 HEAD FIGURES APPLY TO WORST CASE
2 THIS HYDRAULIC GRADIENT HAS BEEN DRAWN FOR A CW FLOW OF 14.77 m^3/s
3 TP DENOTES TERMINAL POINT

FIG. 2.65 Typical hydraulic gradients

18.2 Cooling water pumphouse

The cooling water pumphouse houses the pumps, valves, pipework, strainers and auxiliary plant associated with providing cooling water to the station condensers and possibly auxiliaries. The design of the pumphouse is fundamentally different depending on whether the station is direct cooled or indirect tower cooled. The factors to be taken into account in the design of a pumphouse for a direct cooled station are described and then the differences which occur at a tower cooled, or indirect cooled, station are discussed.

18.2.1 Direct cooled stations

In addition to the main cooling water pumps, the CW pumphouse on a major direct cooled power station houses the screens, stopgates, valves, strainers and pipework which allow debris-free water, usually sea water, to be passed to the turbine condensers, auxiliaries and, on nuclear stations, the reactor auxiliaries cooling systems. These cooling systems are required during all station generation. The reactor essential service water systems — also called variously; reactor sea water (RSW), essential services water (ESW), essential cooling water (ECW) — are also required to remove decay heat from the reactor when the station is shut down for maintenance. Figure 2.66 shows the layout of the CW system for the 3 × 660 MW oil-fired power station Littlebrook *D*, and Fig 2.67 shows a view of the pumphouse and intake screens.

Because of their importance for station operation, the systems are engineered to high standards of reliability by consideration of:

- Stable pump suction conditions.

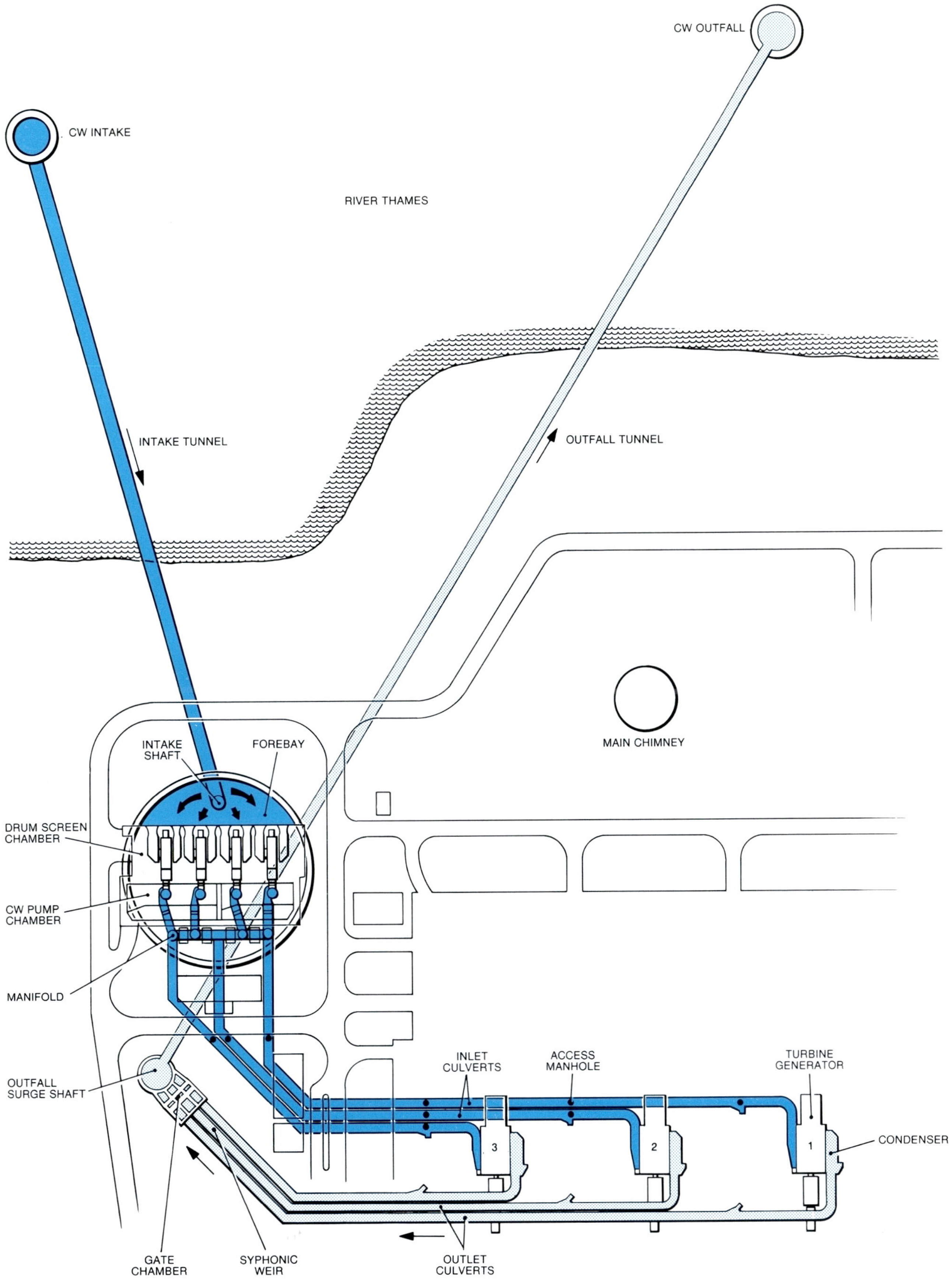

FIG. 2.66 CW system layout — Littlebrook *D*

FIG. 2.67 View of pumphouse and intake screens — Littlebrook D
(see also colour photograph between pp 66 and pp 67)

- Plant redundancy.
- High reliability components.
- Protection from hazards (e.g., flooding).

The pumphouse designer uses his skill to produce a plant arrangement which provides the best possible hydraulic shapes and also meets the design criteria.

The calculation of total cost of a scheme is complex and involves not only the capital cost of the works, but also the operational cost over the life of the station. It is also weighted to cover the knock-on effects of pursuing options which have an implication on the power station project as a whole, such as the excavation methods.

Pumphouse design within CEGB is a collaborative affair where the hydraulic designer merges the best principles of hydraulic design with the civil engineer's structural design to obtain a stable and economic arrangement. The final layout must take account of the variety of plant which must be fitted in, the requirements for high system and component reliability, and the civil engineering methods which are appropriate for the particular ground conditions. The final solution is always a compromise.

Ideal hydraulic shapes tend to be very expensive to construct, so hydraulic model studies are used to demonstrate that the non-optimal hydraulic solutions are at least workable. Mathematical modelling is able to deal with the closed part of the system but at present, full 3-dimensional modelling of fluid flows, especially with a free surface, are only possible at great expense. Physical models are therefore preferred if they can be executed in the available time. One area where a mathematical model is at a great advantage is in the modelling of the transport of various contaminants in the flow (e.g., air, sediment), where the scaling laws do not allow a good representation in a physical model.

The designers of pumping stations around the world produce many different layouts of plant and the majority of them work quite acceptably in terms of reliability and cost. As the quantity of water to be moved increases, the capital cost of each item of plant and the power requirements for operation increase, and it is worth giving the design much more detailed attention.

For direct cooled power station applications, CW pump powers are typically between 1 MW and 2 MW. This is in contrast with the 4 MW to 5 MW requirement on a tower cooled design.

In order to provide the required continuity of flow to the condensers, the CEGB has adopted a number of design policy guidelines. These have evolved over many years and are based on practical experience of the problems which occur in operation. The guidelines include:

- Use of drowned suction pumps, installed with unitised intake screens and hydraulically operated discharge valves.
- Unitised culverts which are interconnected at the pumphouse so that a pump could serve any of the culverts. The interconnection, or manifold, is installed with sectioning valves to allow the individual pumps or culverts to be isolated without shutting down the whole system.
- Splitting the pumphouse into two or more physical sections so that a flood in one part of the pumphouse cannot lead to a complete system shutdown.
- The head and duty of the combined pumping capacity will equal the rated requirements at the 'mean low water spring' (MLWS) tide level. Additionally, at least one pump will be installed for each main steam turbine with a minimum of four pumps to cover for situations of pump breakdown or maintenance.

The layout process

The layout begins by consideration of the ground conditions to see whether a circular or rectangular pumphouse is most appropriate. The mechanical plant is then added in various arrangements. The first plant to be inserted are the main screens, CW pumps and discharge pipework. A major consideration here is whether or not to allow handling of the pumps so that the pumphouse is more symmetrical about its centreline. On recent stations, this has not been done because it increases the number of different components and hence the spares holding required.

On circular pumphouses, the diameter of the outer wall is set mainly by the size of the screening plant, which in turn is determined mainly by the tidal range. This factor makes standardised pumphouse design impossible because even fairly small changes in geometry can upset a good hydraulic arrangement.

The segmental area upstream of the screens forms the forebay. It is almost always smaller than would be desired hydraulically and so model studies are usually required to ensure satisfactory distribution of flow to the screen chambers. The entry to the forebay can be either vertical or horizontal, and depends on the method of construction of the intake works. If the culvert/tunnel approaches the forebay horizontally, the problem for the hydraulic designer is to diffuse the higher velocity flow and distribute it evenly across all the pumps. Diffusers with half angles greater than 20° can cause the flow to become detached from one or other wall and fluidic switching can occur with very disturbed flow patterns. Diffusers with half angles less than 20° are very expensive to construct and it is sometimes necessary to install pillars or similar barriers to break up the high velocity jet.

In poor ground, intake tunnels driven to avoid the use of compressed air are necessarily deep, so vertical shafts at inlet and outlet are sensible. Littlebrook *D* power station located on the Thames near Dartford used this design and has a vertical entry to the forebay (see Fig 2.68). The flow arriving at the main CW pump is conditioned by the velocity gradients in the screen chambers upstream and in the draught tube.

Recent experience at Heysham *2* has shown that the flow is more tolerant to poor conditions in the forebay if the draught tube is long and a balance connection is added just ahead of the drum screen side-channels (Figs 2.69 and 2.70). The space above and between the draught tubes for each of the four main CW pumps has been used for the reactor seawater and screen washwater pumps. This arrangement was possible because the auxiliary CW pumps are located in the turbine hall and draw water from the main system inlet culverts.

18.2.2 Tower cooled stations

The lack of tidal variation in tower cooled, or indirect cooled, systems allows the pumphouse design to be achieved without such deep excavation as on coastal sites and this usually allows the pumps to be placed fairly close to ground level. Pump suction conditions are still determined by the approach flow but this is

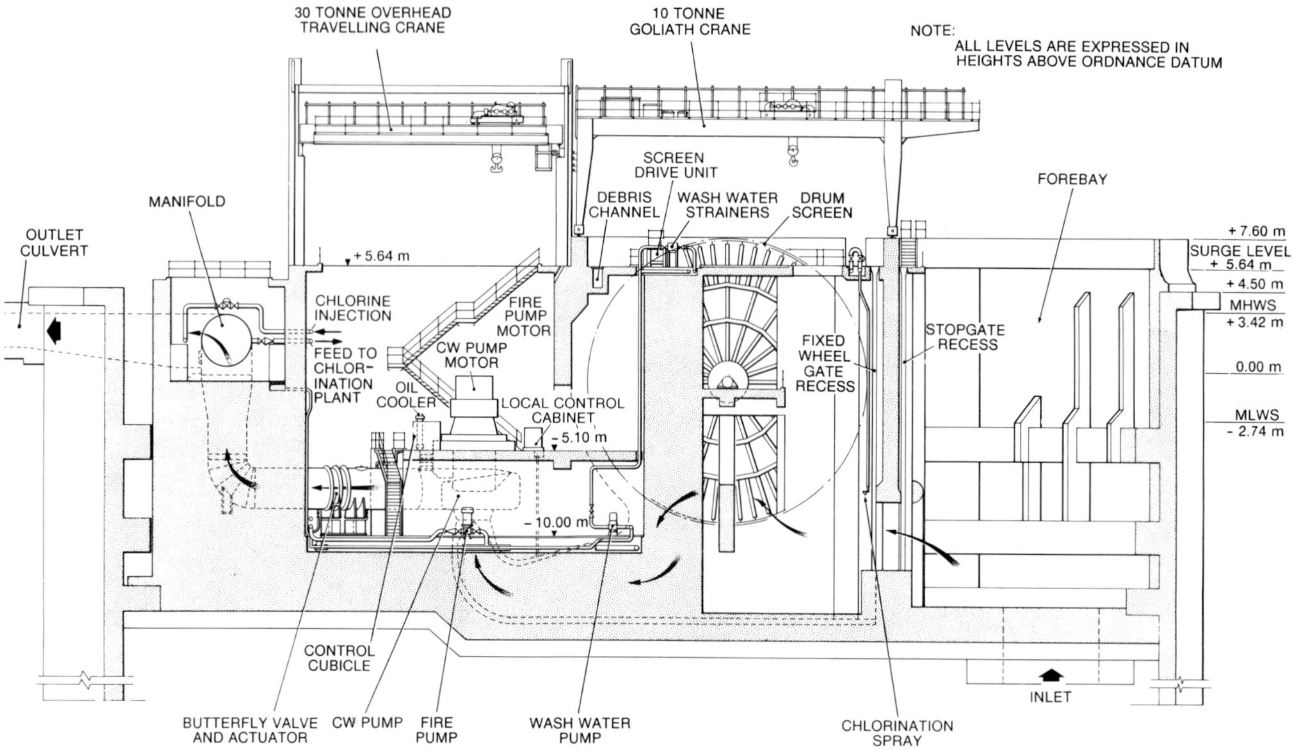

FIG. 2.68 CW pumphouse elevation — Littlebrook *D*

now totally within the control of the designer. Velocities are in the order of 2 m/s in the channels approaching the forebay (often called the *suction dock*). Levels are set by the locations out of the cooling towers. These will normally be designed so that the maximum still water level in the tower pond is at or about site level. Theoretically the water in the tower pond only needs to be deep enough to provide the necessary gradient to make it flow towards the pumps, but in this case the system is quite highly tuned. With continuous loss of water through purge and evaporation a failure in the make-up system would rapidly lead to loss of pump suction. For this reason, most modern stations have a substantial reserve capacity in the cooling tower ponds (up to 2 m working range) and this reflects on the lowest operating level in the pump forebay, which also has to account for the hydraulic gradient losses in the tower flumes. Figure 2.65 (b) shows the hydraulic gradient for Drax power station.

The requirements for screening are much reduced on the tower cooled design but coarse-raked screens are now being considered to deal with the substantial problem of wind blown debris, such as plastic bags, which otherwise may end up on the condenser tube plates.

Isolation arrangements are required to allow removal of a pumpset without draining down the forebay, which is always common to all units.

The pumphouse may also house auxiliary pumps for duties such as compressor cooling, hydrant systems and auxiliary cooling water. Figures 2.71 and 2.72 show one of the two similar pumphouses at Drax power station.

18.3 Main cooling water pumps

The CEGB has standardised on vertical spindle, concrete volute, mixed flow pumps for the largest CW applications, and these have proven to be extremely reliable over many years of operation. The CEGB has a policy of site testing equipment to ensure that it meets its specification. This is especially important for concrete volute pumps because it is impossible for the pump manufacturer to works test the pumps and so a scale model is made and tested as part of the pump contract. During commissioning, the prototype is carefully tested for conformity to the model characteristics of torque, head and flow. Brief details of the testing methods are given here because they have layout implications in the pumphouse.

The concrete volute pumps are driven at low speed (often less then 200 r/min) through reduction gearboxes by 11 kV synchronous motors. The CEGB is examining the possibility of direct drive, multiple pole motors to dispense with the gearboxes, but so far, the slight gains in efficiency and reliability have not warranted the substantial increase in motor cost.

Other power generation authorities have had considerable success with the vertical-spindle bowl pump design. This design allows a more straightforward

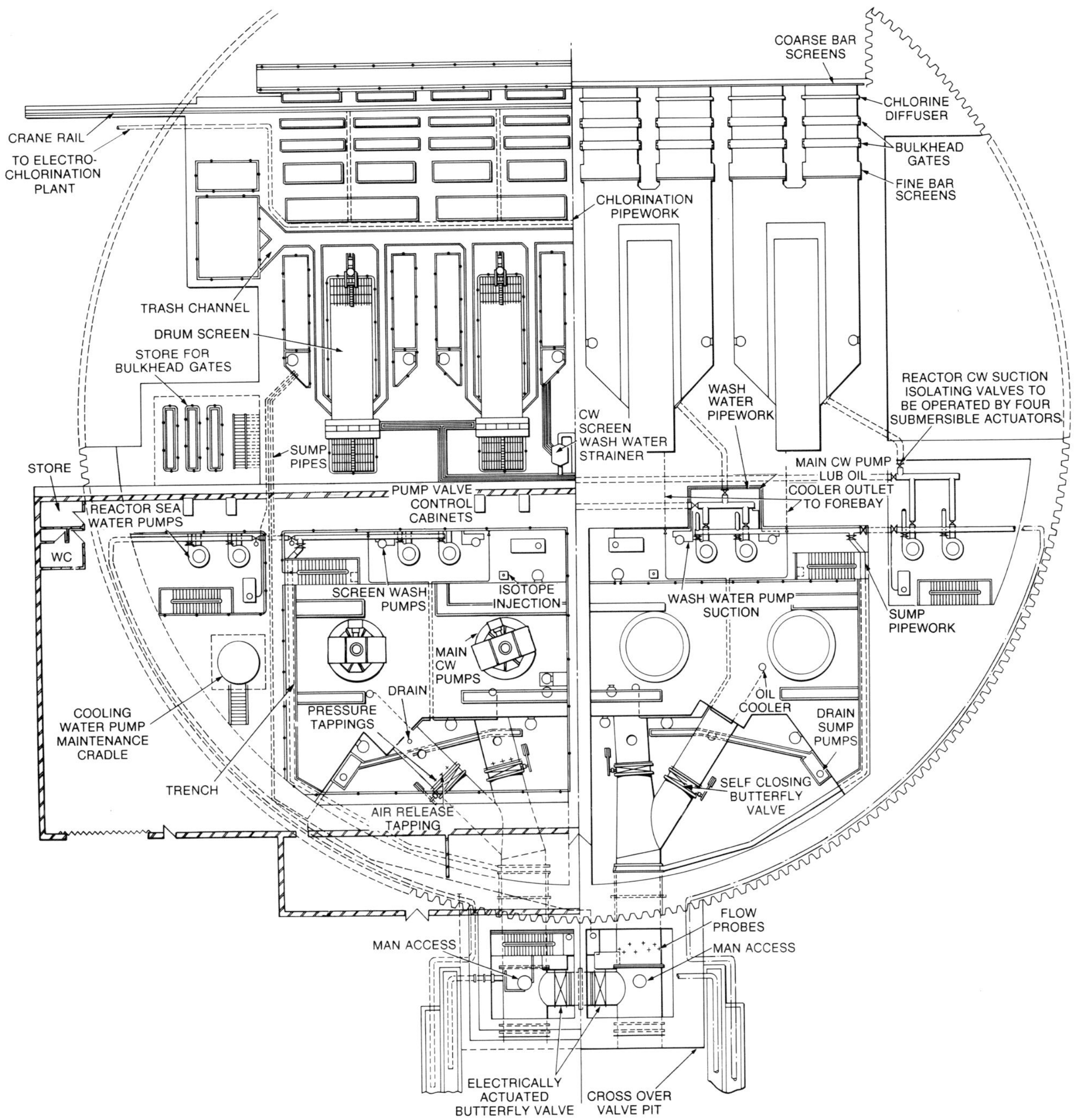

FIG. 2.69 CW pumphouse plan — Heysham 2

suction arrangement but has the disadvantage of very long pump shafts in the flow. These are subject to vibration, leading to gland and bearing wear. The metal casings of the pump also suffer from corrosion unless special measures are taken to avoid it. Corrosion is completely avoided by the concrete volute design. CEGB is therefore unlikely to use bowl pumps for main CW applications, although they are under consideration for lower flow rate applications such as cooling tower make-up systems on indirect cooled stations. Metal casing volute pumps are also used when the head and flow are low enough to give an acceptably small thickness for the pump casing; the changeover from metal casing to concrete volute occurs at approximately 7 m^3/s. The changeover is not abrupt, but pumps below 6 m^3/s would certainly be metal casing, and pumps above 10 m^3/s would certainly be concrete volute type.

18.4 Screening plant

The CEGB uses three types of screening plant at CW pumphouses, coarse screens, fine mesh screens and pressure strainers. Coarse screens of bars are provided at the inlet to the system, which may or may not be at the pumphouse, in order to prevent the ingress of large

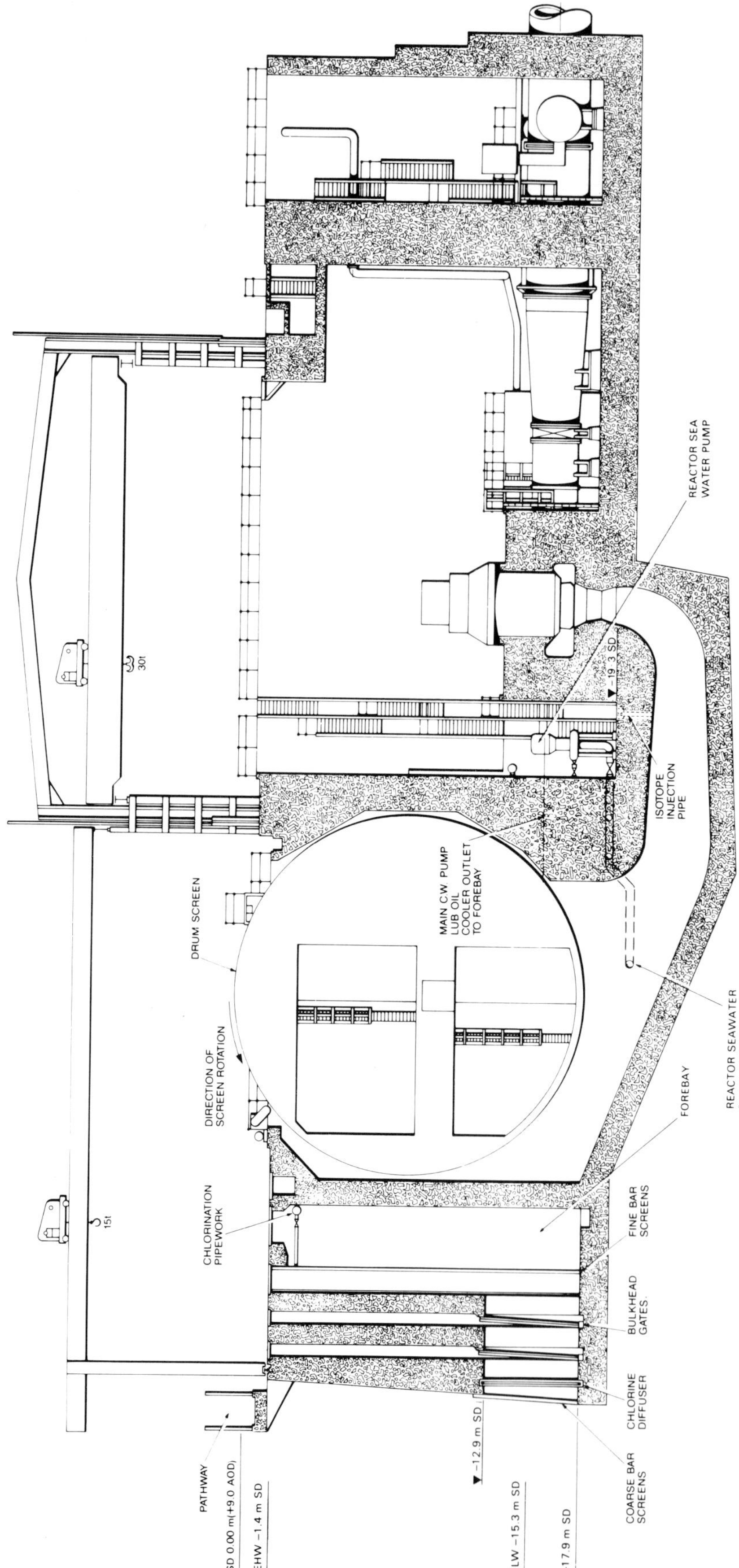

FIG. 2.70 CW pumphouse elevation — Heysham 2

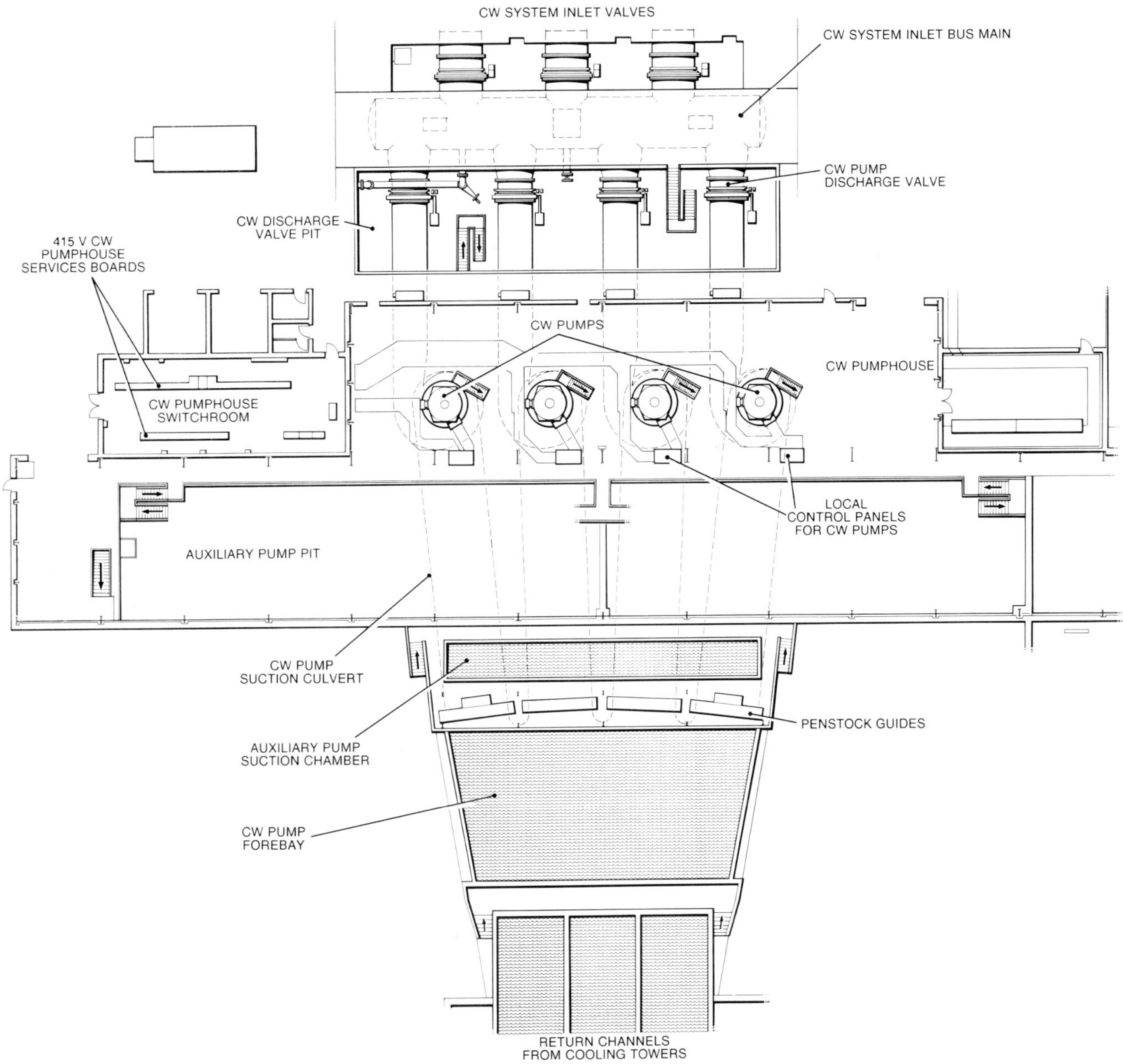

FIG. 2.71 CW pumphouse layout — Drax

baulks of timber which could damage the finer screens. If the system has an offshore intake, this coarse screen is likely to be made of 50 mm bars on a 200 mm pitch. Where the coarse screen is at the pumphouse, the bar pitch is much less, typically 50 mm, and in this case it is necessary to examine the need for permanent raking of the bars. Current practice is to provide room in the pumphouse civil works for the provision of a raking screen. This is usually achieved by enlarging one of the bulkhead gate slots. However, the screens are not fitted until a need is proven, or the station is sited in an area of known debris ingress.

Fine mesh screens are provided to stop the passage of weed and fish into the CW system where they could cause a blockage of the condenser tube plates. The majority of the fine screens are of the moving, self-cleaning, open type such as band or drum screens, although pressure strainers have been installed, downstream of the pumps, at a number of stations. Drum screens are the preferred type because they are substantial steel structures which can be designed to withstand the differential water pressure which could occur if the screen became completely blocked by debris (see Fig 2.67). They are relatively cheap, reliable and the only recurring area where particular attention is required is in the repair and reinstatement of protective coatings. They have lower head losses than other types of screen of similar duty. The only disdavantage of the drum screen is that it needs to extend both above the highest tide level and approximately 2 m below the

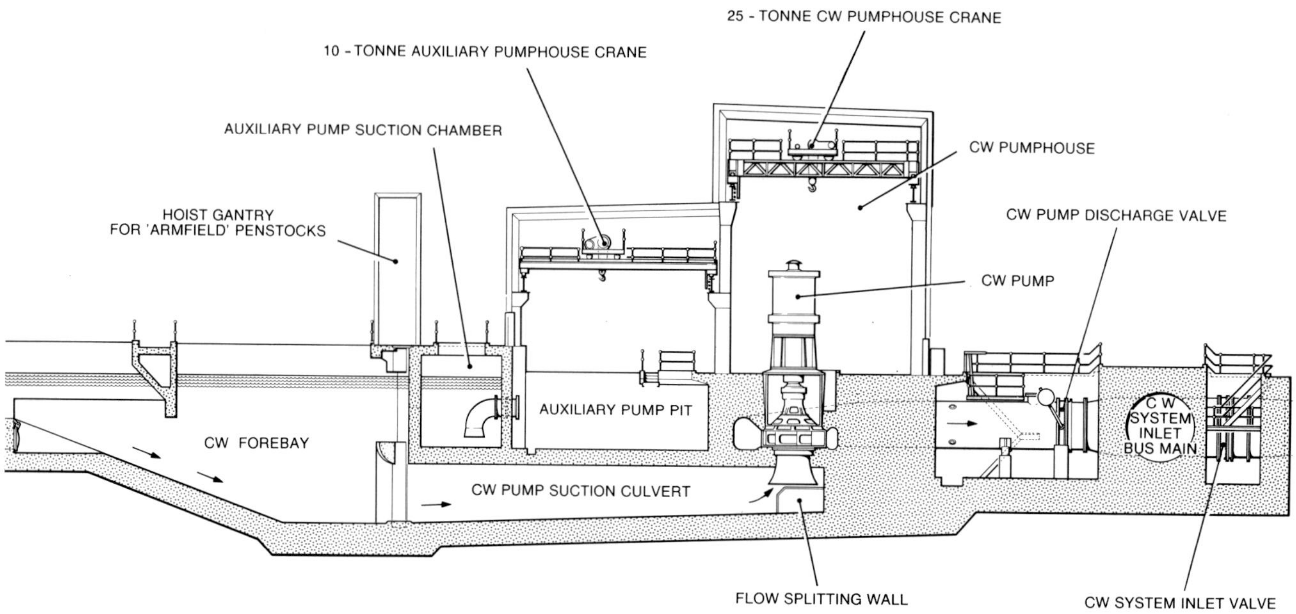

FIG. 2.72 CW pumphouse section — Drax

lowest tide level. Usually, they are about 1.5 times the tidal range in diameter, which leads to very large civil works when the tidal range is large (e.g., Hinkley Point *B* screens are 21.5 m diameter with a tidal range of 14 m). When the tidal range is small, the diameter is set by the need to limit the velocity through the mesh to approximately 0.75 m/s and for a given throughput of water and allowable screen width, the submerged periphery at lowest water level fixes the diameter. Band screens, on the other hand, offer a more compact civil structure, provided that the full flow can be achieved in a single band, but they suffer from the disadvantage of a very large number of moving parts. Greater maintenance and their lower ability to withstand differential pressure make them less attractive for large plants.

Pressure strainers and debris filters offer the most compact civil structure as they can be located vertically above the main pumps. Unfortunately, they have several disadvantages. Being on the high pressure side of the pump, the shell has to be designed to pressure vessel standards. This is complicated by the very large inlet and outlet connections which weaken the shell. The pressure vessel has to be made as small as possible and the compact design gives a much higher head loss than the open type screens. Experience with this sort of strainer in the presence of heavy weed burdens has not been good, although on the continent, they have been very effective in dealing with shell fish (mainly mussels). For this reason, it is unlikely that the CEGB would repeat the pressure strainer design employed on Grain (5 × 660 MW oil-fired station). However, the more compact debris filter could be used in series with the open type screens, especially where a condenser tube ball-cleaning system is employed.

18.5 Pump discharge valves

The CEGB uses hydraulically-operated pump discharge valves to protect the pumps against pressure surge and reverse rotation when more than one pump is supplying water into a common manifold and a trip occurs. On large modern stations, these are exclusively butterfly type. Most designs employ a large drop weight to assist the closure of the valve when the pump is shut down. For reasons of standardisation, the drop weights are all fitted to the same side of the valve body and this means that a pumphouse design cannot be symmetrical about its centreline. On recent stations, these discharge valves have been designed to 6 bar rating and are 2.4 m diameter.

18.6 Section valves

Manifold section valves are provided to allow the isolation of a single culvert for maintenance without shutting down the whole system. These valves may be up to 3 m diameter but on the most recent two-unit stations, the manifold is sized to pass the flow of one CW pump and the valves are therefore 2.4 m diameter.

18.7 Discharge pipework

The main CW discharge pipework and crossover manifold can be constructed in mass concrete or in coated mild steel. If steel pipes are chosen, the design of suitable restraints and expansion joints, and the difficulty of construction of pipe flanges up to 3 m diameter make the mechanical design awkward. Pipework layout which includes flanged stub pipes which are then welded onto plain pipe sections (Fig 2.73) are the preferred methods of construction in metal.

The disadvantage of metal construction is that the shape of the manifold is determined more by constructability and transport considerations than by the objective of low hydraulic loss. Furthermore, the application of protective coatings, which is essential in a sea water environment, needs special attention.

If the manifold is to be constructed in mass concrete, the layout problems are eased but difficulties in interfacing civil engineering with mechanical, and expensive form work are then involved. For pumphouses constructed in difficult (e.g., wet) ground conditions, construction in concrete helps increase the self-weight of the pumphouse and reduces the risk of flotation.

The size of pipework in the pumphouse is based on the use of standard flanges, and the water velocities are kept in the range 2 m/s to 4 m/s. These velocities have been found by experience to keep solids in suspension without causing significant erosion of the pipework. Recent designs are nearer to the lower end of the velocity band from economic and pressure surge considerations.

18.8 Auxiliary systems

Several designs of turbine hall auxiliaries cooling and reactor essential services systems are possible. It is often possible to design an auxiliary cooling water (ACW) system so that pumps are not required (the current single pass, underslung condenser designs make this unlikely), or so that they are only required during plant start-up and shutdown. Where a pump is required, it can either be placed in the pumphouse, in which case a separate pipework system to the turbine hall is required, or it can be placed in the turbine hall drawing water from the main CW culverts and returning it to the discharge culverts or the siphon seal-pit. There are advantages to both schemes and the balance is found by optimising the through-life cost of providing and running a separate system as opposed to utilising the head loss across the condenser to help force water through the ACW system.

Reactor seawater pumps are always located in the pumphouse. These pumps, up to 1 m^3/s capacity, are preferred to be of the close-coupled, in-line, vertical spindle centrifugal type, although bottom-entry pumps are also acceptable. They draw water from the clean side of the fine screens. Because the flow passages in the auxiliary coolers are often smaller than in the main condensers, especially if plate type coolers are being used, secondary fine mesh screens are installed (typically 1 mm to 3 mm mesh). Automatic self-cleaning strainers are used and these are either installed in the pumphouse, which gives an easy debris discharge route, or nearer the heat exchangers. The advantage of having them in close proximity to the heat exchangers is that biological debris build-up in the delivery pipework cannot reach the heat exchangers.

Other auxiliary components and systems in the pumphouse include the control cabinets, small pumps such as screen washwater and sump pumps, main pumpset bearing cooling and seal flushing systems and pipework, flow and head measurement tappings and means of access such as ladders, stairways and gantries. All heavy plant must be located within easy access of lifting equipment.

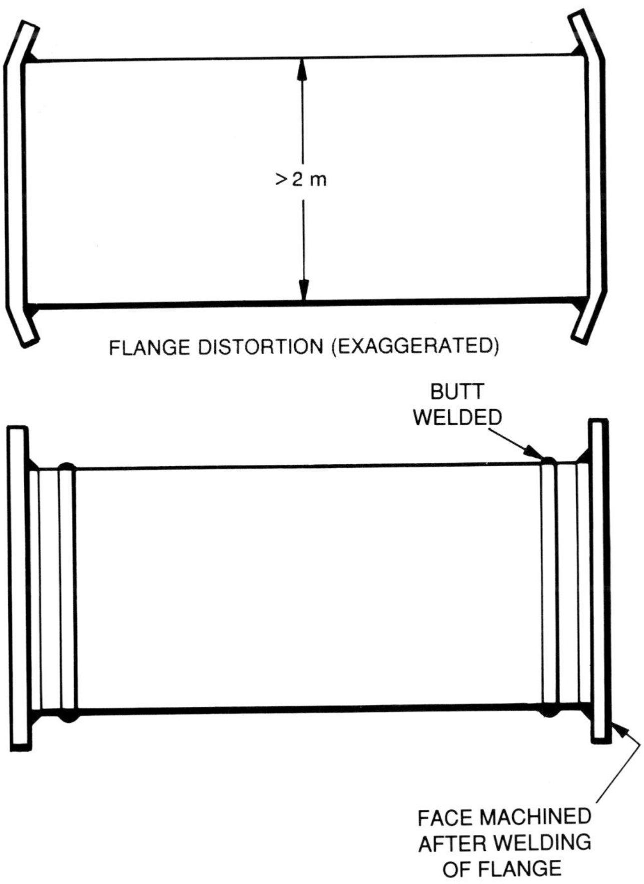

FIG. 2.73 Use of stub pipes in CW pipework

18.9 Gates

CEGB practice is to provide double isolation for maintenance which involves entry into water spaces. Twin bulkhead gates are provided at the entrance to the side channels for this purpose.

18.10 Flow measurement

The difficulties of works testing large pumps were mentioned earlier in this section. In the case of concrete volute pumps, physical testing can only be done on site. Pressure measurement is relatively straightforward. Discharge pressure is measured just upstream of the pump discharge valve by a piezometric ring. Suction head is measured on the clean side of the screens and the pump manufacturer is responsible for the net generated head of the pump and its draught tube.

Flow measurement is much more difficult. Until recently the only reliable method was the so-called 'isotope injection' method where the dilution of a radioactive isotope of tritium is measured as it passes through the system. The isotope is injected through special pipes into the pump draught tube and is then thoroughly mixed with the water as it passes through the pump.

Another method which is gaining favour is the multi-beam time-of-flight ultrasonic method. This method is cheaper than the radioactive method and much quicker to use. The disadvantage is that it relies on being able to fit a number of transducers around the circumference of the discharge pipe, constraining the pump-house designer to provide a length of straight pipe about 0.6 diameters long and preferably more than 3 diameters away from obstructions such as junctions or valves. This seems a small requirement but in practice it is often difficult to find. Where the discharge manifold is in concrete, the isotope method is usually used.

19 Chlorination plant

Chlorination plants, although still in evidence in some power stations, are being phased out throughout the CEGB. Although not relevant to modern practice, a brief outline of the chlorination plant is given here to cover those stations where such plants are still in existence.

Chlorine is a poisonous gas and stringent safety measures are always adopted by the specialist contractors engaged in this work. Chlorine in the liquid state is stored in bulk tank installations, refilled from road or rail tankers. The liquid chlorine is fed to electric evaporators and then in gaseous form to chlorinators which produce the chlorine solution.

The chlorination plant is housed in a weatherproof building, with clear access in and out for personnel. The building must be provided with forced ventilation via a plenum duct at floor level. The chlorine dosing solution is piped to the injection points in the CW system in rubber-lined pipes, the dose rate and periods being programmed through a master clock.

The product used to replace chlorine is a sodium hypochlorite solution, produced on site in the case of coastal stations by utilising a seawater electrolysis plant, and by being purchased as a commercial product by inland stations and stored in bulk. Inland stations could also employ an electrolysis plant, but would have to provide a salt water feedstock by means of a brine producing plant.

For coastal stations the electrochlorination plant is housed in a *dutch barn* type structure adjacent to the CW pumphouse, seawater being taken from the fore-bay area. It is preferable, however, that consideration is given to obtaining a water supply from the CW outlet system and the electrochlorinator power consumption/output performance improves with warmer water. Figure 2.74 shows the basic stages of the electro-chlorination process.

A detraining tank facility is provided in the solution circuit to permit release of hydrogen gas produced by electrolysis, and for this reason special consideration is given to the location of the plant. An open environment is chosen and the same factors apply as those relevant to on-site hydrogen generation plants.

For inland stations, consideration is given to bulk storage in the form of storage tanks, together with the dosing pumps, etc. The size of the installation may vary depending on dosing needs, the fitting of condenser mechanical cleaning equipment, etc.

The sodium hypochlorite solution from the electro-chlorination or bulk storage plant is dosed into the CW system in the same controlled fashion as chlorine.

All installations are provided with road access.

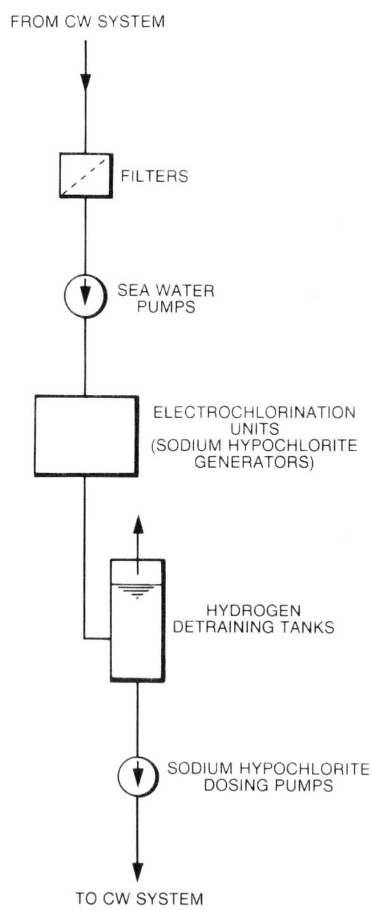

FIG. 2.74 Electrochlorination process

20 Coal handling plant

20.1 Rail-borne reception and discharging

The annual coal consumption of a 2000 MW station, depending on its load factor, amounts to approximately 5 million tonnes, i.e., 20 000 tonnes daily over 250 days. To achieve this delivery rate of coal, trains to large modern power station utilise hopper-bottom wagons on a *permanently coupled* basis. Permanently coupled working means the keeping together of the main line locomotive and its train of wagons as a single unit, from entering to leaving the site, including discharge over a track unloading hopper.

The advantages of this system are a fast turnround of the train at the power station. It is expected that trains will achieve at least 3.5 trips per week between colliery and power station, thus giving greater wagon utilisation together with capital economy in track work and associated savings in operational costs.

A design of hopper wagon has now been adopted having rapid self-discharging characteristics. This wagon is the 32 tonne capacity, 2-axle wagon with six bottom doors operated automatically by lineside equipment. It discharges the coal whilst travelling at a speed of approximately 0.24 m/s.

The largest train at present envisaged by British Rail is 2070 tonnes gross load, with a payload of 1440 tonnes. The size of the train for this payload will be 45 wagons plus locomotive giving an overall length of 425 m. Accommodation in the sidings layout should therefore be 500 m to provide working clearance. Where possible it is prudent to allow for greater lengths, if site space allows, to provide for longer trains should BR wish to introduce them at some later date. Introduction of light wagon bodies offering lower tare weights, and hence higher payload potential, is also currently being considered.

Depending on the space available and the volume of traffic to be handled, the choice lies between the merry-go-round (MGR) *loop* system or the *run round* system. The latter, although necessitating the main line locomotive running round its train, is accepted by BR as permanently coupled working. Figure 2.75 shows the merry-go-round system at a 2000 MW coal-fired station and Figure 2.76 shows the run round system at a 2000 MW coal-fired station.

The CEGB now accepts the weekly delivery of coal over 5 days. Consequently, two sevenths of the daily coal intake is put out to store for reclamation at weekends, when coal is not being delivered. Normally the stations are coaled over two shifts (i.e., 16 hours). However, BR have the right to request the CEGB to

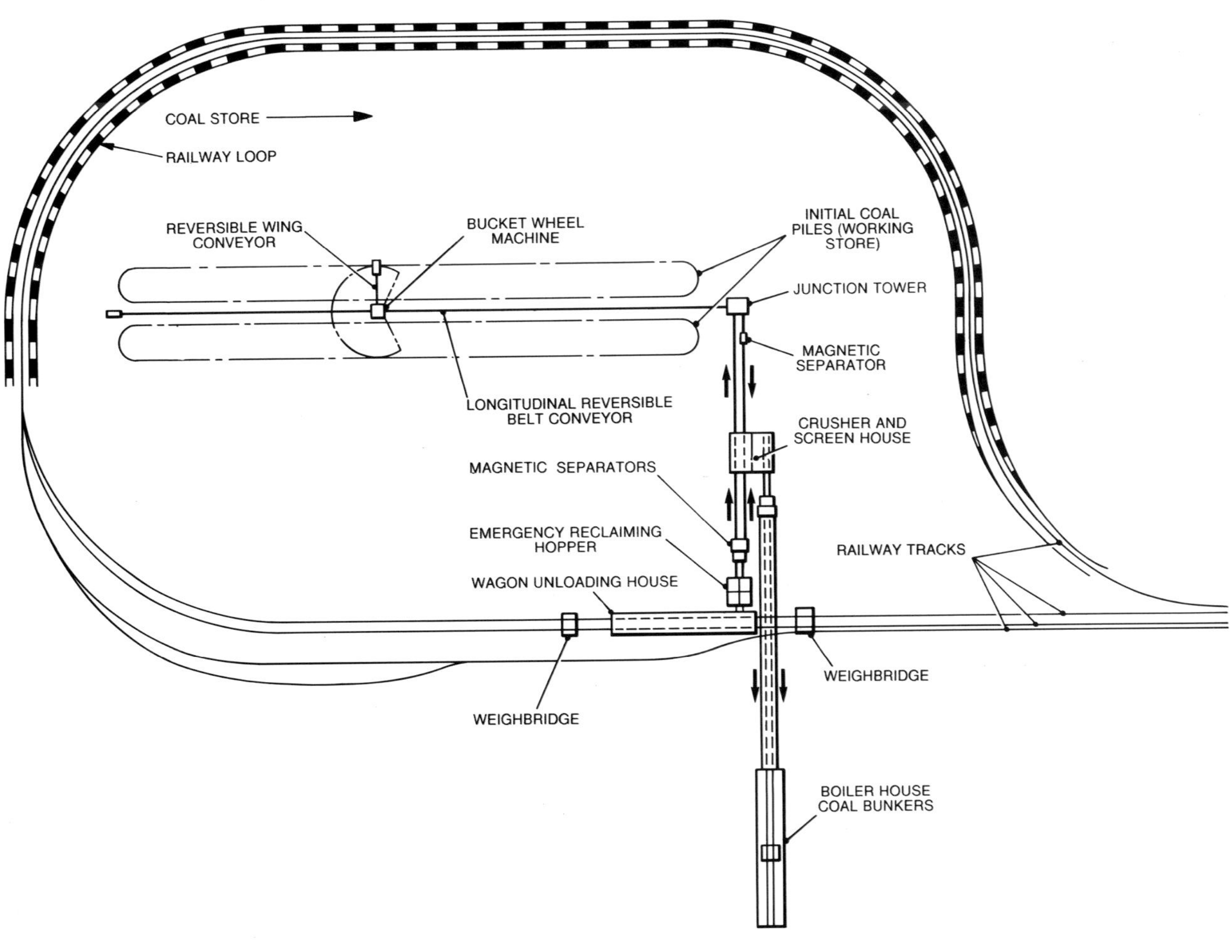

FIG. 2.75 Plan of coal handling plant with merry-go-round system for a 2000 MW station

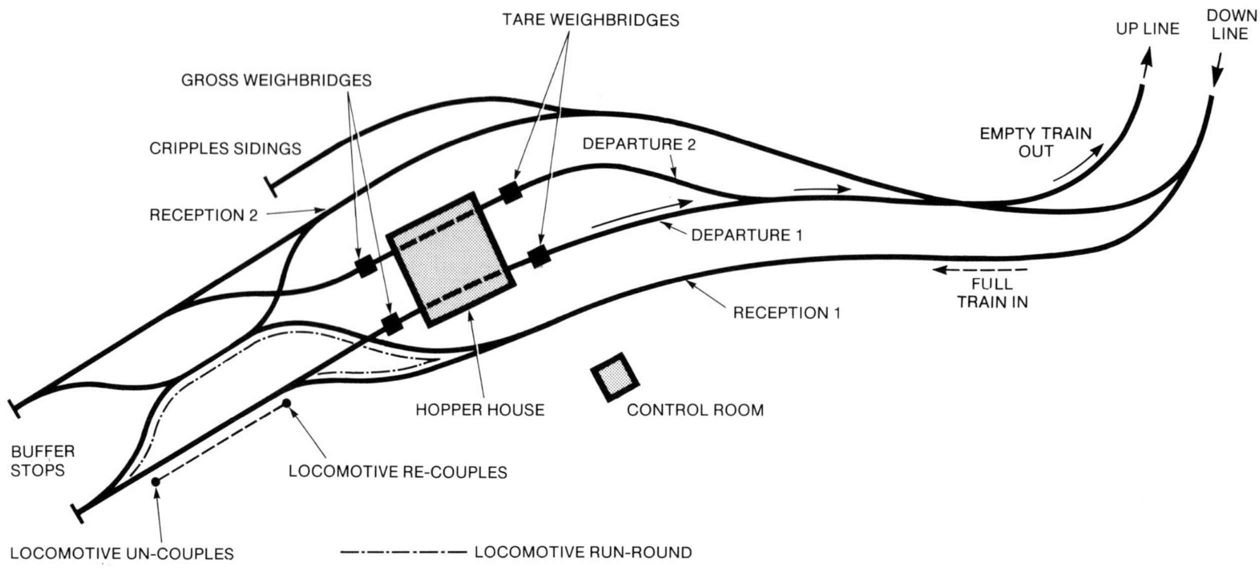

FIG. 2.76 Plan of run-round coal train delivery system

accept trains round-the-clock for which they, in return, offer a reduction in freight rates.

20.1.1 Typical coal handling system

The coal train arrives at the station and proceeds through signals to wagon discharge hoppers in the track hopper house (Fig 2.77). The wagons are discharged and weighed automatically with gross and tare weights recorded before and after the unloading hopper. The coal is removed from the wagon discharge hopper by paddle feeder onto a system of belt conveyors and is sent either direct to the boiler bunkers or to stock via a stock-out machine. When coal is required from stock it is reclaimed by the reclaimer and conveyor system to the boiler bunkers — it is normal practice to avoid a double handling system. To help keep a check on how much coal is in stock, belt weighers are included on the conveyors to and from the coal store and also preceeding the boiler bunkers. Quality sampling is necessary for efficiency monitoring purposes. Provision is also made for trash screening equipment to prevent coal hold up and damage to the pulverising equipment and for magnetic separators to remove tramp iron from the coal.

20.2 Water-borne reception and discharging

Water-borne coal is received at CEGB stations in two ways:

- By sea-going colliers.
- By river or canal barges.

Whilst the second method has been used for deliveries on smaller stations, the coal consumption of large power stations and the limitations of the inland waterway network within the UK makes complete reliance on this method impractical. Only one of the CEGB's 2000 MW stations is equipped to accept part of its coal intake by barge, for which a tippler has been installed.

The use of sea-going colliers is practical for stations located on the coast or on navigable river estuaries. However, each site must be evaluated individually to determine the size of collier which can be utilised since such features as water depth and river width may limit ship size and therefore the practical coaling rate which can be achieved for the station.

In locating the jetty for coal discharging, navigation authorities usually require that the mooring facilities avoid interference with main shipping channels. It is, however, necessary to provide sufficient water depth to enable berthing at all states of the tide. Depending on specific site constraints, this may require an offshore location or extensive dredging of the berth and its approach and departure channels.

Coal unloading facilities at the jetty head may consist of grabbing cranes, grab unloaders or continuous handling equipment. The choice will depend on the size of station, the size of collier and such interrelated aspects as capital costs, assessed overall through ship unloading rate, equipment availability, etc., evaluation which for a given combination of factors will identify the preferred option.

20.3 Road-borne reception and discharging

Road vehicle unloading facilities are arranged for standby or supplementary purposes and, where provided, normally comprise no more than a single ground unloading hopper into which road vehicles can dis-

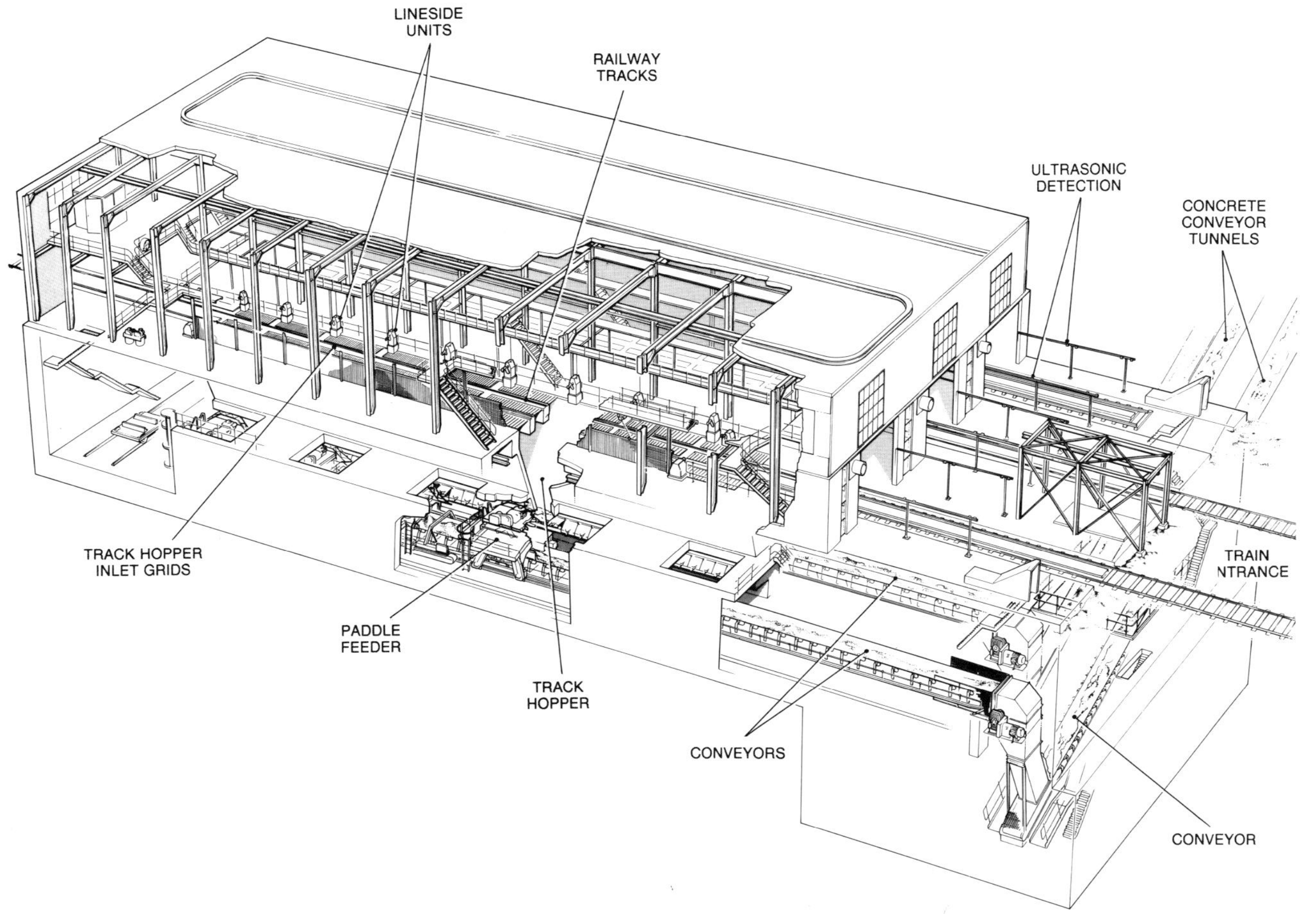

FIG. 2.77 Rail wagon unloading track hopper house

charge and from which coal is conveyed by belt to some convenient point in the main coal handling system.

20.4 Coal storage

20.4.1 Stockpiling

It is now common practice on CEGB stations to provide for a store of capacity equal to a quarter of the annual coal burn, over half of this capacity may be regarded as a strategic reserve for use in emergency. It is consequently unnecessary to incur heavy expense in providing fixed handling equipment for the more remote parts of the store which are usually served by mobile plant. Tests have shown that coal of any quality now in use may be stocked to a considerable depth provided it is properly consolidated, and it is now assumed for planning purposes that nearly all coal stores can be at least 15 m deep. The store is usually accommodated inside the rail sidings loop already described.

20.4.2 Stockpile working

Various methods of reclaiming coal from the stockpile to the station boilers have been examined, covering mobile equipment into ground reclaim hoppers or bucket wheel stocking out/reclaiming machines, to see if a common policy could be adopted. It has been found that many features affect individual sites, including the station capacity, delivery schedules and the developing nature of the reclaim plant that this has to be resolved for each project. For example, an optimum mobile plant scheme for one 2000 MW project comprises three scrapers for long haulage and two bulldozers for a short haul to the reclaim hopper, with working distances varying between 200 m and 510 m (see Fig 2.78). The alternative of a bucket wheel stocking out/reclaiming machine supported by mobile plant was the optimum choice for several stations (see Fig 2.79).

The figures also show the strategic position of weighers for checking freight deliveries and estimating the station efficiency, coal sampling equipment, crushing and screening plant and conveyor interconnections.

20.4.3 Bucket wheel stocking out/ reclaiming machine

For some power stations the bucket wheel consists of a rail-mounted vehicle which operates in conjunction with a fixed belt conveyor running the length of the coal stocking area (see Fig 2.80). The machine, which either

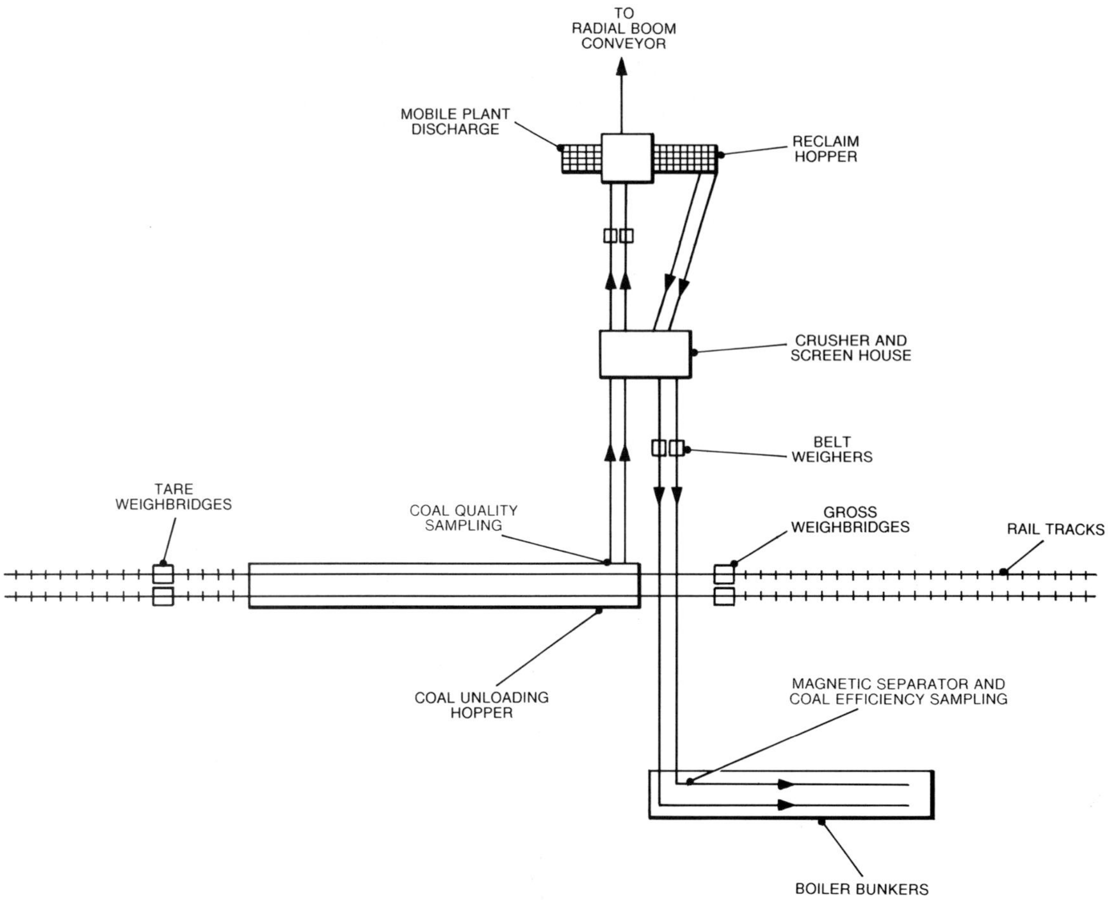

FIG. 2.78 Mobile plant coal reclaim arrangement

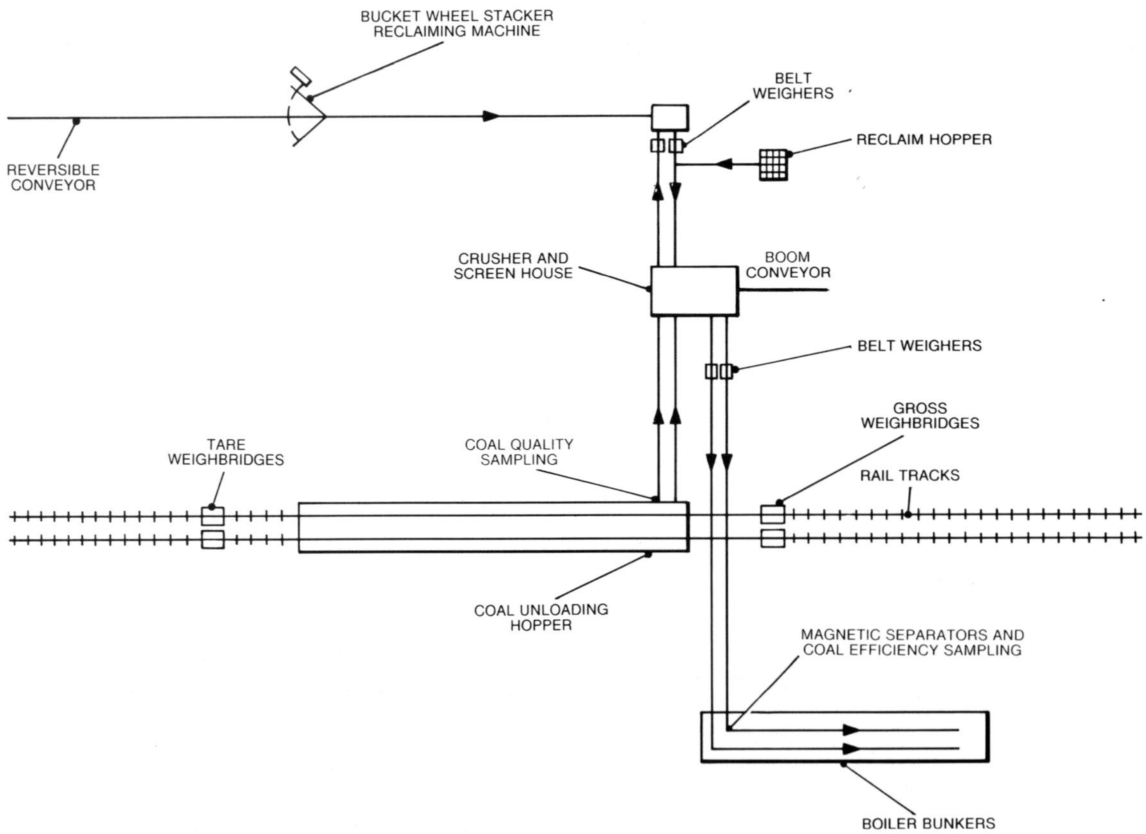

FIG. 2.79 Bucket wheel stocking-out reclaiming machine arrangement

FIG. 2.80 Bucket wheel stocking-out reclaiming machine (see also colour photograph between pp 66 and pp 67)

straddles or runs alongside the fixed conveyor, is essentially a mobile platform upon which is mounted a reversible wing conveyor for stocking out in a continuous heap on one or both sides of the fixed conveyor. At the outboard end of the conveyor boom a bucket wheel is mounted and is used only when reclaiming is taking place. Means are provided for transferring coal continuously between the wing and fixed conveyors during stocking out and reclaiming. The wing conveyor is capable of being luffed and slewed.

20.5 Conveyance from unloading point to station bunkers or coal store

The main line of the coal handling plant runs between the coal unloading point (be it wagon hoppers or jetty) and the boilerhouse, and for maximum economy this line should be as short as possible. The branch line used for stocking out and reclaiming to and from the coal store should also be kept to a minimum length. For a station of 2000 MW, deep-troughed twin conveyors of 1500 t/h each are usually provided for the main line, one being a standby for the other. The branch line conveyor is a single 3000 t/h reversible belt.

Figure 2.81 shows a perspective schematic of the Drax power station coal handling system.

20.6 Plant control

Because the coal handling plant is remote from the station, it is the usual practice within the CEGB to provide a coal handling control room local to the rail unloading hopper, which itself is usually adjacent to the coal store. This control room would contain all the switching, sequence control and annunciators for the complete coal handling system, including the stocking and reclaiming conveyors and paddle feeders, with the exception of the bucket wheel stocking-out reclaiming machine and control of lineside wagon opening and closing equipment. The control room also contains all the necessary track signal switching equipment,

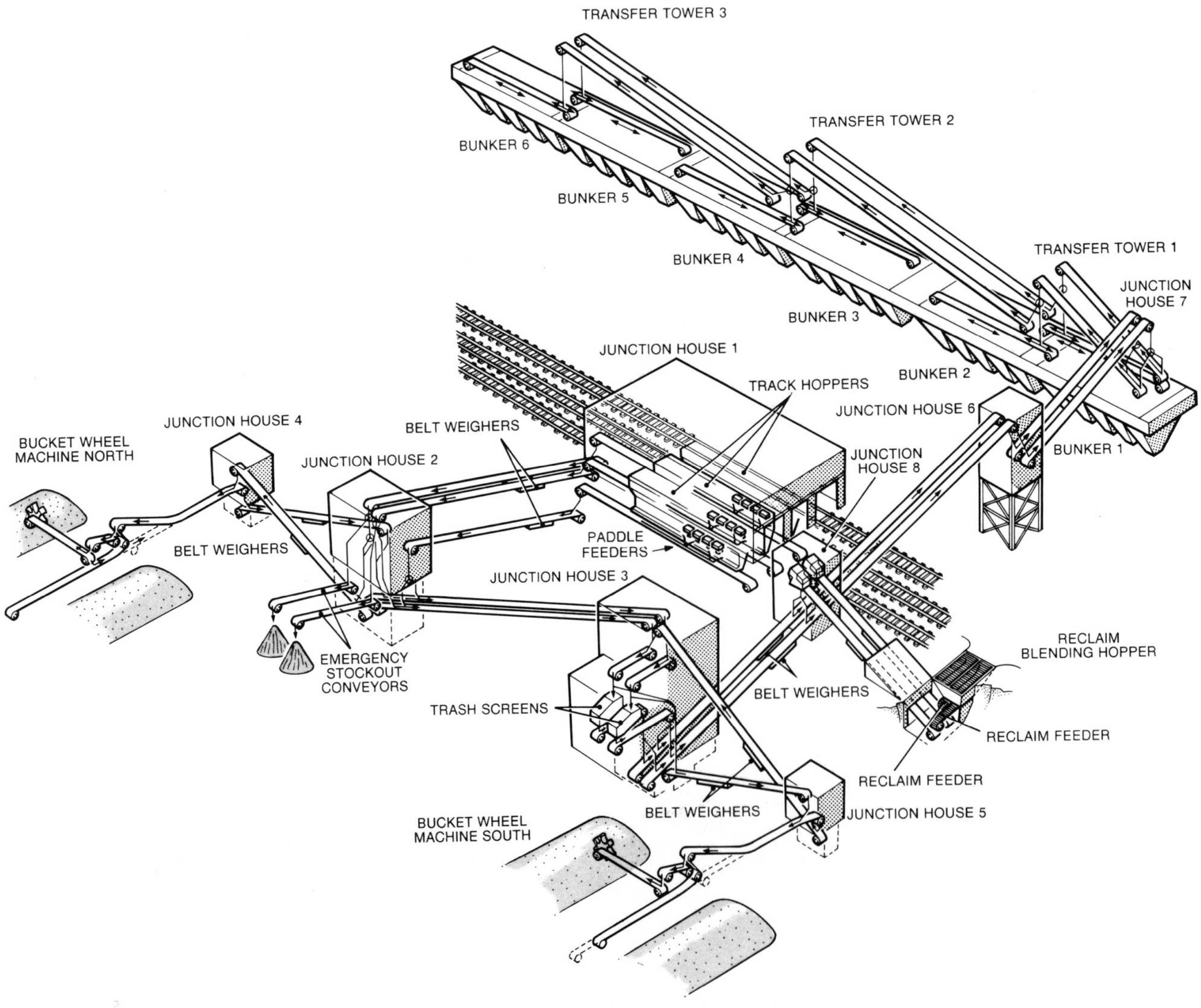

FIG. 2.81 Drax coal handling system

together with a closed circuit television monitoring system for providing supervision of transfer chutes, paddle feeders, rail unloading hoppers, etc.

21 Ash and dust handling plant

Pulverised fuel (PF) boilers now account for all the coal burnt by the CEGB. The coal is milled to a fine powder before being blown into the furnace through burner nozzles. It burns as it passes up through the furnace space, producing very fine molten ash particles, most of which are carried through the boiler passes with the flue gas, solidifying and cooling as they go. Some particles impinge on furnace wall tubes or pendant superheater tubes where they fuse or sinter together forming clinker. As the ash builds up it either falls off or is periodically blown off with steam or air by the sootblowers, falling into the ash hoppers at the base of the boiler as furnace bottom ash (FBA). This normally accounts for 20% of the total ash make. The size of this ash can be very large, up to 0.25 m to 0.5 m in length. The remaining 80%, which has a particle size of less than 80 microns diameter, is conveyed on the gas stream and is known as dust, or more precisely as pulverised fuel ash (PFA), and the very fine dust continues in the gas stream until it is either collected in the economiser hoppers or arrested by electrostatic precipitators and collected in the precipitator hoppers (see Fig 2.82).

A modern 2000 MW coal-fired station will produce about 250 tonnes of ash and dust per hour and this poses a major disposal problem. It is best to consider the requirements of ash handling (i.e., furnace bottom ash and clinker) separately from PF dust handling since two separate systems are involved.

21.1 Ash handling plant

On CEGB stations furnace bottom ash is typically collected in a hopper of welded steel construction and firebrick-lined. The hopper is static and has a water trough around its top periphery into which a seal-plate welded around the furnace bottom outlet dips to prevent air ingress into the furnace. Because the water in this trough evaporates it has a float-operated make-up supply from a convenient clear water source.

Figure 2.83 shows a schematic arrangement of one type of FBA handling system. The FBA falls into the boiler ash hoppers which are shielded as far as possible from direct furnace radiation by the shape of the boiler wall tubes. As it enters the hoppers it is quenched by water sprays; this leads to heat loss from the boiler, but the spray is necessary to prevent the formation of large pieces of clinker.

The hopper is typically sized to accommodate up to 24 hours production of ash at design rates, so providing storage to cover short term problems on the disposal system.

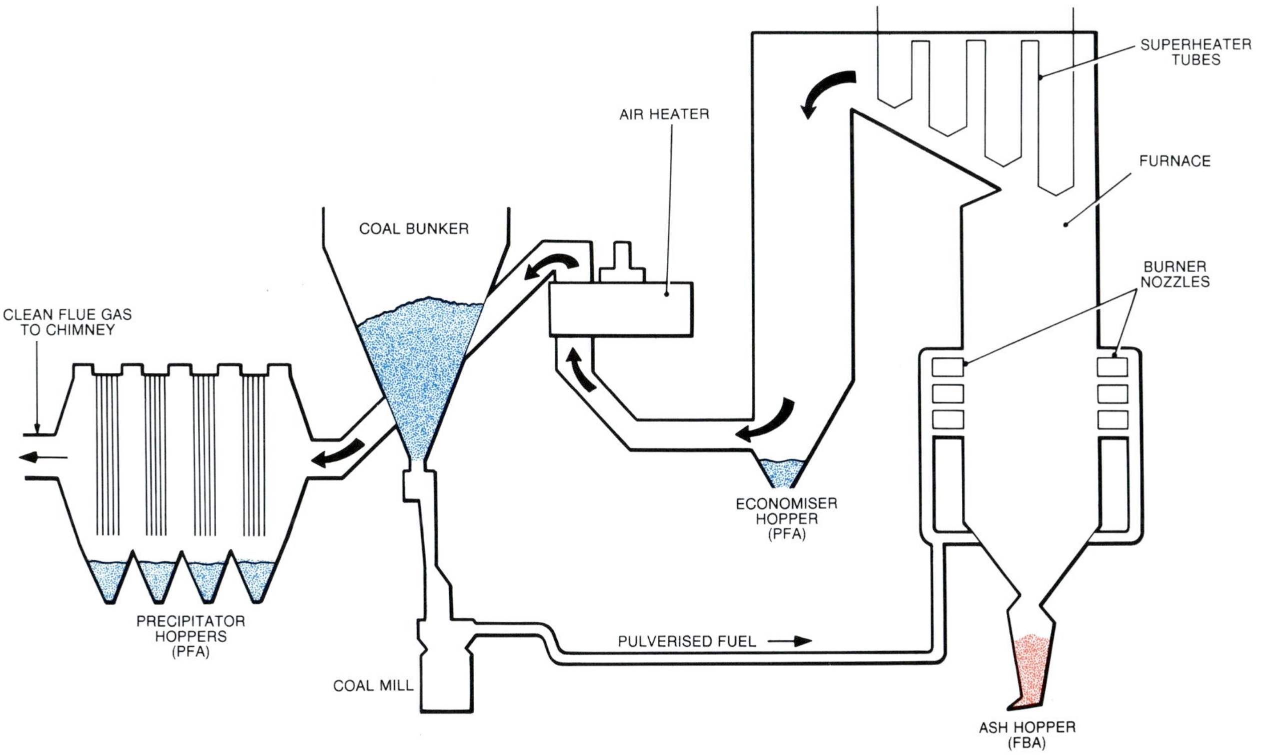

FIG. 2.82 Boiler schematic showing ash collection points

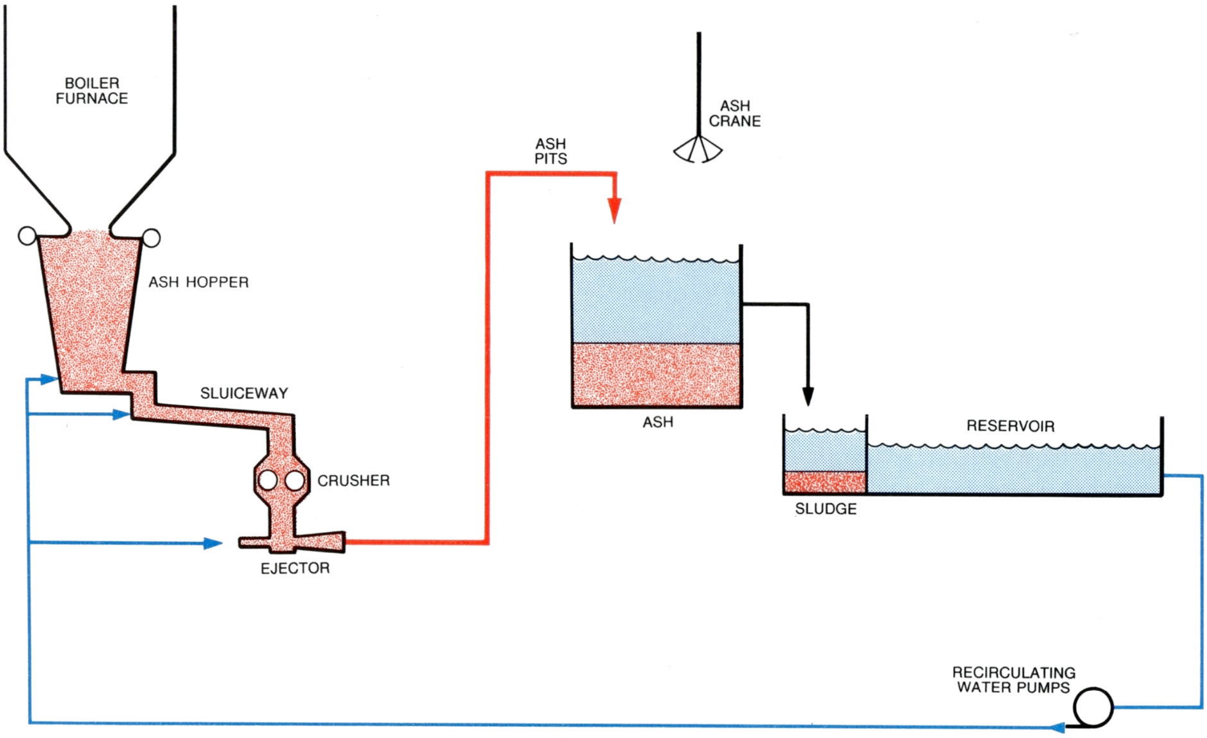

FIG. 2.83 Furnace bottom ash removal arrangement

Water jets are used periodically to discharge the ash from each hearth in turn and propel it along a sluiceway to the end of the hopper. Here the ash drops through an ash crusher which crushes it to less than 50 mm and then into an ejector, where the ash is entrained in a high velocity jet of water from a supply at about 10 bar pressure. The jet discharges through a diverging throat and the consequent pressure recovery to approximately 3 bar drives the ash and water mixture along a pipe to the ash pits. Recently installed ejectors have a 100 mm diameter throat and with a water flow of 200 litres/second they are capable of conveying approximately 70 tonnes of ash per hour. Earlier systems used open sluiceways for the entire journey from ash hoppers to ash pits, but this had the disadvantage on large stations of higher civil engineering costs and of using much greater quantities of high pressure water.

At the ash pits, the ash settles to the bottom and the water is decanted off via a sludge settling pond to a reservoir. When ashing is complete the ash is allowed to drain and the finer sludge particles allowed to settle from the water before it is re-used. Each ash pit is normally designed to accommodate 24 hours ash production from several boilers, three ash pits being used in rotation, one filling, one draining and one being emptied. Drained ash and sludge is taken from ash pits by grabbing crane direct to road vehicles or deposited on the apron for further draining.

After drying, the ash is again picked up by the grabbing crane and loaded directly into purchasers' lorries or into a concrete storage silo from which it may be discharged into purchasers' trucks as and when required.

The cycle of operations must be carefully worked out, starting from the total ash make per ashing operation, settling time in the pit and drying out time on the apron. The storage should ensure that purchasers can be served on a continuous basis during the day shift period.

On a 2000 MW station it is usual to provide two drying aprons (one on each side of the settling pit) for alternative loading, draining and pick-up cycles, and two travelling overhead cranes.

Where ash is made surplus to sales, provision is made to crush the ash in roll crushers so that it can readily be pumped in water to a disposal area which may be at any reasonable distance from the station. The design power of the ash disposal pumps will consequently depend on the distances involved.

Generally the sluice pumps, roll crushers and ash disposal pumps are housed in a chamber below ground level and adjacent to the settling pit. These chambers can be liable to flooding and therefore are provided with sump pumps having a capability equal to or better than the flow capacity of the largest pipe in the chamber which could fail, for example, in freezing winter conditions.

The main characteristic of FBA which concerns the system designer is its abrasive nature. This is important in selecting materials for use in storage hoppers and transport systems. No method has been derived which gives quantitative indication of abrasive wear, and material selection is accomplished from operating experience.

21.2 Dust handling plant

The dust handling system is typically arranged to collect the dust (PFA) in a dry dust bunker as shown in Figs 2.84 and 2.85.

The dust handling plant is associated with the ash handling plant but is considered as a separate layout problem. The dust is handled in its dry state throughout the plant by pneumatic air slides which remove dust from the economiser hoppers and precipitator hoppers.

Consideration is also given to avoiding low trap points in boiler gas ducting to prevent any build-up of dust within the ducting.

The first step in the operation is to collect the dust together from the boiler precipitator and economiser hoppers, which can have typically 36 precipitator dust hoppers and six economiser hoppers. The hopper contents are discharged by gravity into a low pressure dust blower which uses air at about 0.5 bar to blow the dust to a buffer hopper. A dust pump is used to transfer the dust from the boiler hopper to the station dry dust bunker.

Air slides are used as an alternative to the dust blowers. They consist of a rectangular duct at a slight downwards angle divided into upper and lower chambers by a porous membrane.

The dust handling plant is usually entirely automatic requiring no manual intervention except in cases of failure and hence operates on a 24 hour basis.

21.3 Ash and dust disposal

There are several different methods of ash transport and disposal used by the CEGB, each suited to different site conditions. Most of the FBA is sold, and whilst as much dust as possible is sold, a large proportion has to be disposed of elsewhere. In all of its work the CEGB has a statutory duty to environmental conservation, and in the application of this obligation to ash disposal, the CEGB aims, wherever possible, to improve or maintain the existing environment. Some examples of ash disposal schemes are as follows.

Barlow land — Drax

Surplus ash from the Drax power station is being used to construct an artificial hill on land adjacent to the

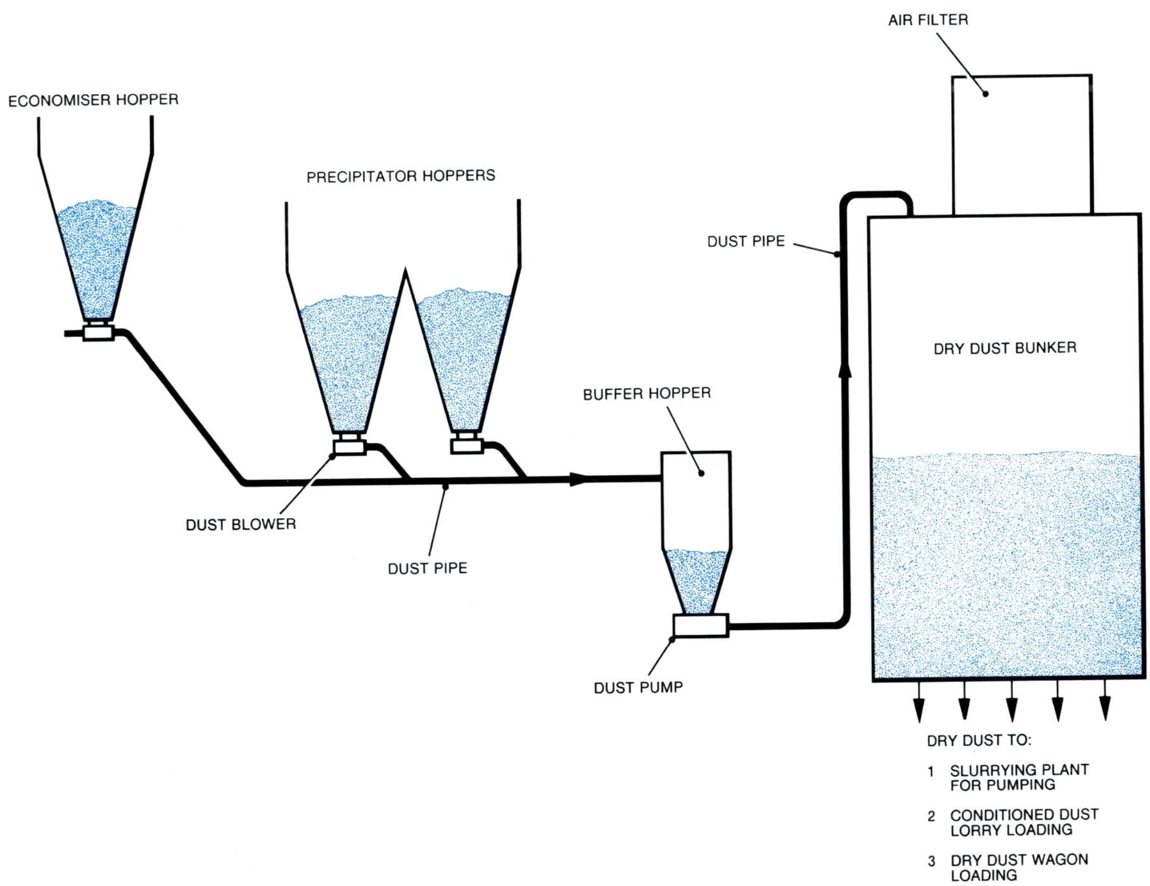

FIG. 2.84 Dust removal arrangement

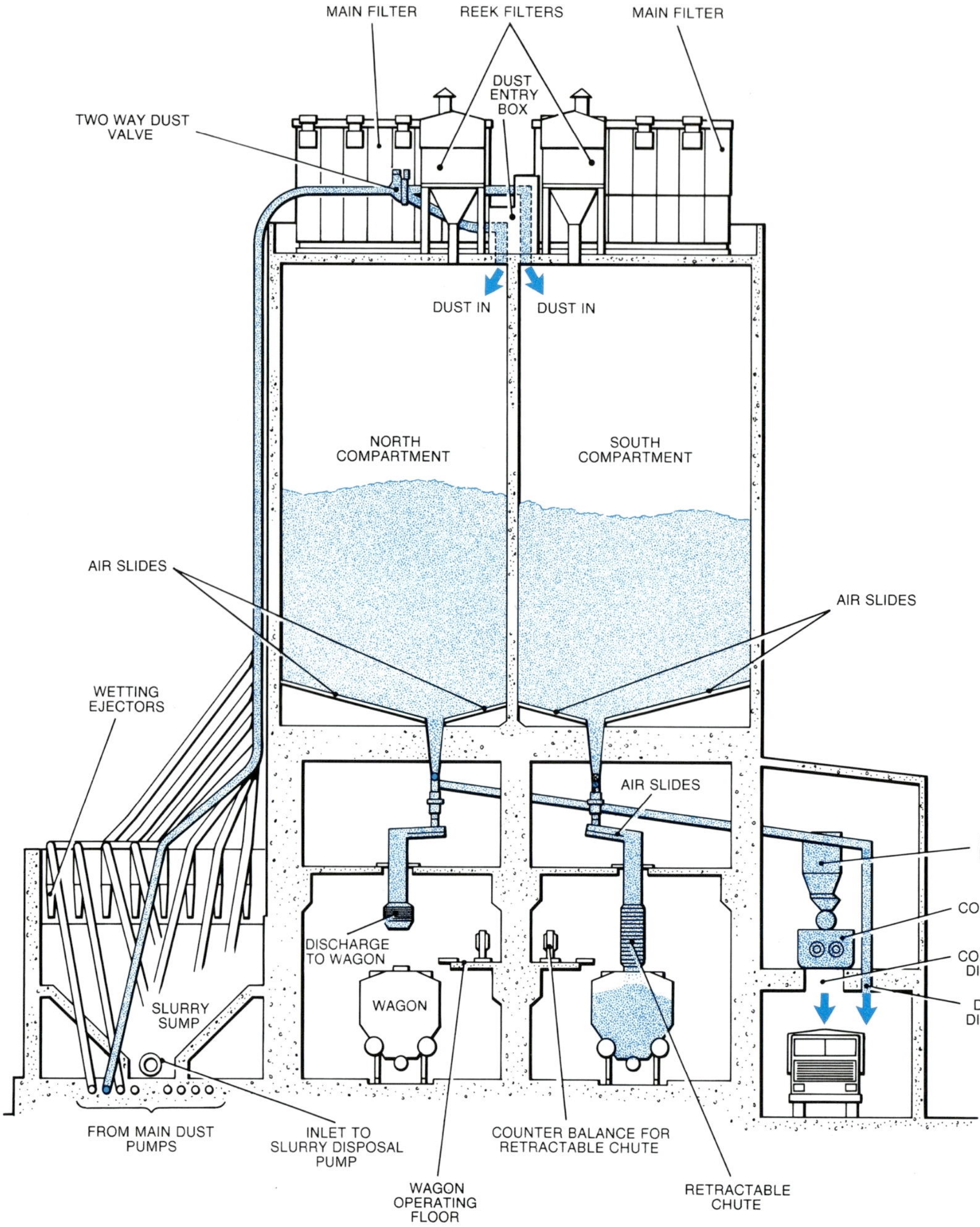

FIG. 2.85 Typical dust bunker

power station. The short distance from the power station enables the ash to be transported by belt conveyor in a conditioned cake form, which offers both economic and environmental advantages (see Chapter 1 Fig 1.12).

Gale Common

Gale Common ash disposal plant in Yorkshire receives ash in slurry form from two 2000 MW power stations, at a rate of up to 35 000 tonnes per week. Approximately half of the ash slurry is dried and used to build the embankments of an ash disposal lagoon, the remainder of the unprocessed slurry being pumped to the centre of the lagoon where it is allowed to settle out. If insufficient dry ash is recovered to build the embankment, the shortfall is made up with waste colliery shale.

Peterborough land reclamation scheme

At this location 400 hectares of disused clay pits are being filled with PFA transported by rail from three power stations in the Midlands. The pits will hold at least 30 million tonnes of PFA possibly rising to 50 million tonnes as further clay pits are excavated by the brick-making companies.

Radley

The PFA is slurried into a 70% water and 30% PFA mixture and pumped along a pipeline. This system has to handle 3600 tonnes of PFA daily in a 12 hour period. The disposal area comprises four separate lakes which settle out the PFA from the slurry mixture. The surplus water from the lakes is returned to the River Thames under strictly-controlled conditions.

Figure 2.86 shows the general disposal arrangements and lake filling sequence.

22 Auxiliary boilers

The auxiliary boiler plant is required to provide auxiliary steam for the main generating plant and auxiliaries. Typical services include:

- Heating for the de-aerator on start-up.
- For atomising the fuel oil burners.
- For fuel oil pumping and heating services.
- Hot water and space heating services for station buildings, including the administration and welfare buildings.

For a 2000 MW fossil-fired station the plant would typically comprise five identical oil-fired boilers supplying superheated steam at 17.5 bar and 228°C. Normally four boilers are required for peak demand with one on standby. Typically the boilers would consist of the multi-tubular, wet back, three pass type, having twin furnaces each fitted with a single rotary cup burner (see Fig 2.87). The burners are fitted with a *voluvalve* system which controls the volume of fuel oil passed either to the rotary cup atomiser or recirculated back to the pump suction. Air is supplied to the burner from a forced draught fan mounted adjacent to the burner. Some of this air is supplied to a primary fan mounted on the burner shaft. A fuel/air ratio motor controls the quantity of air and fuel required for combustion. This motor is operated by electrical signals from a position controller, which receives a desired value from the steam range pressure controller or from manually operated pushbuttons. Initial ignition of the fuel oil is from a flame of propane gas ignited by an electric spark. Each boiler is fitted with shell and superheater safety valves which discharge through silencers to atmosphere.

The boilers are normally located in a separate auxiliary boiler house as shown in Fig 2.88 and are arranged to each supply steam into a ring main distribution system; condensate is returned via a separate system to recovery tanks adjacent to the auxiliary boilerhouse.

As the auxiliary boilerhouse serves several areas on the station site it is difficult to optimise its location. Generally it is positioned adjacent to the station main buildings. Other factors that can affect its position are whether or not the auxiliary boilerhouse and gas turbine house share the same chimney for visual reasons. This is the case in the typical example shown.

Ideally the boilerhouse should be located centrally to the areas it serves, to reduce the extent of steam and condensate distribution main pipework.

23 Gas generation and storage

Gases are used on power stations as essential components of the electrical power generation system and also as part of safety and protection systems. Some-

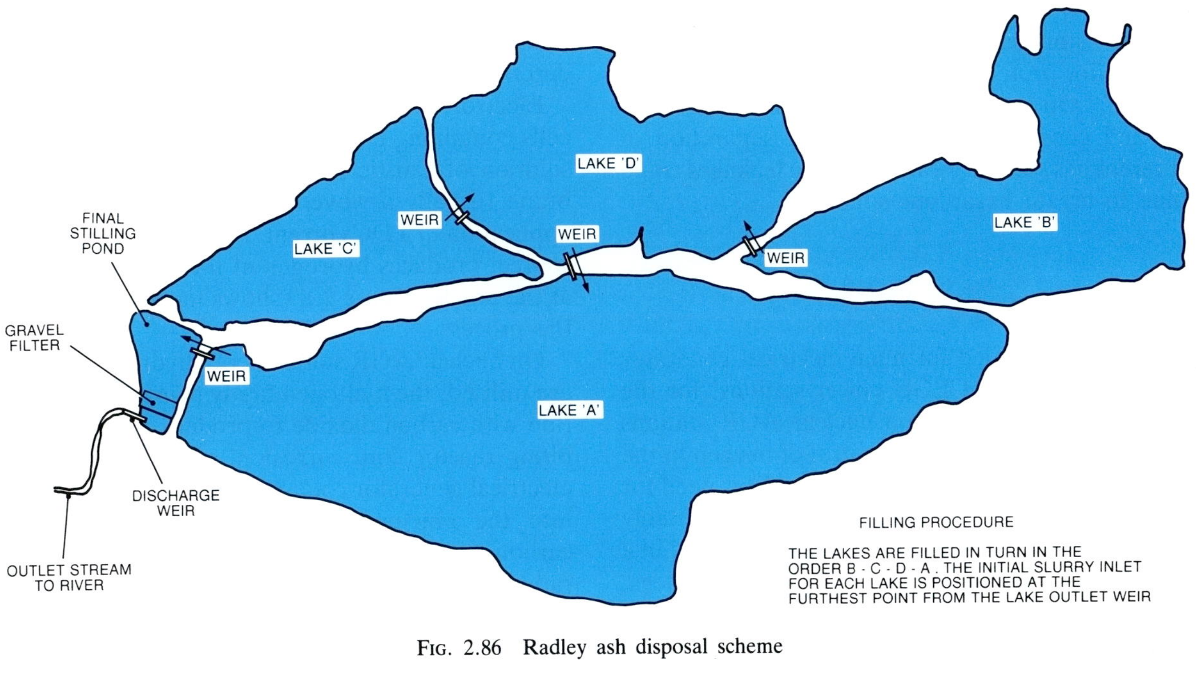

FIG. 2.86 Radley ash disposal scheme

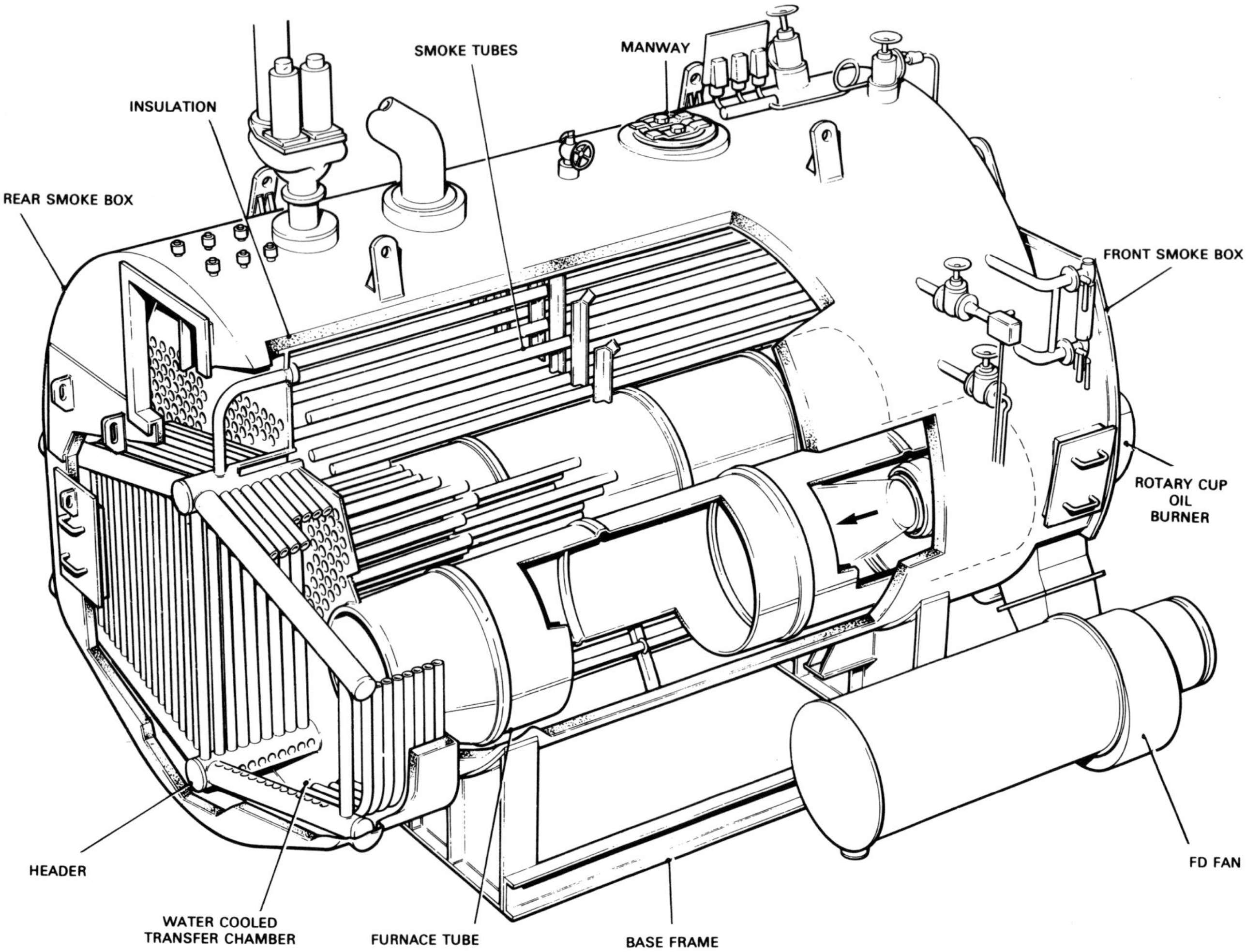

FIG. 2.87 Typical oil-fired auxiliary boiler

times they are generated on-site and sometimes they are imported in cylinders and vessels.

In all cases of on-site gas generation and storage the location of storage tanks, pipelines, road delivery access and fire protection measures are of paramount importance. Open-to-atmosphere environments are chosen and consideration given to the separation of the different gas storage locations where leakages could combine to create hazardous situations.

23.1 Hydrogen

Hydrogen gas is needed for main electrical generator cooling on all modern CEGB power stations, for the production of methane gas on nuclear AGR stations and, amongst other things, for control of oxygen in the reactor coolant on nuclear PWR stations. This need for hydrogen gas is satisfied by the provision of a suitably sized on-site hydrogen gas generator plant, housed in a simple building consisting basically of a roof on legs, i.e., a *dutch barn* type of construction. Depending on the site, local decisions may require the addition of fully-louvred walls to the building, thus affording plant and personnel protection in exposed locations.

Two types of gas generator plant are currently utilised, electrolytic and methanol-cracking and are described as follows:

Electrolytic type gas generators consist of a series of cells containing electrodes suspended in an electrolyte solution of caustic potash, and separated by a membrane (usually a woven type of asbestos mesh). The application of a DC current from a transformer/rectifier source, produces hydrogen at the cathode and oxygen at the anode. Figure 2.89 shows the general features of the process.

On nuclear AGR stations both hydrogen and oxygen are utilised, the hydrogen firstly being used in conjunction with carbon dioxide to produce methane for inhibiting reactor core corrosion, and secondly for main electrical generator cooling. The oxygen is recombined into the reactor cooling circuit to re-establish the carbon dioxide from carbon monoxide produced by radiolysis.

On nuclear PWR and conventional stations, the hydrogen only is utilised, the former requiring gas for

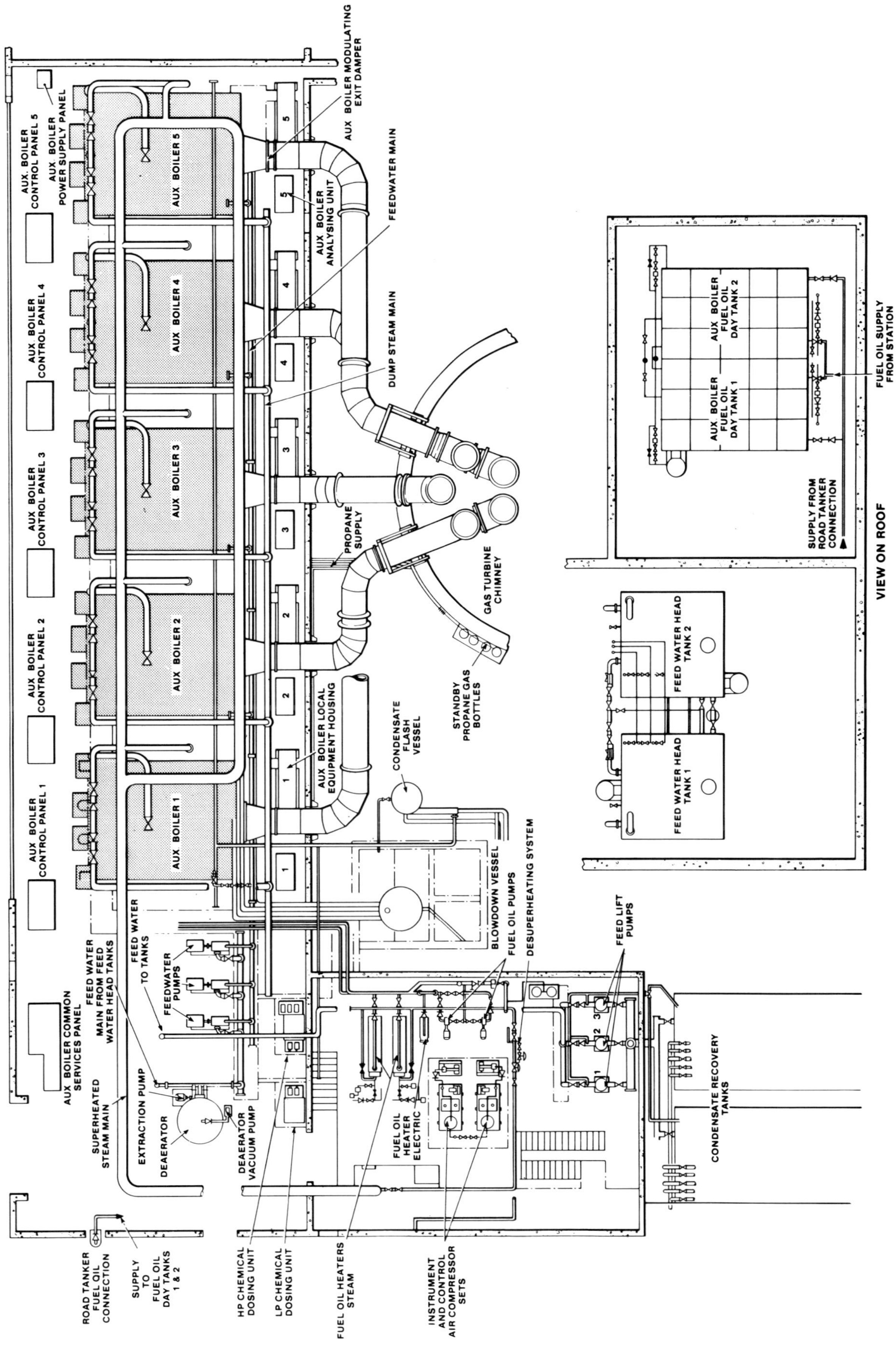

FIG. 2.88 Typical auxiliary boiler house

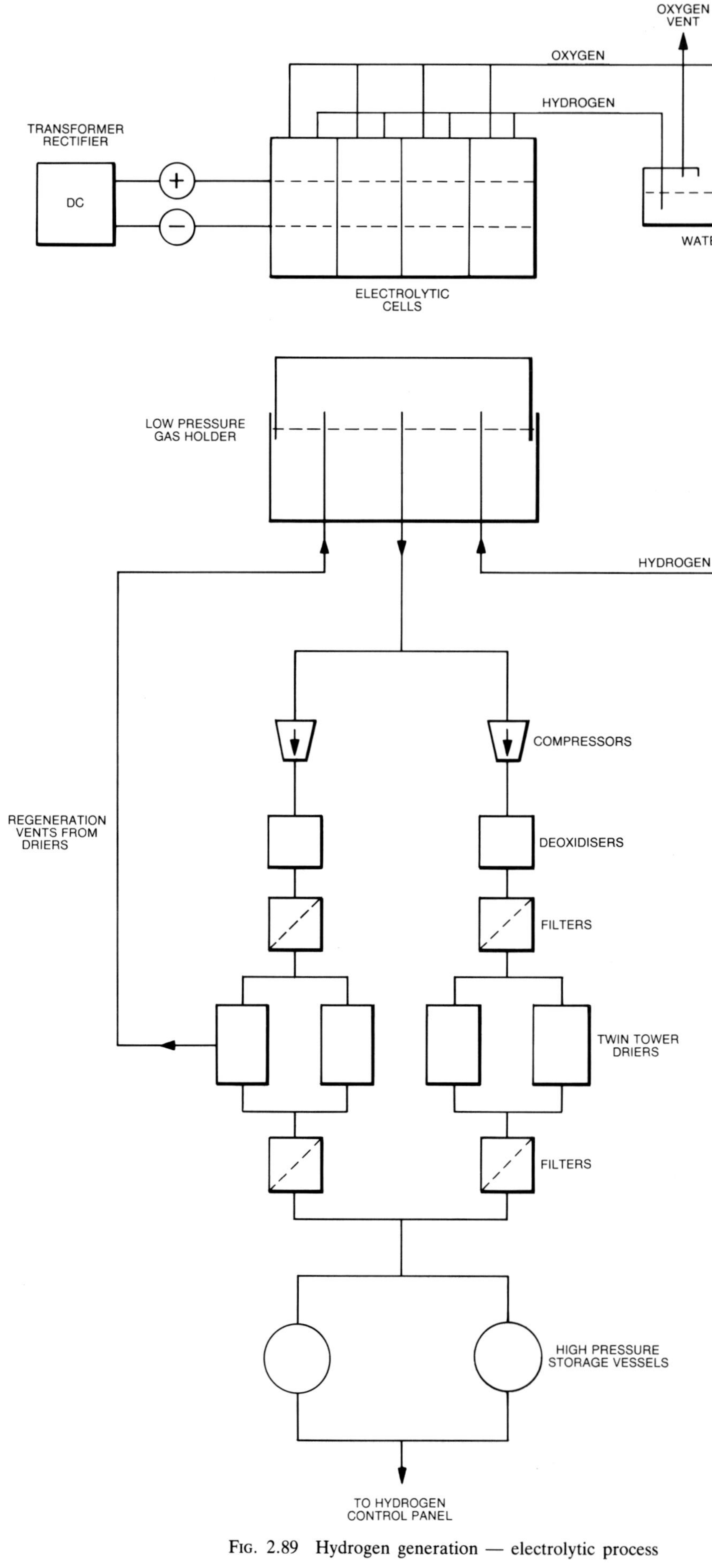

FIG. 2.89 Hydrogen generation — electrolytic process

oxygen control, etc., and main electrical generator cooling, the latter for generator cooling only.

Hydrogen gas produced by electrolysis is wet and contaminated, and therefore is subjected to filtration, deoxidation and drying stages before being stored at high pressure.

Methanol type gas generators consist of modules containing vapouriser, catalyser and diffuser sections. A methanol liquid feedstock is fed into the generator module and is vapourised before being introduced to the catalyser which breaks the methanol vapour into two main components, hydrogen and carbon dioxide. The resulting gas mixture passes through a diffuser section which allows hydrogen gas to pass, but rejects carbon dioxide and other impurities. Figure 2.90 shows the general features of the process.

Hydrogen gas produced by methanol-cracking is extremely pure and dry and in the normal course of events does not require any further treatment before being stored at high pressure.

The methanol type system does require methanol feedstock and this must be stored adjacent to the hydrogen generation plant in suitable storage vessels.

In both systems described, the high pressure gas storage facility is served by compressors from the gas generators, and can take the form of large vessels or smaller cylinders, positioned adjacent to the generator plant and sized to accommodate two main electrical generator charges.

From the storage facility the gas passes through a control panel which regulates pressure and flow in two distribution pipelines to the electrical generator cooling systems. One pipeline allows flow of normal make-up gas, the other allows recharging of a generator in a short period of time.

The whole hydrogen plant is located in a suitable open environment position on the site. Risk factors taken into account are those presented by both the methanol store and hydrogen store as well as the need to conform to the guidance given in documents produced by the Health and Safety Executive (HSE) and the gas producing industry in terms of separation distances, hazardous zoning and general safety matters. Road facilities are provided for tanker access and fire fighting appliances.

23.2 Carbon dioxide

Carbon dioxide (CO_2) is used as a purge gas for main electrical generator purging in both nuclear and con-

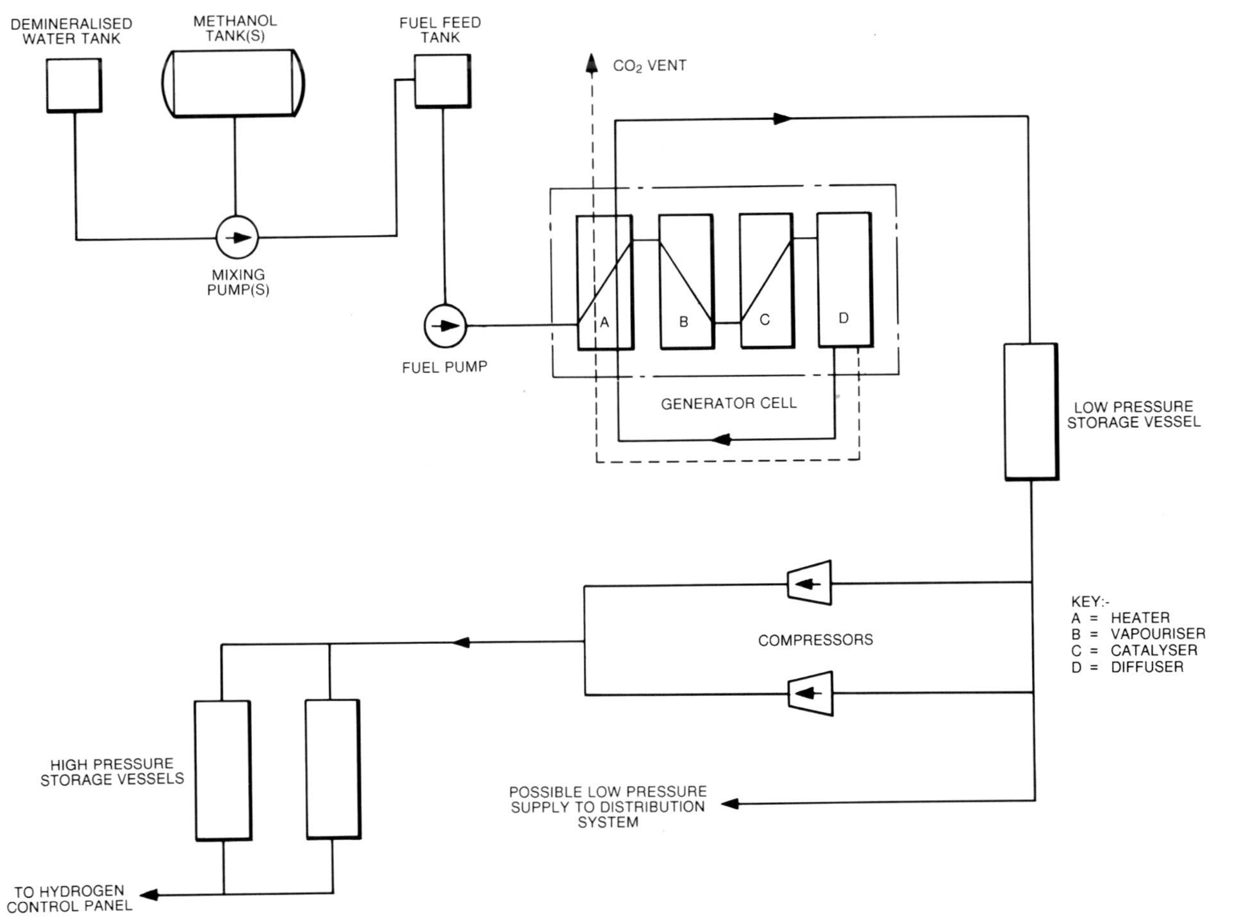

FIG. 2.90 Hydrogen generation — methanol process

ventional stations, and for the reactor cooling circuit and various purging requirements in nuclear AGR stations.

Storage is normally in bulk liquid form, large quantities being required for the nuclear stations usually in the region of six 100 t vessels, whilst a typical fossil-fired station requires far less — one 6 t vessel on average. The storage facility is normally made up of thermally-insulated vessels, each provided with a refrigeration unit to maintain the liquid state. Vapourising equipment is provided to enhance the gas flow rate at times of high demand. In both nuclear and conventional stations the refrigeration units are located adjacent to each storage vessel with condensing coils mounted within the vessel vapour space. The vapourising equipment is a common installation and serves all vessels, being located in the discharge line from the storage facility. The distribution pipework system is routed throughout the station to points of usage.

The whole plant arrangement is contained within an open environment adjacent to the main buildings together with the necessary road tanker access and unloading facilities.

23.3 Nitrogen

Nitrogen (N_2) is used as a purge gas and sometimes for preservation purposes in steam and water spaces, although the latter examples are not common at the present time.

In nuclear AGR stations it is used for the secondary shutdown system, this being the need for an immediate supply of nitrogen gas in the event of a demand for nitrogen injection into a reactor.

In nuclear PWR stations it is used for back-up, purge and cover gas systems, the former being a means of operating say, essential valves in the event of loss of initial means of actuation. Cover gas relates to a blanketing function within tanks or vessels to avoid oxygen pick-up.

In conventional coal-fired stations nitrogen is used for mill loading, i.e., pressurising the grinding facility rams on the mills.

The storage facility normally takes the form of a vacuum insulated liquid nitrogen storage vessel complete with road tanker delivery connections, followed by a liquid nitrogen pump discharging through a vapourising section into a nitrogen gas store. The store is made up of large cylinders which discharge through the distribution pipework system to points of usage. The whole storage arrangement is contained within an open environment adjacent to the main buildings and provided with the necessary road tanker access and unloading facilities.

23.4 Miscellaneous gases

Oxygen (O_2)

The reactor coolant gas circuit on nuclear AGR stations is continuously producing carbon monoxide by radiolysis and oxygen is required to oxidise this back to carbon dioxide.

On AGR stations, oxygen is produced by means of the electrolytic hydrogen generation plant previously described, and stored by means of compression into cylinders. From the store, oxygen is discharged via a pipework system to a recombination unit in the reactor cooling circuit.

The storage arrangements, including the oxygen compression facility, are normally in an open environment adjacent to the hydrogen generation plant. Road access is provided for unloading facilities.

Propane (C_3H_8)

Propane is used for burner ignition purposes on both main and auxiliary boiler installations.

In nuclear stations, due to the relatively small demand from auxiliary boilers only, storage of propane may be in the form of cylinders arranged in banks. They are collectively connected up to manifolds discharging via a pressure control facility and distribution pipework system to points of usage.

In conventional stations the demand for propane gas is created by both main and auxiliary boilers for burner ignition purposes. Storage takes the form of large vessels containing liquid propane which are supplied by road tanker delivery. If necessary, vapourisers may be fitted to each vessel to enhance gas flow at times of increased demand. The vapourisers are sited adjacent to the storage vessels, taking their supply from the liquid space and discharging into the gas space. Propane gas taken from the storage vessels passes through a pressure control facility before being distributed to points of usage through a pipework system.

The location of the propane store is chosen carefully. Due to the nature of propane, consideration must be given to hazardous zones, separation distances, etc., as laid down in HSE guidance documentation. Road facilities are provided for tanker access and fire fighting appliances.

Methane (CH_4)

Methane gas is used in nuclear AGR stations for injection into the reactor cooling gas circuit to inhibit corrosion of the graphite core. It is produced in an on-site methane gas generator plant served by supplies of hydrogen and carbon dioxide, the former being produced by means of the electrolytic hydrogen generation plant described previously, the latter being supplied from the carbon dioxide system described in Section 23.2 of this chapter.

The two gases are combined in a mixing facility before being introduced into the catalyser section of the methanation plant. The catalyst reacts with the hydrogen/carbon dioxide gas mixture to produce methane gas, which is passed on via a cooler to a compressor. The compressed methane gas flows through drying equipment before being stored in cylinders.

Some power stations do not operate an on-site methane generator plant, but prefer to hold methane gas in cylinder stores. In these cases, delivery of the gas is by road trailer and the storage facility is provided with the necessary trailer access and means of unloading. From the storage cylinders, the gas is passed through pressure control equipment before being distributed through a pipework system to the point of usage. The storage area is located in an open environment and care must be taken to ensure separation from sources of oxidants, for example, oxygen stores.

There are other gases which are used in relatively small quantities (e.g., for laboratory purposes) which also require assessment of storage and safety requirements, some of these are:

Argon	(Ar)
Ethane	(C_2H_6)
Ethylene	(C_2H_4)
Helium	(He)
Nitric oxide	(NO)
Nitrous oxide	(N_2O)
Sulphur dioxide	(SO_2)
Sulphur hexafluoride	(SF_6)
Oxy-acetylene	(O_2/C_2H_2)

These gases are normally stored in cylinders and may be portable.

24 Pumped storage plant

The plant required for pumped storage power stations is in principle much simpler than on more conventional power stations because there is no prime requirement for boiler and fuel handling plant. However, because of the potential for providing system reserve and the need to maintain sufficient suction head during the pumping mode, other features and design methods are required which are not found on conventional stations. The principal features affecting the design and layout of the station plant are:

- Station rating.
- Operating response on start-up.
- Duty cycle (daily or weekly).
- Plant operating regime.
- Cost.
- Environmental impact.
- High availability and reliability.
- Flooding hazard.
- Fire hazard.

The last two items in the list are more important if the main plant is to be located underground. Flooding could be avoided by either complex isolation arrangements such as bulkhead doors, or by qualifying the integrity of the hydraulic system. Combating the fire hazard must concentrate on the arrangements for smoke venting, particularly from PVC cable fires, and on the safe means of escape for personnel.

For the purposes of this description it is assumed that the plant is located underground. In the case of plant located at ground level, the plant items required will be the same but the layout task is eased.

In designing the main plant layout some contingency must be made to cover the possibility of encountering severe faulting in the rock. For this reason the initial design will try to use the minimum possible excavation which has an impact on arrangements for initial construction and subsequent maintenance of the plant. A balance is required which aims to minimise the construction programme.

One strategy for minimising the risk of the project if poor ground is encountered is to use a number of smaller chambers rather than one large chamber. During the development of the layout, consideration will be given to the most economical arrangement of multiple caverns to give flexibility of construction, so that if for any reason work is held up on one excavation, work can continue on the others. This can be achieved by arranging multiple access routes to the main excavations. This strategy was adopted with great success on the Dinorwig project; Fig 2.91 shows the arrangement of tunnels and chambers in more detail. The underground complex consists of the machine hall, main inlet valve (MIV) gallery, draft tube valve (DTV) gallery and the transformer hall. The 400 kV transmission substation is located on the upper floor of the transformer hall. Three major galleries, one for each pair of generator-motors, connect the machine hall with the transformer hall, and accommodate the generator-motor busbars and 11 kV electrical switchgear (see Fig 2.92). Within the machine hall there is a 9-storey building accommodating control equipment, electrical plant and welfare facilities. At each end of the DTV gallery there is a chamber housing starting equipment for the generator-motors. Above the main complex a system of high level tunnels provides ventilation.

On discharge from the turbines, the water flows through the DTVs which isolate the pump-turbines when dewatering. The conduit between the DTV and the pump-turbine is the lowest part of the system and the arrangements for dewatering and drainage sumps are located at this point.

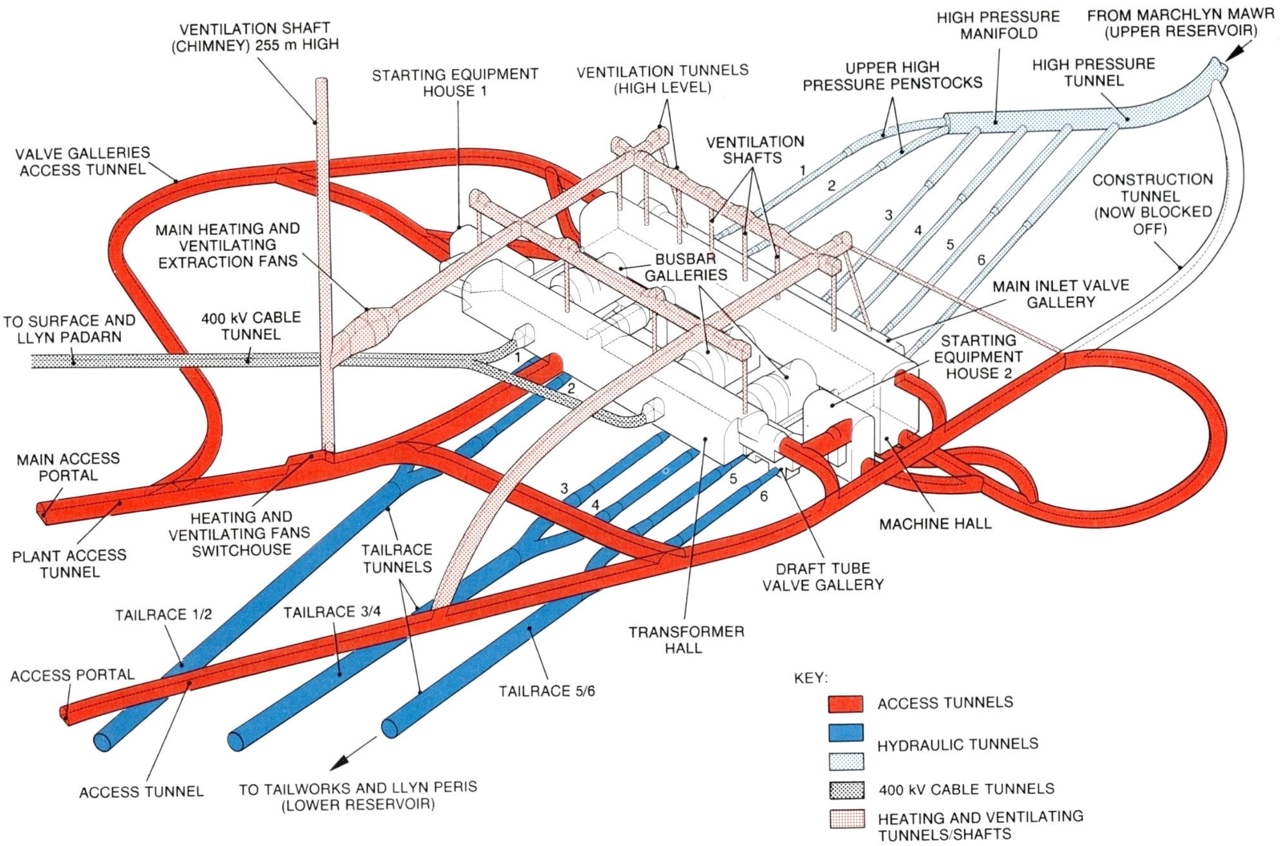

FIG. 2.91 Dinorwig station tunnels complex

24.1 Hydraulic machines

Pumped storage requires the installation of two hydraulic machines, a pump and a turbine. The two functions can be either separate, as at Ffestiniog, or combined into a single reversible machine called a pump-turbine as has been used at Dinorwig. Figures 2.93 and 2.94 show the two arrangements. For reasons of economy the highest practicable running speed is necessary. Large hydro-electric plant is limited in running speed by the ability of the generator to withstand the maximum runaway speed of the turbine. For reversible pump-turbines the ratio of runaway speed to normal running speed is about 1:4 whereas for turbines the normal speed selected can be higher than for a turbine of the same output and head. However this higher speed requires a deeper submergence when operating in the pumping mode to avoid cavitation. In the case of underground power stations the economic penalty for this greater submergence is not very great since, apart from the small extra length of access tunnels and penstocks, the total excavation is unchanged. Penalty in the civil works is more than compensated for by the lower machine costs. Design considerations for the generator-motor also influence the speed selected. At Dinorwig a speed of 500 r/min was selected with a minimum submergence of 60 m at the runner centreline.

The pump-turbine structure is subjected to very large hydraulic forces and partial or complete embedment in concrete is sensible to physically restrain movement and to keep noise levels to acceptable levels.

24.2 Generator-motors

The generator-motors can either be reversible as at Dinorwig, or have a single operating direction as at Ffestiniog. The latter is only possible when the hydraulic machine modes of pump and turbine are realised in separate components. The factors affecting the choice of generator-motor are:

- Unit size and speed.
- Cost.
- Cooling and reliability.
- Physical size.

The specific speed of the pump-turbines dictates the generator-motor speed and many combinations are

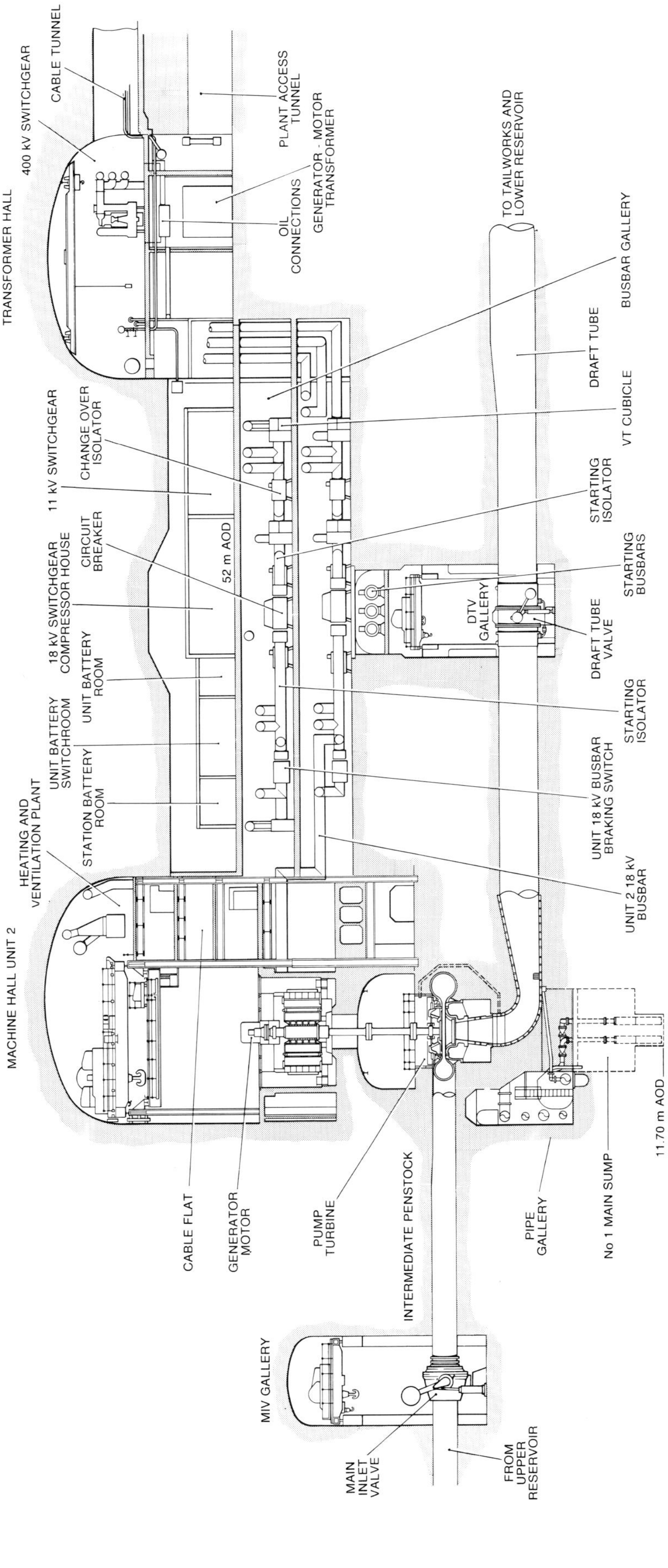

FIG. 2.92 Section through Dinorwig main plant

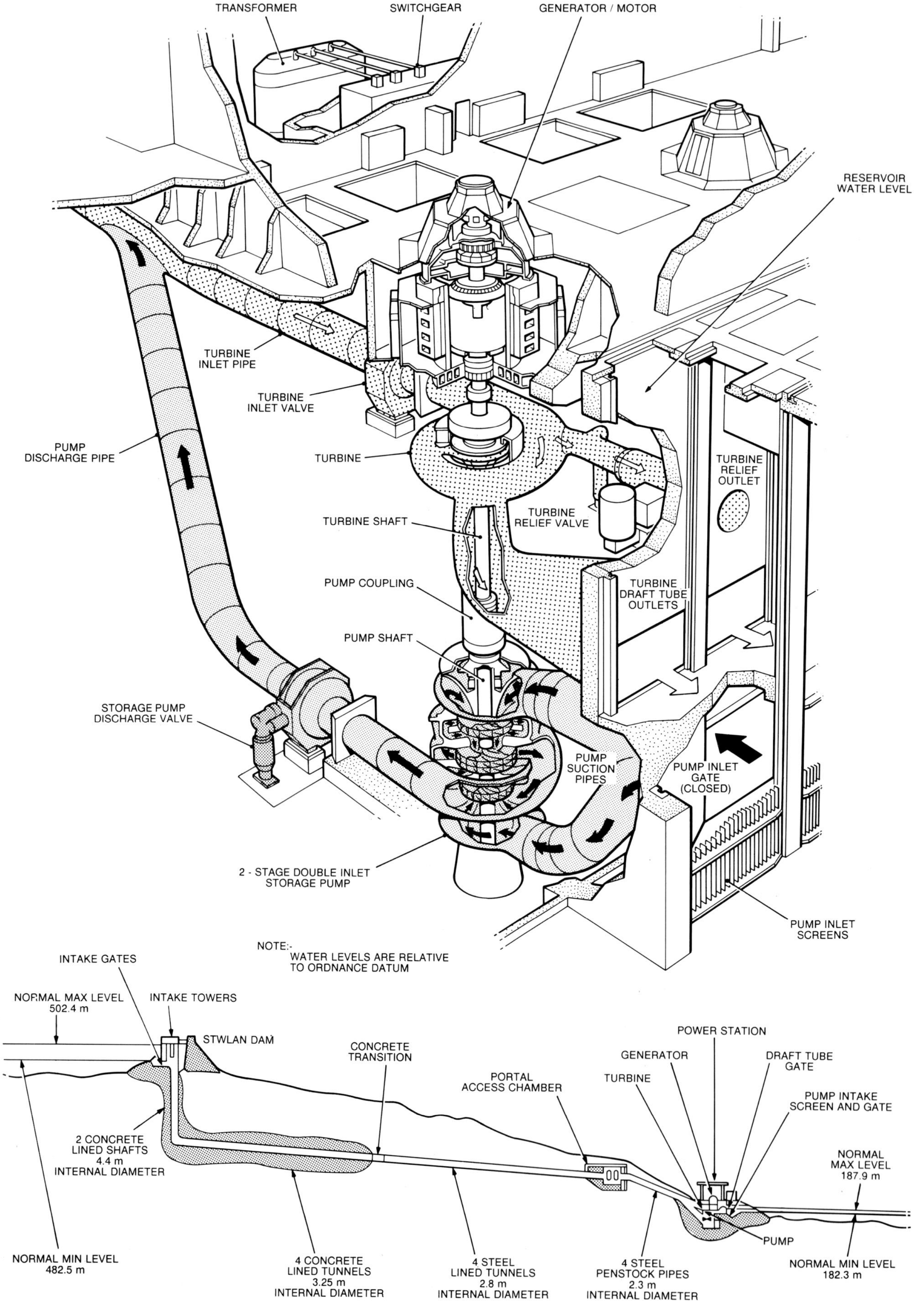

FIG. 2.93 Ffestiniog hydraulic machines

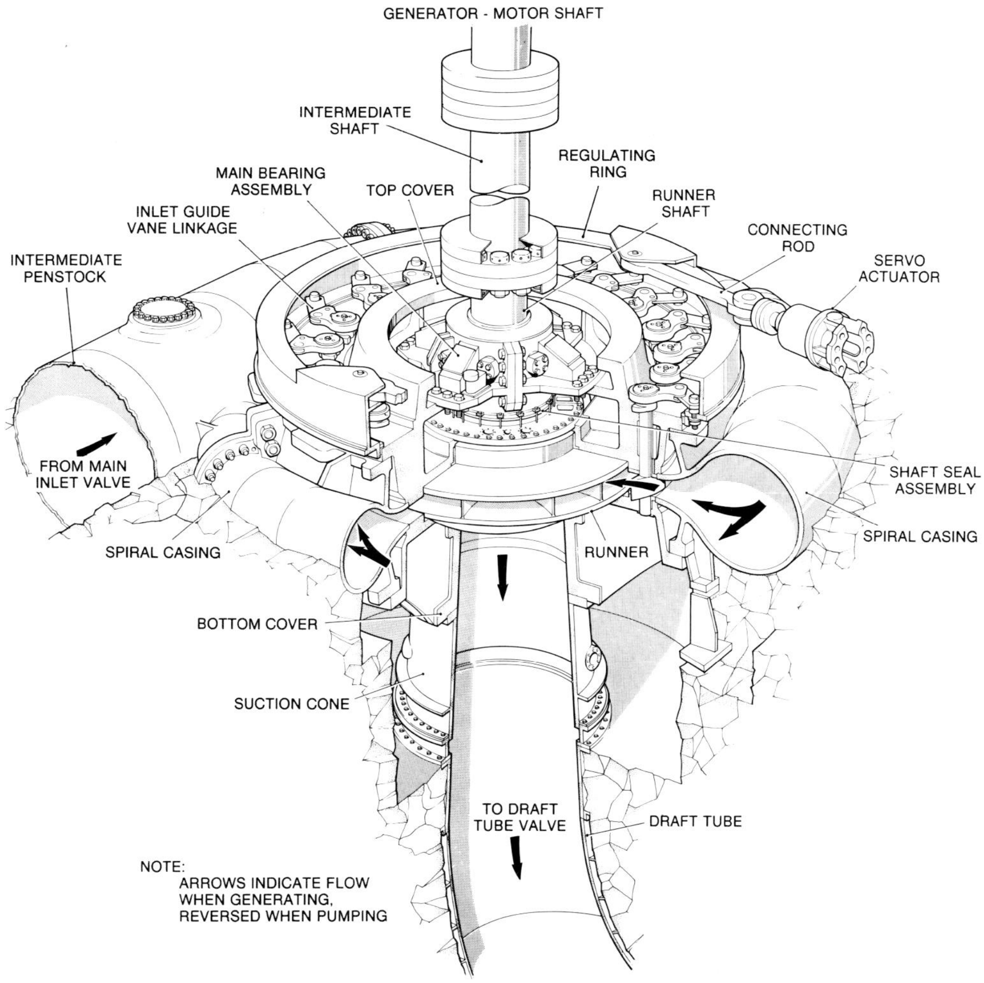

FIG. 2.94 Dinorwig reversible pump-turbine

possible. The choice is usually made on economic grounds without resorting to large extrapolation from existing practice.

Costs usually favour a small number of large machines. This is because the size of the caverns increases greatly with many small machines, whereas the plant costs stay reasonably constant because the optimum speed of operation decreases with unit size, leading to more expensive electrical machines.

Generator-motors may be cooled either by air or water. Air cooling is currently just feasible up to 450 MW unit size but this requires the extrapolation of all the critical design features of the cooling system. Water cooling would offer a more secure engineering design but the reliability of the water-cooled machine may not be high enough. At Dinorwig, air cooling was used based on the following reservations about water cooling:

- CEGB experience of water-cooled stators.
- No international experience with reversible water-cooled sets.
- Existing and proposed designs for water-cooled sets did not include high air pressures within the machine; water leaks would be critical and cause damage to the stator windings.
- The large number of load cycles and stress cycles specified for the project would give rise to accelerated thermal fatigue problems on water cooling pipes.
- There was no experience of reversible pump-turbines at outputs as great as 450 MW with a high head of 500 m.
- There was no point in reducing the size of the rotor by employing water cooling if additional inertia had to be built-in to maintain the required inertia constant.
- The 300 MW and 225 MW machines were within acceptable parameters for air cooling.
- The lack of experience of any set running at 600 r/min above 135 MW made the 500 r/min machine preferable.

- Experience with fully water-cooled hydro-machines was limited to a few single-rotation units with minimal numbers of cycles per year.

On the basis of the arguments detailed above, Dinorwig was built with six sets of 300 MW at 500 r/min.

24.3 Main inlet valves

The main inlet valves provide the primary means by which the pump-turbine may be isolated from the high pressure penstocks. If the station is to perform an instantaneous reserve duty then these valves will be required to open much more quickly than would normally be required for large valves (greater than 2 m diameter). Additionally they must be capable of closing against maximum possible flow in either direction. Some operation of the MIV will be required during all mode changes and this means that the fatigue life needs to be assessed.

The required reliability of the MIV will depend on the hydraulic system design for the station. Stations with more than one high pressure tunnel will require lower reliability than where there is a single tunnel. Maintenance access will only be allowed when the HP system is dewatered. At Dinorwig dewatering may be postponed indefinitely by arranging for the tunnel inspections to be done using submersibles.

The design of the Dinorwig main inlet valve is shown in Fig 2.95 and the important design features are:

- Rotary valve giving full-throat low headloss in operation.
- Design qualified by fracture mechanics and photo-elastic stress analysis and strain gauging of the prototype during final pressure test.
- Two seals which are jacked into position when closed.
- Mechanical latches to lock the valve in the open or closed position.
- 4 second opening, 30 second closing.

24.4 Draft tube valves

The primary function of the draft tube valve is to isolate the pump-turbine from the lower reservoir to enable the draft tube and pump-turbine to be drained down for maintenance. The DTVs are therefore normally in the

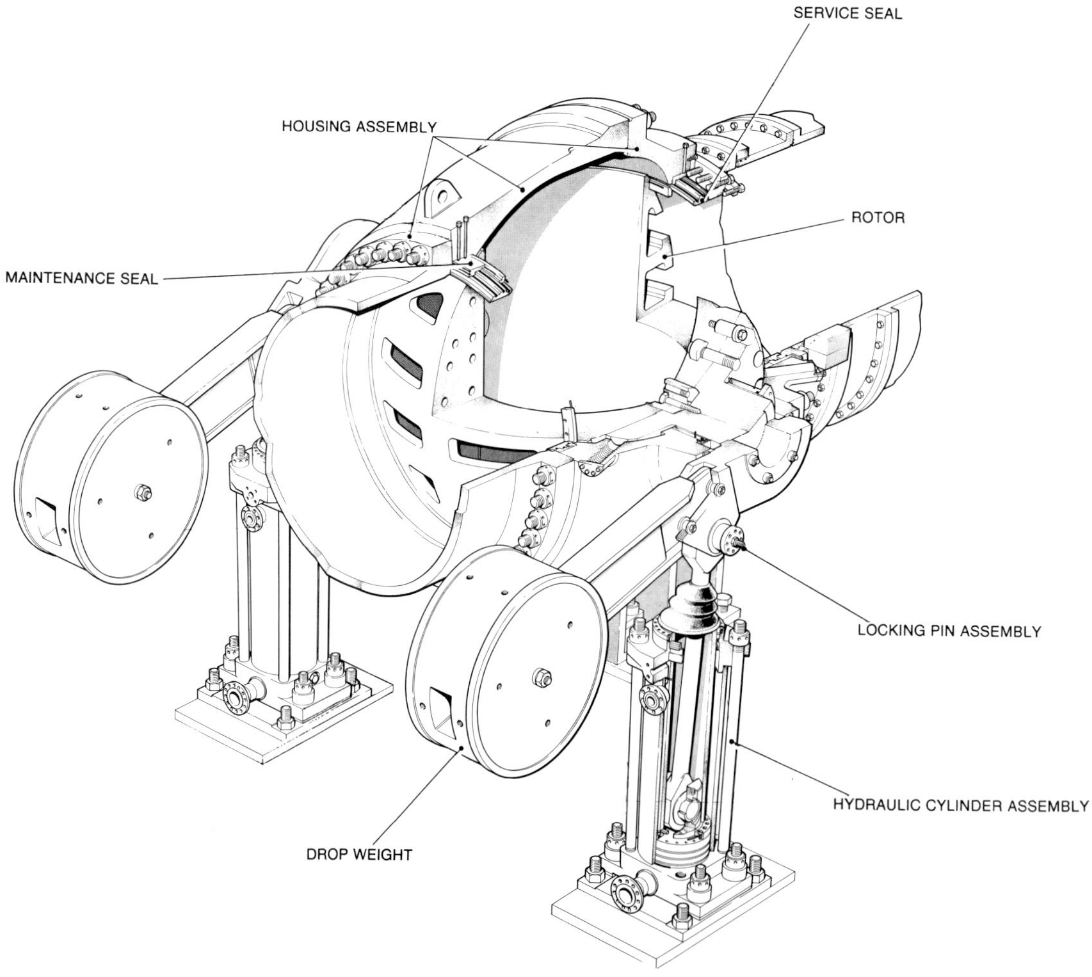

Fig. 2.95 Main inlet valve

open position and not part of the automatic sequence of operation. Additionally they are required to be capable of closing in the event of a major burst to prevent the station flooding. When the station is built at the surface the function of the DTV may be achieved using bulk-head gates. In this case the flooding risk is much reduced.

The DTVs are also required to act as a safety valve to prevent the pressurisation of the draft tube from the upper reservoir. The DTVs are fitted with bypass connections for filling the draft tube, pump-turbines and intermediate penstock so that the valve should normally open in a balanced pressure condition. The Dinorwig DTV is shown in Fig 2.96. It is 3.75 m diameter, weights 80 t and is designed for 60 bar water pressure.

24.5 Gates

Isolating gates are required at both the upper and lower reservoirs. Double isolation is usually provided with one gate capable of operating against a flow whilst the other only operates in still water. Fixed roller gates are preferred for the main gates with sealing faces away from the relevant reservoir. The function of the gates is to isolate the reservoir from the hydraulic system to enable maintenance work to be carried out on the tunnels and turbine isolating valves without draining the reservoirs. They also allow the controlled filling of the hydraulic system following dewatering. Figure 2.97 shows the headworks gate installation.

Additionally the head gates at the upper reservoir provide:

- Initial impounding of the upper reservoir.
- Closure to hold the contents of the upper reservoir in the event of hydraulic system failure and a turbine trip or runaway.

The secondary stop gates are provided to give safety back-up to the main gates when the tunnels are drained and maintenance work is in progress.

At Dinorwig, the gates do not perform an emergency duty. In the event of a gross fracture in the station, whilst closure of the main head gate would retain the contents of the upper reservoir, the tunnels and surge chamber volumes greatly exceed the volume of the

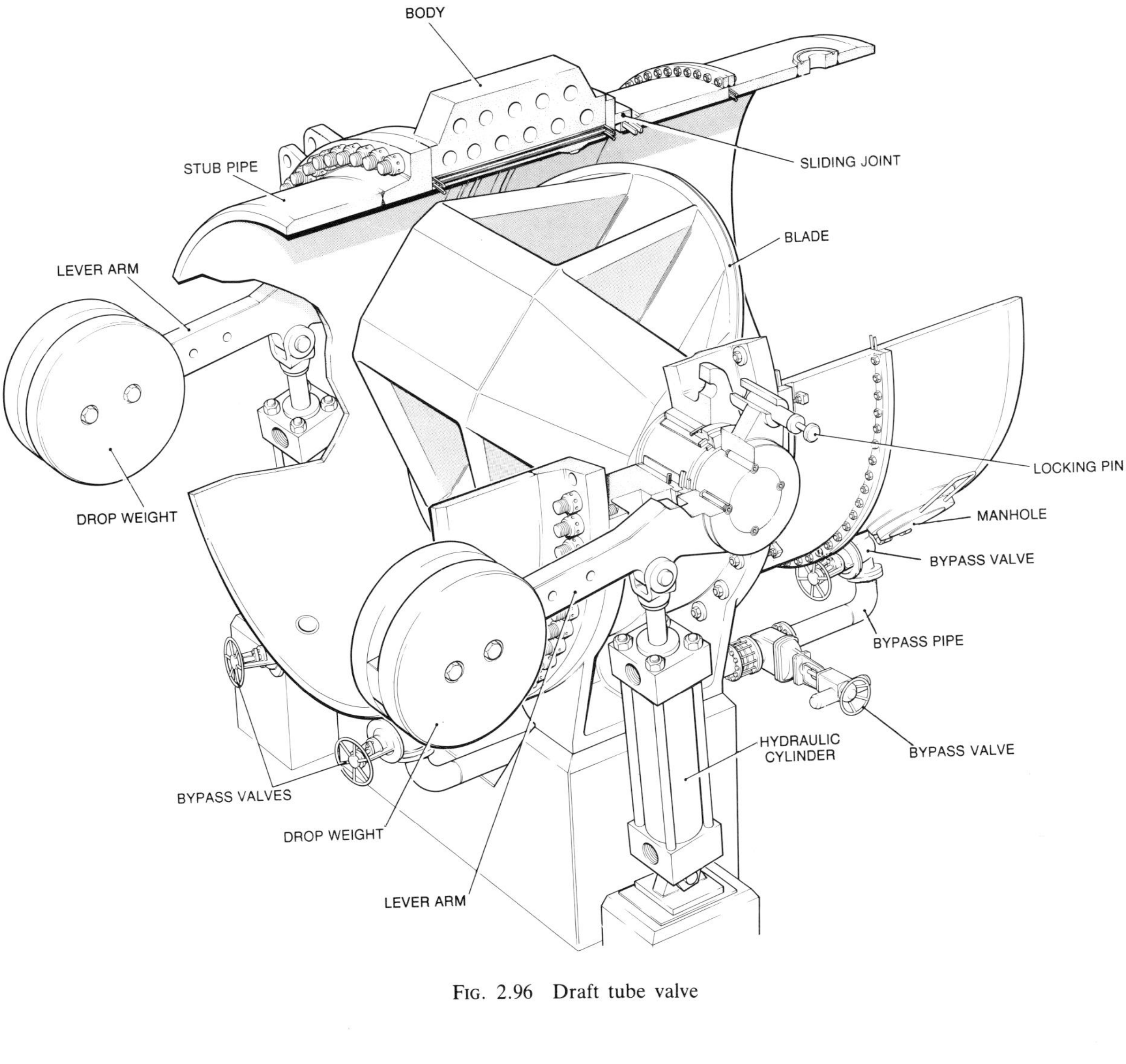

FIG. 2.96 Draft tube valve

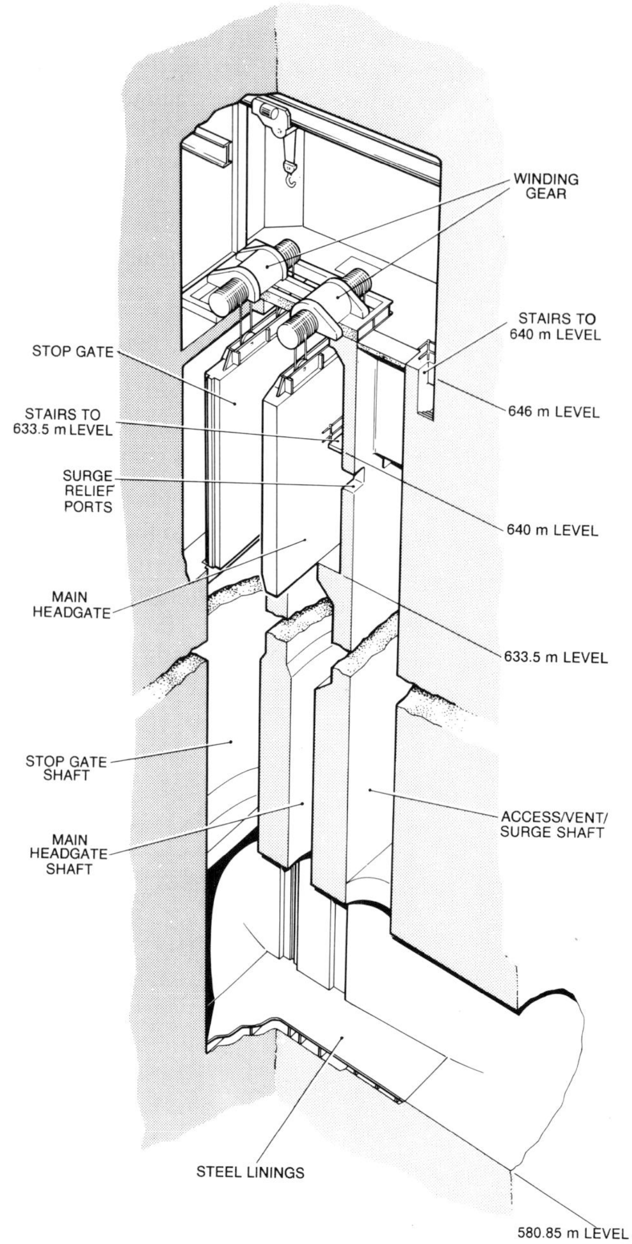

FIG. 2.97 Headworks gates installation

station and flooding of the station could not be prevented. Closure of the head gates takes 30 minutes.

24.6 High integrity pipework

When the hydraulic system is underground, much of it can be concrete-lined rock tunnel. In the sections near the machines, steel linings are appropriate to limit leakage but the majority of the strength of the pressure containment will be derived from the rock. In fact, the steel membrane in the steel-lined rock sections will be designed mainly to withstand a collapse criterion when the tunnels are dewatered.

Within the station, the intermediate penstocks (between the main inlet valves and the pump-turbines, see Fig 2.90) must be capable of withstanding all the mode changes through station life without failing in fatigue. This requires very special design assessment to determine the defect tolerance. Major failure of these components would lead to rapid flooding of the station galleries while the main inlet valve and draft tube valves are closing, unless complicated bulkhead arrangements are incorporated into the design to isolate parts of the station. The choice of materials is very important in guaranteeing the integrity of the critical components.

Over the last 20 years there has been an increasing use of the high yield strength quenched and tempered (QT) steels in the fabrication of hydro-electric penstocks. When designs can be based on the simple *internal pressure times safety factor* approach, considerable reductions are made in the section sizes required. This results in economic savings due to reduced weld volumes and transported weight and in many cases the avoidance of the need for post-weld heat treatment.

The more traditional material for this application would be plain carbon manganese steel plate manufactured by conventional steelmaking routes and fabricated using established welding processes. The major disadvantages with this choice is the transport problem involved with moving the thicker sectioned penstocks and the volume of in-situ welding required.

At Dinorwig, after considering the operational and safety requirements, the decision was made to abandon the usual approach used in penstock design of applying a safety factor to the maximum internal pressure. Instead, a fitness for purpose philosophy was adopted based upon state-of-the-art fracture mechanics. This philosophy recognised the potential for crack-like defects appearing in the penstock material and evaluated the contending materials against their ability to tolerate such defects under the extremes of the loading conditions expected. This appeared to be the first time such a philosophy had been used in penstock design and it required a considerable amount of experimental and analytical work.

The basic philosophy is now commonly used and consists of the following steps:

- Assume the presence of a pre-existing crack in a weld with dimensions immediately below that which can be sized by ultrasonic non-destructive testing (NTD) techniques.
- Predict the amount of sub-critical crack growth that would be expected by the fatigue loading over the full design life of the station.
- Demonstrate that at the end of life, the crack will still be tolerated in the structure even under the most onerous loading condition imaginable.

For such a set of analyses, a pessimistic approach is adopted to ensure that failure is avoided by adopting upper-bound crack sizes with lower-bound material properties. This also requires the worst case crack locations to be considered and includes consideration of stress concentration effects and such like.

25 Gas turbine plant

25.1 Introduction

In a gas turbine, the continuous combustion process of the fuel is harnessed to drive a turbine shaft, which is in turn coupled to an electrical generator. The overall simplicity of the fuel system and its associated thermodynamic cycle has several advantages for power generation. There is no requirement for a steam cycle, thus keeping down the capital cost of the plant. Correspondingly no condenser cooling is required, enabling a greater freedom of plant siting to be exercised. In addition, gas turbines can be quickly started and brought on load, automatically or under remote control. However, against such advantages, comparatively high fuel and running costs have traditionally restricted gas turbines to low load factor operation on CEGB stations.

25.2 Operational requirements

Gas turbines are currently operated on the CEGB system to satisfy both auxiliary power generation and peak load generation requirements. At several of the major conventional and nuclear power stations emergency or auxiliary gas turbines based on aero-engine technology have been installed. These units serve two requirements. Firstly, on conventional stations they are able to contribute to the overall output of the main station during periods of high demand. Secondly, they play an important role in maintaining the system frequency and voltage and thus adding to the overall security of the system. During certain operational situations the outputs from major stations are progressively reduced through the lowered output of their frequency-conscious auxiliaries. Ultimately if this situation is allowed to continue, cascade tripping of the stations can occur.

The gas turbine's capability of starting and quickly running up to full load permits it to be synchronised to the 11 kV unit board, even when the system frequency is falling. If the system frequency continues to fall, the unit board can be isolated from the rest of the system, the gas turbine can then supply the auxiliary plant at normal frequency and so keep the auxiliaries at full speed and maintain full output from the main unit.

The major role of the CEGB's six gas turbine stations is to satisfy the peak load requirements placed upon the system. Their output is important in controlling and stabilising system frequency and in meeting sharp or unexpected rises in consumer demand. The gas turbines installed at these peak load stations have the turbines coupled to the generator via a clutch in accordance with a CEGB need for power factor compensation under certain conditions of system operation. The clutch automatically engages the turbine when generation is required and disengages it from the generator when reactive compensation operation is selected, i.e., the generator may be used for synchronous compensation when required.

25.3 Aero-engine-derivative gas turbines

In the emergency gas turbines installed at major CEGB power stations, the supply of hot gas under pressure is provided by an aircraft jet engine. At the time that they were installed there were two British engines of suitable power available:

- The Avon, capable of producing sufficient gas for 14.5 MW of power.

- The Olympus, capable of producing sufficient gas for 17.5 MW of power.

Significant development has taken place in aero-engine technology since the initial machines were installed. Rolls Royce has expanded its range of aero-engine-based gas turbine units and machines are currently available with a peak output of up to 70 MW for future consideration. The current range uses the Spey, Avon, Olympus C and RB211 engines.

These gas generators supply the hot gas under pressure to a mechanically separate power turbine running at synchronous speed. Figure 2.98 shows the layout of a typical aero-engine-based, peak-load set.

Aero-engine-derivative gas turbines have very quick start times and can be brought up to full load in two to three minutes. Another advantage of such machines is *repair by replacement* of the gas generator units. The relatively light weight of the gas generator (approximately 2 tonnes) is also an advantage in this context and the set can be returned to operation with refurbishment of the gas generator at central stores or works.

25.4 Industrial gas turbines

During the 1970s it was decided to install industrial gas turbines. The incentive for doing so was a practical determination of the comparative economics of these and aero-engine-based units. At the time, the available evidence favoured the industrial unit for reasons of longer design life and the absence of any weight limitations. Also the design of industrial gas turbines produced a machine that was more robust than aero-engine-derivative plant and suitable for extended periods of generation. There was also the possibility that greater cost savings with scale would be shown by the industrial over the aero-engine-based machine, where improved output is obtained through a multiplicity of gas generators. Industrial units also have a greater tolerance of lower grade fuel, thus giving the possibility of reduced operating costs.

Two prototype GEC sets have been installed using EM610 machines at Leicester gas turbine power station. Each unit has an output of 52 MW. Industrial gas turbine units of up to about 130 MW are currently available from the major manufacturers for future consideration. Figure 2.99 shows the arrangement of the Leicester industrial gas turbine set.

Normal start times for industrial units are in the order of 20 to 30 minutes from first rotation to full load.

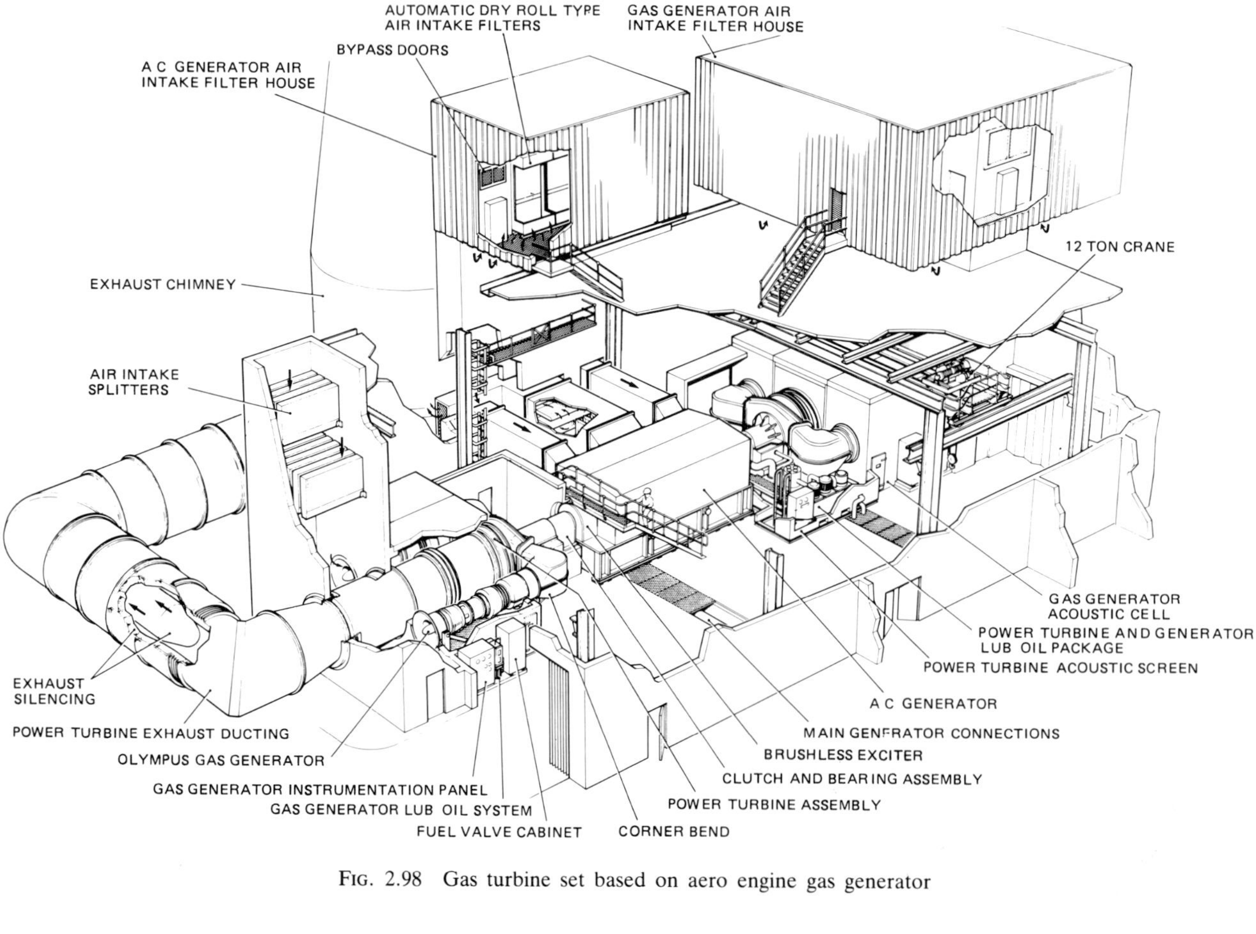

FIG. 2.98 Gas turbine set based on aero engine gas generator

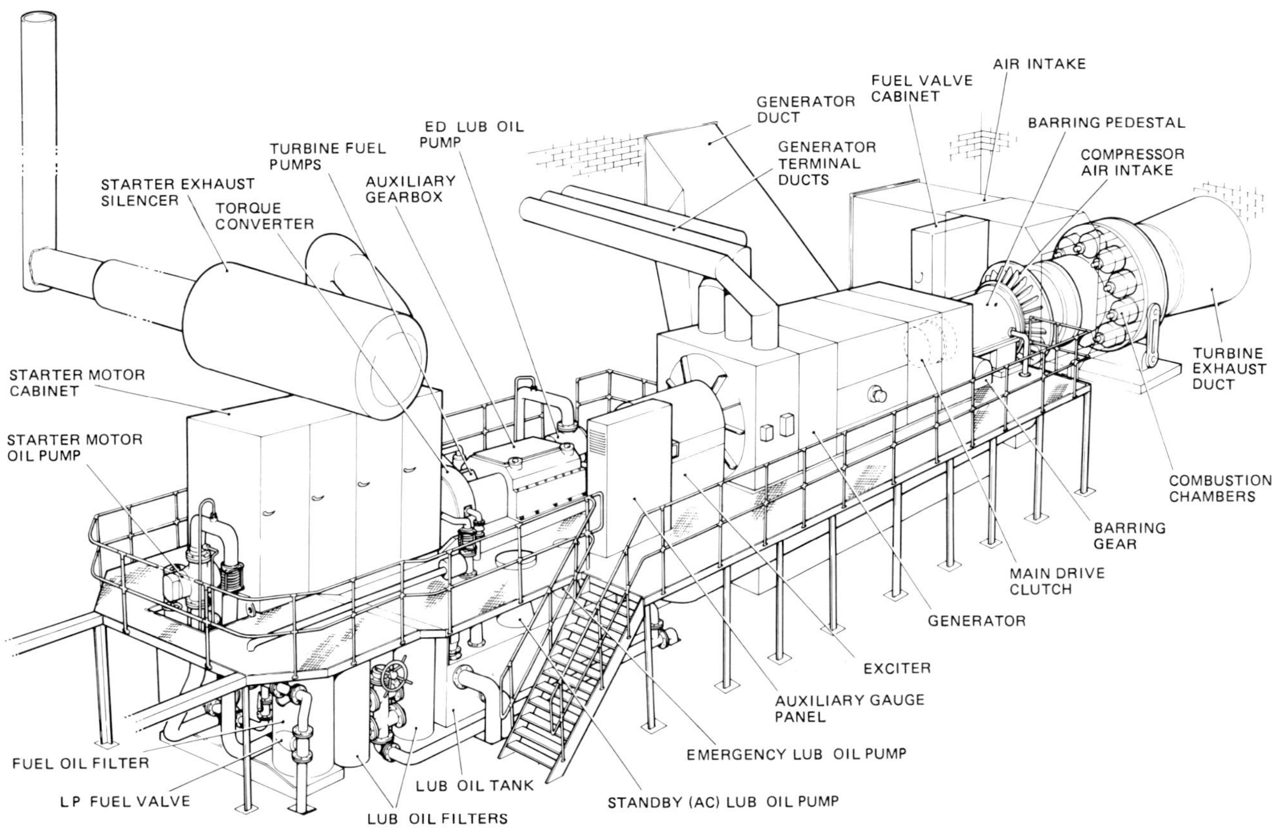

FIG. 2.99 Industrial gas turbine set

25.5 Gas turbine power station layout

25.5.1 Introduction

Gas turbines require relatively few items of auxiliary plant for their operation. They occupy relatively small areas of land and require few site resources, thus enabling great flexibility to be exercised in their siting. The following sections briefly outline the major plant items that have an influence on the overall layout of a gas turbine power station.

25.5.2 Station plant

The major plant requirements for a gas turbine station may be summarised as follows:

Gas turbines

Two optional formats of gas turbines are currently available for both the aero-engine-derivative and industrial type machines. The units may be located in a small power station building representing a traditional CEGB approach of *housing* generating plant. Alternatively the gas turbine units are available from the manufacturers as works-assembled *packaged* units. This format requires no building around the units and offers advantages of lower cost and a shorter lead time from order to commissioning.

Fuel storage and handling

The basic concept envisaged for future gas turbine stations is for natural gas fuel with a fuel oil capacity as back-up. For industrial machines this may include both residual fuel oil (termed ORF) and distilate fuel oil (termed ODF) capacity. To enable start-up and shutdown when firing ORF, an ODF system must also be included to facilitate initial combustion and to prevent problems of residues depositing in fuel oil pipework whilst firing on natural gas. Thus, in addition to a natural gas receiving valve station, both ORF and ODF storage tanks and delivery systems may be required.

Pumping plant for unloading supplies of residual and distillate fuel oil from rail tankers into storage vessels and feeding the gas turbines is essential. Treatment plant for both types of fuel oil is also required. Residual fuel oil heaters are also required to provide fuel oil to the turbines of the correct viscosity if this fuel type is chosen.

Site services

The remaining site area is occupied by:

- Fire protection water storage tanks and pumphouse.

- Workshop and stores buildings.
- Lagoon for contaminated drainage and to receive discharge from the fire protection system and fuel oil tank contents in the event of a fire.

25.5.3 Industrial gas turbine site layout

Figure 2.100 shows idealised conceptual layouts for future gas turbines incorporating industrial type machines capable of running on a range of fuel options. The overall layout shows 'station based' gas turbine units, whilst the lower inset shows the gas turbine units in 'package' form. The layouts chosen provide an economic arrangement for interconnections between the various site services. In reality, however, site specific features and restrictions may have a major influence on the overall layout for any future station. The design concept and orientation of plant items and buildings for these conceptual layouts is briefly explained as follows:

Fuel oil tanks

It is envisaged that four ORF tanks would be installed in total since this would allow flexibility in arrangement and operation. Long term tank outages for cleaning or repair can therefore be accommodated with the station still in service.

ODF contains fewer contaminants than ORF, and due to the absence of tank heaters, an ODF tank outage is a much more remote possibility. It is envisaged that only two ODF tanks would be required.

Fuel oil pumphouse and tanks

The fuel oil pumphouses, including the heating and fuel treatment plant, are situated in the centre of the site to minimise pipe runs to the unloading points, the fuel oil tanks and the gas turbine units. For the same reason, ORF treatment plant is situated next to the ORF tanks.

Fire protection pumphouse and tanks

The fire protection pumphouse and water tanks are again situated in the centre of the site but adjacent to the ODF tanks. This is because ODF is more volatile than ORF and is thus seen to constitute the main fire risk.

Other plant items

The natural gas receiving valve station is situated adjacent to the gas turbines and away from the ODF tanks since they constitute the main potential risk of a fire. This layout also minimises natural gas pipe runs.

The contaminated drainage lagoon is sited adjacent to the fuel oil tank farm, to receive bund drainage in the event of a fire or tank failure.

25.5.4 Cowes gas turbine station layout

Station and site details

Figure 2.101 shows the site layout for Cowes gas turbine station, which is representative of the latest of the CEGB's peak-load gas turbine generating installations. The station is situated on the east bank of the River Medina on the southern outskirts of East Cowes on the Isle of Wight. The station consists of two 70 MW aero-engine-based gas turbine units each powered by four Olympus gas generators, two at each end of the gas turbine houses, which exhaust into power turbines located at each end of a centrally-mounted electrical generator.

The gas turbine generating station, the fuel oil reception and storage complex, and the 132 kV substation have been constructed partly on land previously occupied by the Cowes coal-fired power station and partly on land acquired from British Gas, who occupy the adjoining site. The total site area is approximately 4.75 hectares. The site is bounded to the south by farmland, with the south eastern corner of the site occupied by an 11 kV/33 kV substation. Over the northern half of the eastern boundary, the site is bounded by land scheduled for light industrial development. Along its northern and western boundaries, the site is bounded by the Southern Gas Works and the River Medina.

It is very rare in the UK that power station design engineers are provided with an ideal greenfield site to work with due to the limited land areas that are available for such developments. It is thus often required to be able to design an effective power station layout based upon the limitations and restrictions imposed by any particular site. The gas turbine station at Cowes provides a good example of such a development, where the shape and nature of the land available has imposed certain restrictions on the layout of individual plant items within the station.

Site development

The main elements of the installation comprise:

- Two turbine houses each with exhaust system and chimney.
- Control block linking the two turbine houses with integral station and generator transformer compounds.
- Amenity and workshop building.
- 132 kV substation.
- Fuel oil handling pumphouse and fire protection pumphouse.
- Bunded oil storage area.
- Combined heavy load and oil unloading berth.
- Stores (modified existing workshop building).
- An extended 33 kV substation.

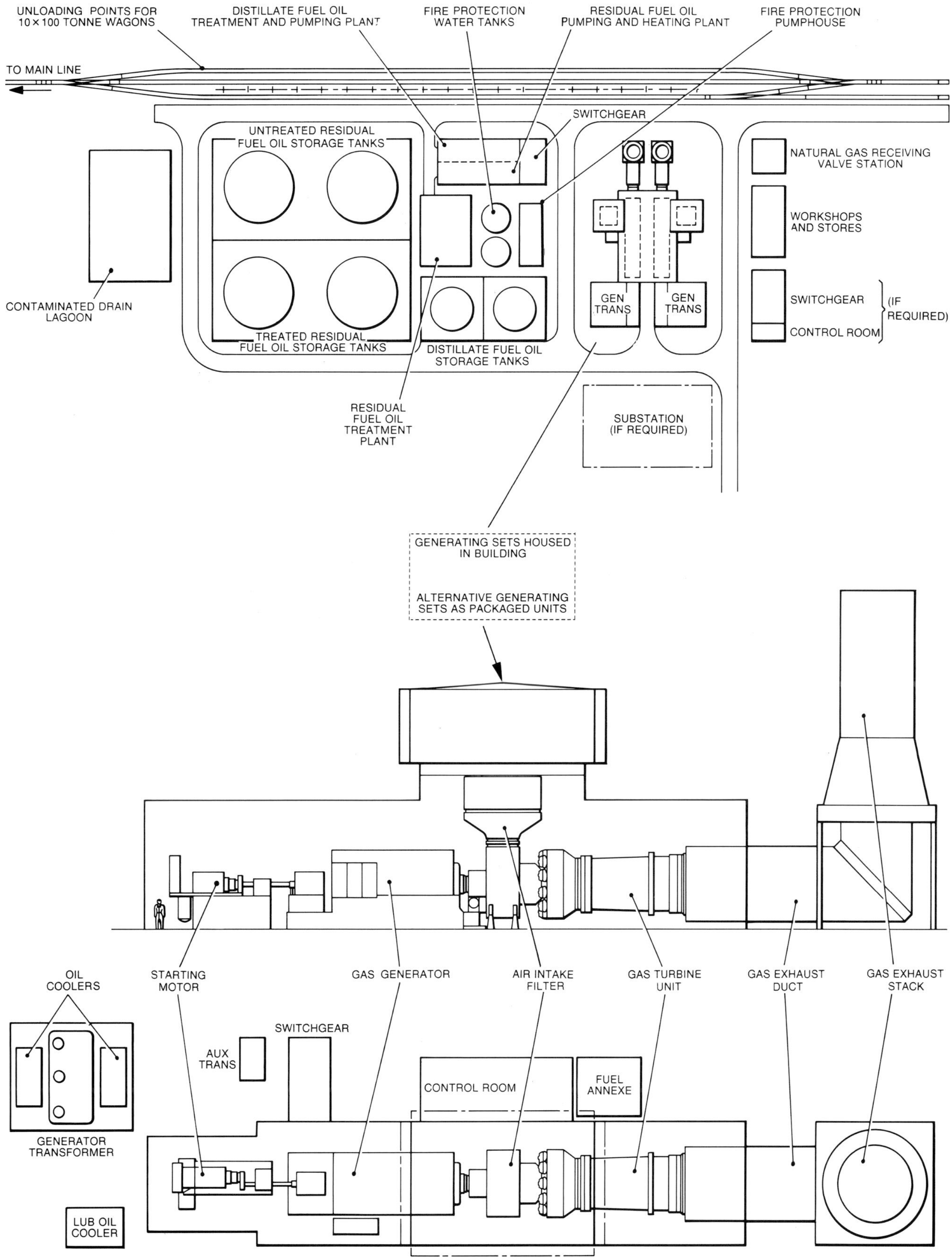

FIG. 2.100 Conceptual layout for two gas turbine package unit generating sets

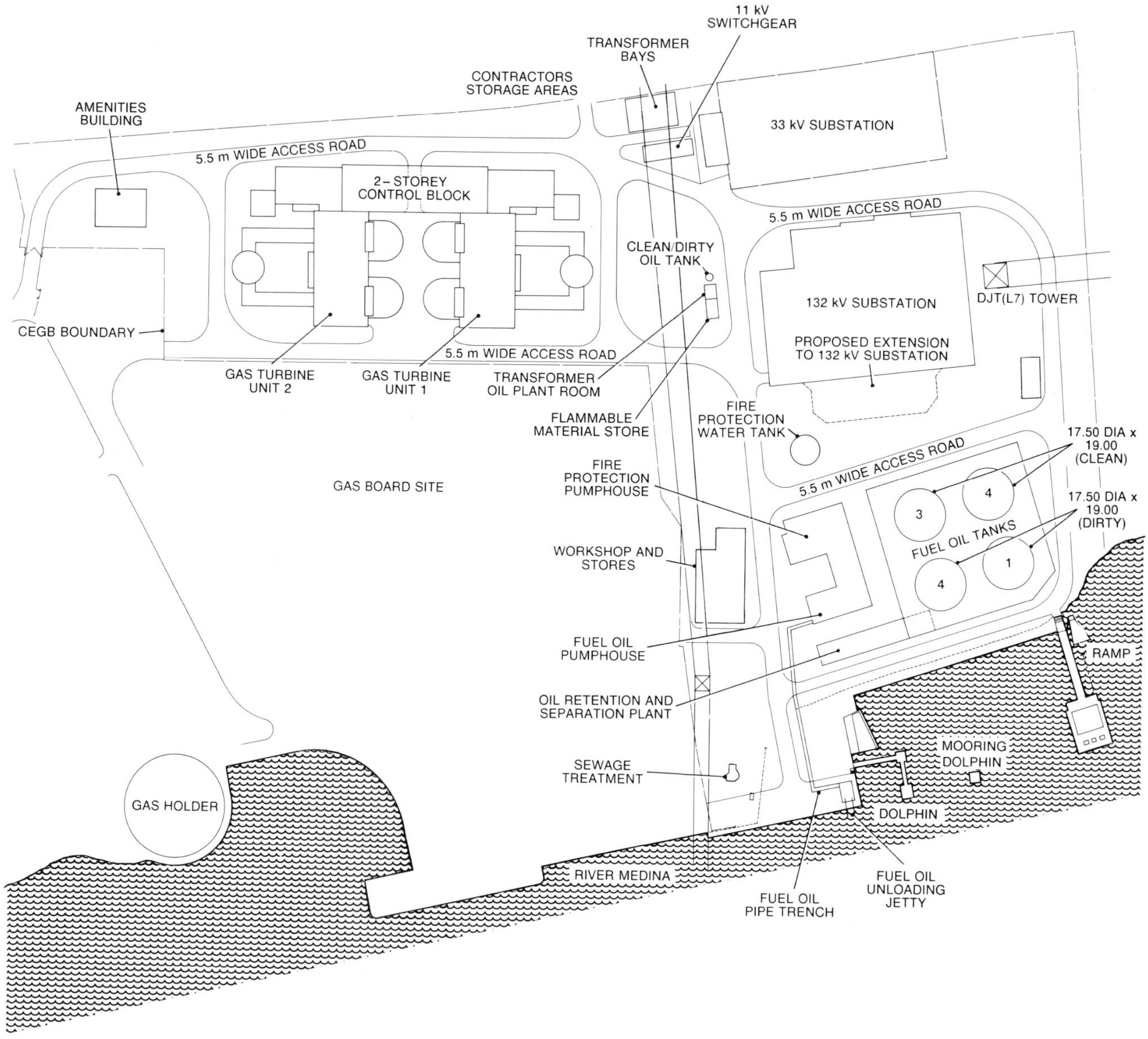

FIG. 2.101 Cowes gas turbine station layout

The location of these various elements has been dictated by the shape and size of the land area available, the slopes and levels of the ground and the intention to make the route of the oil from reception to pumphouse to storage vessel to turbine houses as simple as possible.

The width of the terrace of land purchased from British Gas was insufficient to accommodate the turbine houses and control block with service roads while retaining the existing access road. It was therefore necessary to reconstruct the access road at a lower level to provide a combined access and service road. The ideal of a flat site from east to west could not be achieved as this would have entailed either additional excavation at the access road or additional fill over most of the site, and especially at the west side of the terrace, each of which would have given rise to stability problems. The turbine house site therefore slopes 1.5 m from east to west. The ideal of a flat site for the 132 kV substation likewise could not be achieved. The compromise is a flat top terrace and a lower terrace sloping at 1:30. Any other arrangement would have created problems with the road from the wharf, the gradient of which is critical for handling the heavy loads, or would have entailed a very substantial depth of fill, creating stability problems on the slope.

Due to difficult foundation conditions, the only reasonable location for the fuel oil storage tanks is at the southern end of the old power station site.

To overcome the various problems with site levels and slopes, the fire protection pumphouse and the fuel oil handling pumphouse have been combined, linked by a switchroom and cable basement.

25.5.5 Littlebrook *D* auxiliary gas turbine installation

As mentioned previously, many of the CEGB's major conventional and nuclear power stations have been installed with auxiliary gas turbines. Figure 2.102 shows the layout of the installation at Littlebrook *D* power station. The three units, nominally one for each of the main generating units, are housed in a small building in a transverse orientation.

The spacing between machines is largely governed by the requirement for providing adequate space for access and maintenance activities. The gas exhaust chimney is located along the centreline of the central gas turbine unit in order to minimise the overall length of the gas exhaust ducts.

26 References

[1] BS476: Part 22: Methods for determination of the fire resistance of non-loadbearing elements of construction: 1987.

[2] Miller, D.S: Hydraulic design data for power station cooling water systems: British Hydro-Mechanics Research Association: December 1970.

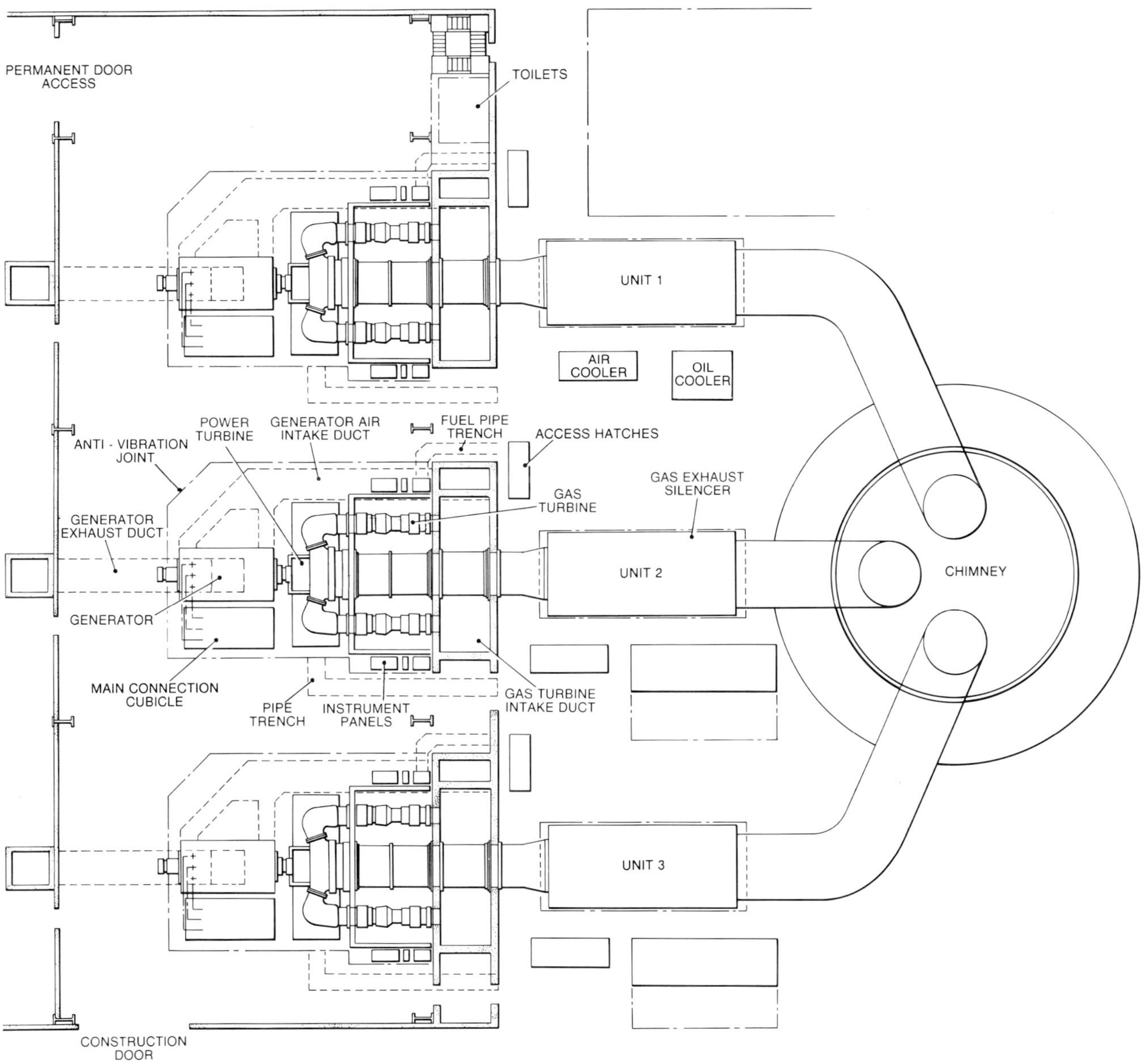

FIG. 2.102 Littlebrook *D* gas turbine installation layout

CHAPTER 3

Civil engineering and building works

1 Introduction

The construction of a power station can be divided conveniently into two components. These are construction below ground level — *foundations*, and construction above ground level — *superstructure*. Accordingly this chapter has been split likewise.

Sections 2 to 10 of this chapter are devoted to matters related to foundations and some structures which are wholly or primarily below ground while Sections 11 to 22 relate to superstructures and construction above ground level. Let us define those matters which are to be covered under foundations. These structures and constructions are very often in reinforced concrete, sometimes at, but more materially below, ground level. Firstly, methods of soil and rock investigation are reviewed and then there are basic descriptions of foundation types in general use. This leads to a discussion with examples of reinforced concrete foundations of typical main and secondary structures. Section 7 considers other works which are not strictly classified as *foundations* but are encompassed by civil engineering and are often of a site preparatory nature. In Section 8 primary construction methods are discussed for land sited works.

River and sea works normally associated with cooling water systems are considered in Section 9, followed by an outline of the works and methods of construction required for wharves and jetties where fossil-fired stations often receive their fuel from colliers and tankers. Steelwork, concrete and building techniques are discussed and the chapter concludes with architectural considerations, statutory regulations and civil engineering contract strategy.

2 Geotechnical investigations

2.1 General and desk studies

The extent and scope of geotechnical investigations vary greatly depending on the object of the exercise. These can range from initial site selection studies, through the investigation of preferred sites to feasibility and design stages. The financial implications vary accordingly from tens of thousands to millions of pounds. There is often advantage in phasing such investigations so that the earlier findings can be utilised to focus later attention on those matters likely to prove crucial to the development.

Geotechnical investigations are often undertaken to deduce the availability and extent of constructional materials from local sources, and in the case of nuclear stations special considerations (usually related to safety against defined hazard criteria) need additional geotechnical evaluation. As all power station construction is undertaken against a background of political overtones it should be recorded that constraints are often imposed, particularly in terms of timing, granting of planning permissions, consents, etc., which can radically affect the nature, timing and even sometimes location of geotechnical investigations relative to the *ideal* investigatory plan.

It is important to state the principles behind the planning of all site investigations and the first element to consider in this context is a desk study. It is essential that all investigation planning commences with a desk study to examine all the available information on topics such as mapping geology, hydrology, ground conditions, site usage (both present and past), geography, etc., so that maximum use can be made of such existing information and the risk of unnecessary duplication is minimised. In the case of a preliminary site selection study this desk exercise may be all that is warranted, but usually it is the forerunner to further work, much of which is site orientated.

It is normal for a desk study to cover an appreciable area around the potential site as well as the site itself. Judgement is required in assessing the area to be included in a desk study, and certain proposals will warrant a larger area or degree of study than others.

For instance a site may be adjacent to a river, whose water levels have been recorded over many years at or near to the site. But to equate those levels to volumetric flows may mean obtaining data from gauging stations many kilometres up or down stream and then interpolating or extrapolating the gauging data to produce equivalent data for the site. If this issue is deemed of particular importance, or if a high degree of accuracy is required, it may be prudent to consult a hydrologist or the Water Authority regarding the figures produced.

A desk study should be set up with a list of objectives and these must include such items as the examination of potential problems in developing the site, making economic use of investigation techniques and the selection of techniques appropriate to the anticipated ground conditions and geotechnical requirements. Such studies will establish guidelines for the scope, depth and layout of investigation measures required in subsequent site investigations.

Unless the desk study is of a very preliminary nature, such as basic site appraisal, a walk-over survey is highly desirable. This may have to be confined to roads, tracks and other public rights of way, but even so limited a walk-over is often most valuable in confirming desk study impressions and gaining further information to supplement published data. A walk-over survey is essential before committing expenditure on items such as physical sub-surface investigations or geophysics. As with the desk study, the area of coverage might be appreciably larger than the site itself. A list of study aspects to be recorded should be compiled prior to a walk-over survey and the possibility that such a survey may warrant extension into mapping exercises (topographical, geological and/or geomorphological) should be borne in mind, particularly for large or complex sites.

In these very early stages of site evaluation, recourse should also be made to the more sophisticated forms of aerial surveillance. As well as vertical and oblique photography using standard film, infra red photography and satellite imagery can highlight surface and near surface features which can be difficult to distinguish at close range. Such methods can be especially valuable where access is difficult in terms of either geography or wayleave permissions, or where structural geology can play an overriding role, such as in the location of the safety related elements of a nuclear station layout.

2.2 Geophysical investigations

A wide variety of methods of geophysical investigation is available. These methods measure the variations in selected physical properties of the ground such as wave velocity, resistivity, density and magnetic susceptibility. These parameters are generally related to geotechnical or geological characteristics of the ground. The methods work best if there are strong contrasts in the measured parameters either with depth or laterally across the site.

The accuracy of geophysical techniques can be very variable and often cannot be predicted in advance. The results always therefore require to be checked against or correlated with subsurface geotechnical or geological information from a limited number of exploratory excavations or boreholes.

If ground conditions are suitable, the use of an appropriate geophysical method can prove very economical in establishing some of the more important features of ground conditions on a large site, such as the depth to rock. This enables the minimum number of expensive boreholes to be employed later. Alternatively, anomalies revealed by geophysical investigations can be used to select the optimum locations for the more expensive subsurface investigation techniques. For example, low seismic velocity zones may indicate very fractured bedrock associated with faulting, while magnetic anomalies can indicate the presence of basic igneous intrusions.

Care and experience are required in selecting appropriate geophysical techniques. If methods inappropriate to the ground conditions are adopted they can be of dubious accuracy or even misleading. Thus it is essential to have some idea of the likely ground conditions from desk studies or other preliminary investigations before employing any geophysical investigation technique. Expert interpretation of the results of geophysical surveys is essential. The primary data are often produced in a form that is largely unintelligible to the practising engineer, but which are capable of remarkably accurate assessment by the relevant professional.

The most common methods of geophysical surveying and their uses are listed in Table 3.1.

2.3 Trial excavations and boreholes

Trial excavations and boreholes are traditional methods of direct subsurface exploration. The former permit visual examination of the ground in-situ. Both methods allow samples to be taken for description and laboratory testing. In-situ tests can also be carried out below the ground surface in either trial excavations or boreholes.

Trial excavations allow the most thorough in-situ examination of soil or rock masses since a relatively large volume of material can be inspected and a large area of exposed surface can be examined. This is important in common circumstances where the ground possesses a network of fissures or fractures or contains a complex pattern of seams or lenses of materials of varying composition. In these cases the behaviour of the ground may be critically controlled by this large scale *fabric* rather than the characteristics of small laboratory specimens which do not contain a representative amount of the *fabric*.

Large trial excavations may also be used to advantage where geological conditions are very complex. Trenches may be particularly useful to locate the position of faults or other linear features known to be present at shallow or even at reasonable depths. Adits can reveal three-dimensional geology of rock masses pertinent to the design of underground excavations or abutments of dams in complex formations. However adits of deep trenches may represent substantial engineering works in their own right.

Machine-excavated trial pits and trenches can prove very economical down to depths of about five metres although shoring or battering of the excavation sides will be necessary if personnel are to descend into them. Trial excavations may prove impractical or costly in ground with a high water table. Rapid groundwater inflow may cause problems in permeable ground. And in less permeable soils such as silts and fine sands groundwater inflow may destabilise the sides and base of the excavation. Base instability can also be troublesome in very soft clay soils.

Trial excavations rapidly become more expensive with increasing depth. Therefore boreholes become a more cost-effective method of investigation as the depth of exploration increases or where groundwater conditions are troublesome.

Investigation by means of boreholes will inevitably form a major part of any investigation for a power station site. Boreholes will certainly be adopted for any overwater investigations associated with cooling water or jetty studies.

TABLE 3.1
Common methods of geophysical investigations

METHOD	MAIN APPLICATION
Seismic refraction	Determination of depth to a hard stratum e.g. bedrock.
Seismic reflection	As refraction but capable of resolving ground conditions to a greater depth.
Resistivity	Identification of simple stratigraphical boundaries or water table level.
Magnetometer	To infer the presence of bodies of iron-rich rocks such as basic igneous intrusions or buried artificial ferrous objects.
Electromagnetic	Identification of simple stratigraphical boundaries.

In British practice, methods of forming site investigation boreholes fall into two categories — cable tool methods and rotary drilling methods. The former, often referred to as shell and auger methods, are normally used for investigating uncemented or weakly cemented materials such as sands, silts and clays. They may also be used to advance the borehole in the upper weathered zones of rock, but the rate of progress becomes too slow as more solid rock is encountered. Frequent samples have to be taken from each borehole to enable a record of the ground conditions to be compiled and for possible laboratory testing. In-situ tests are often performed to provide information about the mechanical properties of the ground and the improvement in these properties with increasing depth.

Figure 3.1 shows a light cable tool boring rig.

Disturbed samples are obtained during shell and auger boring from the tools used to advance the borehole, but these have suffered severe disturbance and are only of use for recognition of the materials penetrated. More representative and less disturbed samples can be obtained in cohesive strata by hammering or pushing a sample tube into the soil at the base of the borehole. The sample recovered in this way is often referred to as 'undisturbed', but its true quality is determined by the details of the boring and sampling procedures, particularly in either very hard or soft materials. Only thin-walled samplers employed in a carefully drilled borehole produce samples really suitable for testing for deformation or strength properties. Piston samplers and special sampling procedures need to be employed in very soft soils. It is always very difficult and often impossible to obtain samples of cohesionless materials (sands and gravels) which retain the original soil structure. In-situ testing is generally relied upon in such soils.

In foreign practice, boreholes in soils are frequently advanced by rotary drilling methods using hollow steam augers or non-coring drill bits. Sampling and in-situ testing can be performed through the hollow stem auger or through the base of the borehole.

Boreholes in rock strata are formed by rotary core drilling methods. These are intended to produce a continuous core of the materials penetrated. The core can be described and tested to establish respectively the geological succession and geotechnical characteristics of the ground. Again, in-situ tests can be carried out in the borehole.

Figures 3.2, 3.3 and 3.4 show a large rotary drilling rig and its application to an upwardly inclined borehole.

In some instances boreholes intended primarily for in-situ testing or instrumentation may be drilled without recovering cores or samples.

Improvements in rotary core drilling methods have now enabled good quality cores to be recovered in many types of uncemented soils with appropriate drilling techniques. These methods yield continuous cores

Fig. 3.1 Light cable tool boring rig
(see also colour photograph between pp 242 and pp 243)

of the material penetrated and thus give a more complete record of ground conditions than is obtained by intermittent sampling in shell and auger holes. They may therefore be used to supplement the findings of shell and auger boreholes on important investigations.

Rotary core drilling in strata consisting of solid, strong rocks is normally a fairly straightforward process, but in weaker or fractured rocks or in soils the drilling process may fail to recover a complete core of the material penetrated. Alternatively the condition

FIG. 3.2 Large rotary drilling rig
(see also colour photograph between pp 242 and pp 243)

of the recovered core may be unsatisfactory. Prudent assessment of zones of core loss or poor quality core may well lead to unacceptable uncertainties about ground conditions and hence to unduly pessimistic designs. Measures can be taken in most cases to maximise the quality and quantity of core recovered. These include using larger diameter drilling tools, changing the flushing medium used to remove drill cuttings from the borehole or using more sophisticated drilling tools. However, the ultimate success of all drilling techniques is critically dependent on the skill of individual rig operators.

A few other methods of direct exploration are also available, although only a few companies own and operate the more specialised items of equipment. Probably the most important such method is the Delft sampler, which is restricted in use to softer, mainly cohesive, alluvial soils. It yields a continuous small diameter core which can be very useful for establishing the presence of a *fabric* of thin seams of silt and sand which is often present in such materials and can have an important influence on their geotechnical behaviour.

2.4 In-situ tests

As noted, it is extremely difficult to take undisturbed samples of cohesionless materials, and samples of cohesive materials have inevitably suffered some disturbance which may affect the results of laboratory tests. These factors have led to the development of tests which attempt to measure the characteristics of the ground in-situ.

In-situ tests range from routine simple procedures to very complicated and expensive methods. They aim to measure many different geotechnical parameters. Tests may be performed in exploratory excavations, boreholes, or even in some cases from the ground surface. The more common tests are listed in Tables 3.2 to 3.4.

It must not be thought that in-situ tests are a panacea for all problems of measurement of geotechnical parameters. The behaviour of the majority of geological materials is extremely complicated, being generally non-linear, anisotropic, time dependent and loading path dependent. No single in-situ test can reproduce all the relevant conditions and thus provide an incontrovertible measure of field performance. The success of field performance made using in-situ test results still depends strongly on engineering judgement and experience.

Other more sophisticated tests may be carried out in adits for example. These may include cross-tunnel jacking tests, which are a variant of place bearing tests, or measurements, by a variety of methods, of the in-situ state of stress in the rocks surrounding the adit. To obtain useful engineering data from such sophisticated methods normally requires the employment of appropriately specialised firms or university research staff.

In certain circumstances there may be advantages in carrying out large scale field tests designed to examine specific aspects of geotechnical behaviour. These may occur where, because of the prevailing ground conditions, unacceptable geotechnical design uncertainties or contractual risks remain even after a full conventional investigation. Alternatively, large scale field tests may be warranted where their cost is justified by the potential savings to a major project resulting from a particularly accurate assessment of geotechnical parameters.

FIG. 3.3 Detail of rotary drilling
(see also colour photograph between pp 242 and pp 243)

Examples of large scale field tests which can be used are instrumented trial embankments or excavations, pumping tests, pile tests and trials of ground improvement processes.

2.5 Groundwater investigations

Groundwater is the most frequent cause of problems in geotechnical engineering. It is therefore essential that the geotechnical investigation provides a sufficiently thorough understanding of the groundwater regime and the consequent pore pressure distribution. Not only does the existing groundwater regime influence the construction of excavation for foundations and the long term performance of a development, but a large development may itself alter that regime. These potential hazards can only be evaluated from a knowledge of the initial hydrogeological conditions.

The first aspect of groundwater investigation is to establish and characterise the spatial variation of water pressures in the ground. They probably vary laterally across the site and may also exhibit a non-hydrostatic distribution with depth. Groundwater observations

FIG. 3.4 Rotary core drilling of an upwardly-inclined borehole in an existing dam (see also colour photograph between pp 242 and pp 243)

should be made routinely as drilling proceeds in each borehole, but such measurements may not be sufficiently reliable on their own because the duration of drilling is too short for equilibrium levels to be measured. A proper distribution of piezometers or observation wells is needed to establish reliable groundwater pressure information. It should be noted that various types of piezometer installation exist, and it is important to select a type of instrument appropriate to the permeability of the strata encountered and the rate of variation of groundwater pressure which is expected.

Observation of groundwater pressures using piezometers need to be made over a significant period (preferably at least a year) because they are influenced by seasonal and climatological factors. Abstraction of groundwater by pumping can also be a major influence on groundwater conditions. On coastal sites tidal influences need to be examined.

For a complete understanding of the groundwater regime, which will be needed, for example, for the design of a dewatering scheme, it will be necessary to establish the coefficient of permeability of the various strata. This will entail in-situ testing, possibly supple-

TABLE 3.2
In-situ tests which may be performed in trial excavations

TYPE OF TEST	DESCRIPTION
In-situ density test	Used in soils. Several methods available depending on soil type. Measures soil density for classification or routine design purposes.
California Bearing Ratio test	Measures the California Bearing Ratio which is a design parameter directly used in designing roads and other pavements.
Plate bearing test	Measures the deformation behaviour of soil or rock within the zone directly beneath the plate to a depth approximately equal to the plate width.
Direct shear test	Expensive test which can be used to measure the shear strength characteristics in-situ. May be useful on important projects involving slope stability where the ground is troublesome to sample and test in the laboratory.

TABLE 3.3
In-situ tests which may be performed in boreholes

TYPE OF TEST	DESCRIPTION
Standard penetration test (SPT)	Probably the most commonly used in-situ test. An inexpensive test empirically correlated with condition and performance of granular soils but also of use in stiffer cohesive soils or weak rocks where sampling or recovery of cores may prove difficult.
Field vane test	Relatively cheap method of measuring the undrained shear strength of soft cohesive soils which are prone to sampling disturbance.
Permeability tests	These measure the rate of water flow into or out of the borehole and hence allow the coefficient of permeability of soils or rocks to be computed. Several different methods are available, each with many variants. Packer tests are a commonly used type of test to measure the permeability of rocks. Permeability tests may also be carried out using piezometers which have been installed in boreholes.
Pressuremeter or dilatometer tests	These tests rely on the expansion of a cylindrical probe against the walls of a borehole. The tests theoretically measure the shear modulus of the material around the probe. Again there are several different types of pressuremeter probe and the quality of the outcome may well be critically dependent on the skill and experience of the operator.

mented by laboratory tests. Pumping tests will be advisable where major dewatering works are envisaged.

Water quality measurements may also be necessary. In the simplest of cases these will entail determinations of the pH value and sulphate and chloride contents of the groundwater. If a site is potentially contaminated by previous industrial use more detailed chemical analyses of groundwater may be needed. Where extensive dewatering measures are required it may be necessary to determine the chemical characterestrics of groundwater to check on possible future dewatering inducing changes in water quality and hence potential environmental problems.

2.6 Ground description and classification

Soil and rock descriptions are made from samples of continuous cores taken from boreholes, faces exposed in trial excavations and natural or man-made exposures. The cores and samples obtained in an investigation may deteriorate or become lost, and exposures become overgrown or covered up. Therefore their descriptions, together with any in-situ or laboratory test results, may form the only lasting record of ground conditions proven by the investigation. They are also more readily transmitted to the designer and the builder. Photographs can be very useful and should always be taken, but they cannot convey the tactile properties of materials. It is therefore vital that the description of soils and rocks are accurate, unambiguous and do not omit any important details.

Figure 3.7 shows the interior of a sample store established for a large site investigation.

A systematic method of description using universally accepted terminology is therefore necessary. Guidance on these matters is given in BS5930: 1981 Code of Practice for Site Investigations [1]. It recommends that soil descriptions should cover the following characteristics in the approximate sequence given:

TABLE 3.4
In-situ tests which may be performed without trial excavations or boreholes

TYPE OF TEST	DESCRIPTION
Static cone penetrometer test	This method involves measuring the resistance to penetration of a standardised design of cone which is slowly pushed into the ground. The parameters recorded have been empirically correlated with geotechnical performance in a wide range of ground conditions. Useful predictions of the type of material penetrated can be made in most cases, making this an economic indirect exploration tool as well as a valuable method of in-situ testing. Figures 3.5 and 3.6 show equipment for conducting electric static cone penetrometer tests.
Dynamic probing test	This method is similar to the static cone penetrometer but less well developed in the UK. It can be used to assess economically and rapidly the depth to a hard stratum and to correlate uncomplicated ground conditions between boreholes. Various methods exist.
Flow-through sampler	Uses and limitations are similar to the dynamic sounding methods. This technique can also recover a small disturbed sample from the horizon at which penetration ceases.
Penetration vane test	This is essentially the same as the field vane test described in Table 3.3 but the vane is advanced from the ground surface, tests being carried out at successively greater depths. This technique is only suitable for very soft soils.
Self boring pressuremeter test	This is a more sophisticated version of the dilatometer test described in Table 3.3. The instrument contains its own rotary drilling head and is drilled into the ground from the surface with tests being carried out at successively deeper levels. This instrument is intended to cause the absolute minimum of disturbance to the ground around the probe which is tested. If hard materials are present the device may need assistance by a drilling rig to penetrate those layers. Again the outcome is critically dependent on the operator's experience and skill.

(a) Mass characteristics:

- Field strength or state of compactness and indication of moisture condition.
- Bedding.
- Discontinuities.
- State of weathering.

(b) Material characteristics:

- Colour.
- Particle shape and composition.
- Soil name, grading and plasticity.

(c) Geological formation, age and type of deposit.

(d) Classification (optional):

- Soil group symbol.

Similarly, description of rocks needs to follow a systematic pattern with unambiguous terminology. Again, guidance on description is contained in BS5930. However, the engineering behaviour of rocks is normally dominated by the effects of rock structure and weathering, rather than by the characteristics of the solid rock types involved. For this reason it is important to distinguish between description of the rock material and description of the rock mass. It may not be possible to appreciate the full characteristics of the rock mass from rock cores from a limited number of boreholes.

BS5930 recommends that the description of rock material should cover sequentially the following information:

- Colour.
- Grain size.
- Texture and structure.
- State of weathering.
- Rock name.
- Strength.
- Other characteristics and properties.

In addition to this list it is vital that records of rock cores contain a quantitative description of the fracture state of the rock cores. BS5930 describes the terminology generally used for this purpose.

Rock mass description should describe the rock material and then furnish information on the following characteristics:

- The description of rock types in the mass with reference to major geological structures.
- The dip magnitude and direction, nature, spacing and persistence, width or opening of any discontinuities and the presence of any materials in the joints or discontinuities.
- Details of the weathering profile.

FIG. 3.5 20 tonne penetrometer truck for conducting electric static cone penetrometer tests

It is axiomatic that field description of soils and rocks should be reviewed in the light of the results of all in-situ and laboratory tests subsequently performed on them.

The classification of soils is intended to group together soils having similar engineering characteristics. By experience it has been found that useful groupings can be established on the basis of simple laboratory index tests for plasticity and grading characteristics of disturbed soil samples, or approximately by visual and tactile examination.

Classification of soils can be useful for the following reasons:

- The results of past experience can be used in a rational empirical manner to infer certain aspects of the performance of soils without resort to expensive fundamental tests.
- It can be used to guide the selection from a large number of samples recovered during an investigation towards the most representative ones for expensive engineering tests.
- It can assist in the interpretation of laboratory test results.

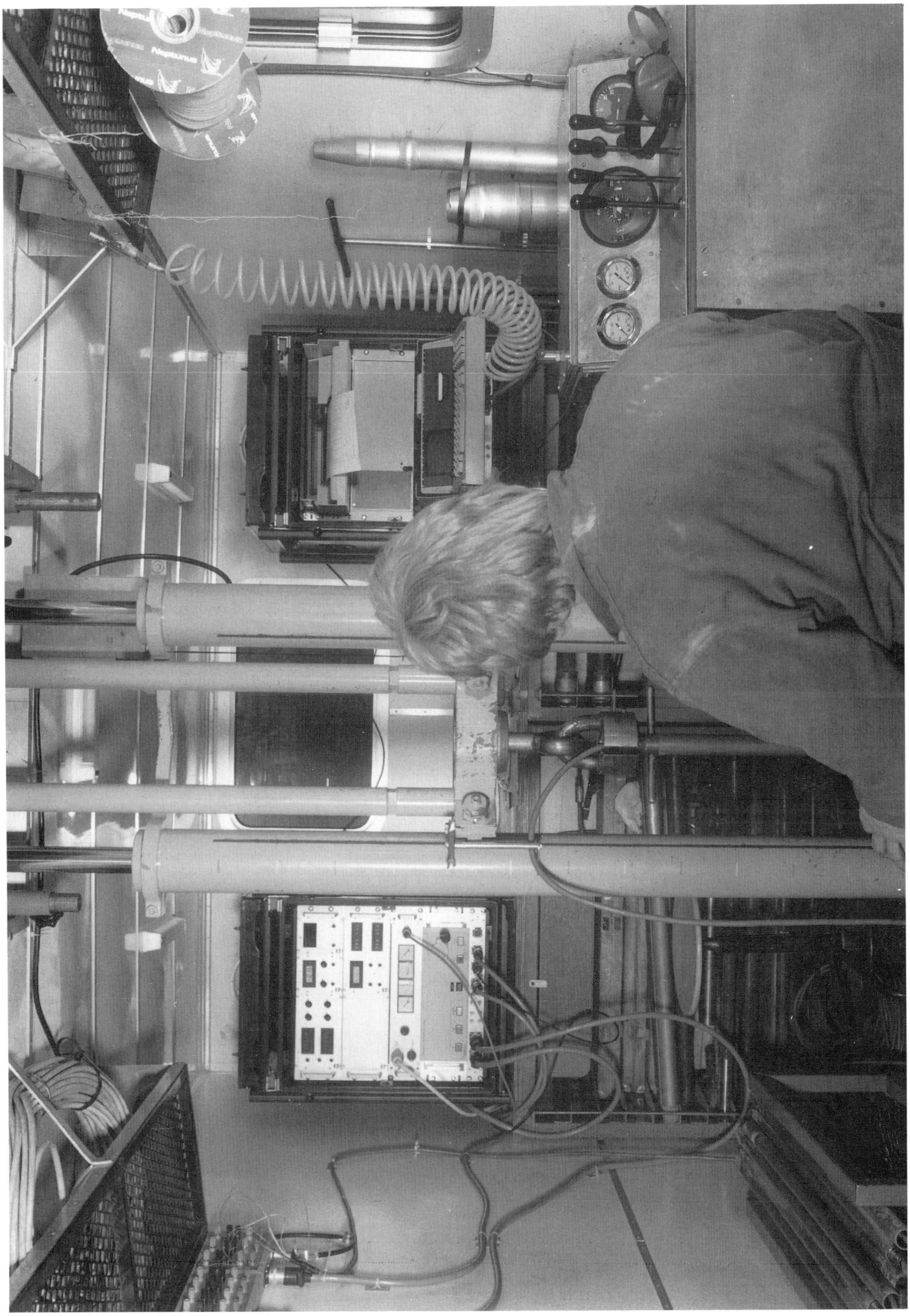

Fig. 3.6 Interior of electric static cone penetrometer truck showing data processing equipment (see also colour photograph between pp 242 and pp 243)

FIG. 3.7 Interior of sample store for a large site investigation (Wimpey Laboratories Ltd.)

In the USA the Unified Soil Classification System is commonly employed. This was developed by the Bureau of Reclamation and the Corps of Engineers. BS5930 defines the British soil classification system for engineering purposes. Both are useful systems and have close similarities. However, neither takes full acount of the in-situ condition of a soil since they are based on index properties of disturbed samples. The in-situ density, strength, moisture content and fabric are not taken account of in these classification systems and are of fundamental importance to many aspects of in-situ geotechnical behaviours. Therefore, classification systems are generally more relevant to the performance of disturbed soil used in earthworks than to performance of the same material in-situ.

Classification systems have also been developed for rock mass characteristics. These are used for the design of tunnels or underground chambers but may have application to other rock engineering schemes. They generally involve an assessment of all relevant characteristics of the rock mass, including strength, fracture state and weathering, according to a predetermined weighting scheme, and yield an index of rock mass quality which can be used for design purposes.

2.7 Laboratory tests

Laboratory tests can be carried out to measure a wide range of soil and rock characteristics. It is therefore tempting to employ the results of tests for a particular parameter directly in soil or rock mechanics formulae to obtain a design solution to specific engineering problems. This should only be done after the test results have been critically examined and possibly modified, using engineering judgement, because there are a number of factors which may limit the ability of laboratory tests to predict field behaviour. The most common ones are as follows.

Samples will inevitably have suffered some mechanical disturbance during sampling, transport and test specimen preparation. This generally reduces the undrained strength and increases the compressibility of a sample. It also reduces permeability. Stress relief is unavoidable, even in the best sampling operations. It has similar but normally less severe effects to disturbance. It can be very important in fissured materials.

Samples used for laboratory testing should be large enough to include a representative amount of the fabric of the material sampled. Many commonly used sample sizes of 100 mm diameter or less are too small to meet this criterion which is particularly important in permeability and consolidation tests. Sample size effects are also important in strength testing of rocks and fissured soils. Finally, and most importantly, the samples tested must be representative of the strata encountered.

The standard soil and rock mechanics tests do not model simultaneously all the factors which may influence ground behaviour under field conditions. Since geological materials exhibit complex inelastic behaviour, it is therefore inevitable that field behaviour will differ somewhat from that observed in the laboratory. It is, however, best to conduct laboratory tests according to accepted standard procedures.

There are methods of mitigating or making allowance for some of the above limitations where they are recognised to exist. Also, empirical design procedures often implicitly make an allowance for the influence of standard testing techniques under normal conditions. There, if samples are judiciously chosen, the test methods are appropriate and the tests are carefully carried out, then useful results can be obtained from laboratory testing. However, as previously noted, the results should always be critically reviewed to see that they are consistent with all other data obtained for the site and any published information for comparable geological materials.

Laboratory tests fall into the following categories — index or classification tests, chemical tests may be used to categorise the materials encountered within a standard soil classification system. They may also be used to guide the selection of representative samples for engineering tests and so interpret the results of the latter.

A useful summary of the common laboratory tests on soil and rock is contained in Table 4 of BS5930.

2.8 Factual reports

It is imperative that a good quality factual report is compiled as a record of all the investigations made for any specific site. In terms of the desk study this might be a fairly brief typewritten account of the various matters examined and the more important facts which emerged. Copious data records should not be discarded but rather consigned to references or appendices where they will remain available for inspection by the few people who may require to study them in detail. It is important to remember that some projects and sites do not immediately proceed to further development stages but they may re-emerge at a much later date. Then the factual report of the earlier work is the only data readily to hand, and is consequently of great value at that time.

If the geotechnical site work has been undertaken by a contracting firm, it is common practice to specify that it produces a factual report of all its operations, including the laboratory testing. Site plans are an important element of reports as it is often significant to know exactly where certain data were obtained. Commonly the site plans should be at two scales — at a small scale to define the site area relative to its surroundings, and at a sufficiently large scale to be able to pinpoint the locations of certain items, e.g., a borehole or the buried foundations of a demolished building.

2.9 Interpretation of site investigations

Whereas a factual report of any site investigation should always be compiled for record purposes, there will be occasions when interpretations of the factual findings are required in addition. Such interpretive reports are usually published in a separate cover from the corresponding factual data. They may be compiled by one or more staff members from a contracting, consulting or client organisation. Naturally the authors should be fully qualified and highly competent people who, from their specialist experience, are fully able to translate the facts into sound geotechnical advice. It is also of prime importance that the subject matter of the interpretive report is agreed by the parties concerned before the report is commissioned. Depending on the report content, the authors may have a ground engineering, geotechnical or geological specialism — the all important issue is that the authors are selected for their eminence in the matters requiring interpretation.

The subject matter of interpretive reports varies considerably, depending on the terms of reference for the site under consideration. But a typical report might give advice on the following matters:

- Safe bearing capacity of soils and rocks.
- Settlement criteria relative to major structures — immediate settlement, long-term settlement, creep considerations and differential settlement.
- Dewatering requirements and solutions.
- Design slopes for cuttings and embankments.
- Compaction and settlement of back-filling and embankments.
- Stability calculations for structures and earthworks.
- Geotechnical design safety factors.
- Piling — loads, design and construction methods.

This list is only intended to be a typical example, and the contents of any interpretive geotechnical report must be tailored to meet the engineering design needs of the site in question.

3 Seismic hazard assessment

Because of the risks involved, modern nuclear power stations are among the few structures in Britain designed specifically against the effects of earthquakes. The assessment of suitable hazard levels and design criteria for any site requires an extensive investigation of the geology and seismicity of the region of the earth's crust which surrounds that site out to radii in excess of 60 km. This investigation should aim to identify and assimilate, not only all recorded earthquakes but also all known sources of additional data which could support the understanding of seismicity. Such data concern the development and structure of the crust around the site and its state of stress and current dynamics, and extend from the surface to the greatest depths permitted by the evidence. The investigation may include the monitoring of local seismic events on a specially-commissioned microseismic network.

The data absorbed into the study are of numerous types and drawn from many sources, but resolve into two principal categories — the record of the earthquakes themselves (as recovered from historical and instrumental sources), and the evidence of crustal movement, past and current (as inferred from geological and geomorphological studies). One major task is to reconcile these two, very different, types of information.

Using established empirical correlations relating isoseismal areas with relevant instrumental determinations of surface wave magnitudes and focal depths, these latter parameters are estimated for each earthquake.

The completeness of any catalogue of earthquakes is dependent on the production and preservation of contemporary accounts, and hence not only on the disposition of recording centres and the state of communications around them, but also on the present day survival and availability of their records. Thus some gaps in the historical earthquake records may simply be cultural. Therefore, the historical dataset is analysed to assess its level of completeness, which, in principle, must be a function of the severity and extent of the effects of the earthquakes that occurred. On the basis of historiographical research, two thresholds are chosen to define through history those earthquakes which, had they occurred, must have been reported in surviving accounts (Set 1), and those which are likely, but not certain, to appear in the record (Set 2, which includes Set 1). Small earthquakes falling below the lower threshold (Set 3) are taken to have been recorded fortuitously and, although possibly significant in terms of their location, have little statistical importance.

A list of instrumentally-recorded earthquakes is compiled from available catalogues. Very few such events prior to 1969 are known but after this date many more records are available, mostly from seismometer networks operated by the British Geological Survey. Like those in the historical record, the instrumentally-recorded events are graded according to their probable accuracy of location and separated into completeness sets according to improving levels of detectability.

Special significance is accorded to faults since they are manifestations of former seismogenic rupture, with the potential, in some cases, to remain persistent sources of crustal weakness subject to renewed movement. The implications of neighbouring faults and their dates of last movement are recognised in several international regulations and codes governing the siting of nuclear installations.

3.1 Geology

Knowledge of the surface geology across the region of interest is compiled from British Geological Survey maps at various scales, specially commissioned studies of satellite imagery and local field studies supplemented by aerial photographs. The subsurface stratigraphy is known from deep boreholes and, in combination with geophysical data, helps to map subsurface structure. An exhaustive review of the geological and geomorphological literature is supported by fieldwork in pursuit of evidence for recent crustal movement. Finally, at locations indicated by the field evidence, radiometric and palaeontological dating is carried out to constrain last dates of fault movement.

3.2 Earthquakes

The seismological database incorporates all earthquakes known from historical and instrumental sources. The historical record is critically re-evaluated from the primary sources using a data searching technique designed historiographically to minimise the chance of 'missing' any earthquakes or of failing to recover any reports of significant felt (*macroseismic*) effects. Secondary sources are analysed to identify and explain shortcomings (particularly multiple entries or incorrect dates) in earlier catalogues. Descriptions of felt effects of earthquakes are assessed with regard to their contemporary cultural contexts and 'intensities' are attributed on the basis of comparison with standard scales of diagnostics such as the Modified Mercalli Scale, most of which extend a scale of I to XII.

Each macroseismic earthquake is treated by mapping the assessed intensities and drawing isoseismals, i.e., intensity contours enclosing common levels of severity. In many cases only the 'felt area' of the earthquake, as enclosed by isoseismal III, can be represented, but for a few larger events isoseismals are drawn for higher intensities. From the mapped data, the 'macrocentre' (the centre or peak of the intensity field) is located and graded according to the inherent uncertainty of this location, which depends on both the quantity and the quality of the data available to constrain it. Four grades are used: better than 5 km uncertainty; better than 10 km, better than 20 km and worse than 20 km uncertainty.

3.3 Crustal dynamics

Because the historical record of seismicity offers only a brief and incomplete picture of contemporary crustal dynamics (and hence the controls on future seismicity) supplementary sources of information are researched.

Evidence of relative movements of the surface, determined over a variety of timescales, can indicate local and regional concentrations of crustal deformation. However, geological observations of crustal movement are only of direct relevance to the contemporary dynamics if they represent a response to the same tectonic regime (it can be demonstrated for Britain that the Current Tectonic Regime has not persisted for longer than about 8–6 million years, since the Upper Miocene). The impact of the ice-ages and the associated marked variations in sea level through the past 2 million years has removed much potential evidence for recent tectonic movements so that, in general, only information for the past 10 000 years of the post-glacial Holocene remains. Observations from tide gauges and from repeated geodetic levelling surveys should provide some insight into contemporary vertical movements, although the duration of accurate monitoring means that any tectonic signal can scarcely be resolved from the inevitable determination errors.

The implications, in terms of direct measurement of deformation patterns, from all three of these data sources is compared with the strain rates inferred from the historical record of seismicity.

Measurements of the state of stress from in-situ borehole observations are not widely available but observations from southern and southwest England, as well as more widely across northwest Europe, confirm a regionally consistent northwest to southeast principal horizontal compressive stress.

The record of seismicity is examined for any correlation with crustal provinces defined by the geological data. Also, in seismic hazard assessment a primary uncertainty is whether or not a specific fault is active; determination of this 'active status' is by criteria, utilising both geological and seismological evidence, which have been developed specifically for use in Britain.

The seismotectonic constraints identified by these studies are incorporated in the construction of a model for the computation of the seismic hazards at the site.

3.4 Ground motion hazard

Earthquake 'ground motion' describes the phenomenon of shaking as the earthquake waves arrive at, or travel across, the surface of the earth. By virtue of geometrical spreading and material damping, the severity of shaking reduces with distance from the focus of the earthquake until eventually wavetrains are so reduced in energy that they can be observed only by extremely sensitive instruments. Although engineering significance is attached only to the area within which so called 'strong ground motion' is experienced, this can still extend over considerable lateral distances depending on the size of the earthquake and its depth. Ground motion is conventionally described in terms of peak free field horizontal acceleration expressed as a percentage of gravity.

A numerical formulation has been developed for treating the inherent random uncertainty in frequency and location of earthquakes within sources and in the attentuation of the seismic waves propagated from them to the site.

Variabilities in model source parameters are incorporated, in a Bayesian sense, by the assignment of appropriate weights to discrete value representations of these parameters, the weights being determined by a consensus of expert opinion. The weighted parameter functions are attached to branches of a computational logic tree for calculating hazard on which the probabilities at each node are disjoint and exhaustive of all possible choices. On the logic tree there are main branches for each area and fault source on which individual probability distributions for geometry, earthquake recurrence, and strong motion attenuation are defined.

The boundaries of area sources are adjusted to reflect tectonic understanding of the region but remain conservative with regard to their hazard implications for the site. Boundaries between the area sources are tested for statistical significance.

Having statistically tested that the Set 1 seismicities of area sources can each be represented by Poisson processes, the next step is the derivation of their respective activity rate, b-value, focal depth and maximum magnitude distributions. Because of their impact on the hazard assessment, these assignments integrate the available information from different historical periods into overall distributions which are weighted by a consensus of expert opinion.

Of the many faults found in the region, very few will be significant by meeting the criteria for modelling as a discrete seismic source and by virtue of their proximity to the site. These faults are modelled with appropriate geometries with respect to the site and with parameters that are consistent with the available evidence. Compensatory adjustment is made to the activity rate of an area source which surrounds any particular modelled fault.

The final computational model, therefore, consists of a combination of nearby area sources and, perhaps, specific fault sources surrounded by a zone which is assumed to be subject to average British seismicity. The expected ground motion hazard at the site is then calculated using an appropriate computer program. The results are tabulated and plotted for a wide range of probabilities together with their corresponding curves at various levels of confidence. Sensitivity studies are carried out to allow the evaluation of the effects of any remaining uncertainties in the final mood.

3.5 Ground rupture hazard

Earthquake 'ground rupture' describes the phenomenon of displacement at rockhead or at ground level. If it occurs at all, it is very localised.

The emphasis in the assessment of the hazard it poses at any site is therefore on the present and active status of known faults at or very close to the site. Great attention is paid in the on-site geological investigation (see Section 2 of this chapter) in order to understand as fully as possible the geometry, relationships and date of last movement of any faults which are encountered. If any are found which cannot categorically be shown by geological evidence to be effectively extinct, it may be necessary either to abandon the site or to compute the ground rupture hazard (using methods comparable to those for the ground motion hazard) in terms of probability levels for the exceedance of various displacements, and then demonstrate the acceptability of the resulting hazard at the lowest levels of probability. This latter route is best avoided or at least ameliorated by adjustment to the layout of safety related plant.

4 Types of foundations

4.1 Isolated column foundations

Any columns may be founded on a pad foundation, provided that the size of pad required to spread the concentrated column loading is compatible with the allowable bearing pressure under that foundation. However, pads are only used where they are shown to be economical in excavation and construction, such as where rock occurs near ground level.

With other soils and rocks the use of such foundations may also require consideration of the acceptability of the settlement which may occur under the stipulated loadings. In general the recommendations of BS8004 [2] for shallow foundations should be followed in design.

Low rise structures with light frames are normally founded on unreinforced or nominally reinforced concrete pad footings. Here the footing supporting the frame should be so proportioned that the angle of load spread from the pier or baseplate to the outer edge of the ground bearing does not exceed 45°, hence minimising tensile stresses within the footing.

Buildings with heavier column loadings should be designed to the requirements of BS8110 [3] for reinforced concrete.

4.2 Strip foundations

The basic requirements for a strip foundation are similar to those which apply to an isolated pad foundation. It is the most suitable form of foundation for both interior and exterior walls of all low rise buildings on reasonable ground.

On sloping sites strip footings should be on a horizontal bearing stepped where necessary to maintain adequate depth.

For heavy column loads on weaker soils the size of adjacent isolated footings may be such as to make them touch. In such cases some economy may be achieved in excavation costs by providing a strip foundation and some further advantage may be gained in limiting differential settlement.

4.3 Raft foundations

Raft foundations are essentially large plan area foundations built either solidly or in cellular form. They may be employed for lightly loaded structures on soft natural ground or on other ground where weak zones exist. In this case the raft is designed to span across such zones with tolerable bearing pressures and acceptable settlements.

Possible applications are discussed in BS8004. Of particular interest is the case where heavier structures are supported in ground conditions which are such that there is unlikely to be significant differential settlements. Isolated footings, under these conditions may occupy such a large proportion of the available area that they may profitably be united to form a raft, usually of reinforced concrete.

The raft comprises a layer of reinforced concrete covering the whole area of the building, or even of several buildings. It may be strengthened by increased thickness of concrete in areas of heavy loading, such as under columns, or by a system of main and secondary beams.

Simple solid raft foundations may be easily and quickly constructed, provided that a suitable stratum exists fairly close to the finished site level — say within 3 m. In cases where the depth of excavation to firm ground exceeds the structural depth of the raft foundation, the excess depth over the raft can be filled with weak concrete or compacted broken stone. Figure 3.8 shows a raft foundation strengthened by main and secondary beams, and with filling placed over to carry the floor at the appropriate level.

4.4 Cellular raft foundations

Where excavation to reach the foundation stratum is in excess of 3 m or where loads become so great as to require a simple solid raft of excessive thickness, a cellular raft foundation may be considered. Herein the resulting structure forms a basement to the building as shown in Fig 3.9. The basement is constructed throughout in reinforced concrete. The floor and ceiling, spaced apart by the intervening walls, together form a structure of great strength and rigidity, which is very resistant to differential settlement, and spreads the total load more effectively than an equivalent solid raft foundation. The use of large quantities of weak concrete fill is avoided and the ground is relieved of the weight of soil which once occupied the void in the basement. For this reason cellular foundations generally give a lower intensity of loading on the ground than solid raft foundations, and so can be used on weaker soils.

A foundation of this kind, in which the weight of the structure and sub-structure, including all imposed loadings, is balanced by the total weight of the excavated soil, and any hydrostatic head resulting from a natural water table resulting in no net applied loading on the foundation, is described as *fully compensated:*

Generally a cellular foundation will be over-compensated or partially-compensated.

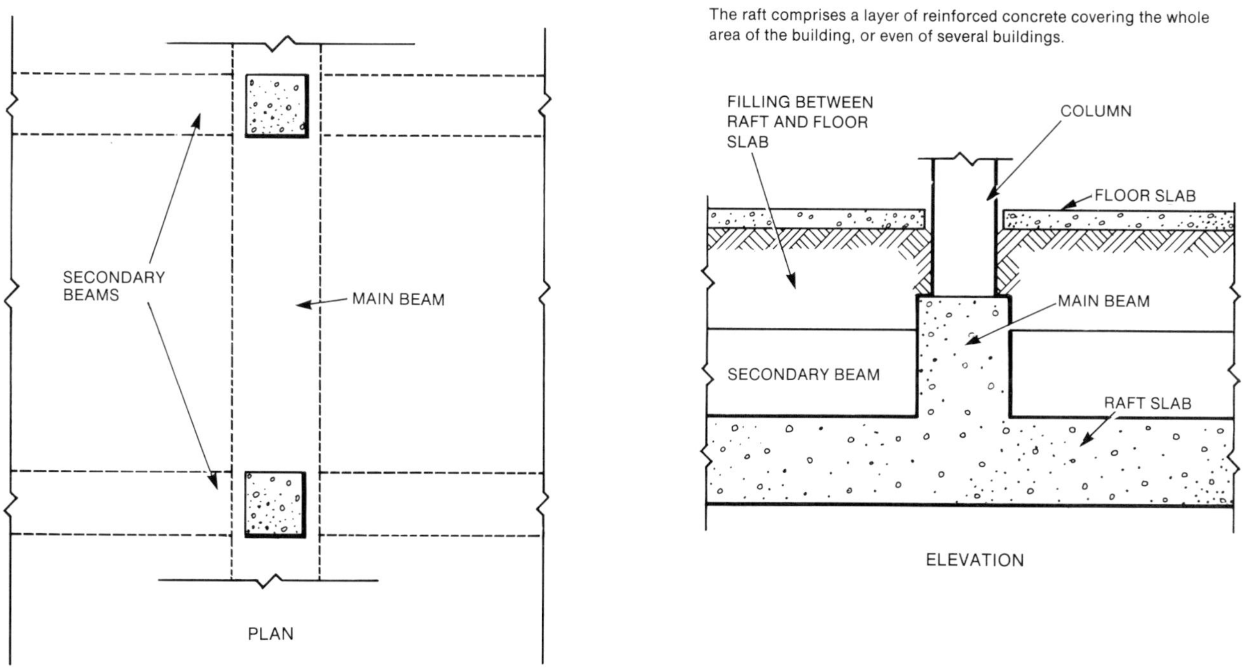

Fig. 3.8 Raft foundations

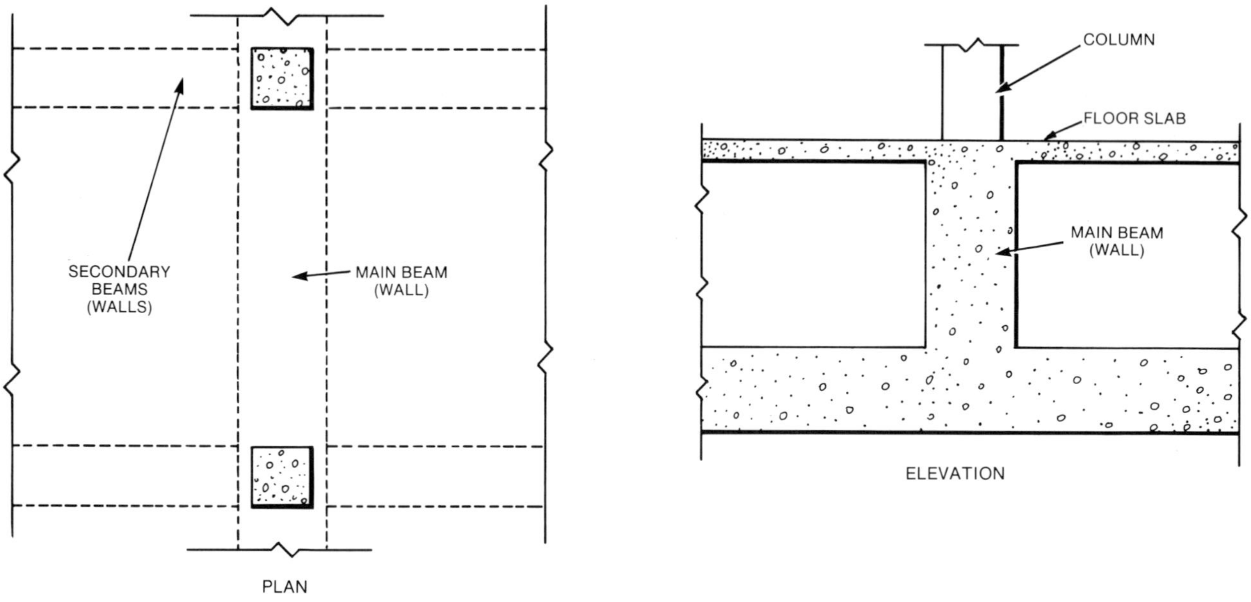

FIG. 3.9 Cellular foundations

The designer must check that a cellular foundation constructed in water-bearing ground will not float due to the hydrostatic head exceeding the imposed loading at any stage of construction or usage. This danger may prevail in the case of over-compensated foundations, such as empty submerged storage tanks, where loadings vary due to operational demands. In a similar manner the effect of ground heave when such foundations are supported on cohesive soils is an equally vital design condition requiring careful study.

4.5 Piled foundations

When a stratum sufficiently strong to support the imposed loading does not exist fairly near the surface, piled foundations may be used. A pile is a 'column' in the ground which is used to transfer the load to a stronger and deeper stratum.

Piles may be of timber, hollow steel section, prestressed or reinforced concrete; the last named being the most used for foundations of buildings. The load carried by a pile can vary from 300 to 10 000 kN depending on the size, type and ground conditions. The load is transmitted from the pile to the ground either by end bearing of the pile toe on a hard strata underlaying soft ground, or by 'skin friction' between the surface area of the pile and the surrounding soil, or by a combination of both.

There is no one pile type which is best in all conditions and all types have their advantages according to circumstances. Piles can be sub-divided into two main types — displacement piles and replacement piles.

4.5.1 Displacement piles

Displacement piles are sub-divided into *preformed* (precast, in the case of concrete) piles and *in-situ* piles.

The precast piles in general use on power station sites are made either as reinforced or prestressed concrete usually of square or near-square section. Reinforcement consists of heavy longitudinal bars with square links or helical binders. Prestressed piles are most usually pretensioned. In both types, the spacing of links and binders is decreased at the toe and the head to resist driving stresses; the longitudinal reinforcement or prestress provided is greatly in excess of that required to take the structural load, but is needed to resist the stresses of lifting, handling and driving. The toe may be strengthened with a pointed cast-iron shoe, although this may not be necessary when driving in clay.

The casting, storing, transporting and driving of the precast piles for a major contract, which may call for some 20 000 piles, is a considerable undertaking. A large area of the site has to be set aside and levelled and accurately paved with concrete to provide a casting bed, and further areas are required for storage. Travelling derricks are required for casting the piles and lifting and handling them, a batching plant is needed to produce the large quantity of concrete required, and a light railway system to transport the piles to the driving frames.

Piles precast in this manner have the advantage that they can be inspected before driving, and good control of the quality of concrete can be maintained. They are driven only after sufficient time has been allowed for the concrete to mature, and are thus resistant to attack from harmful substances in the soil.

Some wastage of concrete is necessary as the tops of the piles have to be cut away to expose the pile reinforcement for inclusion in the foundations. An accurate knowledge of the level of the bearing stratum is required to avoid further wastage through excessive cut-off. It is possible to drive piles below ground level using a wooden 'dolly' as a temporary extension to the pile. Permanent extension of the pile and subsequent redriving is a tedious business as the head has to be stripped away, a new length of pile cast on *in-situ* and left to mature before redriving can be undertaken.

A mobile pile frame — a tall steel structure on rails fitted with diesel-driven lifting gear — is employed to drive the piles. This hoists the pile into position, holds it and supports the pile hammer during driving, which continues until sufficient resistance is encountered. A variety of pile hammers are in use. Drop hammers varying in weight up to some 4 tonnes, dependent on the weight of pile, are still used, but diesel-driven hammers, which are much quicker in operation, are nowadays more common. The pile head is protected during driving with a steel helmet lined with a wooden or hard plastic packer.

Some of the problems related to uncertainty concerning pile length can be addressed by jointed reinforced concrete piles which are offered as proprietary systems by some specialist contractors. Such piles can be precast in appropriate lengths (up to 10 m) and combinations. It is not desirable to have a joint just below ground level when the pile is being hardest driven to its set.

Hollow steel piles, generally of hexagonal shape and with pointed toes, are frequently used to support jetties, especially where long pile lengths are required. Although more expensive than concrete piles, they are much lighter and do not require such careful handling. Lengthening, when required, is a simple matter of trimming the head and welding on a further section. A protective external coating of bituminous compound, or protective paint is required before driving. Once driven, the piles are sometimes filled with concrete or with inert water.

A number of systems for forming in-situ displacement piles are used, each specialist firm having its own method. A heavy steel tube fitted with a detachable cast iron or concrete shoe, or a concrete tube formed of hollow precast sections and fitted with a concrete shoe, is driven using a special piling frame usually mounted on a crawler-tracked excavator. A cage of reinforcement is lowered into the tube and the void concreted. The precast concrete shells are left in position and have the advantage that the outer skin of mature concrete is resistant to chemical attack. Where steel tubes are employed, these are withdrawn as concreting proceeds (great care must be taken to ensure that voids are not created by pulling sections of concrete with the tube). Piles of this type need be only lightly reinforced to take the structural loads, and can be accurately finished to the required level, thus avoiding wastage of concrete. They are less certain in quality than the precast pile, particularly where a steel outer casing is used, as the ground pressure may cause the pile to close up after the casing is removed. A few systems employ a lighter steel permanent casing which is internally supported during driving by a mandrel in a similar way to the precast concrete shell type.

The greatest part of the loading on power station foundations is vertical, but some horizontal forces due to wind and operation loads have to be resisted. These forces are resisted by driving some piles at a rake of up to about 1 in 3. Raking piles are commonly used on such structures as jetties, cooling towers, chimneys, coal-handling structures and transmission towers.

The very high column loadings now experienced in power station work cause the spacing of piles to be reduced to a point where the increase in ground pressure and consequential 'heave' caused by displacement piles can become a serious problem. Great care has to be taken if ground movement is not to occur, and piles be displaced. Pre-boring at the location of each pile is sometimes undertaken to solve this problem. This procedure is also adopted to enable driven piles to penetrate thin hard layers of soil overlaying softer material.

4.5.2 Replacement piles

Replacement piles are constructed by specialist contractors, with various methods of boring. It is rare for a single pile to be used on its own; piles are normally driven in groups at a spacing of 2.5 to 3 times their diameter and their heads incorporated into a thick capping of reinforced concrete designed to transmit the column load evenly to the piles. An arrangement of precast piles to support a column load of about 2000 kN is shown in Fig 3.10 and one to support about 30 000 kN from a major column using large diameter bored piles is shown in Fig 3.11.

If the piles are too close together, there will be considerable overlap of stress from adjacent piles in the soil or rock on which they rely for end bearing. Interaction is a particularly important consideration between piles which rely on skin friction. The bearing capacity of a group of piles is therefore, often less than the capacity of one pile multiplied by the number of piles in the group. This reduction in capacity is greatest for piles in cohesive soils, and in large groups can be as much as one third.

4.6 Caisson foundations

For larger concentrated loads it may be necessary to use large diameter caissons either singly or in groups. The concrete shell of a cylinder is built from a steel cutting edge at ground level, and supported vertically by temporary guides. The walls are in the order of 150 mm thick. Precast sections are used for cylinders between

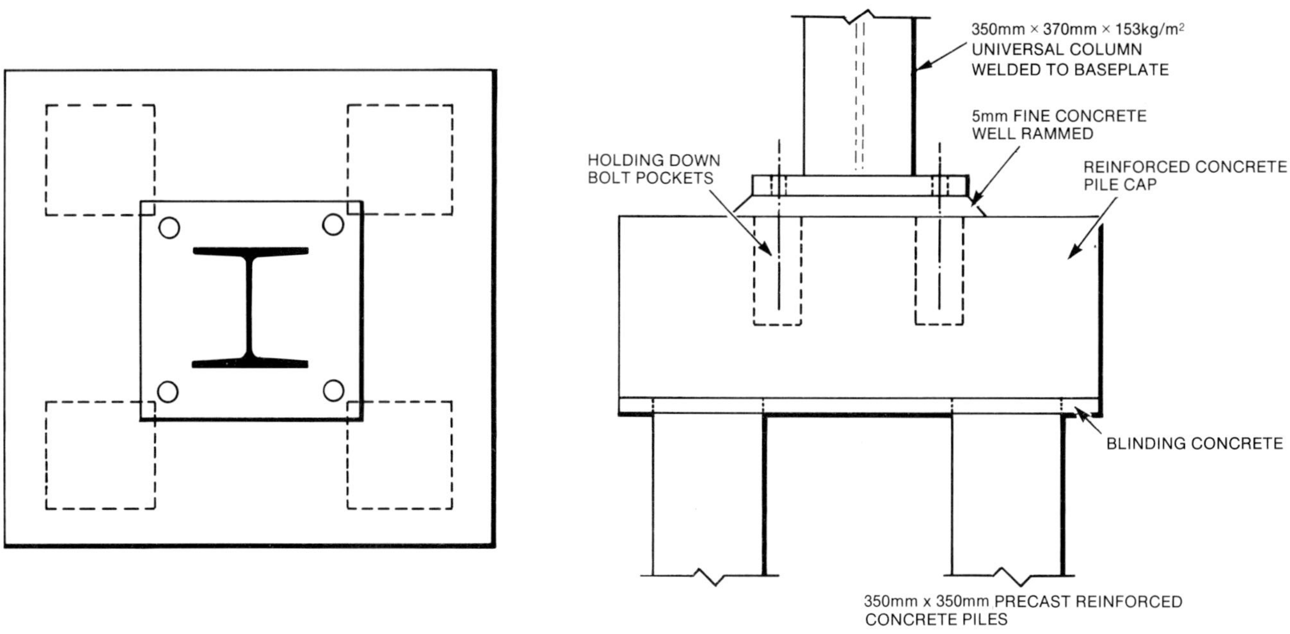

FIG. 3.10 Piled foundations — precast concrete piles

2 m and 3 m diameter, but above that size, in-situ concrete is used. The soil from inside the cylinder is removed with an 'orange peel' circular grab, and the cylinder is forced to sink by loading with kentledge. Grabbing can continue underwater if water-bearing strata are encountered, or an air lock chamber fitted and excavation continued by hand in compressed air. Further precast sections are added at intervals, or the in-situ walls extended until foundation level is reached. The bottom of the cylinder is plugged with mass concrete either placed under water or in compressed air, depending on the method of construction, and the remainder of the core concreted. Reinforcement is added if required. Cylinders are frequently used as the supports for jetties.

Construction through unstable ground is facilitated either by the use of temporary (or sometimes permanent) steel casing or by filling the boring with a thixotropic bentonite suspension.

Small diameter piles (up to 600 mm diameter) are traditionally constructed by a tripod rig which cuts the boring by cable-mounted cutting tools. In recent years very small piles (micropiles) bored by small rotary drills have appeared. Both systems have the advantage of being operable in conditions where the working space is restricted. Larger diameter piles up to about 2 m are bored by heavy lorry or crane-mounted rotary augers. In suitable conditions such piles are sometimes pivoted with an enlarged base using under-reamers.

On completion and cleaning up of the boring, concreting of dry holes is usually by direct discharge. Where the boring is full of water or bentonite an adjusted concrete mix is placed using a trevice pipe. Depending on the quantity of reinforcement required, the steel can be inscribed either before or after concreting. Considerable care has to be exercised during the concreting operation to avoid forming objects in the pipe, particularly with respect to the withdrawal of any temporary casing.

A recent development in the field of replacement piles is the continuous flight auger system where the soil is not removed until the flight auger has penetrated to the required depth. The auger is then withdrawn as the space beneath is continuously filled with pumped sand cement grout which passes down the centre of the auger.

4.7 Anti-seismic foundations

Anti-seismic foundations incorporate engineered devices to isolate the main building structures and plant from the damaging effects of earthquake ground motions.

The commonest form of isolator is similar to a bridge bearing pad made up of alternate bonded horizontal layers of neoprene rubber and steel plates. Bridges and other structures may also be isolated with devices containing a solid lead core which, in plastically deforming, absorbs vibrational energy. The isolators are placed between the structure to be isolated and the ground foundations (see Fig 3.12).

Isolators dramatically reduce the peak horizontal accelerations suffered by the structure, particularly the amplification caused by building resonance in the damaging 1 to 10 Hz frequency range, as illustrated in Fig 3.13.

Because of the complexities involved in designing connections between isolated and non-isolated plant,

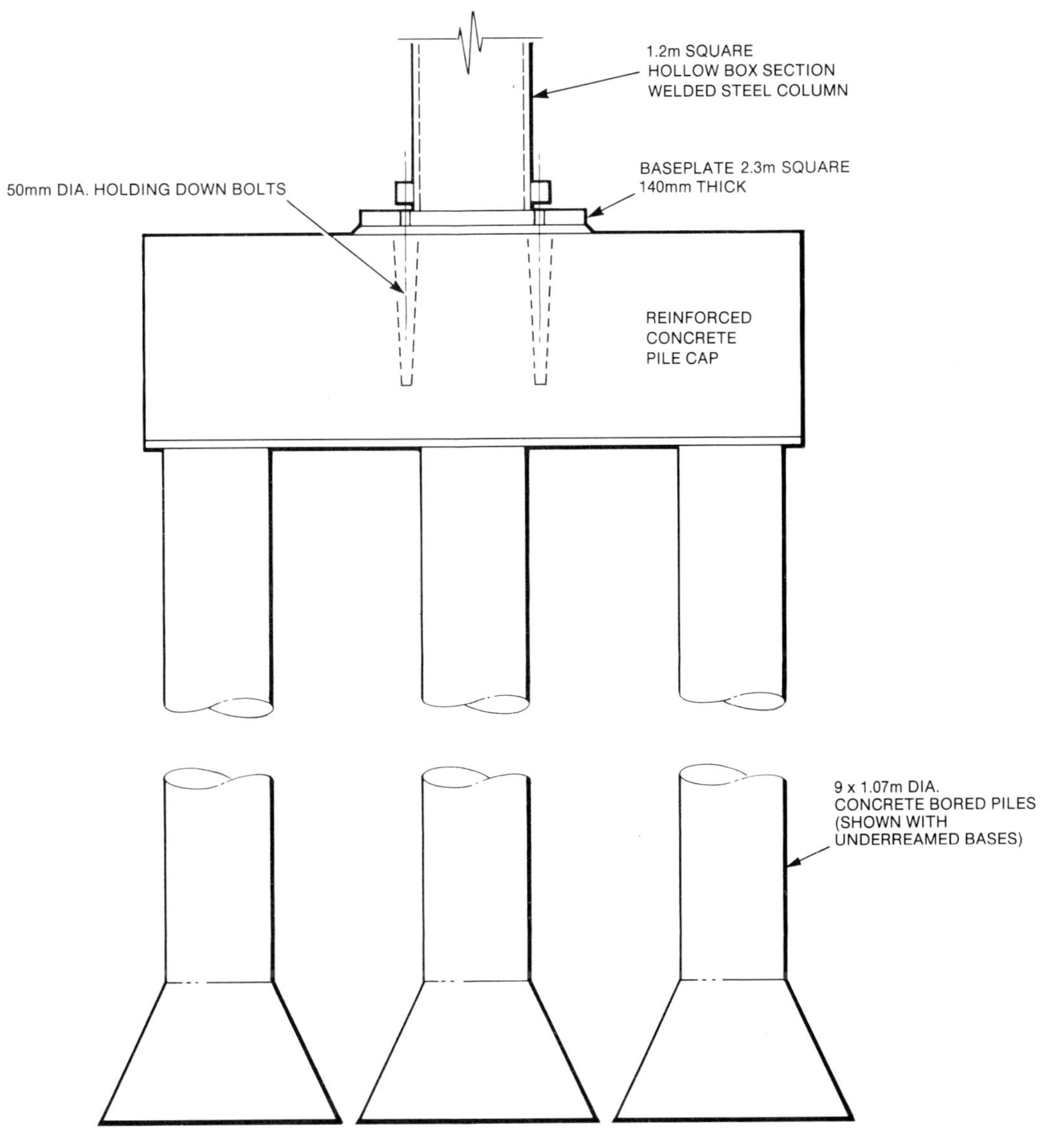

FIG. 3.11 Piled foundations — large bored piles

anti-seismic foundations are only used when it is wished to site a standard plant design in an area where the seismic hazard is greater than that for which the safety-related plant would normally be qualified. In these cases, anti-seismic foundations for safety related plant may be an economic solution enabling the standard plant to be retained, rather than redesigning and requalifing to higher seismic acceleration levels than those for which the standard design has been qualified.

To date, only two nuclear power stations have been equipped with anti-seismic foundations. These are at Koeberg, South Africa and Cruas, France.

The CEGB's Heysham *2* power station and SSEB's Torness station are unusual in that the concrete reactor pressure vessels are founded on a special radial annulus filled with bitumen. This has the opposite effect to seismic isolation, since the bitumen allows slow expansion and contraction of the vessel to take place during prestressing and pressure loadings during their service life, but 'locks' the vessels in place under rapid dynamic seismic ground motion.

More complex devices which will isolate against vertical as well as horizontal motion are currently being developed for fast reactor plant. See 'Seismic Isolation of Nuclear Plants' by A H Hadjian [4].

5 Foundations design and construction

5.1 Concrete

As concrete is the most widely used material in the construction of foundations some elementary notes on its constituents are given. Concrete is an artificial stone manufactured by mixing together coarse and fine aggregates, cement, water and often other components such as cement replacement materials, e.g., pulverised fuel ash, and chemical admixtures, e.g., superplasticisers.

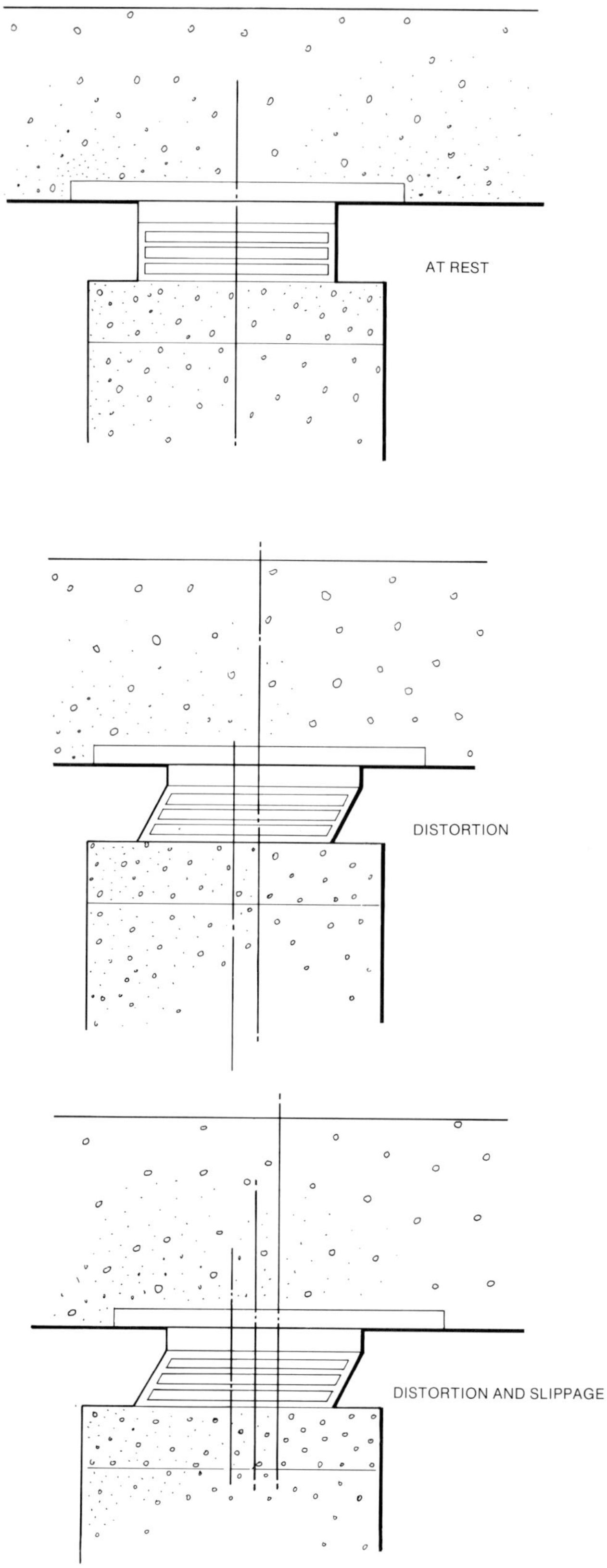

FIG. 3.12 Dynamic behaviour of sliding elastomer bearing pad

The grading of an aggregate defines the proportions of particles of different size in the aggregate. The size of the aggregate particles normally used in concrete varies from 37.5 mm to 0.15 mm. BS882 [5] places aggregates in three main categories: fine aggregate or sand containing particles the majority of which are smaller than 5 mm, coarse aggregate containing particles the majority of which are larger than 5 mm and 'all-in' aggregate comprising both fine and coarse aggregate. 'All-in' aggregate is not normally allowed in high class concrete work unless specifically authorised by the engineer.

The grading of an aggregate has a considerable effect on the workability and stability of a concrete mix and is a most important factor in concrete mix design. In practice, fine and coarse aggregates are batched separately, their proportions being governed largely by their respective gradings. The coarse aggregate may be 'continuous graded' or 'gap graded' — the latter being where one or more intermediate size fraction is omitted. It is best practice to specify a combination of 'single sized' aggregates for a mix design provided that they are combined in such proportions that the resulting grading falls within the grading requirements of BS882 [5]. Fine aggregate has to comply with the grading limits as specified by Table 5 of BS882 and is incorporated into the mix as delivered to site without regrading or recombination.

Aggregates should be clean and free from deleterious substances such as organic matter and clay. Their crushing strength should be greater than the specified compressive strength of the concrete. Whilst most aggregates are generally regarded as being chemically inert some, such as opal and chalcedony, are potentially reactive with the alkalis produced from within the cement matrix and it is possible for adverse reactions to occur such as alkali-silica reaction (ASR). In ASR, reactive silica from the aggregate reacts with free alkalis within the concrete to produce a gel. The gel material will hydrate with any water trapped within the cement matrix. If the supply of water is plentiful, such as at the surface of a frequently wetted structure, then the gel products may expand thus inducing high internal stresses which may eventually damage the concrete structure. Serious ASR effects remain rare in UK concrete practice.

Aggregates such as gravels may also contain chlorides particularly if they were deposited in a marine environment. Such aggregates must be thoroughly washed if they are to be used in either reinforced or prestressed concrete foundation structures, as the chlorides might otherwise cause the corrosion of the steel. Aggregates should not be porous or be liable to susceptible volumetric change with varying water content or be subject to damage by freeze-thaw action.

The term *cement* covers a wide range of materials all of which have the property of setting and hardening, through the application of water. The cements of importance in the construction of foundations are Portland cements, the most frequently used of which are Ordinary Portland cement (BS12 [6]) and Sulphate Resisting cement (BS4027 [7]). The type of cement used in concrete foundations is determined by the concentration of sulphates in the groundwater.

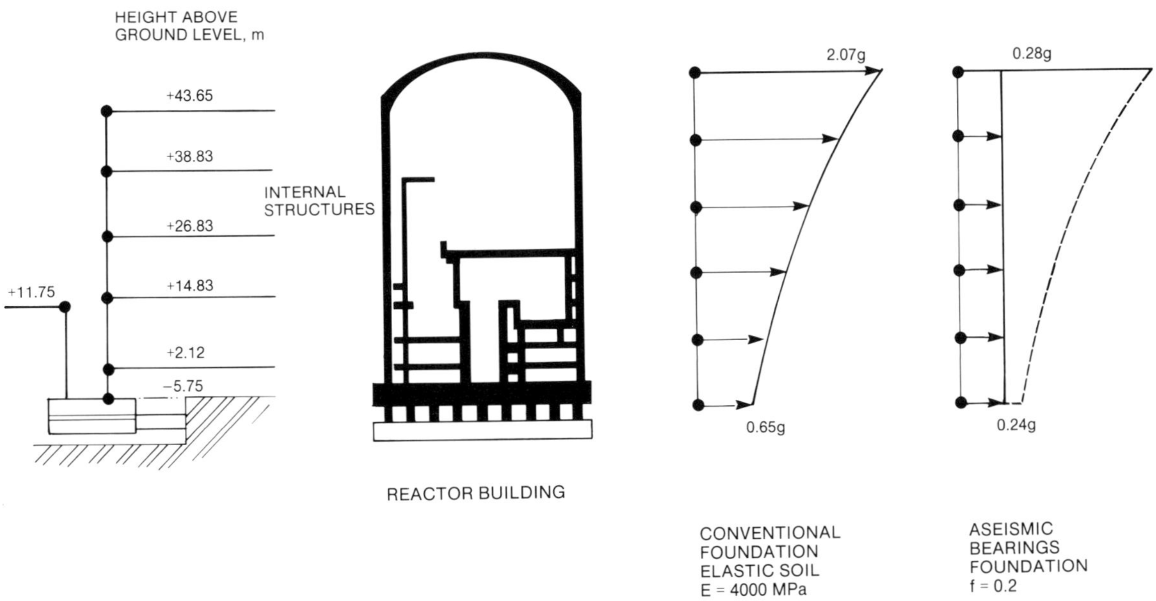

FIG. 3.13 Maximum horizontal acceleration levels in reactor building under 0.6 g El Centro quake

Portland cement is manufactured by grinding limestone or chalk with clay or shale, mixing the ground materials in specified proportions and then burning the material in a kiln to a temperature of about 1400°C. The material resulting from the process is known as clinker which, when cooled, is ground to a fine powder. The powder is blended with gypsum to form Portland cement.

Water used in the production of concrete should conform to BS3148 [8]. All potable water is suitable for making concrete. In general, concrete should be made with the minimum quantity of water that allows the mix to be fully compacted. The CEGB specifies the use of a cement replacement material, pulverised fuel ash (PFA), for all major construction projects. Pulverised fuel ash is a fine-grained material which is extracted, via electrostatic precipitation, from the flue gases of power station boilers fired by pulverised coal. The use of PFA, conforming to BS3892 [9], as a cement replacement material offers considerable advantages both to the performance of the plastic and hardened concrete. The addition of PFA to a mix enables a more durable and higher strength concrete to be produced. This is primarily due to the water-reducing and pozzolanic properties of the material thus producing a dense concrete of low porosity and permeability.

Chemical admixtures may be used, on the written permission of the Engineer, to improve the performance of either plastic or hardened concrete. Admixtures include air-entraining agents, accelerating agents, retarding agents and water-reducing agents. Admixtures containing any significant quantity of chlorides are not permitted to be used. Admixtures are required to be specified in accordance with BS5075 [10].

Traditionally, concrete has been specified by the volumetric proportions of the major constituents together with the concrete compressive strength. However, it is current CEGB practice to specify a concrete mix in terms of its characteristic compressive strength, the maximum aggregate size, the maximum water/cementitious ratio, the maximum and minimum cementitious content and the allowable range of PFA (%) to be used in the works. The specification of the aforementioned will give greater confidence to the production of a durable concrete with the required strength.

The exact choice of proportions for a concrete mix, within the limits of the specification, is left to the contractor. Such a mix is known as a *designed* mix whereas in the alternative, a *prescribed* mix, the client specifies the proportions of the constituents and is responsible for ensuring that the proportions will produce a concrete with the required performance. Other factors that have to be taken into consideration in a mix design include its workability, i.e., the ease with which it can be handled and compacted and if necessary the suitability for transporting by concrete pump. The mix must also be suitable for placing around the proposed reinforcement arrangements, in constricted sections of the work and for producing the desired surface finish.

Apart from being sufficiently workable, fresh concrete should have a composition such that its constituent materials remain uniformly distributed in the concrete during both the period between mixing and compaction and the period following compaction before the concrete hardens. Due to differences in the particle sizes and constituent materials there exists a natural tendency for them to separate. The two most common features of this separation are segregation and bleeding. Segregation may be defined as separation of the constituents of a heterogeneous mixture so that their distribution is no longer uniform. Bleeding is a feature in which some of the water in the mix tends to rise to the surface of freshly placed concrete. Bleeding may result in weaker concrete at the surface and is often associated with shrinkage cracking and the production of a 'dusty' concrete surface.

The contractor's mix design is subject to the approval of the CEGB, and final adjustments may be made following testing of the trial mixes.

Due to the large quantities of concreting involved in major power station construction works, it is found economical to establish site batching and mixing facilities. This also has additional benefits as both the contractor and the CEGB's site engineers have full and ready access to the source of concrete production and thus may ensure that strict standards of quality control are maintained.

All batching is undertaken on a weight basis and the equipment is calibrated on a monthly basis.

To consistently produce high quality concrete it is essential that good site control is maintained throughout the duration of the contract.

5.2 Bearing pressures and settlement

Foundations invariably settle under load and a limited amount of settlement can be designed for and tolerated, particularly if it is uniform over the foundation area. The allowable bearing pressure is the ultimate net pressure divided by a factor of safety. For structures supporting or containing power station plant the safety factor is usually set at a minimum figure from 2 to 3 according to the importance of the plant and structure and taking appropriate safety considerations into account. Earthworks are often allocated lower factors of safety — even down to figures not much above unity, e.g., 1.25 — for those where some form of failure would not have serious consequences.

Safety factors are also set having proper regard for the sensitivity of differential settlement. For example, a structure having rigid beam-to-column connections would be subject to high framework stresses if its column bases settled unequally.

In the following cases it is assumed that ground conditions equate to the commonly found case that the foundation is bearing on a uniform stratum overlying further strata which are stronger or as strong as the founding stratum.

For granular soils above the ground water table level the allowable bearing pressure for a three metre wide footing and a safety factor of 3 varies approximately linearly from about 500 kN/m^2, when N in the standard penetration test is 47, to about 100 kN/m^2, when N in the standard penetration test is 12. As the foundation width is increased there is a modest decrease in allowable pressure; the extent of this decrease is debatable but, as a guide, the largest raft foundation would have an allowable bearing pressure not more than 20% lower than that for a three metre wide foundation.

For shallow foundations in cohesive soils above the water table, the ultimate theoretical bearing pressure for a long strip footing is 5.14C where C is the shear strength of the founding soil stratum. For a square base the ultimate value is about 6.7C and the factor varies between 5.14 and 6.7 depending on the length-to-breadth ratio of the foundation in question. Furthermore this bearing pressure may be increased by a further wD where w is the soil density and D is the foundation depth of embedment in metres. However deep the foundation, the ultimate allowable bearing pressure must not exceed 9C, and if calculations infer a factor greater than 9 a limitation must be applied to regulate the bearing pressure. Hence, with a safety factor of 3, the allowable design bearing pressure will never exceed 3C.

The design concept of allowable bearing pressure is in reality a gross simplification of a series of complex problems, and should be employed only after careful assessment of the subsurface conditions. In reality the load distribution under a foundation is rarely uniform, being greatest at the foundation centre in granular soils and greatest at its edge in cohesive soils. The loading effect from a foundation extends beyond (below) the foundation for a considerable distance normally to the plane of the foundation. It is commonly assumed that such a zone of increased stress has a *pressure bulb* distribution and extends to a depth of 1.5 times the width of the foundation. Under a structure such as a boiler house it is common for the *pressure bulbs* of adjacent column base foundations to interact to form a single *bulb* of much greater depth. All the strata within this stressed zone require consideration when deducing a figure for the allowable bearing pressure at foundation level. This is the prime reason why loading tests on relatively small soil areas at foundation level can give misleading results.

Assuming that a proper design has been undertaken to avoid shear failure of the soil, settlement of the foundation can occur for a variety of reasons, three of which are described in the following.

5.2.1 Consolidation settlement

This is brought about by the expulsion of pore water from the soil (or rock) interstices resulting from the increased pressure being exerted from the foundation. On granular soils (sands and gravels) this occurs quickly and is usually complete by the end of construction. With cohesive soils (clays and silts) it is a much more gradual process usually resulting in settlement over a number of years, depending in turn on the thickness and compressibility of the stratum being consolidated and on the degree of disturbance induced by the new loading.

5.2.2 Plastic settlement

This is caused by underlying soft plastic cohesive materials gradually being displaced or squeezed out from under a foundation. This can happen at loading levels appreciably less than that which would cause simple shear failure of the same stratum. Also it can occur where a plastic soil is sandwiched between layers of harder material. Unless revealed in the site investigation beforehand such behaviour can be puzzling to design engineers and, even if forecast, can still prove difficult to analyse accurately.

5.2.3 Settlement due to changes of conditions

Settlement can be caused by conditions which are largely or entirely independent of the new foundation loading. Instances of this phenomenon are:

- Drying and shrinkage of cohesive soils brought about by lowering the water table.
- Saturation of granular soils brought about by raising the water table. This has the effect of reducing the safe bearing capacity of the granular soils.
- Long term seepage having the ability to wash away fine materials.
- Temperature changes, such as under a boiler house, where appreciable long term temperature increase produces shrinkage resulting from moisture loss of the stratum having contact with the foundations.

Over and above the preceding reasons, differential settlement of foundations arises from three common causes:

(a) Spacial variation of soil conditions having varying capacities to carry the uniform imposed foundation load.

(b) Variation in thickness of a compressible stratum under the foundations.

(c) Net foundation loads may vary around the structure, thereby inducing uneven ground loading. If at all possible this should be eliminated by careful design.

Mixed types of foundations are also to be avoided if possible as they invariably increase the likelihood of differential settlement. It should be borne in mind that settlement rather than bearing capacity is the critical parameter in the design of successful foundations for most major structures.

5.3 Test piling

A theoretical estimate of the load which a particular size of pile is capable of carrying may be made when the soil properties are known. Alternatively the load which a driven pile can carry may be determined, fairly reliably under favourable circumstances, from pile driving formulae when the *set* or penetration resulting from a known hammer blow is measured (this method is notoriously unreliable in conditions where dynamic and static loads produce dissimilar reaction in the soil). Examples of these calculations are provided in Appendix A of this chapter.

However, the most accurate method of determining the capacity of a pile is to apply a test load to the head of the pile after it has been constructed and this provides the most reliable source of information on which to base a design. Full information from any test piles should be made available to companies tendering for piling contracts.

The test load is usually applied to a pile by:

- Building an independent structure loaded with kentledge around the pile head and jacking from the structure down on to the pile.
- Jacking down on to the test pile from a crossbeam; each end of the crossbeam being connected to one or more piles which carry the uplift reaction in tension. The uplift piles must be sufficiently remote from the test pile to prevent interaction.

Load testing of piles can be carried out for various reasons; on preliminary piles to derive an adequate design, as described; on working piles, to verify the adequacy of this design across a large site where conditions may vary, and again on working piles, to check that standards of workmanship are such that acceptable performance can be expected from all of the piles. The need for a number of preliminary pile tests will be dictated by the uncertainty or safety factors in the design process. The number of working pile tests will depend on the size of the contract, the variability of ground conditions, and the difficulty of construction (typically on a major piling contract one working pile in every 200 is load tested). While working piles are rarely tested to more than 150% of the working load, preliminary piles should be loaded very much higher, to 250% working load or to destruction.

The purpose of a pile test is not only to assess load carrying capacity but also to observe the settlement characteristics of the pile particularly up to its working load. The applied load is measured by either a load cell or a proving ring and the displacement of the pile by extensometer gauges mounted on a separate stationary support structure (see Fig 3.14).

The pile is loaded either in increments (some fraction of the working load), each increment only being added when settlement from the previous increment has sensibly ceased, or, in the case of preliminary piles, continuously such that the pile moves into the ground at a constant rate of penetration (CRP).

For a CRP test to be worthwhile sufficient test capacity must be provided in order that the pile can be failed. To obtain the maximum information from a pile test the cycle of loading should include monitored performance under incremental unloading since this

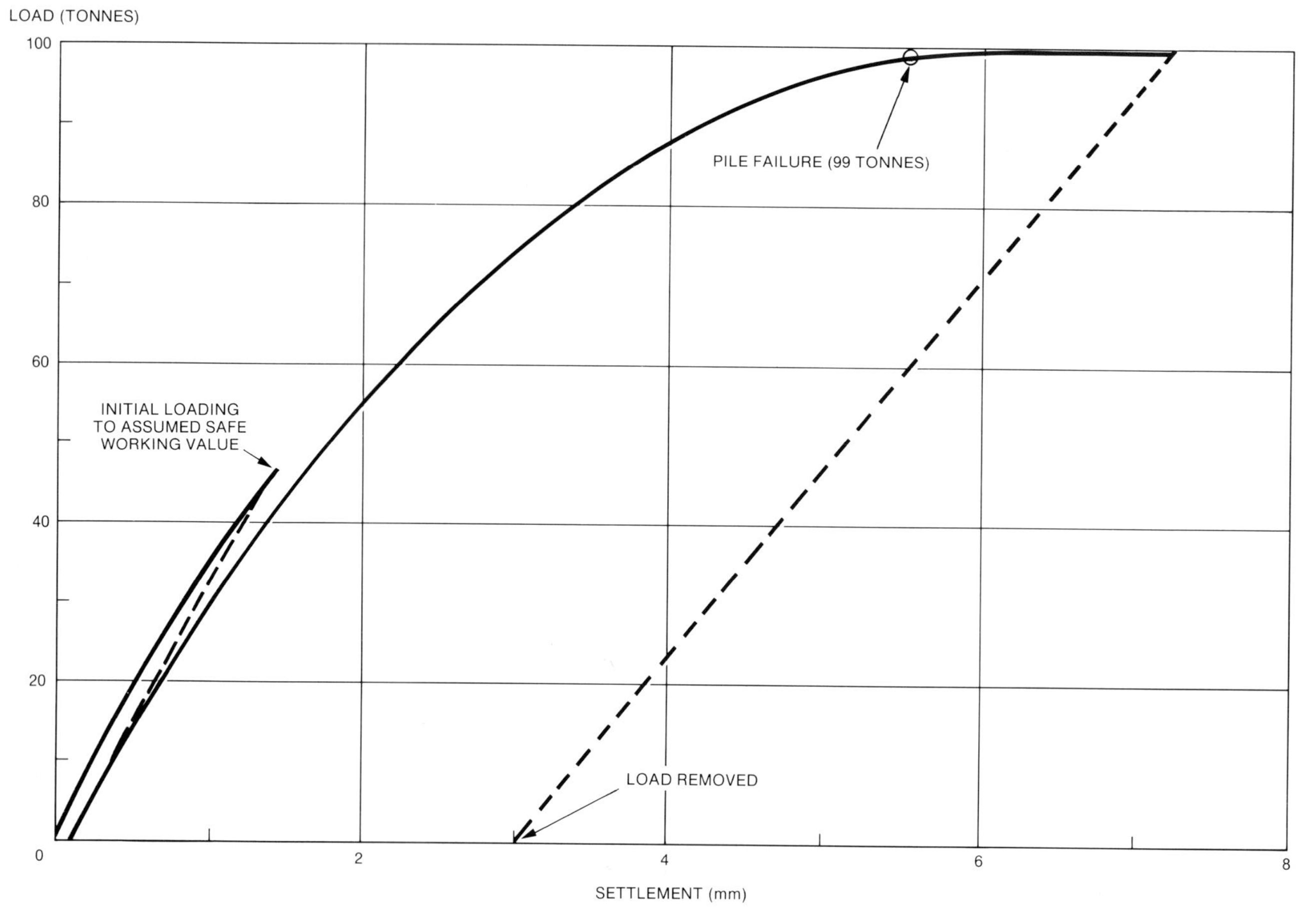

Note: Pile test on 0.4m diameter bored pile 12m long in clay (test carried out continuously over a period of 72 h)

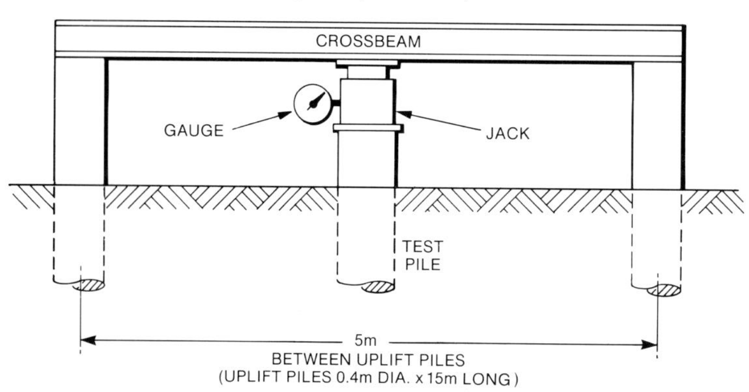

FIG. 3.14 Pile test arrangement and results

allows a refined estimate to be made of the elastic shortening of the pile.

Therefore, a typical working pile load test consists of two full cycles of incremental load to working load and then to 150% working load.

For preliminary pile tests it is often desirable to start with the same procedure and then add one or more incremental cycles to maxima and perhaps finish with a CRP load test.

The results of load tests should be recorded on standard sheets where all the relevant information is noted including full cycles of incremental load to working load and then to 150% working load.

For preliminary pile tests it is often desirable to start with the same procedure and then add one or more incremental cycles to maxima and perhaps finish with a CRP load test.

The results of load tests should be recorded on standard sheets where all the relevant information is noted including full details concerning the construction of the pile.

6 Foundations for main and secondary structures

6.1 Boiler house foundations

Boilers are the largest items of plant, and in modern power stations they are suspended from a framework of heavy steel beams just below roof level. The boilers, with their ancillary plant and associated coal bunkers, comprise a very high proportion of the total load. The foundations provided are largely required to carry loads arising from the plant, the loads due to the buildings being only a relatively small proportion of the total.

Since it has been CEGB practice for many years to build completely integrated boiler/turbine-generator units it follows that the foundations layout is repeated about successive centre lines of the boiler and turbine layout, say at 100 m centres for the current coal-fired reference design.

Pile caps or other types of separate foundations are provided for structural steelwork and specific heavy plant items such as the fans, mill foundations, air-heaters and the ash removal system. In particular the heavy inertial loads generated by the mill operation require careful detailing to provide isolation of the other foundations and plant from dynamic loading.

In general, the floor plant area will be predominantly occupied by the discrete foundations described. As infill will be required to facilitate construction and subsequent operation, this can take the form of a reinforced concrete slab, say 300 mm thick, designed to withstand heavy wheeled traffic and laid on well compacted fill. A typical layout of a power block foundation for a 2 × 900 MW station is shown in Fig 3.15.

6.2 Turbine hall foundations

Whilst substantial loadings will result from the main structural steelwork forming the turbine hall framework and supporting the main overhead cranes, other substantial loadings arise from the turbine-generator blocks. The layout adopted for the turbines whether transverse, longitudinal or angled, will affect the loadings arising from the building itself in several ways. Hence layout needs to be established before serious foundation design can start.

Other turbine hall plant items can be supported economically on a simple combined pile cap of uniform thickness surrounding, but separate from, the massive discrete foundation blocks supporting the heavier plant and from the pile caps holding the building's structural steel frame columns.

In the CEGB's coal-fired station reference design it is proposed that the turbine hall floor should consist of a pile-reinforced concrete slab 750 mm thick, surmounted by a 200 mm mass concrete topping. Site drilling of plant fixings into this upper layer then can be undertaken without hazarding the structural slab.

In contrast the turbine-generator block sub-foundation and main steel column bases would be separately supported on pile caps some 2.5 m thick.

The turbine hall foundation layout is considerably simplified by minimising CW culverts and cable tunnels below the building, hence avoiding the complicated forms which these items necessitated on previous stations.

Figure 3.15 shows the turbine hall layout for a 2 × 900 MW station and its relationship to the boiler house.

6.3 Turbine-generator blocks

A turbine-generator block provides support for the machine in its static and cold condition and in its hot and rotating condition. That support extends the full length of the shaft at its base level but is normally separated to support particular shaft bearings individually at machine operating level.

The height of a turbine-generator block is dependent on the type and disposition of the condensers, the requirements of the operators and the costs or savings involved in constructing a basement in relation to the capitalised cost of pumping cooling water to a greater height.

The block has openings within it to accommodate plant and pipework and is itself carried on a sub-foundation.

Turbine-generator blocks are made in reinforced concrete or steel with the condensers often placed under or alongside on adjacent plinths. This arrangement gives a maximum basement height of about 12 m. Some units are constructed with condensers not located under the machine. This allows the height of the block to be less, consequently reducing both the height and capital cost of the turbine hall. One such arrangement is for the condensers to be placed on each side of the machine, these being known as pannier condensers.

Most turbine-generator blocks used to be built in reinforced concrete but an alternative is to construct in steel. This reduces the foundation load, and being more slender permits markedly better access and easier layout beneath and around the machine for cooling water pipes and other plant. In the case of a concrete block it is an advantage if it is built ahead of the machine erection in order to allow hydration thermal shrinkage to occur, whereas with a steel block this problem does not exist. As the material properties of steel are more consistent and more accurately known, the analysis of differential settlement problems is less difficult. The compatibility and better control of the properties of the construction material enables the dynamic design of the block to be done as part of the overall machine design.

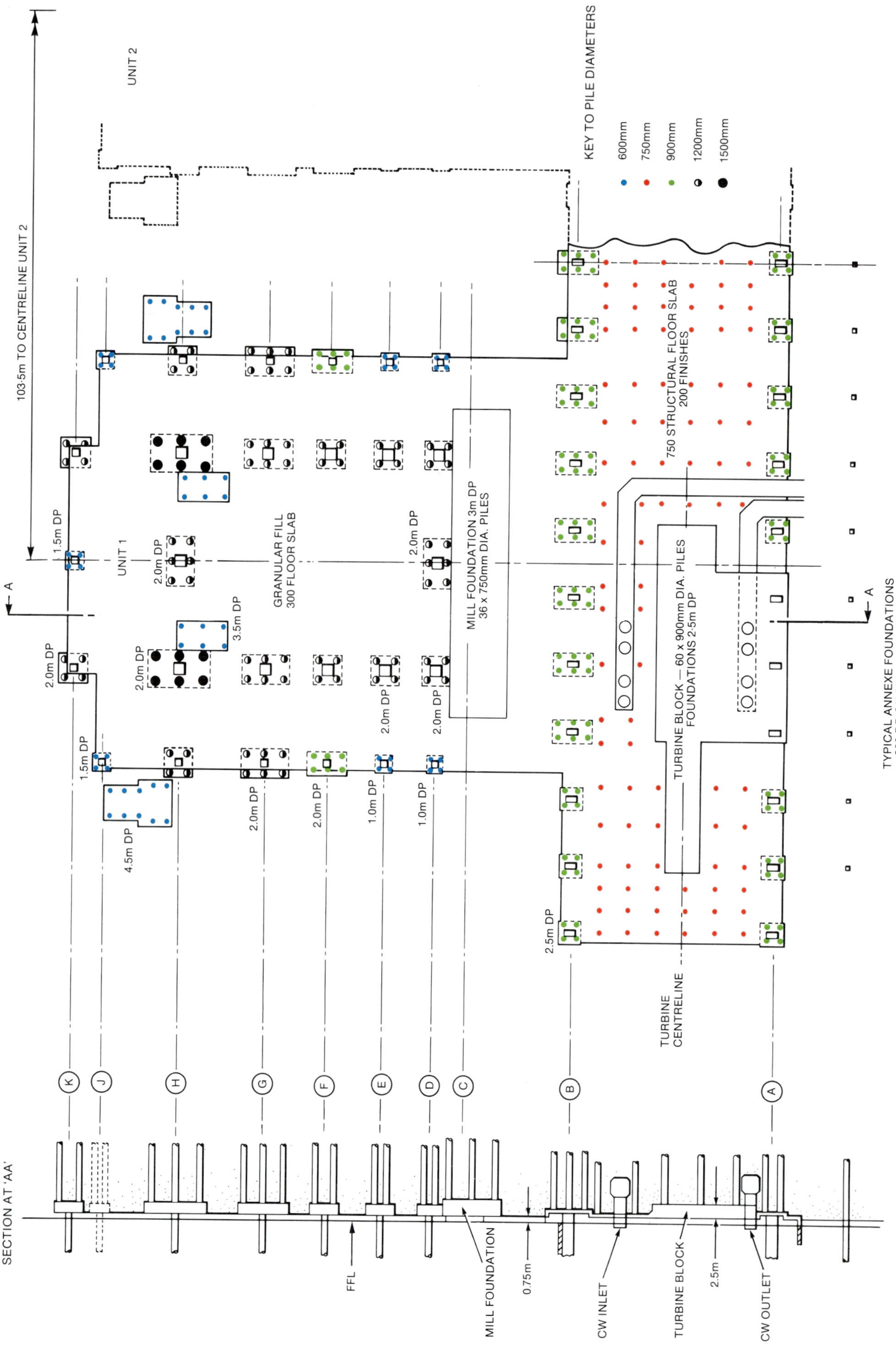

Fig. 3.15 2 × 900 MW coal-fired station power block foundations

A static loading diagram for a 500 MW unit is illustrated in Fig 3.16. This shows a concrete block with axial condensers beneath the machine.

A design of a 500 MW unit on a steel block and with pannier condensers is illustrated in Fig 3.17. The reduction in height of the unit made possible by the choice of condensers should also be noted.

Whichever type of turbine-generator block is ultimately adopted the requirements and basic principles of the design are similar. The plan of the block is determined by the machine designer who provides the civil engineer with the position of bearings and the loads and tolerances cold and hot. Specified differential movements subsequent to alignment are remarkably low, being in the region of 0.01 mm reducing to 0.005 mm for points close together, though actual values may vary with different manufacturers. In this respect it is an advantage if all bearing supports can be mounted over piers or columns rather than on beams. Certain geometrical modifications may be permitted by the manufacturer in order to facilitate construction of the block, to assist the designer to keep deflections within the prescribed limits or, to ensure that resonance will not occur at or close to the machine's normal running speed, thereby minimising operational vibration levels.

With concrete blocks, homogeniety and lack of shrinkage and thermal movement pre-compensation are likely to be more important than rapid setting, finish or strength characteristics.

There are numerous theoretical advantages in adopting special concrete mix designs for turbine generator blocks, but insufficient numbers are built to justify firm conclusions. One option more suited to concrete blocks than steel is to pre-compensate the levels of the bearing pads so that when the block is heat-soaked the greater expansion of the steam end of the block brings the main shaft into alignment at operating speed and temperature.

At the steam end of the turbine foundation care must be taken to ensure that the concrete is shielded from the high temperature parts.

In some cases it is necessary to provide a reflecting or insulating shield. Special care should be taken to provide reinforcement to take care of the temperature stresses.

Although a good quality concrete is required, a very high strength is not necessary as the concrete stresses are comparatively low owing to the considerable cross-sectional area of most members. The average quantity of reinforcement required is likely to be about 1% of the appropriate concrete cross-section. This should be placed vertically, longitudinally and transversely in all structural members, to prevent any possible cracking due to vibration, even though it may not be theoretically required in all planes.

Some blocks have been constructed from pre-stressed concrete, considerably reducing the weight of steel used, and taking fuller advantage of high strength concrete by obtaining a better balance between effective compressive and tensile behaviour. Shrinkage cracks are reduced, and shrinkage induced more quickly. This often requires even more care in design to avoid overheating causing large losses of prestress.

Concrete turbine-generator blocks are cast in sections in about four vertical 'lifts' in a sequence which gives an approximately balanced load on the foundation. If the programme time permits it is an advantage to allow a period of one month between concrete placing of 'lifts', to allow temperatures in the concrete to return towards ambient levels.

The design of the block's structural form and components has to be such that their natural frequency (or its harmonics) are at least ±20% different from those of the machine at normal operating speed (e.g., 50 Hz for a 3000 r/min machine) in order to avoid resonance. For a given material the frequency of vibration of any member may be changed by altering the dimensions or sections but not by prestressing. Slender cantilevers and thin diaphragm walls are particularly liable to vibration and attention should be paid to their natural frequency. If necessary, the section of these members should be increased.

In the case of the vertical columns, and other substantial members, there may be a choice between raising or lowering the natural frequency. The latter choice may be the more suitable as it would represent a considerable saving in material, providing it leaves the structure sufficiently strong. However, this means that these frequencies must be passed through every time the machine is run up or down.

A massive monolithic foundation is essential in order to provide a stable base for the turbine-generator block and to absorb vibration. The thickness of the foundation should not be less than one-tenth of its length, and a foundation of this type is shown in Fig 3.18. It is not piled in this instance, as it is founded on a firm stratum at this extra depth. Discontinuity between the block foundation and the basement floor is attempted to provide some vibrational isolation throughout the turbine hall. However, cooling water culverts passing through form a direct connection, but these can be made relatively flexible by using suitable joints. Cooling water and the surrounding ground contribute to vibration damping.

Relative measurements are made to determine any change in the level of the block which could cause rough running of the machine as a result of movement of the foundation or of the block itself. Steel levelling plates are cast into the block around both the basement and operating floor level. A separate reference point is provided in the turbine hall, which may have to be carried on an isolated pile driven through a hole in the basement floor and retained completely independent of the floor. Measurements from the reference point and around the levelling plates are carried out using an optical micrometer level or a micrometer water level. An Invar rod is installed to measure down from operating floor level.

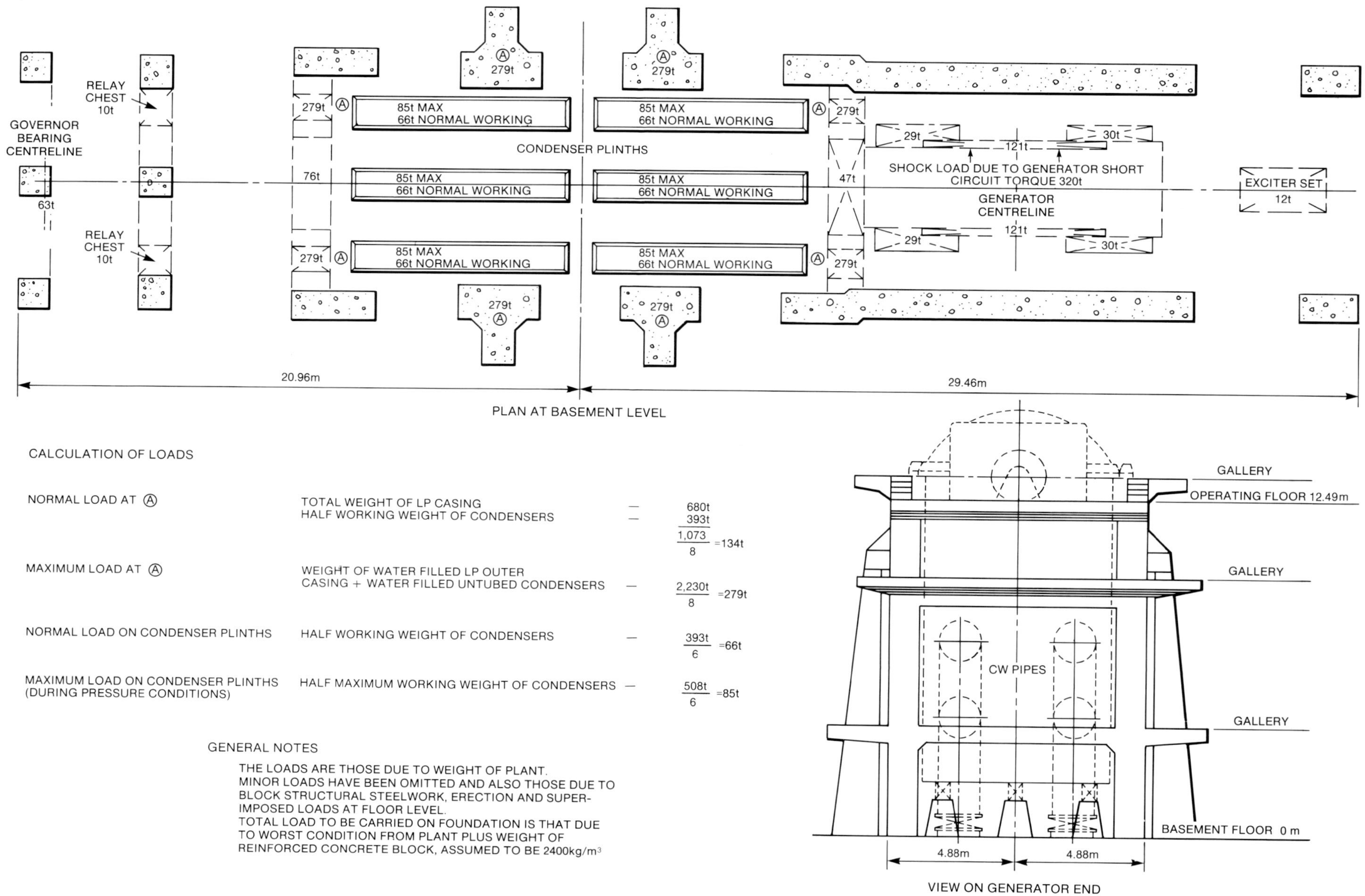

FIG. 3.16 Simplified loading plan for 500 MW turbine-generator

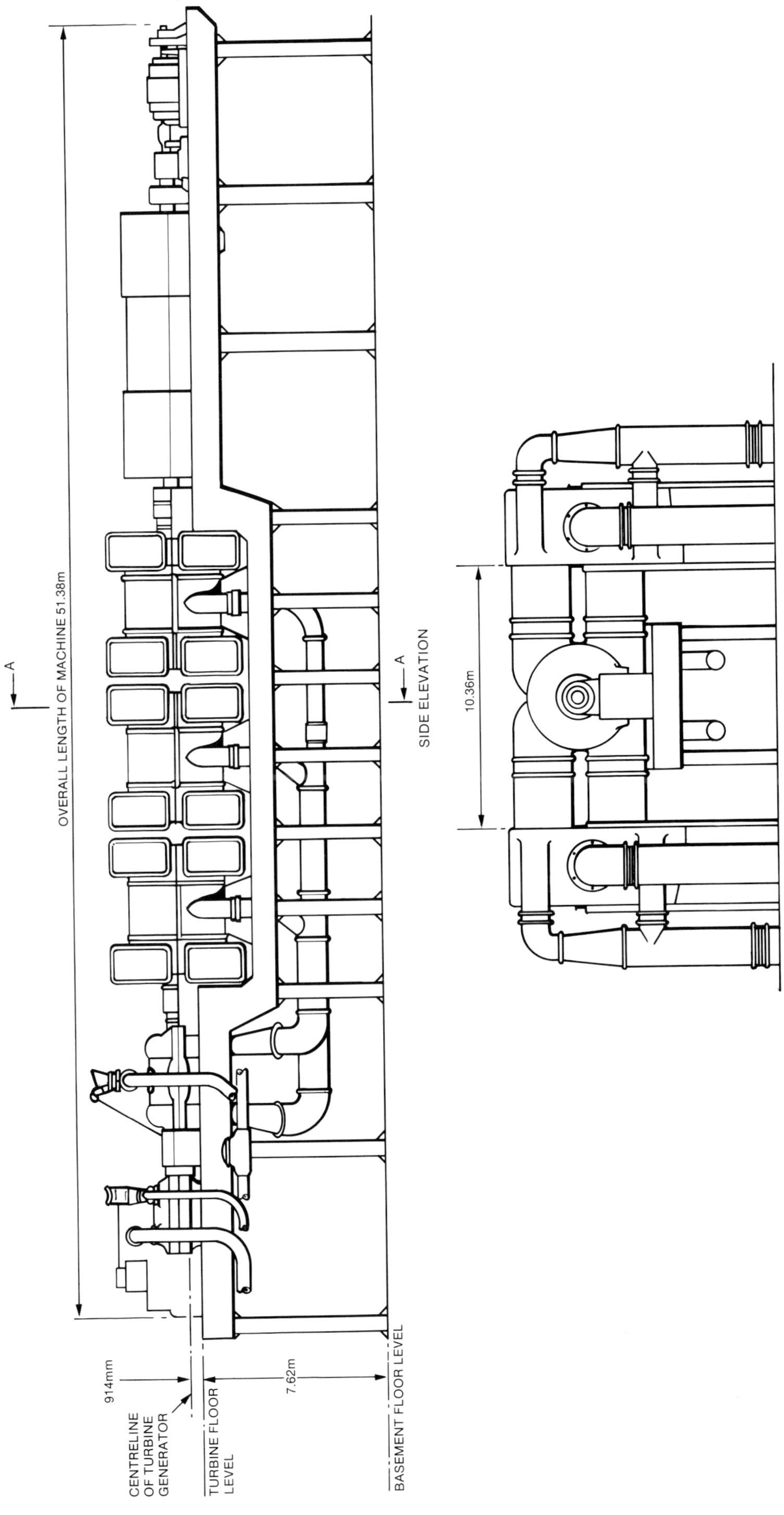

FIG. 3.17 Arrangement of 500 MW turbine-generator on steel block with side mounted condensers

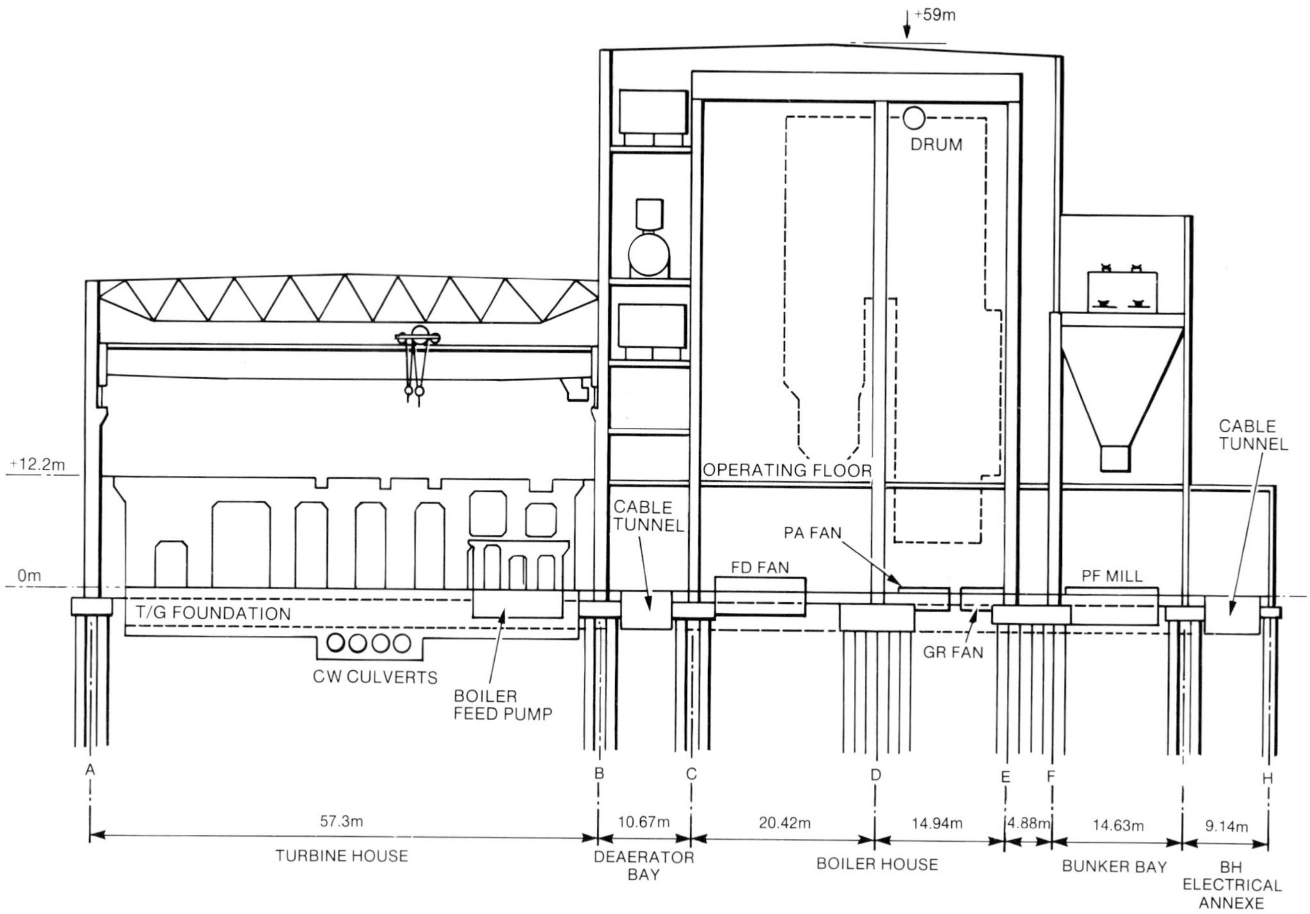

FIG. 3.18 Section through main buildings

In addition to these readings some blocks have been fitted with a large number of remote reading instruments buried at strategic points as construction proceeded. Strain gauges, thermocouples and moisture cells have been used in combination for this, and give some indication of the behaviour of the block under the influence of changing load and long term heating from the turbine-generator itself.

6.4 Basement of ground floor

The choice of level for the basement of operating floor is dependent on various influences, some of which, e.g., hydraulics and layout, are discussed elsewhere in this book. The civil engineer's aim is to provide a suitable foundation for the main plant at the least capital cost, situated not so low as to involve expensive dewatering and not so high as to require imported fill. A cost balance can be arrived at for any given site, which is fine for producing capital estimates for the civil elements. However, other disciplines with a vested interest in station levels show more substantial revenue costs or savings, such as pumping costs and maintenance access. Hence the optimal level to minimise lifetime costs needs to be calculated from a range of levels through progressive iterations.

Building and landscape architects have an input which may also call for planning decisions before the final cost optimisation is done. The final choice from this wide parametric study is unlikely to coincide with the civil engineer's preference and extra costs will inevitably be incurred.

The designer therefore has to accept the layout and levels decided by a range of interests and must tailor his structural design to suit the engineering ground conditions and the construction programme within these external constraints.

Boiler house and turbine hall foundations have been covered in the preceding sections, but the depth of basement will affect the choice of foundation from the range of buoyant, semi-buoyant, raft, piled, cylinders and contact types. For the main foundations it is not easy, or usual, to combine more than two of these methods in similar ground.

If a basement is proposed, the design must take into account the temporary stages when the excavation will heave at the bottom due to being unloaded and later when the whole basement may tend to float. Pressure relief, semi-permanent ground dewatering or ground anchors may have to be introduced.

6.5 Track hoppers

Coal-fired stations requiring some 20 000 t of coal per day need extensive bulk handling facilities for delivery, storage and retrieval of fuel. Track hoppers, into which a train can discharge its load whilst moving, form the biggest single foundation in the fuel handling system, followed by those for the boom stacker and reclaim track.

Figure 3.19 shows typical reinforced concrete hoppers which are about 70 m long and have a throughput of over 1000 t per hour. Coal is removed from the hopper onto conveyors by paddle feeders which can travel the length of the hopper.

The track hopper extends rather more than 10 m below ground level and has to withstand considerable earth pressure. Also, if the ground water level is high, precautions have to be taken to prevent the empty hopper's tendency to float. This may be done by increasing the dead weight of the whole construction. Another way is to extend the width of the base slab, as shown dotted in Fig 3.19(a), and use the weight of soil above the projections to counteract the tendency.

The beams supporting the track over the hopper should be as narrow as practicable so that the discharge of coal from the railway wagons is not impeded. A smooth tile or cast glass lining to the hopper face assists coal flow.

The design of hopper shown in Fig 3.19(b) provides improved handling of coals which have poor flow qualities.

6.6 Chimney foundations

The adoption of tall, multi-flue chimneys in the mid 1960s to reduce ground level concentrations of sulphur dioxide from the main boilers has resulted in structures of 200 m to 260 m in height and weighing 20 000 t to 30 000 t.

Foundations have taken several forms but generally comprise a thick reinforced concrete slab supported in some of the following ways:

- On driven or bored pattern piling with both vertical and raking members.
- On large diameter concrete foundation cylinders. These are adopted when conventional pile driving might require impractically-close pile pitch or deep multiple piling where the as-built geometry would prove extremely difficult to meet the design.
- The slab is replaced by a cellular foundation where only low bearing pressures can be tolerated. The hollow cellular form allows some pre-loading of the sub-base before chimney building commences.
- As a reinforced concrete contact foundation where high bearing pressures are tolerable, such as on a high rockhead site.

Modern design methods require that the dynamic response of tall chimneys to cross wind effects should be established, calling for study of the interaction of the chimney superstructure, the foundation and the underlying soils. Hence the initial choice of foundation scantlings or even forms based on static wind loadings and sized to be within allowable ground bearing pressures, may require modification in the secondary stage of design when the dynamic response is being estimated.

Two types of modern multi-flue chimney foundation are illustrated in Figs 3.20 and 3.21.

6.7 Cooling tower foundations

The foundations for a cooling tower have to be considered in the light of the various components which constitute an operating tower.

Since the intensity of loading will vary in both form and magnitude from component to component, suitable isolation in the form of expansion joints must be provided between them to allow for differential movement and settlement.

6.7.1 The cooling tower shell and shell support columns

The shell and its support columns can be carried on a discrete pier at each lower node point, i.e., the junction of each column pair at their base. Alternatively the columns can be carried on a continuous circular foundation ring beam. The need for piling in either case will depend upon the proximity of a suitable bearing stratum to ground level.

To achieve acceptable load paths from the shell to the foundations, the support columns follow the basic geometry of the lower ring. As such they are raked both radially and circumferentially and the traditional X or W form results from this geometry of structure.

The need to allow in the design for the lateral thrust component of the force in these raking columns will, by analogy, necessitate raking piles where piling is considered suitable or will require a means of tying discrete footings together in the case of contact foundations (see Fig 3.22).

6.7.2 The packing support structure

Since this normally comprises a square grid of columns at 6 m to 9 m centres, it is generally adequate to provide a single pile or isolated footing at the location of each column. Largely shielded by the shell from wind loadings, these foundations are virtually uninfluenced by horizontal loadings.

6.7.3 The pond floor

This generally supports a head of water not exceeding 3 m depth and unless supported on very poor ground

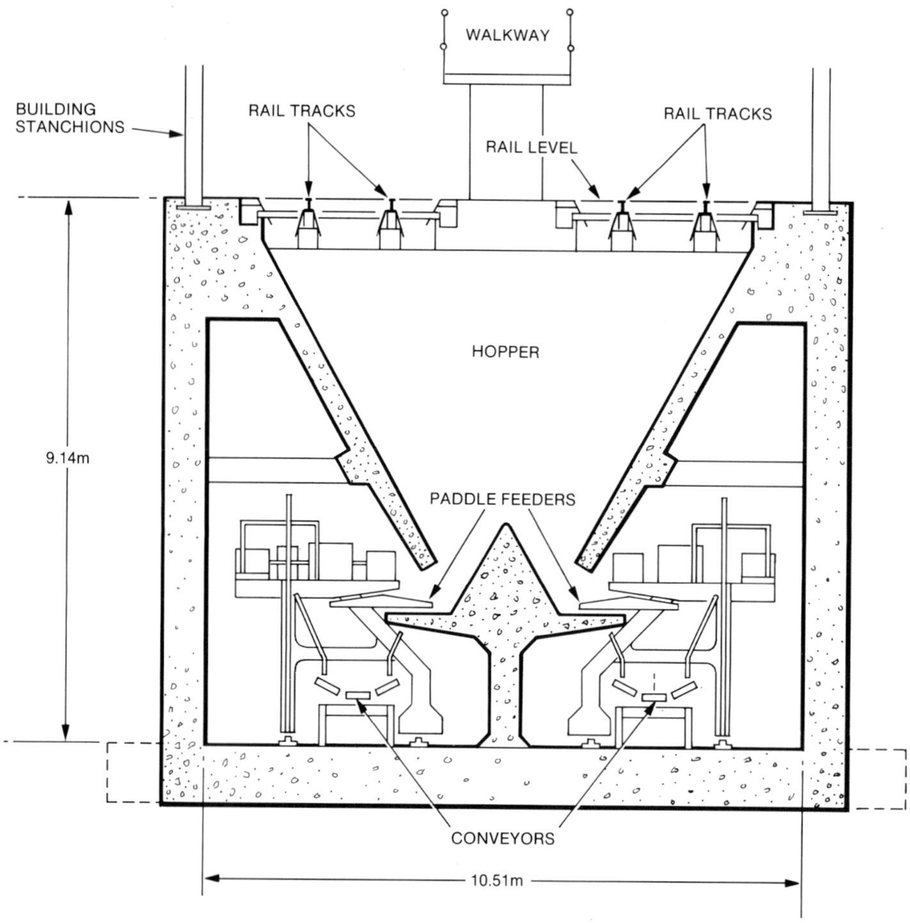

A

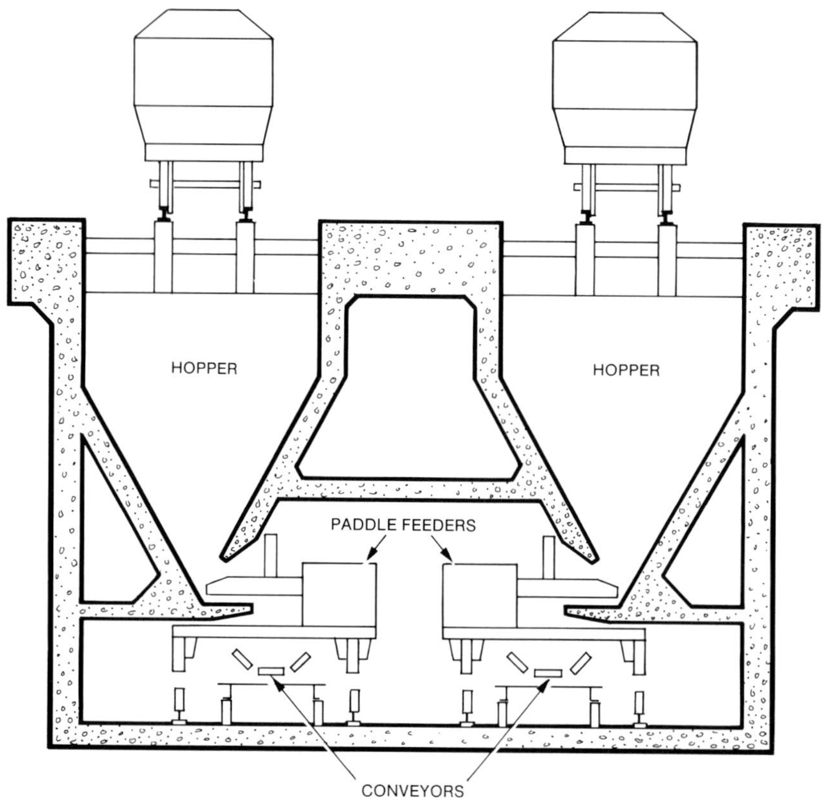

B

FIG. 3.19 Reinforced concrete coal plant track hopper designs

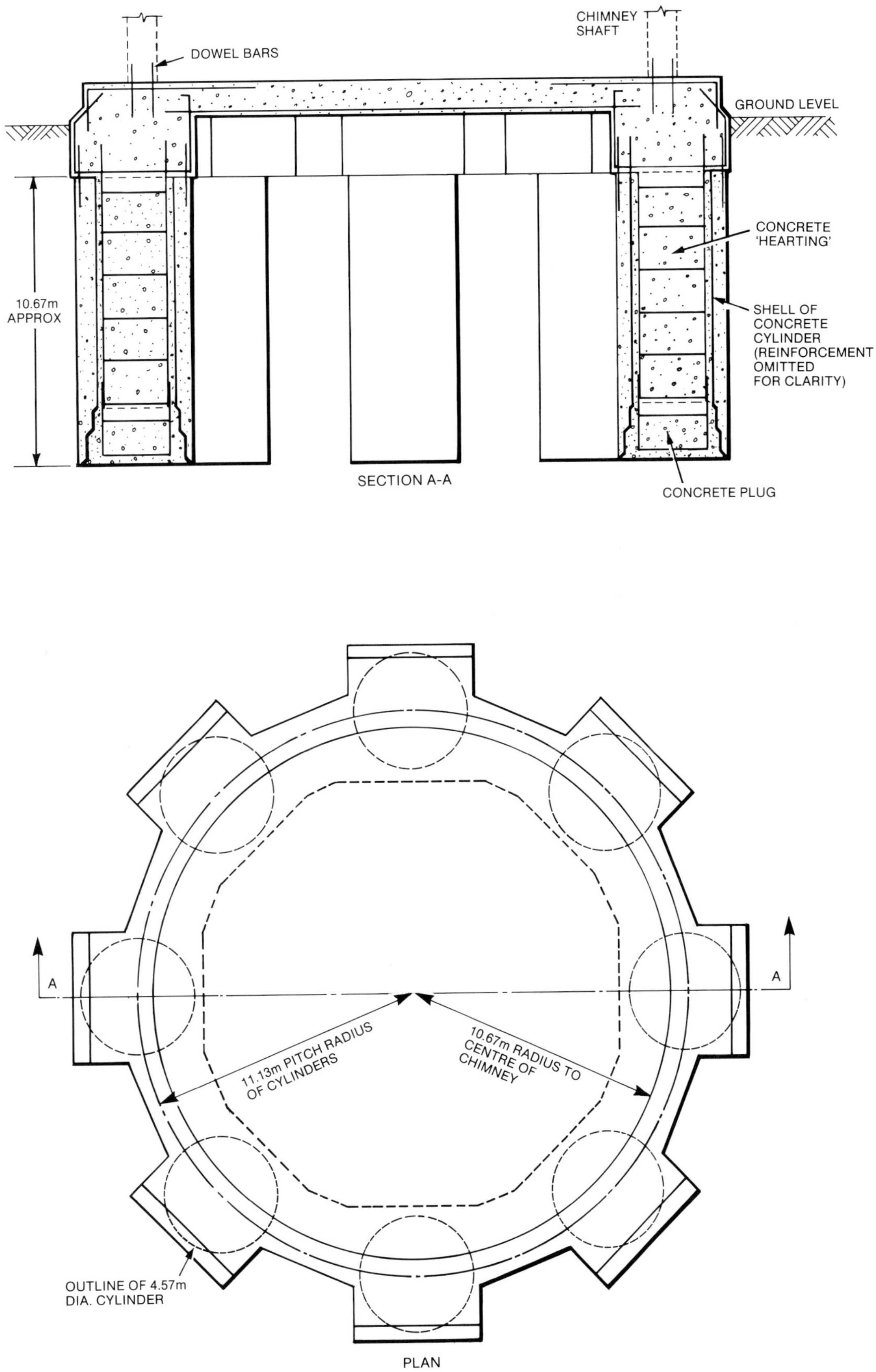

FIG. 3.20 Chimney foundations — cylinder

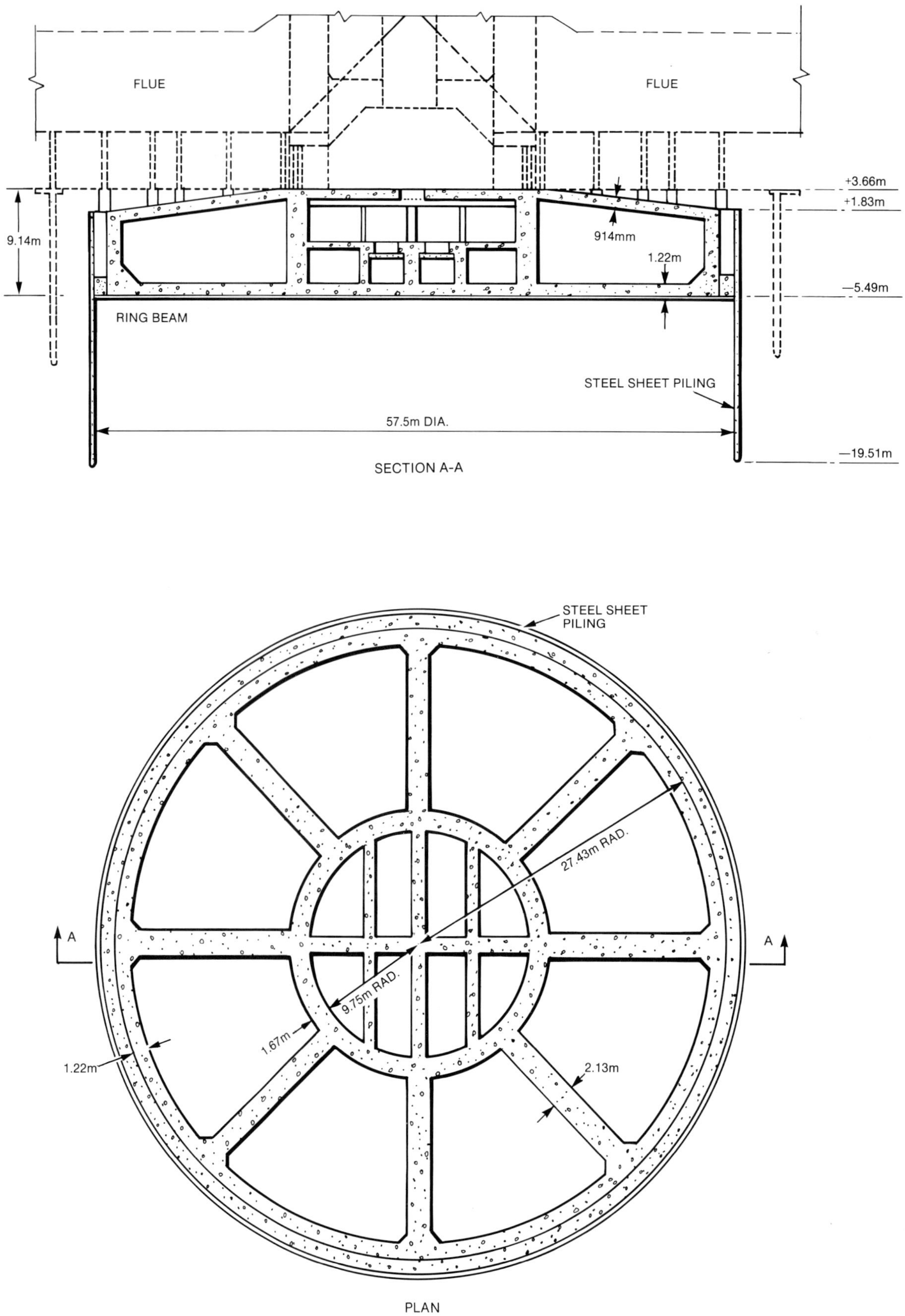

FIG. 3.21 Chimney foundations — cellular

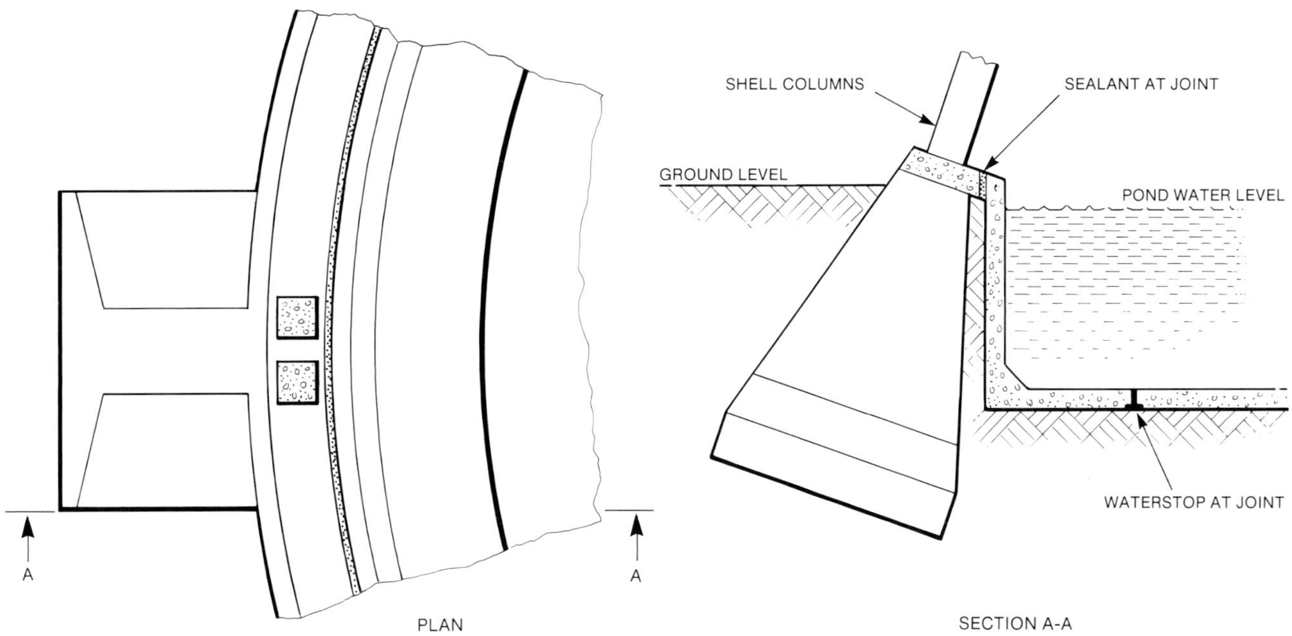

FIG. 3.22 Pond water sealing arrangements

will comprise a simple continuous 180 mm reinforced concrete slab with adequate expansion joints. Construction joints should be carefully detailed and constructed to avoid excess leakage and hence deterioration of the underlying formation. On poor ground the pond floor will require simple pattern piling which has to be fully compatible and probably integral with the packing support piles.

Since operational requirements necessitate the emptying of the ponds on occasions, due attention must be paid in the design to flotation where deeper ponds are built below basic ground level.

6.7.4 The distribution culvert

The culvert that carries water from outside the tower to a distribution system above the packing is a relatively heavy reinforced concrete box culvert of several possible geometries. This culvert presents few problems in the foundation design from its vertical loading, its length will produce lateral forces on the support columns as the structure expands at full operating temperature from a cold start. Hence bending moments will result. Isolation of this foundation from the surrounding pack support foundation is usual.

6.8 Reactor foundations

The foundations for all nuclear reactors are subject to particularly stringent design and construction requirements. The advanced gas-cooled reactors (AGRs) in particular represent exceptionally high concentrated loadings when contained within prestressed concrete pressure vessels and typically weigh 40 000 to 60 000 tonnes per reactor. Such massive single loads have nearly all been founded on deep concrete foundations bearing directly over a large area onto competent soil or rock. This simple foundation concept is relatively easy to analyse and monitor and minimises the effect that hazard factors, such as seismic loading, have on the competence of the structure.

Figure 3.23 shows a cross-section through part of a CEGB prestressed concrete pressure vessel and its foundations. Here the total weight of the loaded reactor vessel is transmitted to the ground through a mass concrete base of 25.9 m diameter. The gross pressure on the underside of the mass concrete base is approaching 1300 kN/m^2, although the net pressure is over 200 kN/m^2 less due to overburden pressure relief as the base is founded over 11 metres below ground level. Long term settlement records are kept by means of precise levelling techniques as onerous settlement criteria are imposed on such critical structures to ensure that critical clearances between reactor components are not prejudiced.

In this case the 3.35 m thick concrete base was poured in three equal height lifts, each lift being subdivided into bays so that the plan locations of all vertical joints were staggered. The pouring of adjacent bays was programmed for a minimum period of four days to allow the major proportion of thermal shrinkage to occur prior to pouring the adjacent concrete. This reactor (as most others) has a surrounding reinforced concrete retaining wall of 0.75 m thickness which carries most of the considerably lighter external portions of the circumferential reactor building. The

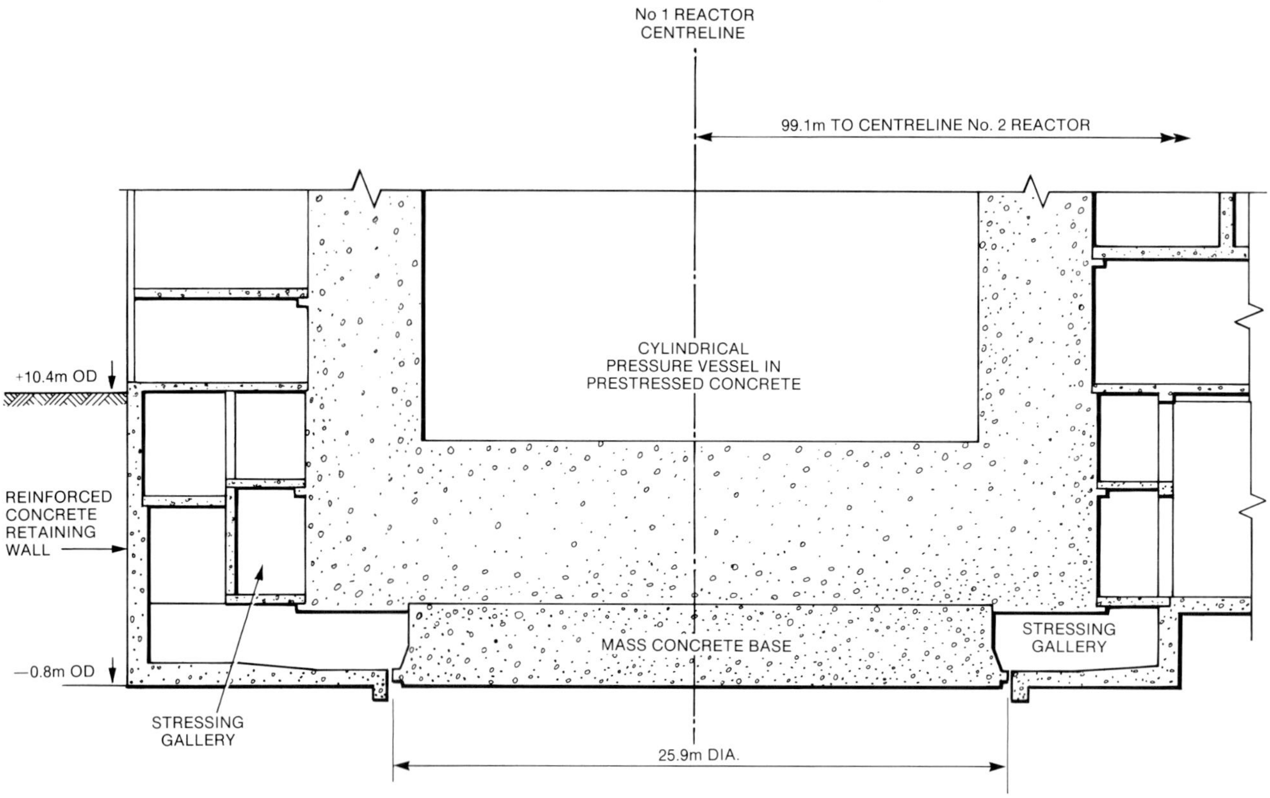

FIG. 3.23 Reactor foundations

retaining wall has a completely separate foundation so as to isolate settlement effects of the reactor base. The only linkage between the two is a continuous horizontal rubber water bar.

7 General site works

7.1 Flood embankments

In order to obtain an adequate supply of cooling water from natural sources many stations are located near to rivers. The locations may be on low-lying land which is subject to flooding. In such cases it may be possible to raise the whole area of the station above flood level by bringing infill from outside sources. If the level has to be raised appreciably, or there is a shortage of fill material in the district, only the area occupied by the buildings may be filled and the cooling towers protected by a flood embankment. Surplus soil arising from excavations for foundations may be suitable to form the embankment, which will vary from some 1 m wide at the top, with slopes on each side of 1 vertical to 1.5 to 2 horizontal, the top being some 1 m above the highest probable flood level. Impervious soils make good flood banks, but sands and gravels need to incorporate a central wall of clay, concrete or steel sheet piling to prevent the passage of water, or be surfaced with an impermeable membrane of clay, concrete, etc.

7.2 Roads

The layout of roads for a power station aims to provide satisfactory routes both for the construction traffic, and for the power station traffic once the station is in operation. Construction traffic is heavy and frequent and provides a severe test for the roads as it includes, not only heavy lorries, but tracked vehicles and large transporters. Permanent traffic may be light on certain sections of roads, but other sections will have to withstand the frequent passage of heavy coal and/or ash lorries, the occasional heavy load from plant transporters or mobile coal handling equipment.

Road widths vary from 3 m for minor roads up to 7 m for the main perimeter road. The layout must be designed with bends capable of accommodating the largest transporters. Clearance is required for these beyond the inner curve of the road at sharp bends in view of their considerable overhang. A transporter carrying a 570 MVA transformer has a total weight of approaching 300 t and is more than 30 m long with six axles at each end and with eight wheels to each axle. A road tractor at both front and rear is generally required.

The design of the road and the depth of construction will depend on the strength of the sub-soil, its suitability for compaction, and the intensity of the anticipated traffic loading. The road foundations must be kept well drained by sub-soil drainage, or the roads built on shallow embankments if practicable. At this stage, allowance should be made for cable and pipe cables and services crossings under the roads, for the roads to cross cable reserves, and the external surface water drainage to be constructed. This drainage will prove invaluable in helping to keep the site dry and clean during construction.

The nature of the sub-soil influences the choice of road design between two types — rigid or flexible. In both types the intention is to spread a high intensity of wheel loading at the road surface, through succeeding layers of progressively weaker material until the bearing load at foundation level is acceptable.

A typical example of flexible road construction would be a sub-base of rolled granular material, such as sandy ballast of say 300 mm thick, depending on the strength of the ground, underlying a base of 200 mm of well-rolled brick hard core. The road surface 100 mm thick, would consist of tarmacadam, bitumen macadam or asphalt, in ascending order of cost, laid in two layers of base course and a thinner wearing course of finer material. Asphalt provides the more impermeable surface and is more resistant to traffic wear. Materials such as cement-stabilised granular soils, pulverised fuel ash, and lean mix concrete are also used for bases and sub-bases, their choice being governed by local availability and economic considerations. Flexible roads are easy to repair where individual areas become damaged or subside, but may need extensive resurfacing after use by construction traffic.

Rigid roads are generally constructed of reinforced concrete up to 300 mm thick; the reinforcement being a welded mesh fabric. The base should consist of up to 150 mm of rolled clinker ash, lean mix concrete or granular material. Rigid roads must be provided with suitable joints at intervals to allow for expansion and contraction. The surface of a good quality concrete road is easy to clean and is resistant to damage, but where uneven settlement occurs the slab will ultimately crack and effective repair is then expensive. This type of construction is often preferable for the heavily used perimeter road.

7.3 Drainage

The water to be disposed of from a power station consists of:

- Surface water, sometimes called stormwater, which is the run-off of natural water from all the various surfaces including paved areas, roofs and unpaved land.
- Foul water, sometimes called soil or sewage, which is the water-borne human and domestic waste discharged from lavatories, ablutions and kitchens.
- Plant water, which is that used for cooling and a variety of other processes inside the power station from where it is discharged to waste.

These effluents are disposed of through separate systems since they require different purification treatments before discharge. Treated effluents from plant or foul water may join the surface water system at or before final discharge or they may be routed through stilling ponds or through the cooling water outfall, the arrangement depending on the station site layout.

Figure 3.24 shows a typical drainage plan where the manholes are numbered and the levels at the bottom of the drain are given.

Surface water is normally uncontaminated and may be discharged without treatment to an adjacent stormwater sewer, river, estuary or the sea. Drainage from vehicle hard-standing areas may also enter the surface water drains via a petrol/oil interceptor. The sizing of the drains at their point of discharge depends on the area of the site and the extreme storm conditions considered relevant. Typically three pipes may discharge into the final outfall each with a diameter of about 900 mm. Any discharge and its quality requires the consent of the statutory water authority.

Foul water pipes are sized to run at a proportional depth of 0.75. Data on discharge quantities from lavatories, etc., is given in BS5572 [11]. The total discharge from a power station can normally be accommodated in a pipe 225 mm in diameter. This is discharged to the local authority's sewer if it is not too distant and has sufficient capacity. When this is not the case, a sewage treatment plant is installed within the site, sized to meet the anticipated constructional quantities, the operational quantities or both.

Discharge from the various items of plant varies greatly in its purity and quantity. The largest discharge of uncontaminated water occurs when condensers are emptied which can be at the rate of 27 m^3/min for 10 minutes. This can be put into the surface water system. Likely areas of oil leakage are restricted by bunding so that oil does not enter the drainage system. However, drainage from floors or paved areas which may contain some oil spillage is passed through an oil/water separator of sufficient size to allow separation to take place. Very acidic effluents from boiler washing and cleaning are neutralised on or off site or discharged to a delaying resevoir, usually the ash lagoon. Discharges of acids and alkalis from the water treatment plant are neutralised in an effluent tank within the water treatment plant area, drain pipes being made of internally coated steel or cast iron or reinforced plastic. All liquid discharges from the site require the approval of the statutory water authority.

Wherever possible, drainage is by gravity at self-cleaning velocities of flow. The slope at which drains are laid is typically such that the flow velocity is not less than 0.75 m/s when the effluent is at a quarter depth in the pipe. The minimum gradient which will produce this varies from 1 in 70 for 100 mm diameter pipe to 1 in 900 for a 900 mm diameter pipe.

Drainage pipes are laid in straight lines and at a constant gradient between manholes. Manholes are

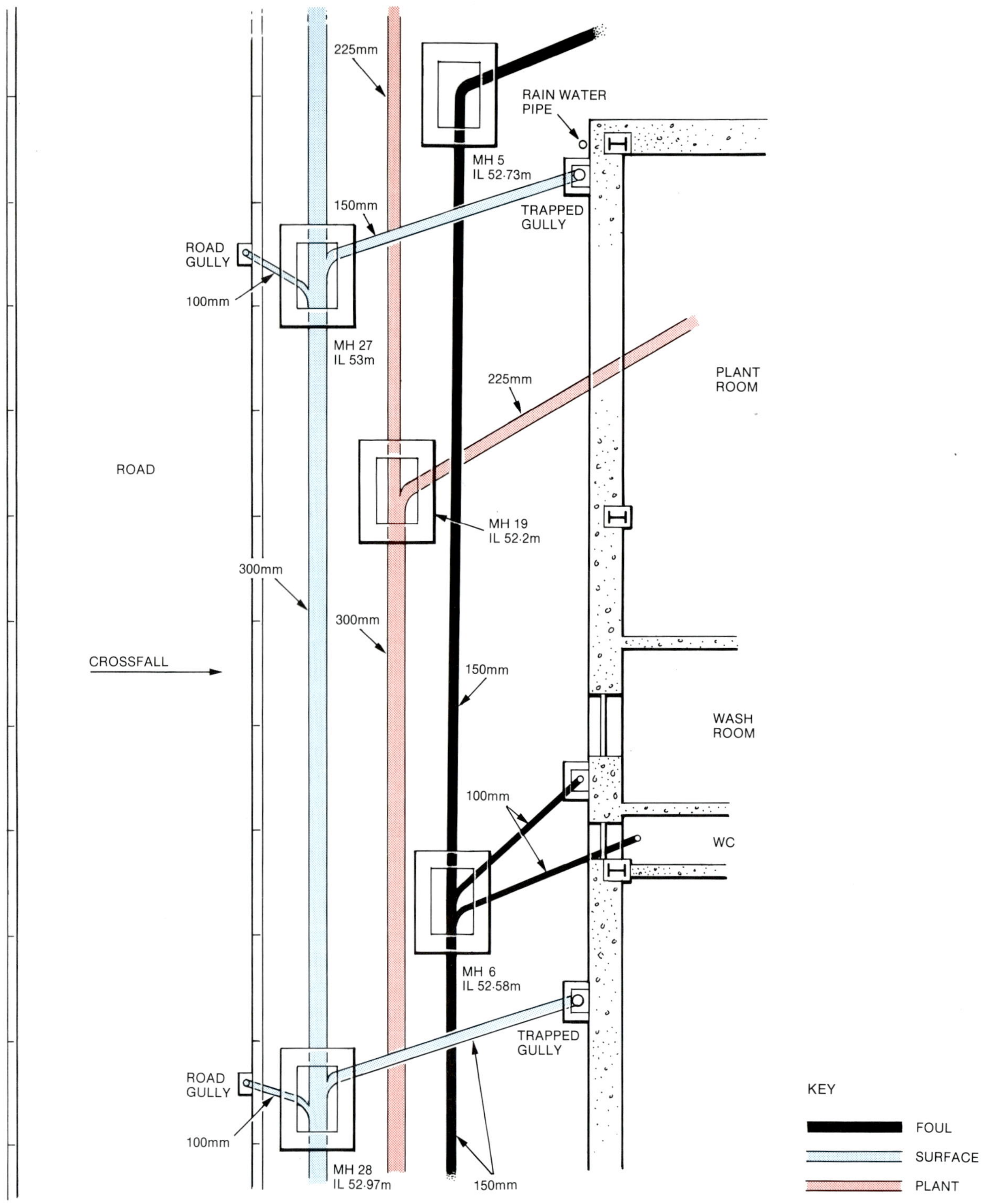

NOTES

1. MANHOLES ARE NUMBERED AND INVERT LEVELS GIVEN AS HEIGHT ABOVE NEWLYN DATUM

2. NORMAL MINIMUM PIPE DIAMETER IS 150mm EXCEPT FOR SHORT BRANCHES WHICH MAY BE 100mm DIAMETER.

FIG. 3.24 Typical drainage plan

provided at changes of direction and at a maximum spacing of 90 m (though a much closer spacing is usual on power stations) to facilitate cleaning of the drains when blocked. Manholes are constructed in engineering bricks, with a minimum wall thickness of 225 mm, or of precast concrete. The latter is usually cheaper, particularly for deep manholes, but brickwork is advantageous where it is necessary to break into a manhole to accept an additional discharge pipe. In either case it is essential that the manhole is sufficiently large to permit cleaning to be carried out. A catchpit is similar in construction to a manhole except that the base is some 300 mm minimum below the invert level of the incoming and outgoing pipes. Its purpose is to trap solids contained in the discharge. They may not, of course, be included in foul discharge lines.

A combination of gravitational and pumped flow is usual in the foul water system. Pumps are used occasionally in the surface water system in remote areas or at the final discharge point depending on the initial and final topography. These higher pressure regions of drainage systems are normally constructed in cast iron or steel pipes. Concrete pipes suitably jacketed at their joints and reinforced have been used for larger sizes. Asbestos cement pipes have been used in the past, but are now not used in CEGB stations.

In nuclear power stations any radioactive effluents from spillages and washdowns are collected by open gullies made of either stainless steel or cast iron. Borosilicate glass or cast iron pipework is used but where pipework is to be encased in concrete borosilicate glass is unsuitable. It is however suitable for both gravity and pumped drains. Secondary containment is obtained by providing an interspace around each drainage pipe and this is achieved in a variety of ways. All potentially radioactive drainage from power stations is held and monitored before final discharge or removal.

Underground drainage pipes are protected from stress by being laid on suitable supporting beds and being protected at the top and sides. The usual power station practice is shown in Fig 3.25. Further general information is given in BS8301 [12]. For the station permanent system, rigid pipes are used with either rigid or flexible joints. Vitrified clayware is used up to about 300 mm diameter with larger sizes in concrete or reinforced concrete. Flexible pipes are restricted to tem-

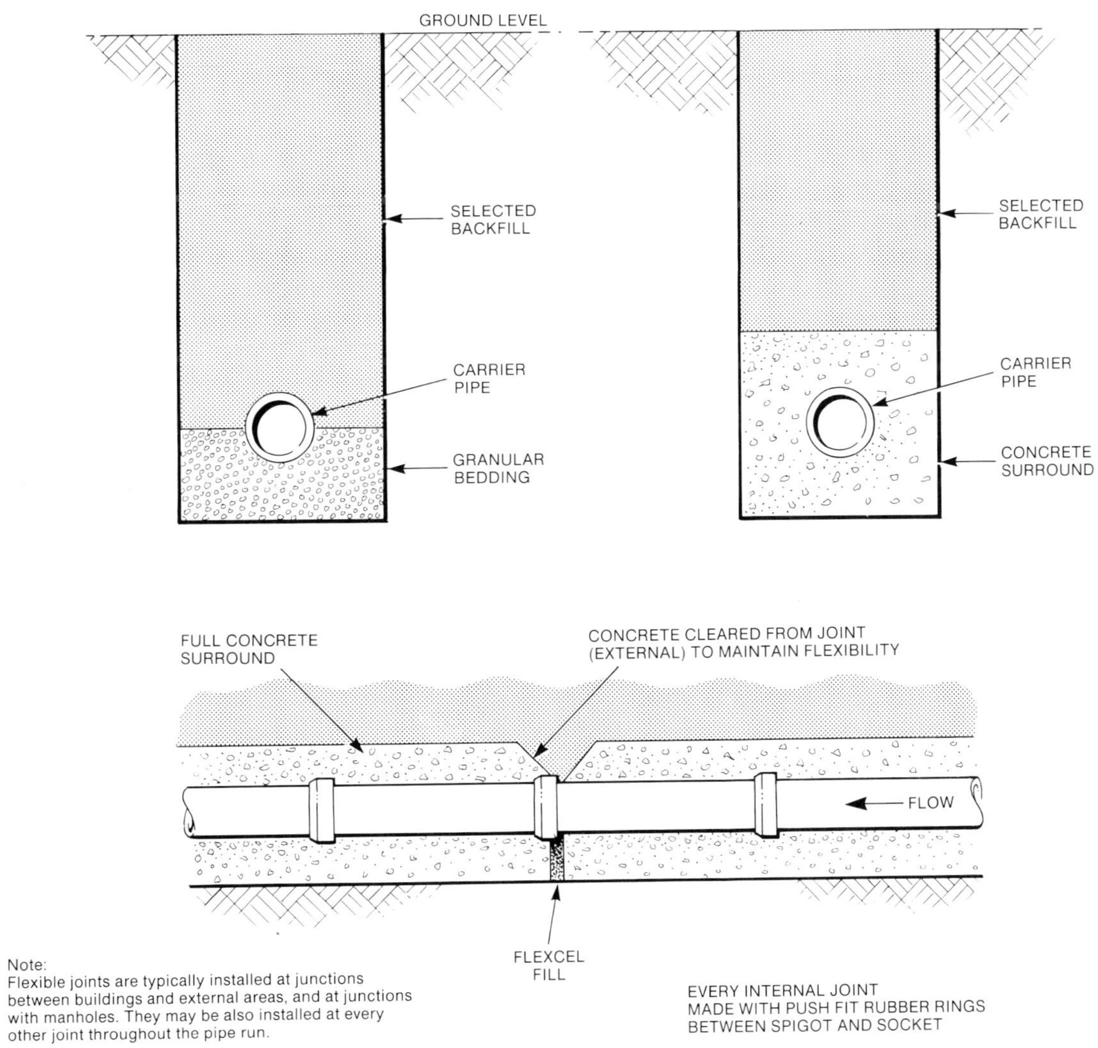

FIG. 3.25 Pipe protection

porary drainage. Underground cast iron and steel pipes are provided with protective bituminous coatings as they are liable to attack by groundwater. Cast iron pipes are used for foul drains in buildings where extra strength is required and for boiler blowdown drains where the discharge may be at high temperature.

Figure 3.26 shows the connection of a road gully to a surface water drainage system. Note that the gully acts as a silt trap and requires periodic cleaning.

7.4 Railways

Until the introduction of close coupled working, large power station sidings were necessary to accommodate coal trains. Today, coal trains with bottom opening trucks are continuously unloaded without uncoupling. Sidings are virtually eliminated except for crippled wagons and construction traffic where materials are delivered by rail. Trains are hauled by British Rail locomotives and the track, loading gauge and minimum radius of curvature have to be suitable for use by the largest and heaviest diesel locomotives.

The adoption of flat bottom running rail to British Rail standards has replaced bull head rail in most modern installations. The track is secured to timber or concrete sleepers by the use of Pandrol clips. Check rails are provided where sharp curves are unavoidable in order to distribute the lateral forces and to avoid excessive wear to the running rails.

Piped sub-soil drainage of rail tracks is frequently required in order to prevent the foundation beneath the ballast becoming partially waterlogged and weakened. This is carried out by laying open-jointed porous clay pipes or other suitable pipes, and discharging these to a main drain run, having catchpits, located parallel to the track.

7.5 Coal storage

The coal storage area is prepared by first levelling off the ground and then laying down a thick bed of rolled ashes or chalk, the level of which is above the highest water table. If the coal store is on an impermeable sub-soil, some drainage should be provided, for example, the centre of the coal store base may be sloped gently to the outside and drains installed at the perimeter. These measures are necessary to prevent coal retaining considerable quantities of water and thus making coal handling extremely difficult.

The coal store is generally stocked out to a maximum height of about 15 m and this gives a ground pressure of about 15 t/m^2. Sometimes the storage height has had to be limited to a lesser figure due to the presence of weak sub-soils, and the consequent possibility of subsidence.

Large areas of concrete paved coal storage have been provided economically at several recent major stations by using high fly ash concrete mixes. At least 50% of the cementitious material in the mix is non-selected PFA, a material readily available at negligible cost at such localities. Such a surface has clear advantages over prepared ground surfaces whenever wheeled delivery or coal moving equipment is used extensively. Again, drainage considerations require close attention.

7.6 Oil tank compounds

Oil storage tanks, which may vary in size from 6 m to 45 m in diameter and by 3 m to 15 m high, are of welded construction. The intensity of loading is due mainly to the weight of the contained oil and is thus moderate for the lower tanks which will generally be founded on a thick reinforced concrete raft, or on a bed of crushed stone hardcore, with a thick layer of sand

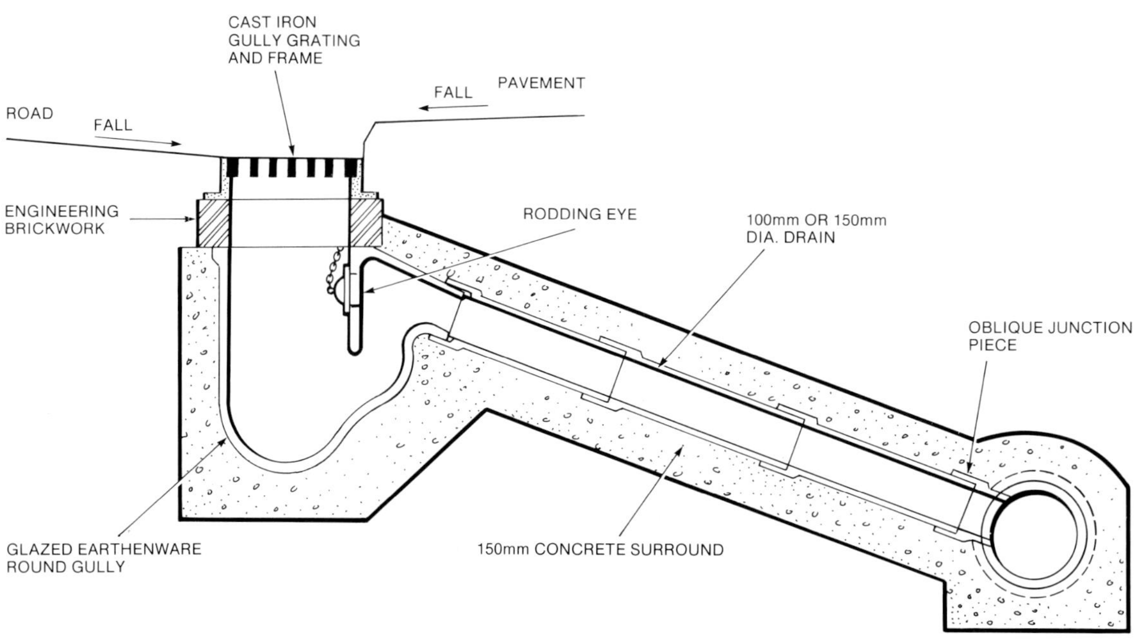

FIG. 3.26 Connection of road gully

topped off by a sand/bitumen mix. For the larger and taller tanks piling is frequently required, as such tanks are frequently sited on poor ground close to a water's edge.

As a precaution against the risk of spread of fire, and against considerable leakage of oil from a damaged tank, tanks are contained within earth retaining walls (called *bunds*) or within reinforced concrete enclosures; these must be sufficiently voluminous to contain the full oil capacity of the tanks. There is no record of fire or collapse of tanks causing a large leakage at any CEGB power station. However, small spillages do occur and these are retained within the bunds.

In general, the larger tanks hold heavy fuel oil and these are enclosed within large unsurfaced earth bunds. Rain falling within these bunded areas either drains into the ground or evaporates, whereas fuel oil solidifies at normal temperature and can be removed or left to decompose. Smaller tanks normally hold light oils and these need to be enclosed within surface impervious bunds which fall to a blind sump within the bunded area. The light oils separate from the rainwater and the water is pumped from the bottom of the blind sump over the bund wall and into the surface water system.

All major tankage installations need to be most carefully considered with respect to the oil and petroleum storage regulations and perhaps the Reservoirs Act [13].

7.7 Ash disposal areas

The amount of preparation required for any ash disposal area will depend upon the final use to which a specific area will be put. If the area is to be landscaped by forming a contoured hill then the main problems will be the formation of terraces such that mechanical plant, e.g., tractor scrapers, can dispose of ash in layers thin enough to ensure adequate compaction and hence avoid subsequent instability, and the provision of sufficient land drainage such that the risk of a catastrophic slide will be avoided.

If the area is to be used as a short term disposal facility from which the ash may be subsequently reclaimed, then it will probably be deposited in artificial or natural lagoons, the former being constructed from the ash itself. In general this form of disposal will be associated with the pumping of ash slurry and hence involves the removal and disposal of the surplus water to a river or other water course.

Large ash disposal facilities are not strictly liable to comply with the Reservoirs Act, but the CEGB has adopted a very similar safety control procedure. Independent expertise is universally engaged for the design, construction and subsequent operational phase of such civil engineering facilities.

8 Methods of construction

8.1 Site clearance, access roads and construction offices

The construction of the power station foundations is carried out in accordance with a detailed programme drawn up to provide the various foundations and general site works, in the sequence necessary to enable the building work and plant installation to proceed in accordance with the overall construction programme.

The first steps in the civil engineering work are the clearance of old buildings, hedges and other obstacles from the site, followed by any general site levelling which is necessary. This may involve only excavating soil from high areas and filling low areas with it, or excavating selected materials to form flood banks, or importing fill to raise the whole or part of the site level. To provide material for landscaping which is almost invariably necessary, the top soil should be carefully stripped off and stock-piled for re-use when construction is complete. Preparatory work for the installation of an electric power supply with ring mains follows.

The permanent main drainage of the future power station is then installed, and arrangements made to dispose of surface water from the system if the permanent outlet or drainage pumphouse cannot be constructed until later. The soil drainage is connected to the local authority sewers or, if none are available, a small sewage treatment plant will need to be built. Lavatories for construction workers will be built at various points around the site at this stage.

A new access road from the public highway to the station will be built, and an access bridge if required. The permanent roads of the station are laid down for use during the construction period, and any temporary roads necessary for the construction of the station are built. These may be removed following completion of the works. Hard storage areas for the plant contractor's material and huts are provided by installing sub-soil drainage and then surfacing large areas with rolled hardcore and ash and making suitable temporary access roads.

A site boundary fence is built to improve security, and a car park built for the construction workers outside the boundary fence.

Site offices are built for the CEGB's site staff and other buildings erected such as stores, garages, concrete laboratory and a site canteen for all the construction personnel. Water and electricity supplies for domestic and constructional purposes are installed. Welfare facilities, such as a site first aid centre, are provided, and in some areas where accommodation is difficult, a hostel or camp may be made available for resident workers.

Completion of the preliminary works will permit later the efficient deployment of the plant and labour required to construct both civil works and plant installation. A separate contract is frequently let for these preliminary works, so that they may be put in hand while the detailed design of the foundations is undertaken and contracts let for their construction.

8.2 Underground construction

All ground is saturated with water below a certain depth depending on natural conditions in the locality. The top surface of the groundwater is called the *water table*, and if a hole is dug below that level water will flow into it until its surface reaches the water table, the rate of flow being largely controlled by the permeability of the ground itself. Before any deep excacations can be carried out it has to be decided how groundwater will be dealt with. Clay strongly resists the flow of water through it, silt is more porous than clay, and ordinary coarse sand is even more porous. Gravels are very porous, and water flows easily through them.

The results of the soil investigation provide the level of the groundwater and the types of soil which lie below. Choice of the best method to be adopted to keep the excavations dry and stable enough to work in can be assessed from this data. For excavations above the water table and for conditions of low flow through clays or silts, it is sufficient to introduce ordinary suction pumps or lift-and-force pumps as the excavation proceeds. This equipment will also remove such surface rain water as finds its way into the excavation. This method is often sufficient for the main foundations excavation for a power station to proceed in open cut, i.e., with the sides sloped back to maintain stability. For deeper excavations this is seldom practicable and the stability of the sides has to be maintained with ring main dewatering on berms surrounding the excavation and deep-well point dewatering to local foundations below the main level.

Other techniques are available for unusual or confined conditions but these are not yet of general application or economic. They include freezing, ground consolidation techniques and grouts, vibro-stabilisation and explosive compaction, and are each suited to particular ground and site conditions.

More generally applicable methods of deep construction are described as follows.

8.2.1 Cofferdams

Figure 3.27 shows a diagrammatic arrangement for the construction of a typical closed sheet pile cofferdam used for a construction in water; a similar arrangement would also be suitable for a construction on land. Sheet piles are obtainable in various sections and lengths and have interlocking clutches along their edges, so that when driven they form a watertight wall. Diesel, steam and compressed air hammers are used for driving sheet piles, and electrically-driven vibrators can be used for driving and extracting piles. A 'silent' hydraulic system is available, albeit expensive, if noise or vibration needs to be minimised. The sheet piles are driven down into an impervious layer, or into a firm foundation such as chalk or hard ballast taking care not to drive so hard as to decouple the clutches at the bottom.

Excavation, or dewatering in the case of an offshore cofferdam, proceeds in stages so that the various levels of bracing can be assembled. In large cofferdams the long struts may need supporting at mid-span by temporary piles called 'king' piles. The bottom bracings are spaced more closely together to take the increased load with increased depth. Excavation and bracing is continued until the foundation level is reached, when concreting of the permanent structure follows in the normal way. Piles may be driven below the excavated level if needed. If the sheet piles are to form part of the permanent structure, concrete is poured against them, and the temporary bracings removed as work progresses. This practice is not recommended, however, for should there be any ground pile movement or any flexing of the sheet piles when removing struts serious damage can be caused to the recently poured concrete structure. The sheet piles are otherwise driven wide of the permanent structure, so that as it rises within the cofferdam, load from the sheet piles can be progressively transferred to the concrete walls by short struts and wedges. The space between the sheet piles and the concrete is finally backfilled or flooded and the piles extracted.

Considerable temporary works design is necessary to place the cofferdam walings and struts so that the main vertical reinforcement is not fouled and the rising shutters can sail past walings until concrete is just below them. Folding wedges between the waling and the sheet piles can then be taken out and the waling and strutting removed. Thus load need not be transferred to green concrete. To achieve this it is often necessary for the permanent works to be substantially re-designed to accommodate the complex temporary works.

This form of construction is widely used for such structures as pumphouse, intakes and outfalls, ash plants, and the deeper parts of large open cut excavations. Conveyors, cable ducts and culverts are frequently constructed within continuous sheet pile cofferdams with long modules of piles, strutting and shuttering being fleeted forward. Intake and outfall structures in calm or shallow water lend themselves to construction within a circular cofferdam of piling or diaphragm walling. In these types cross-bracing can be eliminated by using reinforced concrete circular walings in compression to support the cofferdam wall, leaving the centre free of obstructions.

The diaphragm wall or bentonite trench wall technique makes use of the properties of bentonite and bentonite clay suspensions in water to produce pumpable liquid/gels with specific gravities above 1.0 and some shear strength equivalent. This fluid is used to support the sides of an excavation until digging is complete, when it is displaced by reinforcing cages and concrete carefully tremied in from the bottom. The

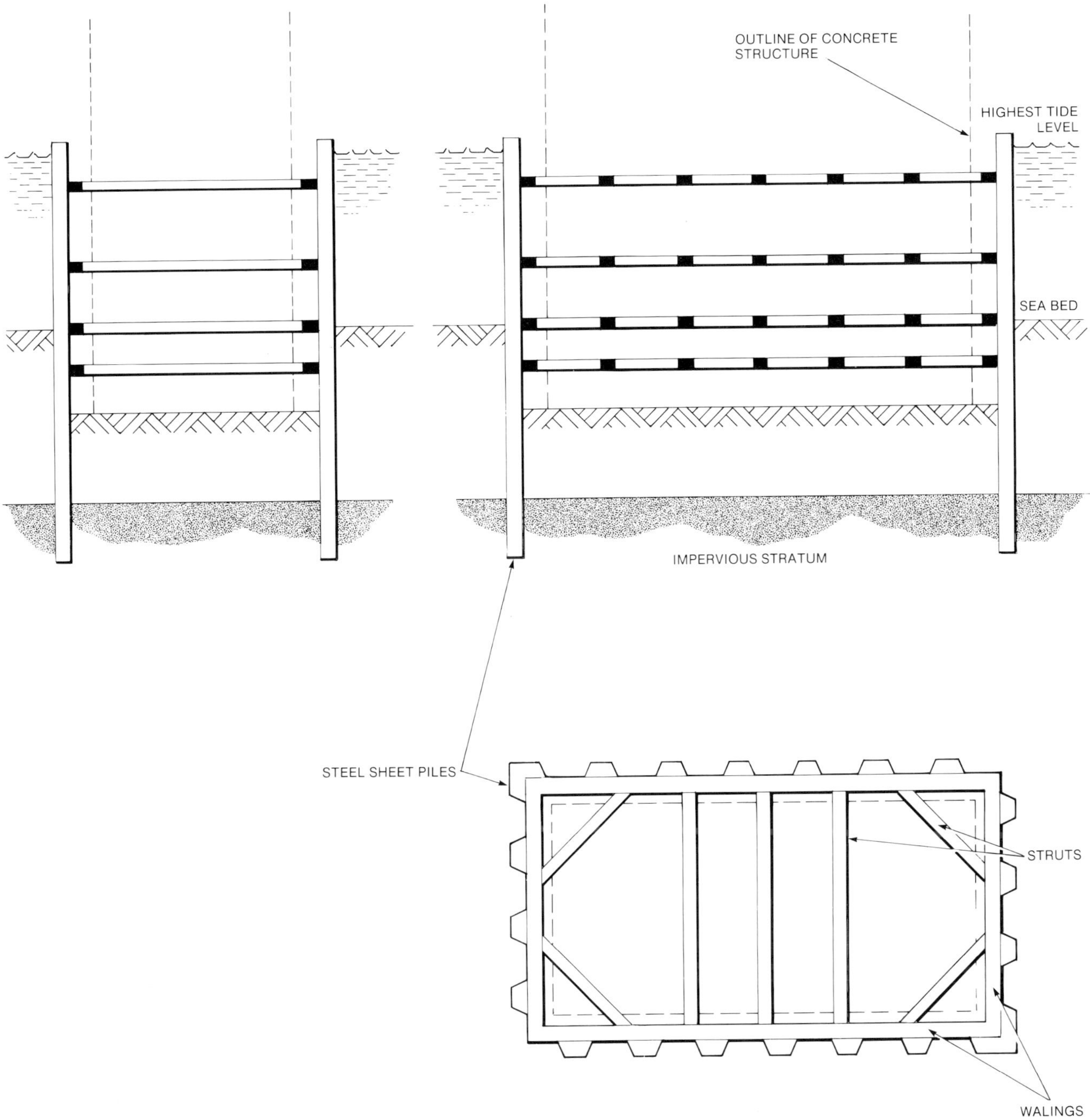

FIG. 3.27 Sheet piled cofferdam

displaced bentonite is then cleaned and pumped elsewhere for re-use. Excavation is done by machine working through the fluid and when working in shallow water it is usual to provide a temporary artificial island from which to excavate. Walls made in this manner have been incorporated in permanent works with little or no surface treatment.

The design of a cofferdam is a difficult exercise because of soil mechanics not lending itself to precise analysis, and failures have occurred. Failure may be due to the mechanical failure of the lower struts, or in a plastic soil the unbalanced earth pressures may cause the sheet piles below the bottom strut to buckle and the bottom of the excavation to heave up. Unbalanced hydrostatic head may cause water and fine material to flow under the toes of the sheet piles, and the bottom of the excavation to 'boil up'. Particular care is needed when there is a large range of tide or groundwater outside which must be balanced during all stages of construction. Groundwater lowering, as described in Section 8.3 of this chapter, may be sufficient to deal with this problem. The out-of-balance forces may also be overcome by flooding the cofferdam, excavating under water with grabs to below foundation level, and placing a layer of mass concrete under water before pumping out for construction to continue. This mass concrete must be thick enough to balance the forces likely to cause failure, and to form in effect a permanent bottom strut and seal to the cofferdam. It may also be tied to underlying rock and anchored down if there is any danger of the cofferdam becoming buoyant.

Injection grouting techniques may be employed to increase the strength of the soils and reduce their permeability. Depending on the nature of the soil, the materials used vary; examples are cement grout, special clays (sometimes mixed with cement), PFA, bitumen compounds, and a variety of chemicals. These are so effective in certain soils that excavation within a grout curtain has been carried out without recourse to sheet piling.

In very large cofferdams, which may be required to enclose the entire area of bulk excavation for a power station, it is impracticable to support struts right across such a large area. In this case the cofferdam walls are tied back with steel rods or cables to anchors in firm ground, or berm is left immediately at the edge of the foundations to support the wall. When the permanent foundation has been built sufficiently close to the sheet piles, the berm is removed and replaced by raking struts braced off the foundations.

The depth to which construction can proceed in cofferdams is limited according to soil conditions and it may be necessary for shafts to use an upper cofferdam to get a seal at rockhead and then sink a second cofferdam through it.

Three other methods for forming deep foundations are now discussed and they are equally suitable for construction on land or through water.

8.2.2 Compressed air or pneumatic caissons

A steel or concrete working chamber is constructed at ground level, its shape corresponding to that of the completed structure. The working chamber walls are sloped to form a cutting edge at the maximum external dimensions and work continues on the permanent reinforced concrete walls above the working chamber roof. Once the walls are high enough to provide sufficient weight and strength, excavation commences under the working chamber roof through access shafts which will later carry the air and muck lock shafts. The ground under the cutting edge is broken out and the whole structure settles and follows the excavation down until water makes it necessary to put the working chamber under air pressure when the air shafts, man locks and muck locks are installed. Where possible they should be installed to full height as the opportunity to lift them may not occur.

Excavation and concreting then continue under air with the same problems as those discussed in Section 8.2.1 of this chapter, until founding level or suitable foundations are reached. Often the weight of the caisson is insufficient to overcome skin friction from the ground and kentledge is added to the caisson as blocks, sand or water. Alternative methods of easing the sinking are to inject a bentonite lubricating film at the cutting edge, or 'blowing down' by removing the workers, raising the air pressure and dropping it suddenly. This, whilst effective, involves some loss of control over where and at what level the caisson is less than the length of its shortest side into the ground. Also, behaviour of the rapidly decompressed ground can be unpredictable, leading to muck being forced up to the air shafts.

When foundation level is reached, the bearing surface is cleaned up, the working chamber filled with mass concrete and then pressure grouted to ensure good contact. The practical depth for caisson sinking is about 25 mm to 30 mm below high water because of the human limitations in compressed air. Final position, level and plumb of a caisson must allow reasonable tolerances due to the inherent uncertainties and so all sluice gate guides, plummer blocks and other plant related items built into the concrete must have adjustment.

Caissons built on tidal river banks always move towards the river (often 1–2 m) and will tend to lean towards it. Allowance should be made for this when setting out the working chamber. For work in water the cutting edge, working chamber and sufficient wall to provide freeboard are floated into position and sunk with as much working plant installed as possible. Other plant must either be afloat or on temporary piled structures outside the area affected by caisson sinking.

8.2.3 Monoliths

Monoliths are not often employed where techniques for ground stabilisation can be used. However they provide simple, but not quick, ways of installing quay walls and harbours which are intended to be dredged later and which may include CW intakes and outfalls.

A cutting edge is assembled at ground level on temporary foundations and the permanent concrete walls built up for a height of three to four lifts. The soil is grabbed out from below the cutting edges and the monolith descends under its own weight. Excavation is halted from time to time while extra sections of wall are built up. The process is continued until foundation level is reached, excavation being carried out under water when water-bearing strata are encountered and kentledge added when required.

The monolith is finally plugged with mass concrete placed under water. Divers using high pressure water jets and air lift pumps are used to clear the slurry from the foundation level. The monolith is then dewatered and the reinforced concrete work completed.

The permanent structure has to be made strong enough to withstand the additional stresses involved in this type of moving construction. In a large structure temporary bracing walls may have to be built and later removed. Pumphouses have been successfully constructed using this method, their 'egg crate' internal structure being particularly suitable from the point of view of strength. In one instance the moving structure was some 40 m × 52 m in plan and 18 m deep and weighed in the order of 60 000 t on completion. Both these types of construction are illustrated diagrammatically in Fig 3.28.

8.3 Groundwater lowering

When major excavations are required the ground often requires dewatering because of the high rate of inflow through sands, gravels or fractured rock. This is achieved by encircling the excavation area with a series of wells from which water is pumped continuously until the construction works within the excavation are completed. Provided water can be pumped from the well system at a faster rate than it can flow through the ground, the water table in that area is lowered. The layout and depths of the wells and pumps require designing so that the water table is lowered beneath the lowest parts of the excavations, thereby leaving the whole area encompassed by the dewatering wells available for dry working.

Suction pumps installed at ground level can only be relied upon to lower a water table to about 4.5 m below the pump level due to vacuum limitations. If the pumps are required to raise water more than about 4.5 m, a two-stage or multi-stage well system has to be adopted. In such cases the sides of the excavation are taken down in bermed stages, each succeeding berm being about

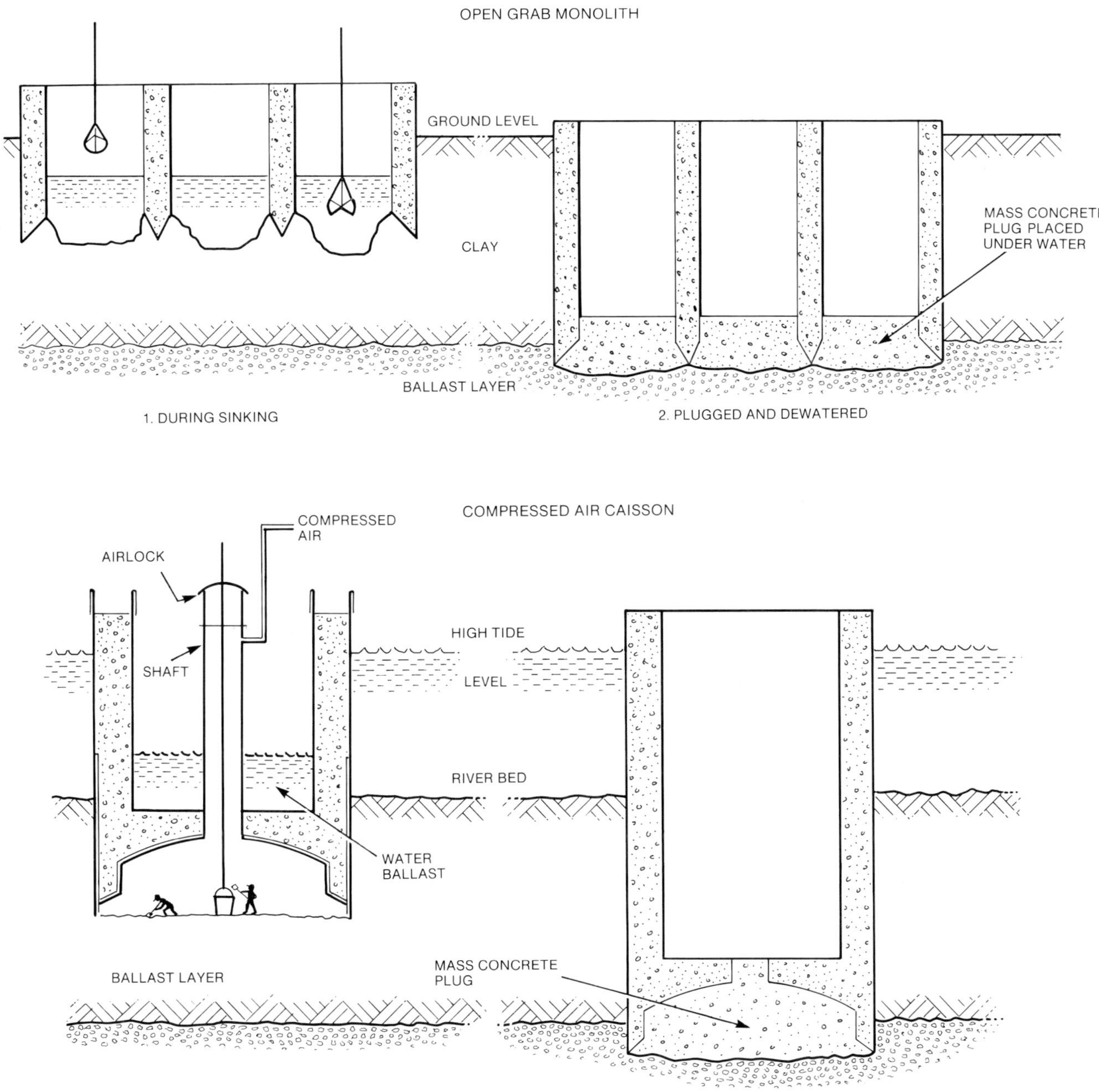

FIG. 3.28 Caisson and monolith construction

4.5 m in level below the previous berm or ground level. It is unusual for more than a three-stage well system to be adopted in practice, and the concept has the major disadvantage that the wells, berms and associated excavation side slopes (if sheet piling is not inserted) take up a large area around the deep excavations. In the planning of power station construction this large surrounding area can rarely be made available for the exclusive use of the well system during the period required for dewatering.

In the system described in the previous paragraph, the dewatering methodology is invariably of the 'well-pointing' type. In this the wells are tubes of about 50 mm diameter fitted with a filter bottom point to minimise fine soil particles being extracted with the groundwater. They are drilled or jetted into the ground at centres often of 1.5 m to 2 m spacing, depending on the ground permeability. The lines of wells are connected to the pumps through common mains laid at ground level or along the berms.

For greater depths and/or larger volumes of groundwater submersible electric pumps are installed at the bottom of larger wells. This is known as a 'deep-well' system and is commonly adopted in major construction. The well is made by driving a perforated tube to the required depth in the ground, removing the encased soils and inserting an inner tube which is also perforated. The annular space between the inner and outer tubes is filled with a graded sand, selected to act as an effective filter to the inflowing water. The submersible electric pump together with its delivery pipe and electricity supply cable is lowered to the bottom of the inner casing and set to work. The diameter of the inner well tube is governed by the size of pump it has to encompass. The outer tube is commonly of 500 mm to 800 mm diameter, the size being selected having regard to parameters such as filter thickness and expected yield. Deep wells are notably more expensive than well points to install, but fewer are required for a given volume of excavation requiring dewatering.

On occasion, recharge wells are also installed in conjunction with a dewatering system. The purposes of these are to maintain an existing ground water table adjacent to an area of dewatering. This is sometimes necessary either to minimise ground movement under a sensitive existing structure or even to maintain water levels for environmental or ecological reasons. If the natural phraetic gradient between the dewatered area and the recharged area cannot be achieved a full or partial cut-off, in the form of sheet piling or a diaphragm wall, has to be inserted into the intervening corridor. Recharge is likely to be an important consideration whenever a new power station is being constructed adjacent to an operating power station or other major structure, especially if that station is nuclear.

8.4 Excavating machinery

Once the method of dealing with the groundwater has been decided and implemented, and sheet piling or other forms of temporary support or cut-off have been installed, the bulk excavation can commence.

A group of machines often used in bulk excavation work are variants of a common machine, and three of these are illustrated in Fig 3.29. A variety of digging arms and other attachments are attached to this basic excavator to enable it to undertake various operations. This basic machine is crawler-track-mounted to allow it to traverse quite soft ground without bogging down. It has a wide radius of operation, having a rotatable body, and can dig and then turn to load direct into transport or dump onto a stockpile. The digging operation is done whilst the tracks are stationary, thereby avoiding churning up the ground surface. However, quite frequent small movements of the machine are necessary for it to maintain its cutting edge at a near optimum position relative to the excavation. Originally these machines were built around a crane body, using winch drums, wire ropes and pulleys to operate the attachments. However, hydraulic technology is now used in the backacter and face shovel modes and also has some application in some of the other machine variants.

The five most common forms of this basic machine have the following forms.

8.4.1 Dragline

The dragline bucket is attached to the crane rope by means of a yoke, to which a hauling rope or dragline is shackled. By using the two ropes correctly an operator can cast the bucket some distance beyond the horizontal reach of the machine jib. Digging is performed by winding in on the dragline. Skill is required to load direct to transport; alternatively the spoil can be side cast to form banks or heaps. This machine can be used for bulk excavation well below its track level in soft or medium ground, or alternatively in well-blasted rock. A special perforated bucket attachment enables it to excavate below water level and it is frequently used for digging large ditches and bank building. It is equally able to load concreting aggregates into batching plant from stockpiles. Although its digging arrangement may seem antiquated and haphazard, such a machine is capable of both accurate work and high outputs when handled by a skilled operator. This machine is manufactured in this format over a large range of sizes, the largest still being the preferred option in open cast coal extraction.

8.4.2 Backacter

The backacter bucket is mounted on a hinged arm, and as its name implies it digs straight back towards the machine and below track level. It can operate to fine tolerances in ground conditions ranging from soft to

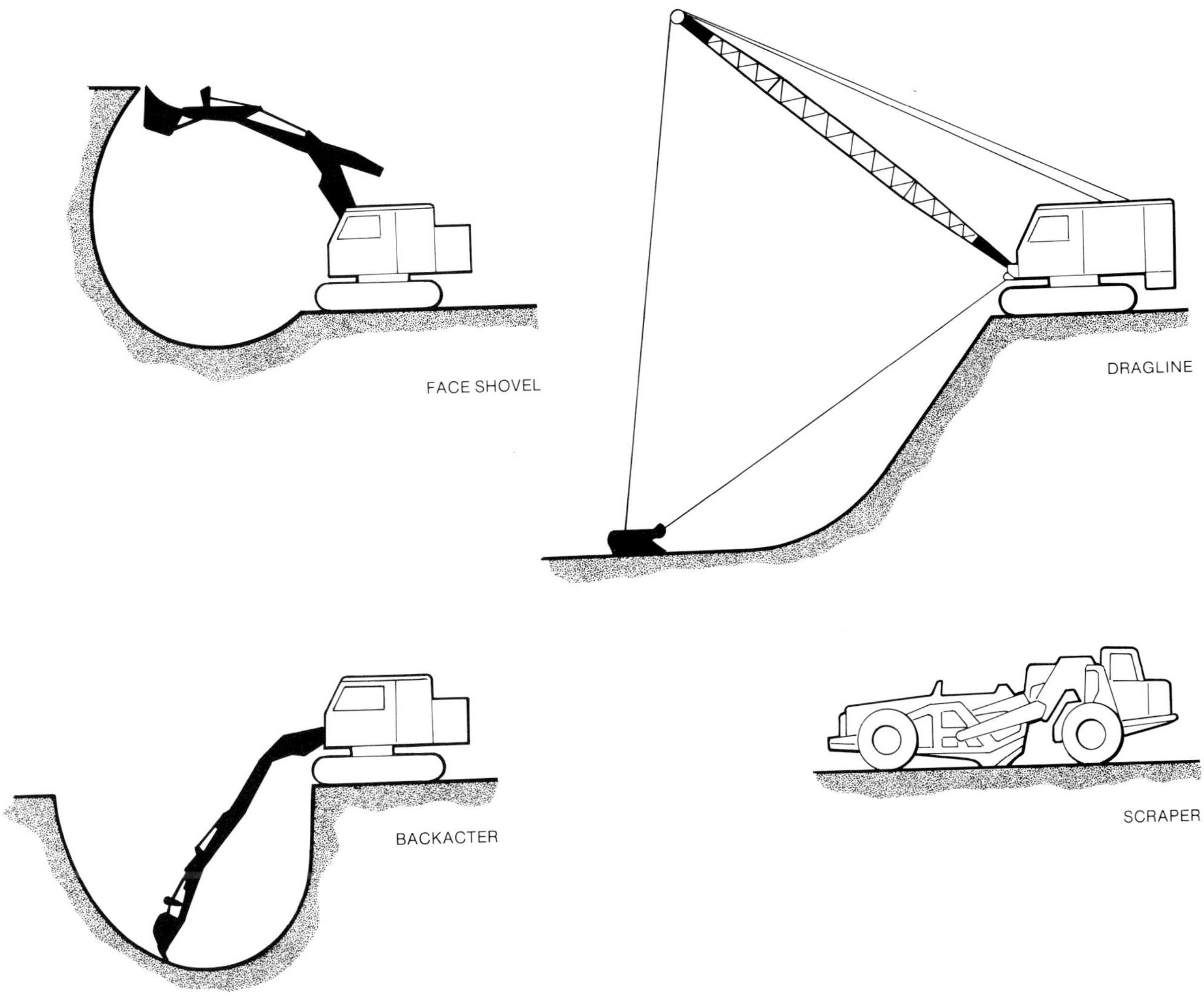

FIG. 3.29 Excavating machinery

stiff and is commonly used in all foundation excavations. It is especially useful for trench excavation (unless the trench is abnormally deep) and for excavating between obstructions, e.g., piling.

8.4.3 Face shovel

The face shovel is mounted on a hinged arm and usually also has the ability to slide. It digs away from the machine and can operate slightly below track level although its most productive output is when digging into a near vertical face above track level. A face shovel is capable of digging into materials firmer and harder than the other variants of the basic machine and consequently it is often used in rock excavation. The bucket has a hinged back, operated by a trip wire or hydraulics for dumping purposes. Even with hardened steel teeth on its bucket, a face shovel cannot cut into sound rock of modest or greater strength. It may loosen and bring down quite good rock if the rock mass has joints and/or fissures that can be prised open by the bucket teeth. Drilling and blasting is required in sound rock excavation prior to the use of a face shovel type machine to excavate the broken rock and load to transport.

8.4.4 Grab

The grab consists of a hinged two-part bucket capable of being dropped down onto material. The bucket closes at the commencement of the lifting operation, thereby removing a 'bite' of the material being excavated. This form of machine is that used for deeper excavations and it is well suited to underwater digging. It is used to form cofferdams of the bentonite slurry variety and has applications for work within all types of cofferdam, sheet piled trenches and other works with closely spaced obstructions. Like the dragline it is also capable of handling aggregates if more suitable plant is not available.

8.4.5 Excavator

Rigged as a crane, this machine has multiple lifting uses on a construction site. It can handle skips of various materials, remove cut-off pile heads, lift structural or reinforcing steel and handle large shuttering units. Its further uses include handling, pitching and driving sheet piles and also boring for and/or driving displacement piles when fitted with further suitable attachments.

As stated previously in the dragline section, these machines are available in a wide range of capacities, but are most likely to be seen on power station sites in the 0.5 m^3 to 3 m^3 range. The size and number required is dependent on both the site and the programme for completion, and can vary appreciably. Excavating production capability is also influenced by the transport capability to the dumping grounds, which may be situated up to several kilometres from site. Off-site dumping is generally performed by road tipper lorries, commonly of 6 m^3 to 15 m^3 capacity. On-site dumping often enables much larger capacity dumpers or tractor-drawn dump trucks to be utilised. Vehicle sizes and numbers have to be related to the size and capability of the excavator buckets, as it is important to ensure that enough transport is available to accept the continuous output from the very expensive unit priced excavating machinery.

As well as the variants of the basic machine reviewed in the foregoing, modern practice is increasingly favouring highly mobile excavating plant based on large tractor units, 'dozer' type machines and motorised scrapers. Equipment mounted on large tractors is often highly specialised and does not merit further description here. But the dozer range of excavating plant and the motorised scrapers contribute greatly to general excavation procedures and warrant further comment in this context.

8.4.6 Tractor shovel and loader

The original bulldozer was a tracked vehicle with a cutting/trimming blade at the front. Developments to this concept happened quickly due to its high degree of manoeuvrability and adaptability and its common successor now has a hydraulically-operated bucket on the front and may be mounted on a rubber-tyred or tracked chassis. Sometimes the bucket is hinged in two parts so that it can additionally act as a grab. These machines undertake a wide range of work, sometimes operating in an attendance capacity on major excavating plant. Typical tasks include trimming bulk excavations, clearing and loading materials, laying temporary roadways and surfacing and minor bulk excavation. It can also be used for spreading and levelling, as could the original bulldozer. Bucket sizes range from 0.5 m^3 upwards, but most are of at least one cubic metre capacity. The smaller ones sometimes have a hydraulically-operated backactor arm mounted at the rear and others carry a rock ripper, which is a hydraulically-operated claw designed to break out fractured rock. The major disadvantage of tractor shovel/loaders are that they load or dig with their main bucket by driving forward and must then be backed or manoeuvred to load transport or a stockpile, thereby churning up the surface they are constantly traversing. They have particular value in closely confined excavation work such as tunnels.

8.4.7 Scraper

Bulk excavation from large areas is normally carried out using scraper equipment (Fig 3.29). A scraper is a large bucket or bowl mounted on rubber tyres and its motive power is provided either by a crawler-mounted tractor which tows the bucket unit or else it is an articulated wheeled machine with engine and cab at the front end. Digging is achieved by lowering a wide cutting edge into the ground while travelling forward; the depth of cut being a number of centimetres depending on the nature of the ground and the work. Such machines are commonly used for topsoil stripping, site levelling, shallow excavations over large areas and constructing earth embankments. Tractor-drawn scrapers have a limited economic distance over which they can travel to discharge as they are relatively slow and have high powered engines to boost the thrust required for scraping loading. By contrast the all-wheeled articulated self-propelled scrapers have less excavating power but offer high travelling speeds and hence can accept longer economic haul distances. If this version of scraper has deficiencies in self-loading (which can be the case in hard ground conditions) the problem is usually overcome by stationing a high powered crawler tractor at the excavation area to assist the loading phase by pushing the scraper at the rear. Bucket sizes can be very large, 30 m^3 being relatively common. Such plant does have operating limitations, being unable to perform successfully on either very hard or soft and/or waterlogged ground. In the range of suitable ground conditions between these extremes, a substantial part (and in a few cases the whole) of bulk excavation for power station foundations have been dug using scraper equipment.

8.5 Construction by diaphragm walling techniques

This technique involves the excavation of an unsupported trench, stabilised by being full of bentonite slurry, which is then filled with the required permanent infill from the bottom upwards. Guide walls about 2 m to 3 m deep are constructed in each side of the proposed wall which can be up to about 1.5 m thick. The trench is excavated in vertical panels up to about 6 m to 8 m long using either grabs or rotary mills with the excavation being maintained full of an appropriately designed bentonite slurry. On completion of the excavation for a panel, which may involve up to three vertical passes of the grab, any reinforcement is placed and the infill material inserted using one or more tremie pipes. For structural walls this material is a suitably designed concrete while for cut-off walls it may be a 'plastic' mixture of bentonite and cement. The method can be an extremely useful construction technique, often for temporary works, but requires close quality control at every stage. To facilitate adequate continuity of the wall, panels are constructed in a non-consecutive sequence and 'stop-ends' are used to produce an appropriate profile of the first and any subsequent panels which are not being constructed against an already cast panel. However, in all cases where the wall is required to act as a structure, a capping beam should be provided to ensure continuity of structural support.

Fairly complicated plan profiles of wall are possible and the technique has been used to provide large capacity support as an alternative to piles.

8.6 Formwork and reinforcement

As soon as an area of excavation is completed and trimmed to the correct profiles, a layer of concrete usually from 75 mm to 300 mm thick but occasionally thicker, dependent on soil conditions, is laid on the exposed surface. This, known as *blinding concrete*, later serves several important functions. It protects the exposed subsoil from the effects of weather — drying out by wind and sun, or softening by rain — and helps to control seepage of groundwater. The subsoil is protected by it from the later activities of construction, such as pile stripping, handling and fixing of reinforcement, timber, pipework, etc., and it can allow light construction traffic to cross. The shape of the underside of the foundation may be accurately formed, and the working area can be kept clean and dry for all the activities of foundation concreting to proceed.

Before concrete can be poured, the reinforcement steel has to be cut, bent and fixed, and the formwork constructed. These are major operations in the foundation construction and employ a great deal of hand labour. The amount of reinforcement used throughout the whole of the foundation construction may be in the order of 10 000 tonnes, most commonly in mild steel bar ranging in size from 10 mm to 40 mm diameter and from 1 m to 20 m long. The bars may be either straight or bent into complicated shapes. The cutting and cold bending is generally carried out on site in a reinforcement yard equipped with power cutters and bending machines, although specialist off-site cutting and bending operations are becoming more common for many smaller jobs in power stations. A tower or mobile crane is generally used to transport the steel around the site either in individual bar lengths or as preformed cages.

Whilst the prepared reinforcement is transported and off-loaded near the place of work, the final lifting into place of individual rods and binding together of their intersections with iron wire is done by hand. There is limited scope for prefabrication, although structural mesh reinforcement is used in roads and for lightly loaded flat areas.

Spacers, in the form of small concrete blocks, are used to maintain the clearance, or 'cover' between the steel and the outside of the concrete to prevent corrosion of the reinforcement. Short bars are used to space one layer of steel from another, and the top layers of steel in a foundation are held up by bars bent into 'stools'. When completed the reinforcement forms a rigid cage. Care must be taken to ensure that space is left to allow concrete to be poured effectively around the steel and for poker vibrators to be inserted.

Buried concrete may be poured against permanent sheet piling, rough timber shuttering or occasionally against the open sides of excavation, but where the finished surface is to be exposed to view, the wet concrete has to be contained by accurately made and closely fitting timber or steel panels. The internal faces of the formwork are made from panels which are braced with walings to withstand the considerable pressure exerted by the plastic concrete. The surface of the formwork is coated with a mould oil before the concrete is placed in order to prevent concrete adhesion to the formwork. Formwork materials are intended for re-use as many times as they remain serviceable. Formwork must not be stripped from fresh concrete until an adequate concrete strength is developed. The shuttering also assists in preventing premature drying out which would otherwise take place due to the heat generated within the setting concrete.

Figure 3.30 shows foundations of a new reclaimer hopper at Drax power station prior to concreting.

Besides using formwork to define the permanent outline of the concrete, a considerable amount of rough shuttering is used to divide the foundations up into convenient sections or 'pours'. The divisions have to take due account of the points of maximum bending and shear in the foundations when loaded. Individual pours in the main foundation areas may contain some 200 m^3 of concrete, depending on the capacity of the concrete mixing and transporting plant. It should be the aim of the contractor to programme the works in order that sufficient pours, of varying size, are planned to keep the batching and mixing plant producing a steady amount of concrete throughout the construction works.

Items such as drains, cable ducts, starter bars, holding-down bolts, and pockets for bolts and steelwork, have to be firmly fixed accurately in position before concreting can take place. It is essential that all bars are free of loose rust, oil or earth (as otherwise a satisfactory bond will not take place with the concrete) and that the inside of the formwork is clean. Before concreting is carried out these points are inspected and the whole arrangement is formally checked by contractor and engineer to see that it complies with the working drawings.

FIG. 3.30 Reclaim hopper under construction
(see also colour photograph between pp 242 and pp 243)

8.7 Mixing and placing of concrete

The amount of concrete now required in typical power station foundations totals over 200 000 m^3 and requires very high average weekly production rates of over 4000 m^3.

The choice of size, and number, of batching and mixing plants depends on the rate at which concrete is required to achieve the programmed construction targets. The choice of plant required to transport the concrete in the most cost-effective manner is dependent on the same factors, and also on the type of foundation being constructed.

The problem of continuously producing amounts of concrete of such large quantity is generally tackled by installing a central batching and mixing plant of high capacity to meet the average demand throughout the contract. This central plant is intended to operate throughout most of the life of the contract. Any extra capacity required to meet periods of maximum demand may be obtained by establishing other plants around the periphery of the works using mobile equipment. This arrangement has the advantage that the individual smaller plants may be moved, from time to time, to follow the construction works without disrupting production. Similarly, a breakdown of plant will not cause a serious stoppage. The travelling distance for the concrete to individual pours may be arranged more economically, and special requirements of mix and aggregates may be more readily accommodated.

The mixers are built into a substantial arrangement containing aggregate hoppers, 'pre-set weight' batching machines, bulk cement and PFA silos and water tanks. Large stockpiles of aggregate are established near the plant and from these the aggregate is loaded into the hoppers by means of an excavator or chain elevator. The materials are pre-weighed and fed into the mixer pan or drum. A pan mixer consists of a large diameter open-top stationary cylinder in which paddles rotate. This type of machine produces a satisfactory concrete mix in as little as 30 s compared with the customary 1 min to 2 min for the rotating vertical drum mixer.

The main building foundations on power station sites cover a large area and are shallow in comparison to their plan dimensions. This type of construction favours distribution by pump, as it is not easily accessible to wheeled traffic. The concrete pipelines vary from 100 mm to 175 mm in diameter; the individual lengths of pipe are designed to be manhandled across scaffold stagings and the joints are quickly made and broken. Bends are available so that the pipeline can reach any part of the pour, and the breaking and re-laying of pipelines is carried out frequently as work proceeds in order to deposit the concrete exactly where required.

Mechanical pumps are most frequently used to push the concrete along the pipelines. Concrete is fed into a hopper from the mixing plant, and gravitates through an automatic valve in the bottom of the hopper into the horizontal cylinder of the pump. A steel piston operated by an electric or diesel engine forces the concrete past a second valve and along the pipeline. The pumps may be single or double cylinder. Pneumatic pumps of various types are also available. Horizontal distances of several hundred metres may be pumped and pumping heads of over 100 m have been achieved.

The types of aggregate and the concrete mix design have to be suited to being handled by these types of equipment; the concrete must be neither too wet nor too dry, and the quality must be controlled to within fairly fine limits to achieve success. A blockage in the pipeline due to badly mixed concrete may result in every joint having to be broken and the concrete tipped from each pipe before it sets. The pipeline is cleared at the end of concreting by forcing a rubber ball through by compressed air. Careful calculation has to be made of the volume of concrete required and allowance made for the concrete remaining in the pipeline if wastage is to be avoided.

Concrete may also be transported about the site in spiral mixer trucks, in front-tipping dumpers, in crane skips, or in lorries fitted with large hydraulically-tipping skips holding up to 5 m^3. Temporary roads and ramps may also be required to reach the places at which concrete is required. It is not good practice to tip concrete from any considerable height as segregation of the materials may occur and the reinforcement may be displaced or the shuttering damaged. With deep foundations, cranes are used to handle skips of concrete to the correct points of discharge; the skips are either brought to the crane full, or are filled at the crane by tipping from the lorries or dumpers. For small foundations mobile cranes are often used, but for larger structures one or more tower cranes would be erected to carry out all the work, from shuttering to concreting.

Concrete for power station foundations is vibrated to fully-compact it into the formwork and around the reinforcement. This should produce a dense concrete free of voids and hence produce a good surface finish. Vibrators may be of the internal poker type, or external type working on the outer face of the formwork. The poker type is more suitable for use in large foundation pours. They vary in size from 25 mm to 100 mm in diameter and are about 500 mm in length. They are operated by electricity, compressed air, or by flexible drive from small petrol engines.

Although an apparently simple operation, some care is required on the part of the operators to ensure that the concrete is properly compacted. The depth of the concrete layers should be kept to not more than 600 mm to allow any entrapped air to escape. When using vibrators care should be taken to ensure against over-vibration thus possibly causing segregation of the concrete.

A contractor often chooses to install stationary compressor plant on site, with a galvanised steel main fitted with tapping points laid over the foundation area. A great deal of compressed air plant is required to operate the vibrators, for cleaning the face of concrete ready for the next pour, and for blowing the formwork clean of dirt and debris prior to concreting.

9 Direct cooled circulating water systems

9.1 Civil engineering structures in direct cooling systems

All major generation plant relies on a 'water to air' or 'water to water' transfer of low grade heat to cool the condensers and increase the efficiency of the plant.

The water to air system is typified by cooling towers or heat exchangers/radiators and is known generically as *indirect cooling* (see Chapter 2).

Direct cooling systems draw water from a natural or artificial source of cool supply, pump it once through the circulating water system and condensers, and discharge it where it will naturally cool before re-entering the system (see Fig 3.31). (The phenomenon known as *recirculation*, is discussed later.)

The source of water is usually the sea or a river, but lakes and artificial reservoirs are also used where necessary.

Once into the cooling water (CW) system the water passes in sequence through the following four main civil structures:

(a) An intake built at such a level in the water source as to ensure that adequate water can be abstracted under all conditions of flow or tide. It must be sited to avoid causing erosion of the bed or channel or drawing an excessive amount of debris, silt or fish into the system. A coarse screen is installed to exclude large debris either at the landward end of the channel or the seaward end of a tunnel.

(b) A screen chamber to house the fine screens which remove from the water the debris which has passed the coarse screen at the intake. These screens must similarly extend from extreme high water to extreme low water, whatever their type, while allowing adequate flow at both. Materials intercepted by the screens are automatically jetted-off into a trough from which they are sluiced into a metal cage for disposal. Any fish arriving at these screens are diverted into by-pass channels and returned to the cold water source.

(c) A pumphouse to contain the pumps at the lowest level in the CW system with the pipework necessary to take water from its natural level in the screen chamber and raise and distribute it to the top of the condensers via the inlet culverts.

(d) An outfall to take the heated water from the condensers back into the sea or river at a low velocity which will not cause erosion of the bed or banks or interfere with shipping. This section of the system may include a syphon recovery chamber or

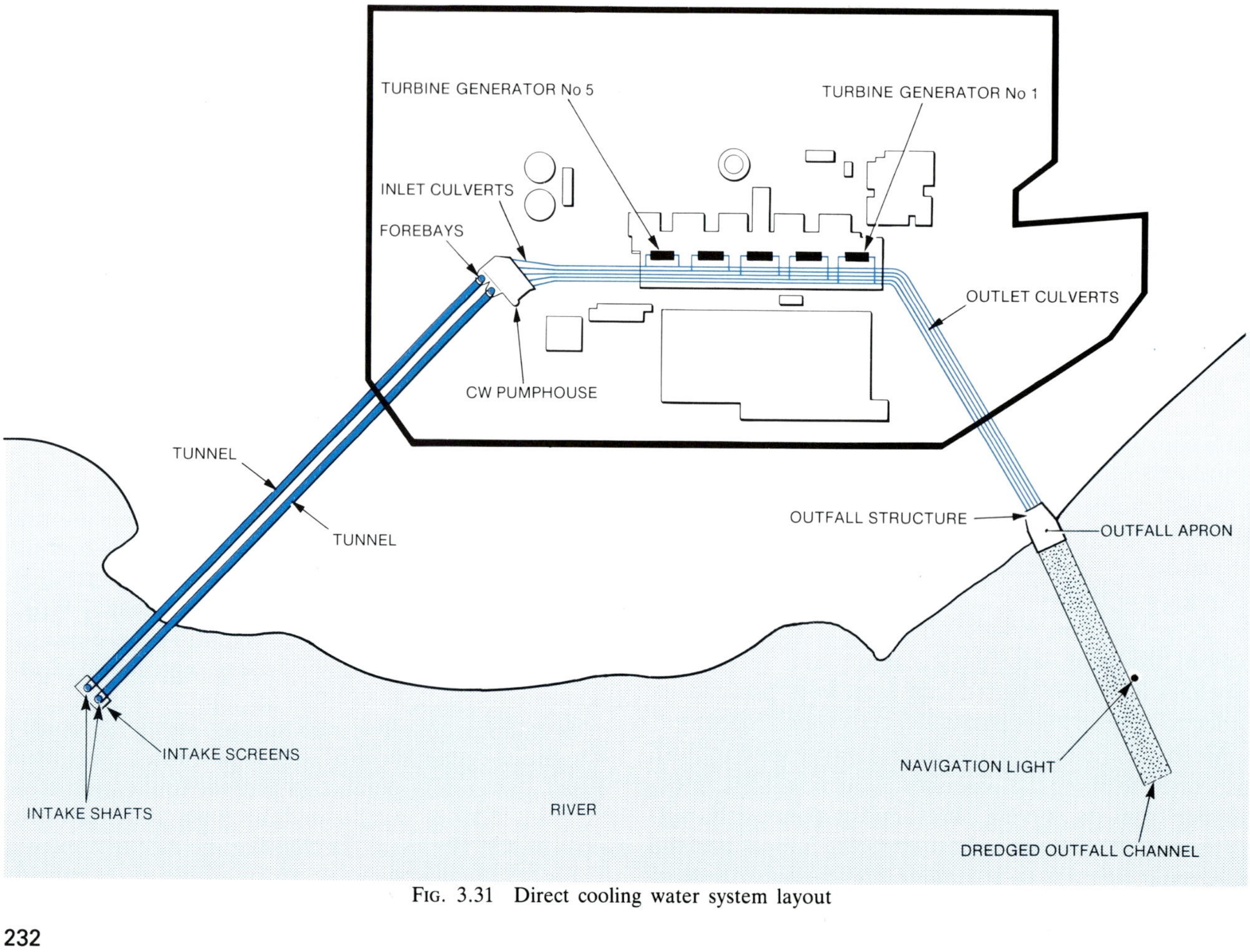

FIG. 3.31 Direct cooling water system layout

seal weir whose purpose it is to keep the syphon head in the condensers and let the water stream break to gravity flow at a predetermined spot. One is required only where tide or flood levels vary substantially.

The cooling water passes through these structures in the order given. It flows via intake culverts of concrete or steel between the intake and the condensers and, after use in the condensers, it flows in outfall culverts to the outfall. Volumes of flow are about 55 m^3 to 60 m^3/s for a 2000 MW station, but can vary depending on environmental constraints and the allowable temperature rise across the condensers.

Care must be taken in choosing the positions of the intake and outfall to ensure that recirculation of the heated outfall water directly back into the intake, and its consequent adverse effect on operating efficiency does not take place, or is minimised. Avoiding recirculation can be readily arranged if the topography is right and water can be taken from a river and discharged to sea for instance. Where no such solution is possible, a physical or mathematical hydraulic model may be necessary to ensure that recirculation is minimised under all tidal and environmental conditions. Several arrangements of intake, outfall and jetty can be modelled until the best is found.

Mathematical models and physical models need data obtained from on-site surveys to produce valid and repeatable results, but once verified can be used to check other environmental changes such as currents and silt. In addition, the models will simulate the results of vertical separation of intake and outfall so as to take advantage of the thermal stratification of the sea or lake water.

9.2 Culverts

The route of the culverts is governed by the relationship between intake and outfall and the axis of the turbine hall and the condensers. Because they require a wide swathe through the site and have little choice in level, they need to avoid crossing other services where possible.

Station lifetime pumping costs are affected by any change in length or level that vary the head required after making full use of syphonic recovery. Other variables are culvert cross-section, material and water velocity which are interdependent with the material governing the maximum velocity and the velocity fixing the cross-sectional area.

Usual velocities are 2.4 m/s to 3 m/s for concrete and 2.5 m/s to 3.5 m/s for metal, although as techniques and finishes improve both are tending to rise. Glass reinforced plastics are also becoming competitive. Culvert areas are usually 3 m to 4 m in diameter or the equivalent in square or octagonal shape.

Care needs to be taken in the design to avoid sharp corners and local changes in velocity which would set up turbulence in flow, to avoid areas of dead water which would encourage marine growth and silting, and to allow for the escape of trapped air from release valves or surge pipes. Man access is required at intervals for inspection and cleaning. The intake culverts also have to be designed to resist the maximum discharge head of the CW pumps on a closed condenser, water hammer from valve failure and external and construction loads.

Outfall culverts have less onerous internal stresses, most of which are transient, partial vacuum, pressures varying with external water levels and construction and traffic loads. All the culverts must be designed for temperature stresses, overburden pressures and not be buoyant when empty. Concrete culverts in particular should avoid thick sections leading to high temperature differential stresses during the curing of the concrete (see Fig 3.32).

Culverts are generally constructed under or within the turbine hall foundations or below ground level in shallow excavations. Each turbine has to be individually served with an inlet and outlet culvert, the total calling for several being built side by side. The massive excavations needed to accommodate this multiplicity of culverting, between turbine hall and CW pumphouse, imposes a severe restriction on many other site activities and needs to be completed as early as practicable in the overall programme.

Steel or metal culverts may be up to 3.5 m diameter and where buried in the open ground require the external protection of purpose-made materials. Their internal protective linings have not proved reliable as they deteriorate with age and tend to peel off and clog the condensers. Concrete linings applied under controlled conditions have been used successfully for several years and are being developed further.

Glass reinforced plastics are becoming available in suitable sizes, either for linings or providing structural properties, again further development is under way.

Reinforced concrete culverts are constructed with walls up to 600 mm thick and are rectangular or octagonal in section to avoid stress concentrations. In the case of inlet culverts operating under pressure, a circular or near circular section is preferred with hoop reinforcement. Where possible they should be concreted in one lift with lengths in-situ of up to 10 m, or longer if precast units are used.

Significant leakage from culverts sited below the turbine hall presents very difficult repair problems and must be avoided by care in both design and construction.

In-situ concrete culverts must be cast in alternate bays to allow shrinkage to take place, and all horizontal and vertical joints require continuous flexible water bars. The main requirements of the concrete are high density and good finish on the water face to cut down leaks and head losses, other requirements are low heat or hydration and good workability, so it is usual to recommend a purpose designed mix for culverts.

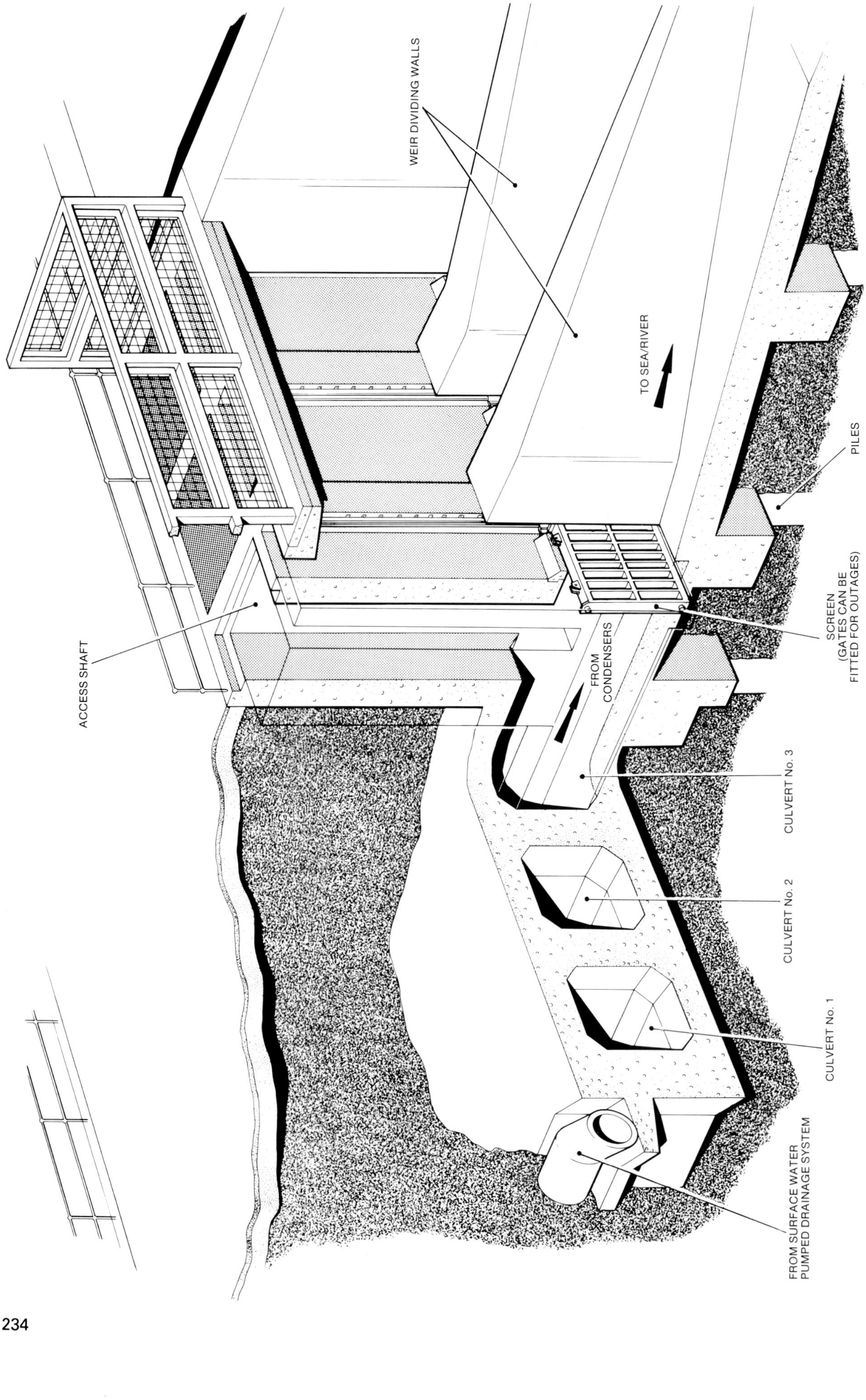

FIG. 3.32 Section through outfall structure

The congestion on site due to excavations for in-situ concrete culverts and the length of time they need to be open (until tested), has led to the rapid development of alternatives in other materials and pre-cast concrete sections built round the use of large mobile cranes.

9.3 Pumphouse and screen chamber intake

In order to draw adequate supplies of water at all states of tide or river flow from an intake situated some considerable depth below site level, the CW pumps are installed in a deep basement (Fig 3.33). This results in very deep excavations for large underground structures and causes problems in construction due to their proximity to water. Techniques for building these structures are discussed in Section 8.2 of this chapter.

The location of the pumphouse, screen chamber and intake varies with the topography, bathymetry and the geotechnical problems associated with construction and operation of the system. The structures may be combined with a jetty, the intake may also be separated from the fine screens and pumps by tunnel or tube. Each site has a series of variations on possible types of structure and the final system is developed from total lifetime cost comparisons.

Construction needs can also affect the design; for example, dewatering is easier from the pumphouse than the intake, and tunnels, are more readily driven uphill than level or down. Soft ground tunnelling carries non-technical risks and may be impossible to drive with adequate safety. Submerged tubes offer an alternative but must not be buoyant if dewatered, and must not represent a permanent hazard by being insufficiently buried in the river or sea bed.

The option of putting the pumps close to individual condensers to save costs on pressure culverts leads to complications with layout and foundations, which often lead to its rejection in favour of a combined screen and pump chamber.

9.4 Cooling water tunnels

Cooling water tunnels are required to carry the water from the intakes to the screen and pump chambers onshore. They can be dispensed with if the pumphouse is close enough to deep water to link them by dredged channel but this is normally not possible. In order to find suitable strata for tunnelling, or one for founding a submerged tube, it is often necessary to go lower than the hydraulic optimum level.

The construction programme usually calls for work on the pumphouse, screen chambers, intake and tunnels to be carried out simultaneously to meet commissioning dates, often in parallel with jetty works.

Providing additional tunnel faces is difficult because until the pumphouse is down to the level, the driving shield cannot be pushed forward at the tunnel level, which occurs late in the programme. It may be found necessary to sink a temporary shaft forward of the pumphouse so that one or two faces can be opened. Rarely is it economic to drive towards the pumphouse. Achieving pumphouse excavation on programme is thus of special importance.

Because of these risks to the programme and the lack of control over tunnel driving rates, alternatives such as submerged tubes may offer a chance to avoid the imposition of fully sequential working from the pumphouse excavation.

A variety of techniques are used for tunnelling CW culverts, depending on the depth and nature of the ground. In hard rock, a pattern of holes is drilled forward, charged, fired and the rock removed. Softer rocks can be dealt with by road headers and full-face tunnel boring machines, with or without shielding. In rock, the tunnel is supported temporarily as necessary with shotcrete, rock bolts or anchors prior to final lining with reinforced concrete pumped and vibrated behind formwork in the normal manner. If needed, final back grouting is done through eyes left in the lining.

Tunnelling in rock at power station sites close to deep water is almost invariably wet. Water enters through fissures and faults and considerable pumping may be required. Weak rock and steeply-dipping rock may require extensive ground treatment or rock anchoring, while close to the sea or river bed compressed air working may have to be adopted.

Where tunnels must be driven through soft ground that is likely to include beds of pervious materials, it may be prudent to consolidate these layers with grouts or other means ahead of the driving face. Again, compressed air working may be required to prevent water entering the works, however thorough the pretreatment of the ground. Nevertheless such pre-treatment will reduce the volume of water entering and the amount of 'fog' in the tunnel. It may also reduce the chances of a major air escape causing a blow-out to the river or sea.

It is not recommended to work in air above the physiological limits which are equivalent to about 25 metres below high water because of long term health risks. The UK imposes such limits which form part of the Work in Compressed Air Regulations [14] and these regulations also lay down decompression times for all workable pressures. Health risks to men working in air, coupled with high costs of plant and labour have led to great efforts to eliminate the need for compressed air working in tunnels.

Where it is not possible to avoid using air, a large compressor plant has to be built on site to supply sufficient air to keep the workings dry. It should have a back-up prime mover, e.g., diesel and electric or gas and diesel, and be fitted with aftercoolers and an effective means of removing products of combustion from the internal compressors. Air dryers are advisable

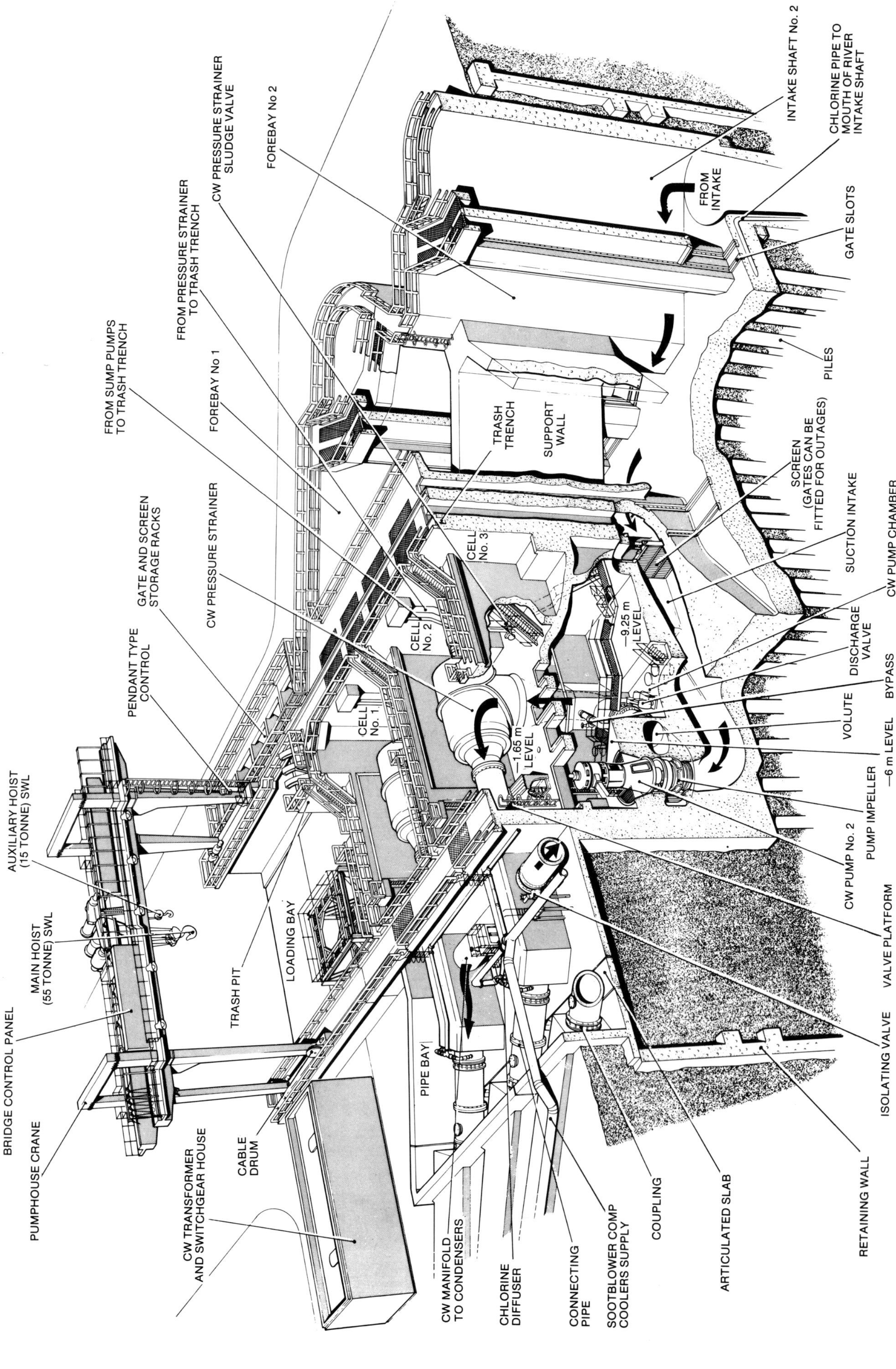

FIG. 3.33 Section through CW pumphouse structure

if air losses are high, otherwise condensation of expanding air cuts visibility in the tunnel and interferes with progress and setting out.

A high pressure supply is required for air tools, boogee pans and grouting. Water, electricity and communications are also needed at the face, together with safety monitoring equipment for methane and carbon dioxide, and Budenberg-type individual air lock pressure gauges.

The passage of men and materials into and out of the tunnel is through air locks, which are chambers with double airtight doors in which the air pressure can be raised or lowered without affecting the pressure inside the tunnel. Air locks are either bolted to the top of the working shaft or are formed with double bulkheads in the tunnel.

Tunnels in soft ground are normally driven using a circular steel shield. This is fitted with a cutting edge and a skin plate and is driven forward by jacks reacting on the most recently fixed tunnel lining. A short heading is first dug by hand or compressed air tools in front of the shield, the shield is moved forward on the jacks and the excavated material is removed. A further ring of lining is erected inside the skin plate when the jacks are withdrawn.

The structural tunnel lining may be formed with ribbed cast iron or concrete segments. The meeting faces are a close fit, but the joints are caulked with lead or plastic to make the lining watertight. The segments have screwed plug holes through which grout is pumped to ensure full contact between the segment and the soil outside. When the lining is complete, caulked and grouted, air pressure is reduced to atmospheric pressure and the tunnel further lined with concrete where required to produce a smooth water flow surface.

As with culverts, the cooling water tunnels and shafts require a smooth surface to the lining, and bends have to be properly designed to eliminate hydraulic losses. The size of tunnel has to take into account the designed velocity of flow, but in some cases may also depend on the size of tunnelling shield available or on an economic decision being taken to standardise the intake and outfall tunnels.

The CW tunnel is completed by forming the shaft at the outboard end connecting it to the source of cooling water. This shaft can be sunk independently, raised from the tunnel or dropped in a cofferdam to meet the tunnel. Numerous ingenious methods of doing this have been devised (see Fig 3.34), but it is difficult to devise a method which allows the shield to be recovered and the tunnel to be dewatered on completion.

As a consequence of efforts to avoid the risks inherent in shield driven tunnels, the development of submerged tube tunnels and pressure balanced drilling mud shields has been rapid, but neither can be said to be an all purpose answer in any ground or water conditions.

9.5 Submersible cooling water structures

In parallel with the incentive to avoid risky, unpredictable and labour intensive tunnelling work has been the growth of the capital plant industry capable of operating in shoaling tidal water. The range of dredgers, tugs and floating cranes available and the experience gained in estuarine road tunnel construction makes their application to offshore structures for cooling water the next logical step.

Plant currently available can dredge to 20 m depth, provide 12 000 kW of towing capacity per unit and lifting capacity of over 5000 t per hook; and the development of more powerful equipment is underway. This means the problem of digging a trench, towing in a precast length of reinforced concrete tunnel, sinking it and back-filling has become much simpler than attempting to do the work piecemeal under the sea, even though the weather is still a major risk to plant, trench, units and men.

Intake and outfall structures can be built in the same way, although the tunnel element design requires detailed analysis or modelling to ensure that they are stable under tow or when being lowered. In the permanent condition, if the tunnel is required to be dewatered, it is often the positive buoyancy which constrains the final design.

With the floating of large capacity plant coming available and dredging costs falling, it is becoming more attractive to consider floating ever larger sections of CW systems, or indeed other modules for station construction, and bringing them to site by water.

9.6 Maintenance considerations

Civil engineering works associated with the cooling water system should be designed to be maintenance free for the first generation of plant. Exposed metal surfaces, screens, gates and valves will need inspection, and if any form of cathodic protection is incorporated it will need regular inspection and maintenance.

Apart from this routine servicing of metal surfaces it should be possible with good design to achieve a balance of water velocity, chlorination and concrete quality control so as to limit maintenance to inspection only, unless the cooling water is particularly aggressive or biologically active.

Access for inspection used to involve the need to dewater, which induces large stress changes to the structure. However, with the use of underwater scanning and film records these can be avoided and the system designed for a more consistent stress regime. This may not be a total advantage for, if inspection shows remedial work is needed, it must then be done underwater and access gained underwater. Again, modern equipment can usually be adapted to do this,

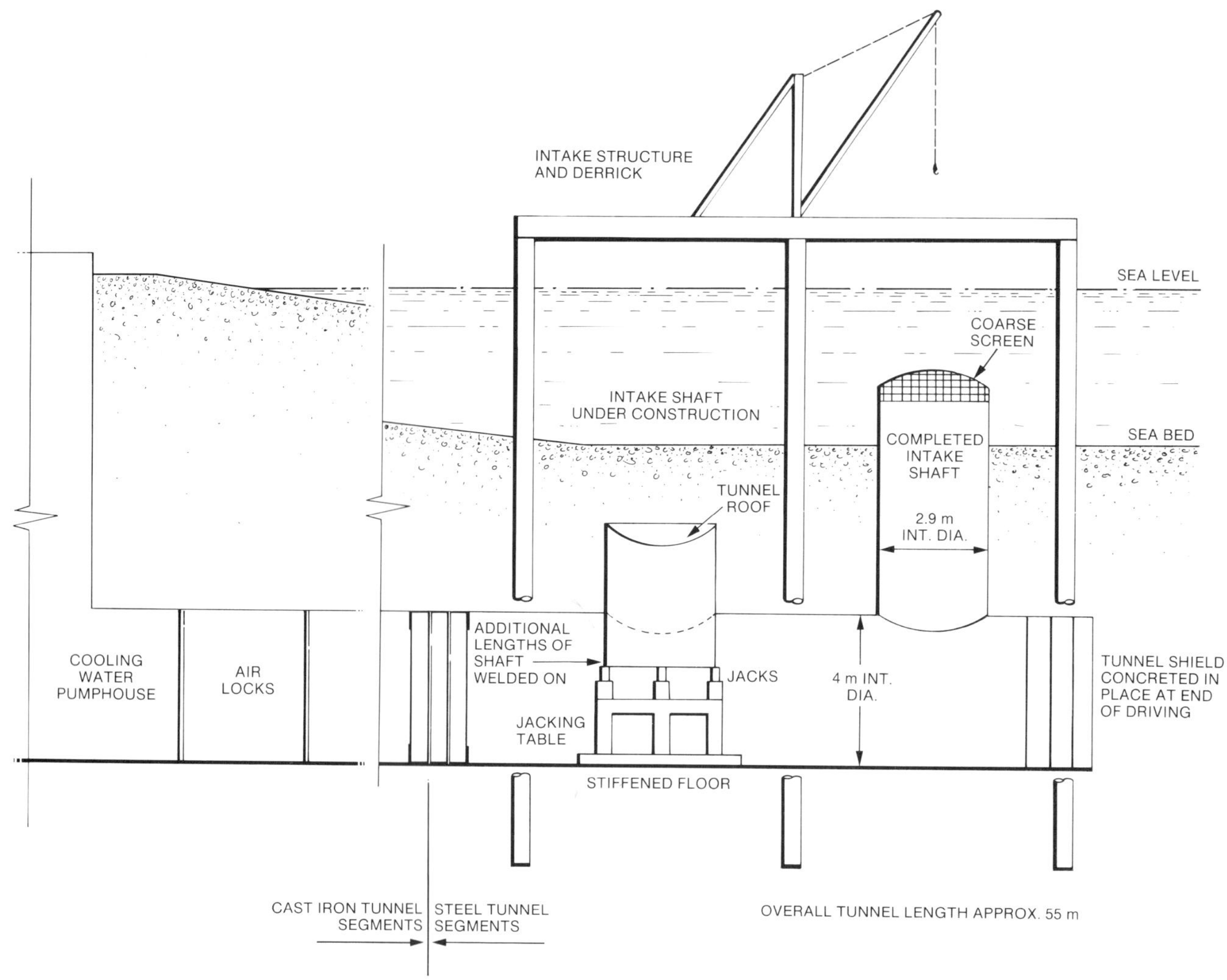

FIG. 3.34 Arrangement of intake shafts constructed from inside the tunnel

but the emphasis must be on control of the design and specification at the construction stage.

A prior knowledge of the characteristics of the water and the environment both physical (silt, etc.) and biological (fouling, etc.) is essential at the planning stage, as well as any changes that a thermal and water flow shock to the area may trigger.

10 Harbours and jetties

10.1 General

All coastal and estuarine stations have seaborne fuel and other deliveries and exports as a valuable option, especially those with either over- or under-developed hinterland infrastructures. Apart from pipelines, sea transport often offers the most economical method of servicing a station; and once the decision is taken to provide a berthing facility, it then becomes a more logical and economic option for other materials. Exports such as ash or unprocessed fuel, sections of the power station and its plant, both incoming and outgoing, can clearly be considered to be potential waterborne loads.

The jetty and its connections to the shore need to carry all cargo handling equipment, conveyors, pipelines, etc., all services needed by vessels and must have space for access and manoeuvring by fire, ambulance and other services. Overall dimensions are governed largely by long term considerations such as source and quantities of fuel, size and types of ship, turnround time and varieties of other anticipated cargo. These factors depend on commercial decisions, but they control the minimum laden draught and hence the possible location of a jetty and its structural form.

Figure 3.35 shows the West Thurrock coal-fired power station where coal is received by collier unloading at the riverside jetty.

Landarms can be over or under water, or non-existent if deep water is close inshore. They may also double as cooling water inlet or outlet pipes if this suits the layout with the intake, outfall or pumphouse forming part of the jetty or quay structure. Interference with shipping movements needs to be avoided by correct

FIG. 3.35 West Thurrock coal-fired station
(see also colour photograph between pp 242 and pp 243)

placing and direction of discharge and intake structures and their flows.

Development of harbour facilities needs to be integrated with other marine works and sea defences, and to be superimposed on the background topography and current pattern so as to cause minimum disturbance to the natural balance of silt and littoral drift movements. Often this can only be done by physical models and by reference to similar sites elsewhere. Mathematical models may be helpful for specific parts. Both types of model need hydrographic information from site to enable building and validation.

10.2 Types of harbours and jetties

Free-standing jetties in deep water present totally different engineering problems to those posed by wharves or quays suitable for shallower-draught vessels. Opportunities for cheap wharf construction are virtually limited to redevelopment areas in Western Europe. Basic types of construction are described here but it should be remembered that with large power stations the logistics of programme construction, plant and techniques available may affect marine works more than pure cost.

The simplest type of wharf is illustrated in Fig 3.36. Here the river bank is supported by a line of sheet piles with their toes driven well below the ultimate dredged level of the river bed, and with their heads supported by walings held back by tie-rods to anchor blocks set well back in stable ground. This type of wharf is most suitable for low ranges of tide and shallow draught vessels, due to the limited earth pressures which can be supported.

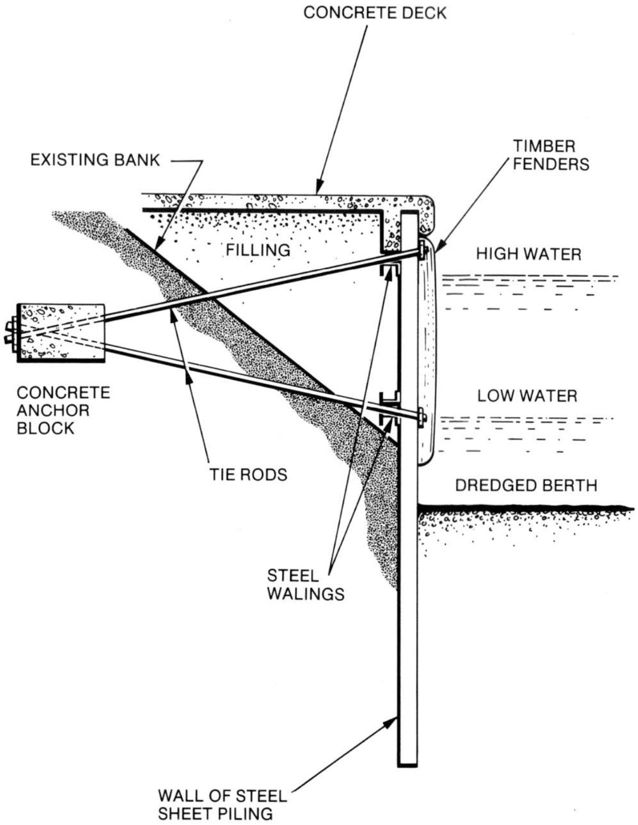

FIG. 3.36 Section of wharf with sheet piling retaining wall

Figure 3.37 illustrates a type of wharf used where deeper berthing is required. The stability of the bank is again maintained by a sheet pile wall and by a reinforced concrete sub-deck and retaining wall supported on piles. The bulk handling plant is carried on a beam and slab deck supported partly on piles, and on its outside edge by a row of cylinders. These are spaced at intervals of about 10 m and besides supporting the front of the wharf, also resist the impact loads from ships berthing. In this instance, piles are driven down through the base of the cylinders into the chalk, and the cylinders are braced from the sub-deck and concrete retaining wall by reinforced concrete trusses. Replaceable timber fenders and rubbing timbers are fixed to the face of the wharf and cylinders to minimise impact loads. The berth needs to be dredged, but the river bed can be left at its natural profile behind the row of cylinders.

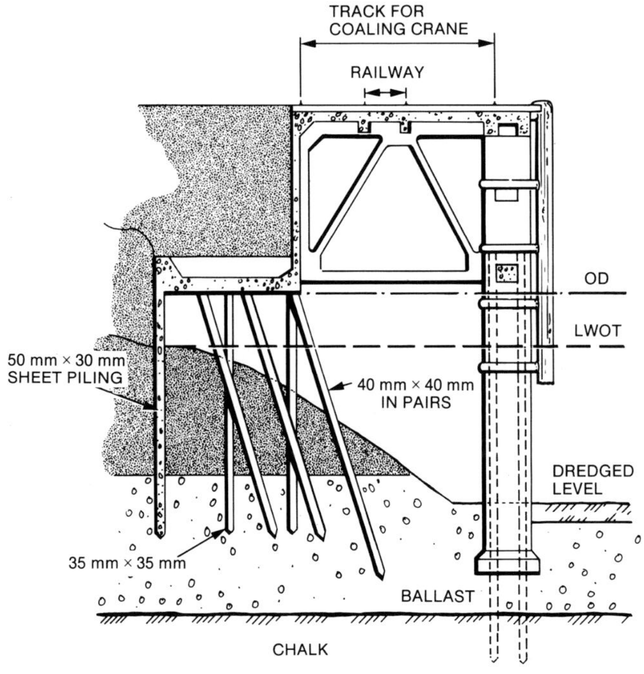

FIG. 3.37 Section of wharf with concrete piles and cylinders

Figure 3.38 shows a cross-section of a wharf taken in a position where the sub-deck has been replaced by coarse screen chambers. The cylinders, fendering and deck are indicated, with one of the travelling cranes. The river bed at the wharf's frontage has been dredged to give sufficient depth of water at all stages of the tide for laden vessels to remain afloat. Consequently they can enter or leave the berths at any time.

Deeper berthing facilities may call for the construction of free-standing jetties far enough off-shore to be on the edge of the deep water channel. A reinforced

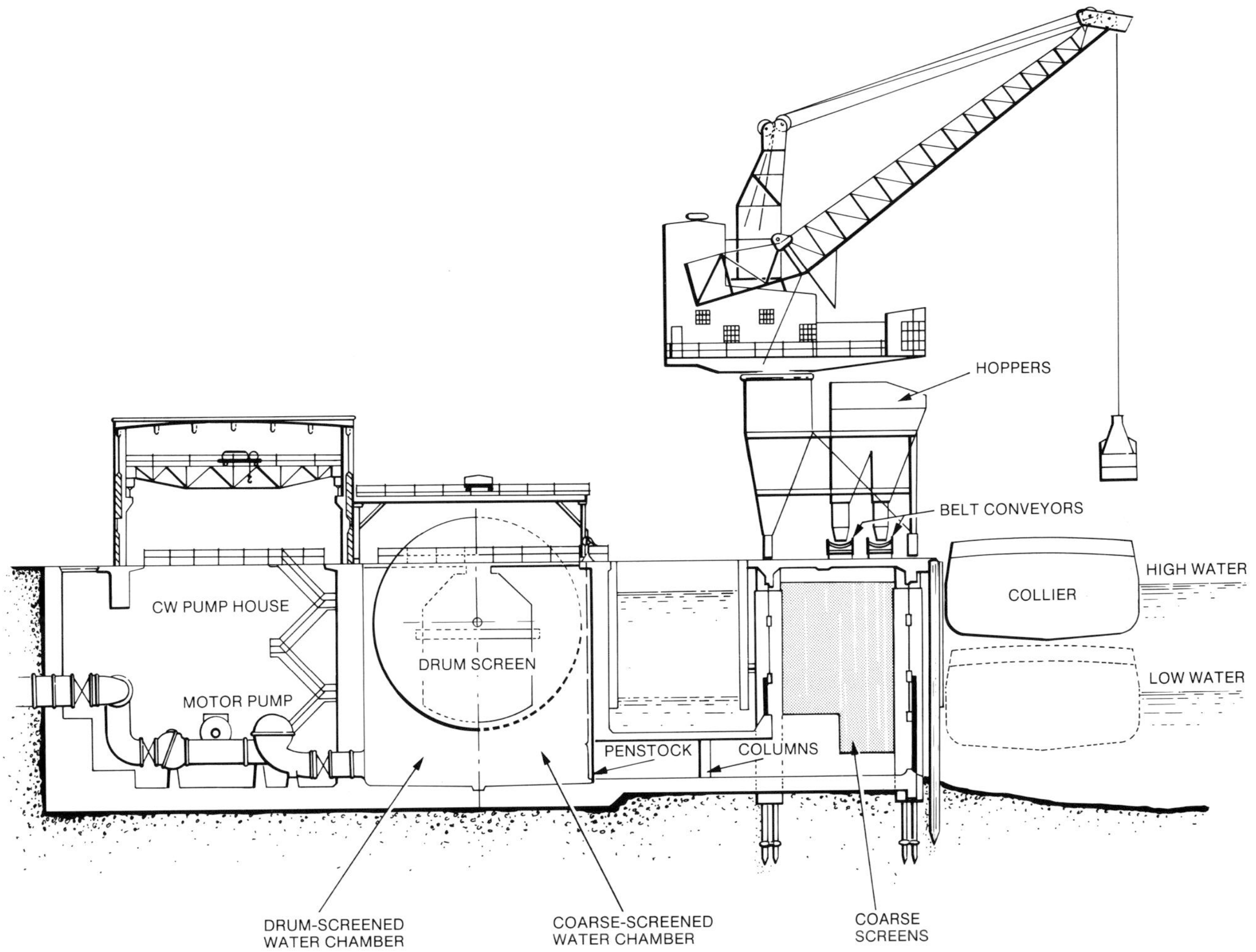

FIG. 3.38 Section of wharf showing intake, screen chambers and pumphouse

concrete deck of beam and slab construction is supported on rows of raking and vertical piles, or on two or three rows of cylinders. The cylinders are connected across the jetty by deep beams. The junctions between beams and cylinders are heavily reinforced, as are the cylinders, so that a rigid structure is produced. This removes the need for deep cross-braces between beams and cylinders, and the consequent difficulty of working between tides. The cylinders may be taken down to final foundation level or may have piles driven down through their bases to take the vertical and berthing loads.

A typical fendering system may consist of hollow steel piles driven between the front row of cylinders and supported off the deck by buffers. Timber protection can also be provided, fixed to the steel piles, and spanning between them. This type of jetty may also have intake or outfall chambers built integrally with it.

The free-standing deep water jetty needs to be connected to shore for virtually everything that passes to or from it. Because of the high cost of this link it is better to limit its capability to essentials, the largest of which is usually fire appliances, and to use conveyor and pipeline housings as structural members.

10.3 Construction of harbours and jetties

Whether the site is in a remote area or an urban area with a crowded infrastructure, there is always pressure to have a working jetty or harbour early in the construction programme to relieve traffic or open up isolated sites for deliveries of construction equipment. This pressure usually means that jetties are designed for quick construction rather than engineering innovation, while construction depends heavily on the weather at site and the floating plant available. Wharves and harbours are not so amenable to the use of flexible or prefabrication techniques and usually take longer to construct and require more site-produced concrete. Hence jetties are usually the first choice if their sole purpose is to service the power station.

The necessary major plant; floating cranes, pile frames, batchers, dredgers and tugs must be known to be available at the site before a specific design is adopted. Alternatively, contractor designs should be invited from tenderers who have known access to suitable plant.

Mobilising marine construction equipment is likely to be a major part of the cost so standard sizing of piles, cylinders and pre-cast beam or deck sections offers the advantage of simplicity and speed. Flexibility of construction and type of pile is also advantageous. Acceptable alternatives should be able to be driven, vibrated or jetted to the level or permission given to use any combination of these methods. Soffit shuttering should be avoided and pre-cast work used where possible. Insitu concrete, where necessary, may even be placed more economically by helicopter from a shore-based batcher.

The decision on how much energy absorption to take in the structure and how much in the fendering depends largely on external factors such as current, wind, type and size of ship and how closely-coupled the vessel needs to be during loading or unloading. A tanker, for instance, requires much less rigidity than a collier or container ship, as all shore connections are flexible.

11 Loadings

11.1 Definitions

When designing any structural element in a building it is necessary to ascertain the imposed loads in addition to the dead load of the frame, cladding and other components making up the building. The imposed loads are the result of static plant, dismantled plant being placed on floors, impact operation of plant, vibration and dynamic forces, foot and wheeled traffic and also wind and snow loads.

11.2 Imposed loads due to plant

Loads due to plant items may vary from less than one tonne on small pipe supports and similar items to those up to 13 000 tonnes for slung boilers, turbine and generator components. If maintenance of the heavier plant is not carried out by station personnel, no provision is made for accommodating plant components on the boiler house and turbine hall operating floors, and the crane is used to move equipment directly to transporters in the loading bay. The loading bay floor in the turbine hall is designed to carry transporters and loads up to 310 tonnes.

11.3 Distributed imposed loads

Although it is necessary to limit the positions on which heavy loads can be set down on floors, it is not practical to do this with smaller items and it is therefore usual to decide on a floor loading per square metre which will cater for this form of loading. Foot traffic and the movement of materials are also taken into account when assessing the imposed distributed load. At the time that the design of the frame is prepared it is not possible to ascertain all the smaller individual imposed plant loading requirements, especially loading on floors due to underslung pipework and similar items, and it is common practice to increase the distributed imposed load per square metre by an amount which will cater for these requirements.

It should be noted that the loadings in Table 3.5 are in addition to individual items of heavy static plant, the positions and weights of which are accurately determined and which in some cases have their own separate foundations.

The recommendations given in BS6399 — Loading for Buildings: Part 1: Code of Practice for Dead and Imposed Loads [15] are normally used to establish the design imposed loads for offices, canteens, laboratories, etc. Typical imposed distributed loads for offices from BS6399 are 2.5 kN/m^2 for offices for general use and 3.5 kN/m^2 for offices with fixed computers or

TABLE 3.5
Typical distributed loads

FLOOR LOCATION	LOADING IN kN m^2
Boiler house operating level	12.5
Boiler house basement	25
Turbine house operating level	12.5
Turbine house basement	25
Turbine house loading bay	50
Control block	10
Electrical annexe	5—12.5
Battery rooms (TH annexe)	10—17
Junction house	5—12.5
Machine shop	25—50
Diesel house/compressor house (400 kV compound)	11
Stores	10—17.5

FIG. 3.1 Light cable tool boring rig

FIG. 3.2 Large rotary drilling rig

FIG. 3.3 Detail of rotary drilling

FIG. 3.4 Rotary core drilling of an upwardly-inclined borehole in an existing dam

Fig. 3.6 Interior of electric static cone penetrometer truck showing data processing equipment

Fig. 3.30 Reclaim hopper under construction

FIG. 3.35 West Thurrock coal-fired station

Fig. 3.60 Lower stressing gallery for AGR pressure vessel

FIG. 3.71 Precipitators

similar data processing equipment. It is usual to add 1 kN/m^2 to allow for partition loads. The imposed loads referred to are used in the design of floors and secondary members. For main beams and columns it is usual to allow a reduced loading as discussed in Section 11.6 of this chapter.

11.4 Cranes

A variety of cranes is provided in power stations, and on conventional stations the largest are those in the turbine halls (see Chapter 2, Fig 2.37).

The start point for determining the design loadings to be applied to a crane support structure is the maximum static wheel load. For conventional design on simple craneage, to allow for dynamic effects, the maximum static wheel loads are enhanced by 25% for design purposes. Horizontal loadings induced by operation of the crane are taken to be 10% of the maximum static wheel loads acting transverse to the long travel rail and 5% acting along the rail.

Loads for heavy or high speed cranes need to be given special consideration, as do the effects of more complex forms of crane gantry construction.

Elements of the crane support structure subject to significant fluctuations of stress are checked for fatigue endurance. The criteria for this is obtained from consideration of the crane's duty and is determined from the number and magnitude of lifts expected over the design life. This may include periods of construction, operation and decommissioning.

Although single cranes are used to lift the heaviest loads, it is sometimes advantageous to use two smaller cranes individually for the smaller loads and linked together with a lifting beam on the two main hooks for the heavier loads. The latter arrangement gives a greater spread of loading on the crane gantry girders with potential reductions in the cost of foundations and structural frame. The use of two cranes for smaller loads improves availability during the erection programme.

Cranage and their support structures in nuclear stations require special consideration. Where failure could lead to a nuclear hazard the cranage and its support structure are designed and manufactured to high integrity standards. This entails the provision of a conservatively designed support structure to withstand the normal operating loads. In addition to these normal loads there is a need to consider the effects of accident conditions and extreme events. The loading applied from such hazards, earthquake, etc., are the subject of special evaluation. The design requirements for the structure under the action of these loads will vary with circumstances but will at least require that it does not fail or collapse and may require that it remains serviceable following the event.

11.5 Wind and snow loads

The allowable pressure and suction on the vertical faces of walls and also sloping and horizontal surfaces of roofs due to wind are covered by the recommendations of building Code of Practice CP3: Chapter V: Part 2 [16].

The dynamic pressure (in Newtons per square metre) of the wind is given by the expression $q = kVs^2$ where $k = 0.613$ and Vs is the design wind speed in metres per second. The design wind speed is derived by multiplying the basic wind speed V by three factors S_1, S_2 and S_3. The basic wind speed (i.e., the 3 second gust speed at 10 metres height estimated to be exceeded once in 50 years) for any chosen site in the UK is determined by consulting the isopleth map in the code. The factor S_1 is a topographic factor, the value of which is usually taken as 1.0. However, in certain very exposed hill slopes and in valleys shaped to produce wind funnelling S_1 is increased to 1.1. The advice of the meteorological office is sought if there is any doubt about a particular site.

Factor S_2 takes account of the combined effects of ground roughness, the variation of wind speed above ground and the size of building or component part under consideration. Most new power stations are likely to be in terrain category 1, i.e., in fetches of open level country with no shelter. Factor S_3 is based on statistical concepts which take account of the degree of security required and the period of exposure of the structure in years. A building life of 50 years and a probability level of 0.636 is usually adopted and this gives a value for S_3 of 1.0.

Having obtained the dynamic pressure q for a particular surface, the surface pressure is calculated by multiplying q by an external pressure coefficient Cp_e, values of which are given in the code for buildings of various shape and proportion. The code also gives advice on determining the appropriate internal pressure coefficient Cp_i. Thus the wind load F acting in a direction normal to the individual structural element or cladding unit is $F = (Cp_e — Cp_i)\, qA$, where A is the surface area of the structural element or cladding unit.

Typical wind loads on cladding and roofing on the main building of a modern power station boiler house are of the order of 2.5 kN/m^2 in both pressure and suction. Boiler house roof corners and edges where wind eddies and vortices can occur are designed for suctions in the order of 4 kN/m^2.

In the cases of important buildings, or components, of unusual shape such as domes, where there is little data in the code and where limited advice is available, wind tunnel tests may be carried out on a model of the proposed building, or arrangement of buildings, to determine the appropriate pressure coefficients directly. This is an expensive and time consuming procedure and the possibility of adopting a very conservative code-based value will suffice in most cases.

Flat roofs are usually designed for an imposed load of 1.5 kN/m^2 where permanent access is provided to the roof and this includes for the snow load. Where no access is provided to the roof (other than that necessary for cleaning and repair) the imposed load including snow is taken as 0.75 kN/m^2.

11.6 Reduced loadings in main beams and columns

Where floors or roofs are designed on a load per square metre basis to accommodate full code-based values of loadings, it is usual to allow a reduction in the imposed load on the main beams and columns. This reduction accepts the logic that while individual areas of floor slabs or roof decks may be fully stressed by these maximum loadings, it is unlikely that the fully integrated maximum design loads will ever be applied to the whole area of the roof or floor on one occasion. One example of these recommendations is in roof construction where the roof decking and purlins may be designed to 1.5 kN/m^2, whereas the main beams and columns are designed for an imposed load of 0.75 kN/m^2. If a turbine hall has main beams which span 60 metres and are at 9 m centres, the total imposed load on the basis of 1.5 kN/m^2 would be:

$$60 \times 9 \times 1.5 = 810 \text{ kN — say 80 tonnes}$$

This is a very high load and by reducing the imposed load to 0.75 kN/m^2 for the main beams and columns, the more realistic figure of 40 tonnes is obtained for the total imposed load on one beam.

11.7 Design loadings

The frame for a building is subject to a continually changing load system (even after completion) due to thermal expansion, varying wind loading, snow loading, dismantling and re-erection of plant, machinery, traffic and other similar varying factors. The frame must be capable of carrying the worst combination of any of the loadings at all stages of erection and after completion. The complications of the distribution of bending moments in rigid joints, structures and continuous members, limiting deflections and shear forces, and the fact that the design must be on a three-dimensional basis, make the computer a most valuable aid for the structural engineer when he designs the frame for the main buildings of a power station.

Figure 3.39 shows some of the more relevant imposed loads on a frame for the main buildings.

11.8 Extreme loadings for nuclear stations

The loadings due to events having a frequency of occurrence at the site of less than or equal to 10^{-4} per year are defined as extreme loadings. Events having a frequency of less than or equal to 10^{-8} per year are of sufficiently low probability to be discounted; for example, a large meteorite strike.

Extreme loadings are split into two categories, namely, internally generated or externally generated, the former being those events which are caused by postulated faults in plant within the site boundary and the latter being due to natural phenomena or from hazards arising outside the site boundary.

The major internal extreme loads which have to be considered are those caused by:

- Missiles, e.g., generated by turbine blade or rotor failure, or failure of pressurised plant.
- Dropped loads, e.g., due to crane or handling equipment failure.
- Pipe whip, e.g., failure of pressurised pipework, vessels or seals.
- Gas or steam release, e.g., from failed pipework, vessels or seals.
- Fire, e.g., conflagrations arising from failure of plant or flammable liquid storage handling systems.

The major external extreme loads which have to be considered are those caused by:

- Wind, e.g., wind speeds with return periods of 10 000 years or more, tornadoes and the effect of tornado-borne debris.
- Earthquakes, e.g., ground-motions with return periods of 10 000 years or more arising from seismic events in the near vicinity of the site (approximately 30 km for the UK).
- Flood, e.g., sea or river levels overtopping flood defences, extreme snow or rainfall with return periods in excess of 10 000 years or more.
- Aircraft crash, e.g., impact of civil or military aircraft on the site.
- Gas cloud explosions, e.g., blast and pressure loadings arising from ignition of gas clouds escaping from rail, road or water-borne tankers or nearby storage facilities.

In some cases, a number of external extreme loads can be discounted following careful assessment of the site location, for example, sea flooding if the site is inland, aircraft crash on sites remote from airfields and gas cloud explosions where no transport routes or storage facilities are located near the site.

The extreme loadings to be considered by the designer have to be specified together with the survival criteria that the plant or structures concerned must meet. The derivation of loadings for wind, flooding and earthquakes are generally the responsibility of the civil engineer and the standards used are described in Sections 11.3 and 11.5 of this chapter.

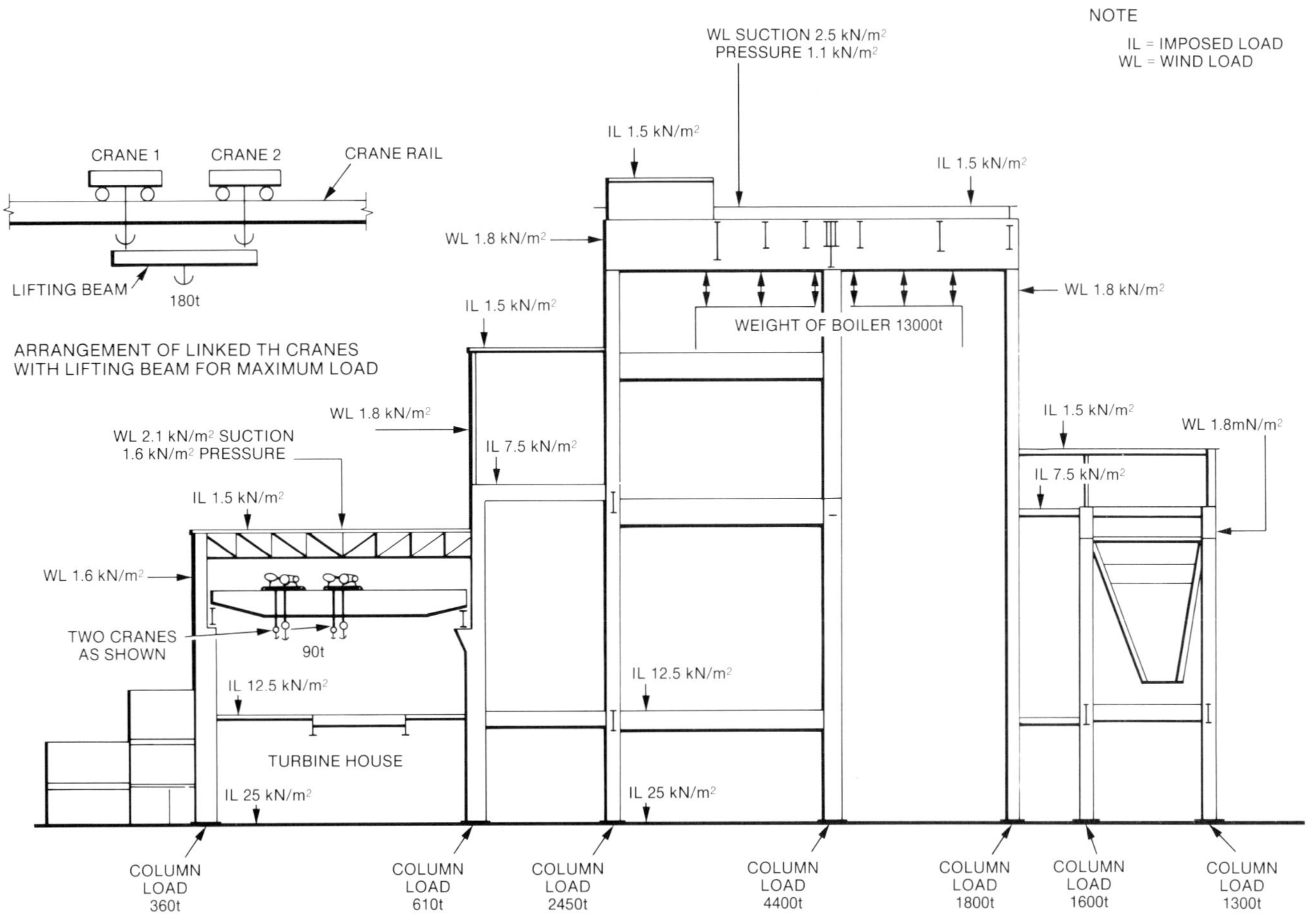

Fig. 3.39 Section through station showing typical loadings

Structural design for extreme loadings generally requires specialist methods, since non-linear analysis or empirical formulae are frequently necessary, and in some cases specially commissioned testing is required to fully validate the design.

12 Steel frames

12.1 Steelwork

Structural steelwork is mainly used for the frames of power station buildings. The beams, columns and other members are fabricated at the works, delivered by road or rail transport to the site and erected on prepared foundations. The structural engineer designs the frame using rolled-steel joist sections, universal beams and column sections, channel and angle sections, flat plates and tubes. In addition to these sections, members such as welded plate girders up to 4 m deep and welded box columns up to 2 m × 1 m in section, with plates up to 90 mm thick, are used to support the heavy plant loading associated with the main power station buildings. Castellated beams (Fig 3.40) which consist of rolled-steel joists or universal beams cut through the centre line of the webs to form a sawtooth pattern, and then welded with the points together, are particularly useful when light loads are carried on a long span and deflection rather than bending stresses is the design criterion (e.g., in roof designs as rafters and purlins to support roof decking). Compound sections, lattice members, box girders and heavy and light trusses are also used.

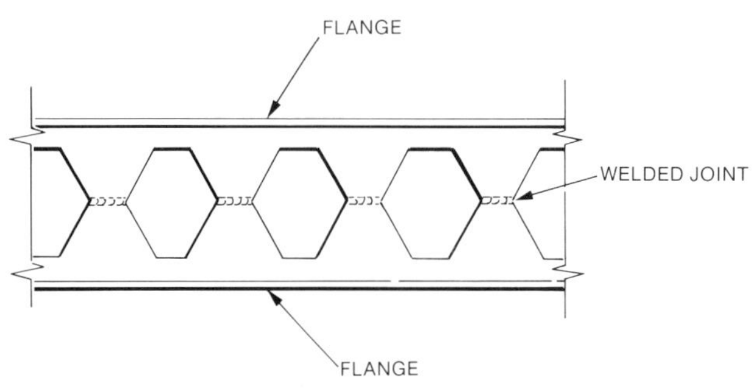

NOTE: THE DEPTH OF A CASTELLATED STEEL BEAM IS APPROXIMATELY 1½ × THE DEPTH OF THE ORIGINAL STEEL SECTION

Fig. 3.40 Arrangement of castellated steel beam

Two types of construction can be used in the frames of the power station main buildings — the turbine hall and boiler house:

(a) *Portal frame construction* This consists of a system of heavy unbraced columns and beams which rely on the bending stiffness of these members and the rigidity of the connections to provide the required stability. The main advantage in this type of construction is that it provides large clear spans that simplify plant access and maintenance.

(b) *Braced frame construction* This consists of a system of heavy columns and beams braced together in the traditional manner to provide the required stability. It is lighter in weight than portal frame construction and consequently cheaper.

A combination of the two types of construction is normally used in modern power stations, portal structures being limited to areas where the need for clear spans is considered essential.

A typical portal frame moment connection is shown in Fig 3.41. The typical box column shown here

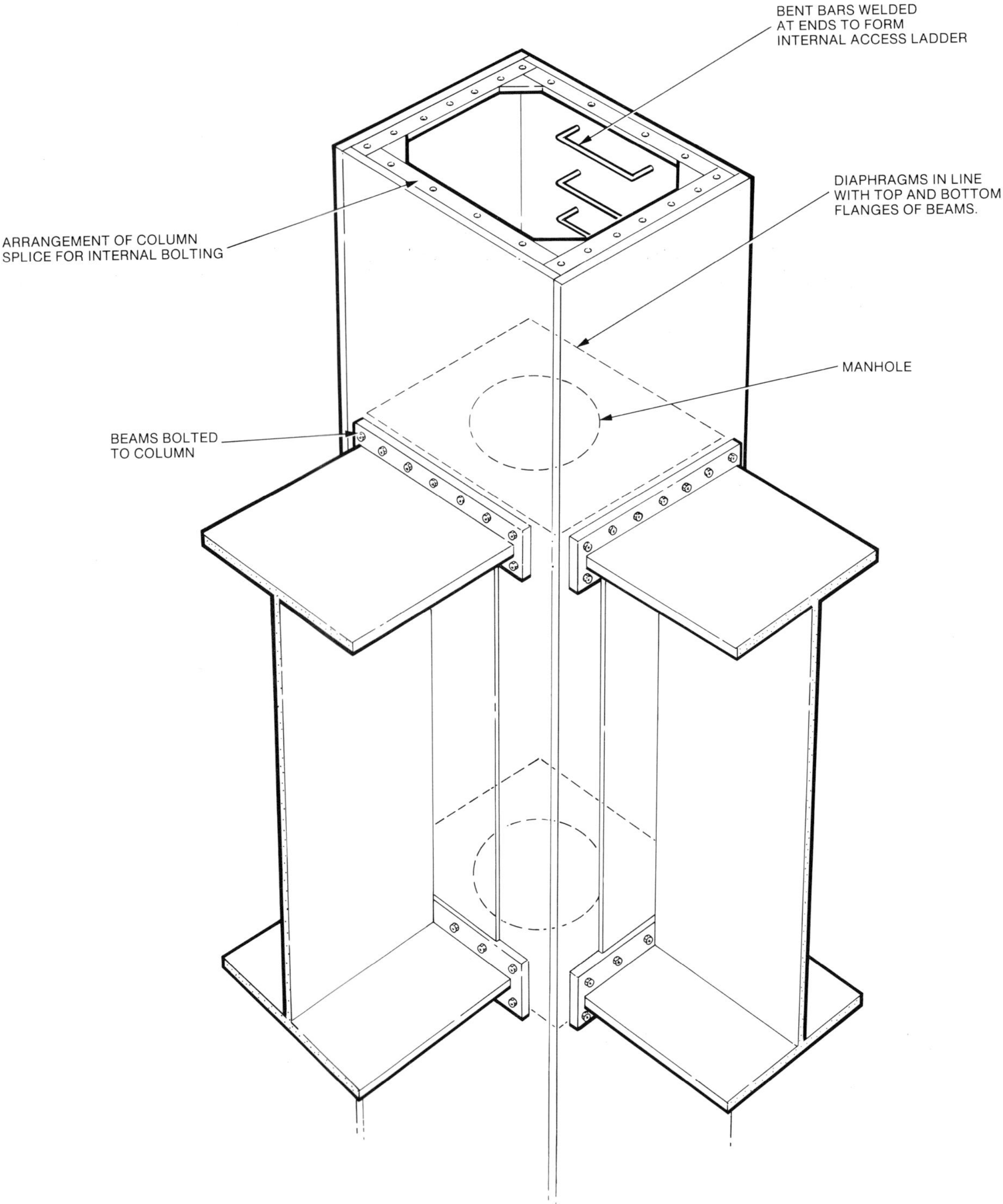

FIG. 3.41 Arrangement of box column

requires internal stiffening by the use of diaphragm plates and/or rolled-steel sections. The diaphragm plates must have manhole access for steelwork erection and maintenance purposes. Manhole access is also provided in the web/flange plates at ground floor level and at a top floor level.

Although steel to BS4360 [17] grade 43 (mild steel) is used extensively, grade 50 (high yield strength) steels are available. The higher working stresses permitted with this material enable lighter sections to be used for the same purpose with their attendant saving in costs, although larger deflections need to be tolerable.

12.2 Design of members

The design of a building frame can be based on simple, semi-rigid or rigid design methods. If the structural engineer uses simple design the distribution of forces may be determined assuming that members intersecting at a joint are pin connected, i.e., freely supported at their ends. In this case the building frame should be adequately braced against the tendency to sway under applied horizontal loads. In most cases considerable saving in weight of steel can be effected by the application of rigid or semi-rigid design or a combination of both methods. These allow for varying degrees of redistribution of primary bending moments between beams or columns, under given loading conditions, throughout the structure.

The total loading on a frame is the result of the imposed loads and dead loads of the structure. As stated earlier it is possible at an early stage in the design calculations to ascertain the imposed loads on the frame. Although experienced structural engineers can make close approximations for the dead loads of the structure it is only after the members have been designed that first assumptions can be checked to ensure that the dead loads have not been underestimated.

The design of steel frames to buildings is carried out in accordance with the requirements of BS5950: Part 1 [18]. The general principles and design methods are based on the *limit state* concept which requires consideration of the limit states at which structures would become unfit for their intended use by applying appropriate factors for the ultimate limit state and serviceability limit state. Examples of limit states relevant to steel structures are given in Section 2 of the standard. The standard comprises nine parts, Part 1 was published in 1985 and its use by the profession is in its early stages. It is intended that design guides will be issued from time to time. To date the British Constructional Steelwork Association has issued a set of design examples entitled 'Course Notes for BS5950, Structural Use of Steelwork in Buildings' which includes examples of beam and column design. For information regarding properties of steel sections and other useful design information reference should be made to the publication of the constructional Steel Research and Development Organisation (Constrado) entitled 'Steelwork Design Guide to BS5950: Part 1: 1985 — Volume 1, Section Properties and Member Capacities' [19].

12.3 Connections

When using rigid or semi-rigid design methods of construction the connections of the beams to the columns must be capable of transferring the bending moments from the beam to the column, which in turn must have sufficient stiffness to withstand the bending moments induced by the beams. A typical detail of a rigid joint between a beam and a box column is shown in Fig 3.41. This type of joint is normally described as a *moment connection*. Friction grip bolts are generally used to transfer the bending moment from the end of the beam into the column. These bolts are tightened up to a specified shank tension such that the shear forces are transferred to the column by friction between the plates and the bending moment is developed by the bolt groups. Slip of the connection is therefore prevented and the bolts are not subjected to shear or bearing forces. The correct shank tension to induce the necessary friction is measured by means of torque wrenches applied to the nuts of the bolts. Alternatively, load-indicating washers or bolts can be used to perform the same function of ensuring the correct shank tension. Load-indicating washers are crimped and the heads of load-indicating bolts are shaped in a manner such that deformation of the washer or bolt head takes place under the load applied when the nut is tightened to induce the specified shank tension. The amount of deformation is related to the induced shank tension. The deformation is measured by a feeler gauge or by observation by skilled steelwork erectors. Another type of friction grip bolt is designed such that the top section of the bolt shears when the required shank tension is reached.

In connections which are required to resist shear and bearing forces, grades 4.4 or 8.8 bolts are employed. The latter grade permits higher stresses to be used.

Welding of connections is permitted if carried out under controlled conditions in the fabrication shop where it is subjected to rigorous non-destructive testing. Site welding is only permitted in exceptional circumstances and requires the permission of the engineer. As a rule all main site connections are bolted by the methods described.

12.4 Protection of steelwork

A disadvantage with steel is its lack of resistance to corrosion. Although non-ferrous metals, mainly of the aluminium alloy groups, have been used successfully in certain instances for structural frames, their high cost does not make them economic for power station frames. When comparing the economics of a steel structure with a reinforced concrete structure the cost of the original protection and subsequent treatments must be considered. As will be seen from other parts of this chapter, other factors must also be considered before the type of frame is decided.

In the past, steelwork, after fabrication at the works, was painted with one coat of red oxide or red lead paint before being despatched to the site. It was found, however, that in many cases the paint was being applied over mill scale and although the steelwork was erected and painted in the normal way, repainting was very soon required because the mill scale and paint film became detached. It has been the practice, therefore, to deliver steelwork to site unpainted and allow it to weather, thus removing loose mill scale before cleaning and painting commences. The paint treatment in this case consists of a coat of rust-inhibiting primer followed by one or two undercoats and a finishing coat of gloss paint.

However, failure to achieve clean conditions of the base prior to painting on site and the high cost of site labour and access, together with the reduction in cost and the general availability of blast cleaning and automatic painting plant at the works, has in recent years made painting at the works a viable economic choice. One finishing coat applied at site, or preferably just touching up damaged or previously unpainted areas is all that is necessary on site.

The provision of large steel frames to indoor substations revealed the need to reduce the frequency of maintenance painting, and grit blasting followed by a good paint sequence is one method of obtaining this finish. One of the best specifications is for a normal paint treatment using a special primer on a galvanised or other form of zinc, or aluminium-dipped or spray coating. Up to 20 years protection before repainting is necessary can be anticipated with this treatment.

In certain areas the steelwork is entirely encased in brickwork or reinforced concrete; this is mostly in offices and amenity areas and provides fire protection to the steel frame and makes it easier to build filler walls and apply decorations.

13 Reinforced concrete

13.1 General

The principles including mixing, transporting, placing and vibration of concrete in foundations, described in Section 5.1 of this chapter, are equally applicable to concrete used in superstructures. A larger proportion of high quality concrete is used on superstructures as most of the work involves columns, beams, slabs, walls and similar structural members, which are more highly stressed than foundation concrete.

13.2 Formwork

In addition to timber and steel panels as used in foundations, shutters lined with plastic and glass-fibre may be used to obtain a better finish to the concrete.

13.3 Reinforcement

Although mild steel reinforcement is used extensively, high tensile steel reinforcing bars and mesh fabric may be used. Due to the higher stresses which may be developed, the saving of steel in a structure using high tensile reinforcement may be quite large. Additionally, the reduced diameter or smaller number of bars of this type may help to simplify complicated reinforcement arrangements and ease the placing of concrete. The design of reinforced concrete allows for the transfer of stress by bond from steel to the concrete and vice-versa throughout the length of the bars. Shrinkage of the concrete, whilst setting, develops the bond between the circular outer face of mild steel bars and the concrete. High tensile steel bars have a deformed outer face which permits them to develop higher bond stresses. In addition to bond requirements the ends of bars may require hooks or bends to ensure sufficient anchorage.

A product certification scheme for reinforcing steel known as the Certification Authority for Reinforcing Steels (CARES) is in operation and the use of CARES material is mandatory in all civil engineering contracts. A typical arrangement of reinforcing bars in a slab, beam and column is shown in Fig 3.42.

Corrosion of steel reinforcement in well-designed reinforced concrete structures is normally prevented by the high alkalinity of the concrete surround. Particularly in very exposed conditions or in acid environments, however, the designer must consider possible reinforcement. Increasing the concrete cover has been the traditional way of protecting such reinforcement, but recently, plastic-coated reinforcement has been introduced for such situations.

13.4 Design of reinforced concrete

The design activity splits naturally into the two distinct spheres of non-seismic-resistant and seismic-resistant aseismic analysis and detailing.

In the UK conventional (i.e., coal, oil, etc.) stations are not designed to resist earthquakes. Therefore the reinforced concrete may be engineered according to normal practice and in particular British Standard Codes of Practice. Nuclear stations are designed to resist an extreme seismic event (with 1 in 10 000 probability per annum), but it should be noted that not all buildings need to be seismic resistant, only those that are safety related require special treatment. Aseismic design is outside the range of normal UK practice and reference is usually made to US Standards and Codes of Practice.

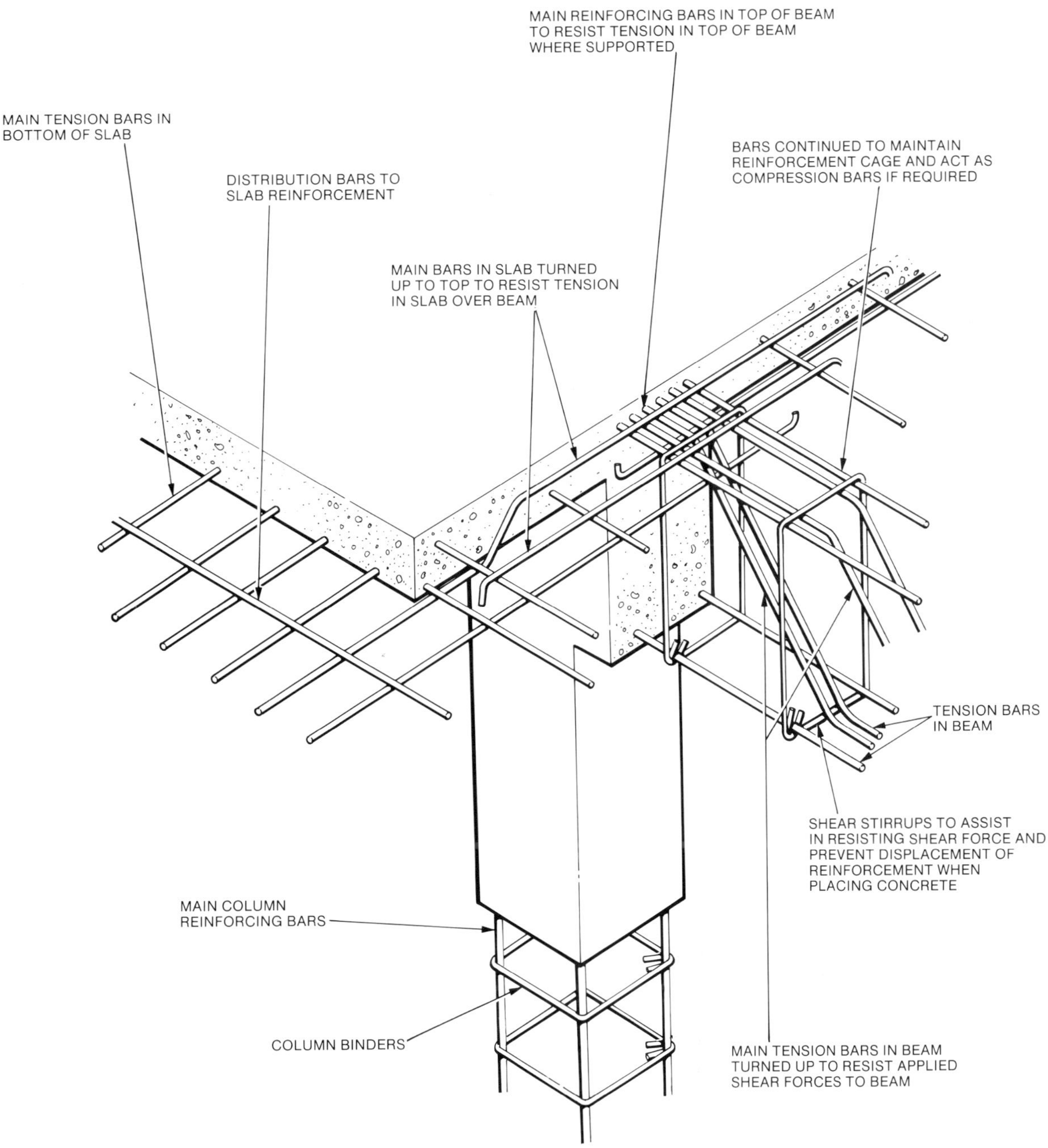

FIG. 3.42 Typical arrangement of reinforcing bars in slab, beam and column

13.4.1 Non-seismic-resistant design

For non-seismic-resistant design it is most convenient to use BS8110 [3] which deals with the structural use of concrete and is published in three parts:

Part 1 — Code of practice for design and construction

Part 2 — Code of practice for special circumstances

Part 3 — Design charts for singly-reinforced beams, doubly-reinforced beams and rectangular columns

This Standard uses the philosophy of limit state design in order to achieve an acceptable probability that the structures being designed will perform satisfactorily during their intended life (about 40 years), sustaining with an adequate degree of safety all the loads of normal construction and use, having adequate durability, and also being adequately resistant to the effects of misuse and fire.

For the purposes of structural design two primary limit states can be identified, the ultimate limit state (ULS) and serviceability limit state (SLS).

An SLS is reached if the structure becomes unfit for use by reason of deformation, cracking, vibration, etc., when sustaining loads no greater than those reasonably expected to occur during the design lifetime. These are the characteristic loads (refer to Section 11 of this chapter).

A ULS is reached if the structure becomes unfit for use by reason of collapse, overturning or buckling. It is expected that this condition could be triggered by unexpected loads greater than the characteristic loads, inaccurate assessment of load effects, unforeseen stress redistribution and variations in dimensional accuracy. The characteristic loads are therefore multiplied by an appropriate partial safety factor and the strength of the structural member is checked against the stresses calculated for the enhanced load condition. When assessing the strength of the structural member the characteristic strengths of materials are divided by appropriate partial safety factors to take account of differences between actual and laboratory values, local weaknesses and inaccuracies in assessment of the resistance of structural sections.

Thus typically a design load combination could be:

1.2 dead + 1.2 imposed + 1.2 wind

and (in checking the section design strength required to resist these loads) partial safety factors for strengths of materials could be:

Reinforcement	1.15
Concrete in flexure or axial load	1.50

The usual design path is to complete the ULS analysis first and then check the SLS. Generally the SLS can be ensured by attention to *deemed-to-satisfy* provisions such as limiting span-to-depth ratios, minimum steel percentages and reinforcement bar spacing during the ULS analysis. If a more rigorous analysis is necessary (e.g., deflections and crackwidths) then BS8110: Part 2: Section 3 [3] contains detailed analysis techniques.

The loading conditions during erection and construction should be considered in design and should be such that the structure's subsequent compliance with the more permanent limit state requirements is not impaired.

Requirements for durability and fire resistance are secondary to the primary limit states in the sense that they require a smaller part of the effort required to develop a design. However, in order to achieve the intended objective (of satisfactory lifetime performance) they are no less important than the ULS and SLS.

The stress analysis can be split into two stages — *global* and *local*:

Global Is analysis of the complete structure to obtain a set of member internal forces and moments that are in equilibrium with the design loads for both the ULS and SLS.

Local Is analysis of the component sections for ULS where inelastic material behaviour is appropriate; analysis of sections for SLS where linear-elastic material behaviour may be used to check the performance expected under the characteristic loads, with appropriate allowance for creep and shrinkage.

It should be noted that a linear elastic analysis is commonly made as a basis for both ULS and SLS, then for the ULS forces and moments to be redistributed with due regard to the ductility of the members concerned. Alternatively an inelastic analysis using yield line theory or Hillerborg's strip method for slabs, and the plastic hinge concept for beams, may be made.

The state of development of digital computers and commercial software packages is now such that the global stress analysis is often best done using a linear-elastic frame or finite element structural analysis program. Inelastic yield line and plastic hinge approaches have also been computerised, but the applicability of these is limited when compared with classical linear-elastic analysis. Such programs vary tremendously in size, capability and cost. They can be run on the full range of micro (personal), mini and mainframe computers. For large analyses, and particularly when finite elements are employed, mesh generation capabilities and graphical output are necessities, while interactive analysis capability can also be very useful for smaller jobs.

Tens, if not hundreds, of programs are available of which the Structural Design Language — STRUDL — family of programs is notable. The original version operated as a subsystem of the Integrated Civil Engineering System (ICES) developed by Massachusetts Institute of Technology, USA. It uses the Problem Oriented Language (POL) for free-format data input and commands, making it much more user friendly than fixed format programs. Many versions and other compatible programs have been developed for use on all types of computer. Early enhanced versions which included the more advanced finite element, dynamic and geometrically non-linear analysis capabilities were unreliable, but now the most comprehensive packages, from IBM and McDonnell Douglas amongst others, are robust and very convenient tools.

13.4.2 Seismic-resistant design

In the UK, the magnox nuclear stations were not specifically designed to resist an earthquake, nor was the first tranche of AGR nuclear power plants. The second generation AGR stations at Heysham *2* and Torness were designed to have an elastic response up to a design basis earthquake (DBE), and to be capable of resisting events up to safe shutdown earthquake (SSE) without prohibiting safe shutdown of the facility. The SSE (which has a probability of occurrence of 1 in 10 000 per annum) has a free-field ground motion twice as intense as the corresponding DBE. This design approach is particular to these stations and is unlikely to be repeated. An overall description of the structural design of these plants, with particular emphasis on the design of the reinforced concrete shear walls surrounding the prestressed concrete reactor pressure vessel, can be found in 'The Aseismic Design of a Reactor Building for the Advanced Gas Cooled Reactor Power Plant' [20].

The next generation of UK nuclear plant starting with Sizewell *B* will have pressurised water reactors (PWRs). Analysis of the seismic response of the main structures for Sizewell is discussed in 'Seismic Design Approach for the Sizewell *B* Nuclear Power Plant [21], which includes the background to soil-structure interaction (SSI) analysis. However, it should be noted that in the early stages of analysis the ground motion is applied directly to the foundations in order to estimate the gross loads for calculation of stresses. This procedure will be shown to be conservative by comparison with the global SSI analysis at a later date. Civil design and analysis work may thus proceed independently of the SSI analysis so that the iterations which would otherwise be necessary between these two parts of the seismic design process are avoided. Whilst the section sizes are determined by adopting a fixed base type of analysis it still remains possible to benefit from a subsequent SSI analysis in the design of reinforcement.

The demonstration that structures can withstand the loading which is imposed on them during the SSE is carried out using analysis procedures in line with internationally accepted standards and practices. As there is a CEGB commitment that the plant will be manifestly licensable in the USA (the country of origin), Sizewell and probably future PWR plant concrete structures (excluding reactor containment) in the UK will be analysed and designed in accordance with ACI 349–80 [22].

There are two categories of plant at a PWR station for which a capability to withstand the SSE must be demonstrated. Plant in the first category is required to function after the SSE so that safe shutdown can be achieved. The design bases are defined to ensure that where relevant:

- The safety related structures or structural members do not collapse or severely distort such that continued operation of safety related plant is prejudiced.
- Structural integrity is maintained to ensure that there are no releases of air or liquid-borne radioactivity in excess of the targets defined, or losses of radiological protection.

This is most conveniently achieved by working to the methods and design criteria of ACI 349 which will ensure that when subjected to various combinations of gravity, wind, thermal, seismic and accidental loads the structures will behave generally in a linear-elastic manner. Elastic behaviour is considered as limited by the yield stress of the effective load-carrying structural materials or ultimate capacity of an element.

Thus using the nomenclature of ACI 349, if:

D	=	dead loads
F	=	lateral and vertical pressure of liquids
L	=	live loads
H	=	lateral earth pressure
W	=	operating wind load
R_o	=	pipe and equipment reactions
T_o	=	thermal effects during normal operation or shutdown
E_{ss}	=	load effects of safe shutdown earthquake (SSE)
W_t	=	loads generated by extreme wind or related internal forces and moments
U	=	required strength to resist factored loads or related internal forces and moments

Then typical design load combinations are:

$$U = 1.4D + 1.4F + 1.7L + 1.7H + 1.7W + 1.7R_o$$

$$U = D + F + L + H + T_o + R_o + E_{ss}$$

$$U = D + F + L + H + T_o + R_o + W_t$$

Although the section strengths determined in accordance with ACI 349 are similar to the ULS strengths of BS8110, appropriate adjustment must be made to account for the different materials and strength testing methods in the UK. Appendix C of ACI 349 deals with special provisions for impulsive and impactive effects.

Plant in the second category comprises items which are not themselves required to function after the SSE but whose failure could imperil other plant of the first category. The design bases have to ensure that where relevant:

- The failure of whole or part of a second-category structure does not jeopardise the integrity of a first-category structure, system or component.
- Radiological protection is maintained as appropriate for structural elements or the structures as a whole.

It is often impractical to provide sufficient separation so that second-category structures would not impinge on first-category structures if they collapsed. Either they must be designed as first-category structures (which is not always economic) or their integrity must be demonstrated by inelastic dynamic analysis to show that under the SSE the deformations are well below those levels that lead to collapse. This second (inelastic) design option is the most demanding of expertise, since it requires rigorous assessment of the damping and ductilities appropriate to the structural behaviour and advanced numerical analysis techniques. The reinforcement distribution and detailing must be correct to achieve the ductility and hence energy absorption implicit in the inelastic analysis. In order to arrive at the proportions and reinforcement of a second-category structure which can then be analysed for resistance to collapse, it is probably most convenient to work within the code requirements of ACI 318M–83 [23], from which ACI 349–80 was developed. The format is common to both codes (though appendices are different as they relate to separate areas) and therefore the methods for the design and analysis of both categories of safety related structures are conveniently similar. The concrete sizes, reinforcement proportions and distribution thus derived are likely to be shown by analysis to be suitably resistant to collapse when threatened by the occurrence of the SSE.

The serious student of the design of reinforced concrete power plant superstructures would be well advised to consult references [23] and [24]. The first of these comprehensively covers the broader aspects of the structural design and analysis of nuclear power stations including impulsive and impactive (missile) loading; the second covers in depth reinforced concrete analysis, behaviour and detailing (particularly with respect to aseismic design and American research and practice).

Commercial (non-safety-related) structures (such as administration buildings and stores) at nuclear power plants could also be designed to ACI 318M, but it is probably more convenient to use BS8110 as for conventionally-planted thermal stations. Here again reference [24] contains very useful information on the behaviour and detailing of reinforced concrete elements; an understanding of this is required to achieve safe, serviceable and durable structures.

13.5 Movement joints

Movement joints are joints which can accommodate relative movement between the adjoining parts of a member or structure. The principal movements which may have to be accommodated are as follows:

(a) Expansions and contractions due to thermal movement and drying shrinkage of the concrete. An indication of the possible magnitude of these movements is given in BS8110 Part 2 [3].

(b) Rotation and/or translocation at a connection.

(c) Long term increase in deflection due to creep of the concrete.

If the movements noted under item (a) are restrained, stresses will build up which can crack the concrete. Where cracking cannot be reduced to acceptable levels by other means, movement joints are introduced.

The following types of movement joint are commonly used in power station construction:

Contraction joint This type of joint is generally used in slabs and walls to permit contraction of the concrete. Initially there is no gap between the concrete on either side of the joint and the reinforcement may be continuous or discontinuous across the joint. Where it is necessary to achieve a watertight joint, a suitable rubber or neoprene water bar and joint sealant is incorporated.

Expansion joint Expansion joints are designed to accommodate both expansion and contraction. They can vary in complexity from a simple gap filled with rot-proof compressible boarding (as used in concrete retaining walls, slabs and similar structure), to the complicated roller, rocker or sliding joints provided at the end of bridge or similar beams.

Expansion gaps in walls and slabs can be made watertight if necessary by incorporating a suitable water bar and sealant. Expansion joints in buildings must be carefully located and made weatherproof by providing cappings at joints in roofs and in the use of copper water bars or similar methods of sealing vertical walls. Care must be taken with the treatment of internal finishes at the positions where expansion joints are located, plates being used on the floors and cover strips or other features being used on walls or ceilings to permit sliding and prevent cracking of plaster or other finishings.

It should be noted that construction joints are not considered to be movement joints. They are introduced to divide the structure into conveniently sized sections for casting. The reinforcement is continuous across the joint and the face of the joint is prepared to encourage continuity of the concrete. In practice, however, shrinkage frequently occurs at construction joints converting them effectively into contraction joints.

13.6 Curing

Curing is the process of keeping the concrete saturated whilst maintaining the temperature within suitable limits during the early stages of hardening.

Concrete hardens as a result of hydration of the cement. Hydration can only take place when the pores in the concrete are adequately wet and the final strength partially depends on the extent to which hydration has taken place. The temperature of the concrete determines the rate at which hydration takes place and hence the rate of strength development.

Keeping the concrete saturated during the early stages of hardening is achieved by preventing loss of moisture by evaporation from the surface. The most common methods of preventing loss of moisture include retaining formwork in position, spraying or flooding the surface with water, covering the surface with a damp absorbent material, covering the surface with a waterproof membrane such as polythene and spraying the surface with a sealing compound. Curing should commence immediately the concrete has set to prevent premature drying out of the surface.

The period of curing required depends on the type of cement, the ambient conditions and the temperature of the concrete. BS8110 Part 1 lays down the minimum curing periods taking these factors into account. For example, the minimum curing period for concrete made with ordinary Portland cement, under average ambient conditions and at an average surface temperature of 5–10°C, is four days.

13.7 Precast concrete

Concrete poured into shutters which retain the concrete during setting into its final form and position is termed *in-situ* concrete. If the concrete is cast in locations which are different from the final positions it is referred to as *precast* concrete. Although on large contracts precasting yards may be established on the site, factory-made units are normally preferred. Factory quality control in manufacture and more economic handling processes can usually more than offset the additional expense of transporting units from the factory to the site.

The principal advantages of *in situ* concrete are that it can be cast monolithic, i.e., there are no joints to make good, and the size of members is not limited by capacity of handling equipment. Some of the advantages of precast concrete over *in situ* concrete are as follows:

- Good concrete of uniform quality is more easily obtained under factory conditions.
- Curing is more easily carried out on precast units.
- Units can be protected against frost more easily during the early stages.
- It is not necessary to delay construction of a further stage of the structure to allow completed sections to gain a strength sufficient to withstand later superimposed loadings.
- Congestion on the structure is reduced by the manufacture of units in more suitable areas and because less support formwork and scaffolding is required.
- There is no delay due to the placing and striking of formwork or the fixing of reinforcement.

Additional requirements imposed by the use of precast concrete, however, include:

- The reinforced concrete design must be adequate to cope with the stresses arising from lifting, handling, transporting and erecting.
- Particular care must be taken at all stages to avoid damage to the units.
- Transport and adequate lifting equipment at site must be provided.
- Detailed planning and programming is necessary to ensure that the units are manufactured and delivered to site to suit the contract programme.
- All units must be clearly identified and non-symmetric elements must be appropriately marked to ensure correct erection.

Precast concrete floors, one type of which is shown on Fig 3.43, have been used for many years but are unsuitable where they may be subject to the movement of heavy plant loads. Precast wall panels have been used successfully on several recent power stations. Attractive architectural effects have been obtained on the face of wall panels by impressing pebbles at the partially set stage, or by water jets to expose the aggregate on the face. Lintels, coping stones and similar items are often made from *artificial stone* or *reconstructed stone*, which is concrete manufactured using stone aggregates and coloured cement, where the colour of the stone is of particular importance. Very often the finished block has only a face of this type of concrete, the backing being concrete mixed from ordinary gravel and sand.

Precast concrete is used extensively in cooling tower construction, for the shell support columns, the ring beam above the air opening and the pack support columns and beams.

14 Prestressed concrete

14.1 Prestressing

When a beam is loaded it deflects and compressive stresses are introduced into the top fibres and tensile stresses into the lower ones. With prestressed concrete, high tensile steel wire or strands are inserted into tubes cast in the tension section of the beam and held in permanent wedge type anchorages; after tensioning by means of hydraulic jacks the tubes are grouted up. This has the effect of compressing the concrete in the lower fibres and introducing tension into the upper fibres. Loading of the beam then has to remove the induced reverse stresses before normal compression of the top section and tension in the lower section commences.

Figure 3.44 shows the stresses which are introduced into the concrete by loading and prestressing in a concrete beam. All the stress diagrams are taken on section A-A at the centre of the beam. Figure 3.44(a) shows the method of arriving at the stresses resulting from the dead load together with the prestress, and diagram (b) shows the final resultant stresses after the imposed load is applied. It may be seen from Fig 3.44(b) that even when the beam is subjected to all stresses resulting from the dead loads, the imposed loads and the prestressing forces, no tension stresses remain in the concrete. Using this method, members of a section much lighter than that required for normal reinforced concrete may in certain instances be used and much larger spans for beams can be obtained. Beams are sometimes tensioned through the supporting columns of a concrete structure to form rigid joints. Prestressing is used for valley beams of barrel roofs, bridges, tanks and pressure vessels.

For precast items such as cooling tower stack members or floor and roof beams, the wires are stretched on a permanent bed, possibly several hundred feet long, and a number of units are then cast around each section of wires. After the concrete has set, the jacks are released and the wires between each unit are cut, the wires remaining tensioned inside each unit.

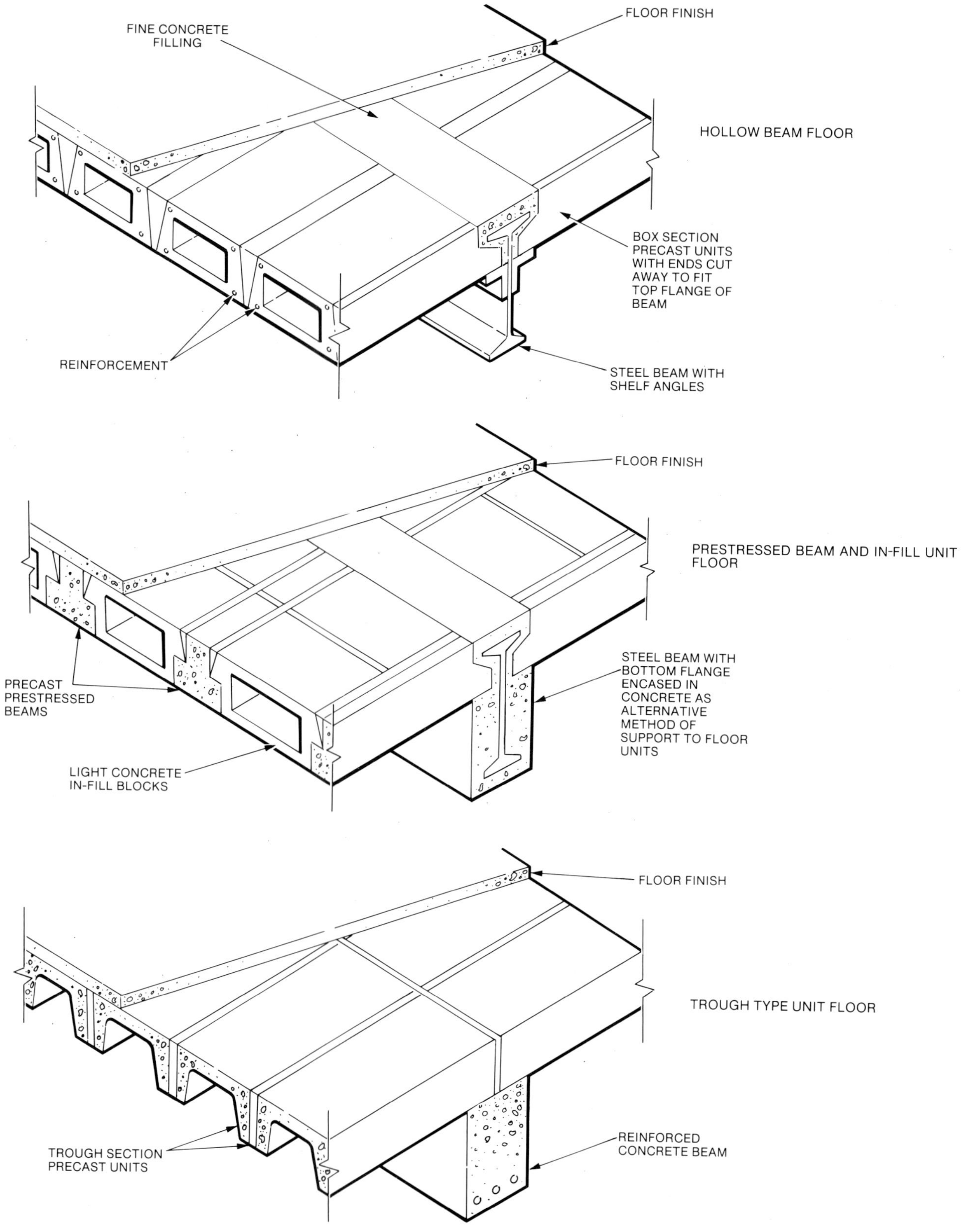

FIG. 3.43 Precast concrete floors

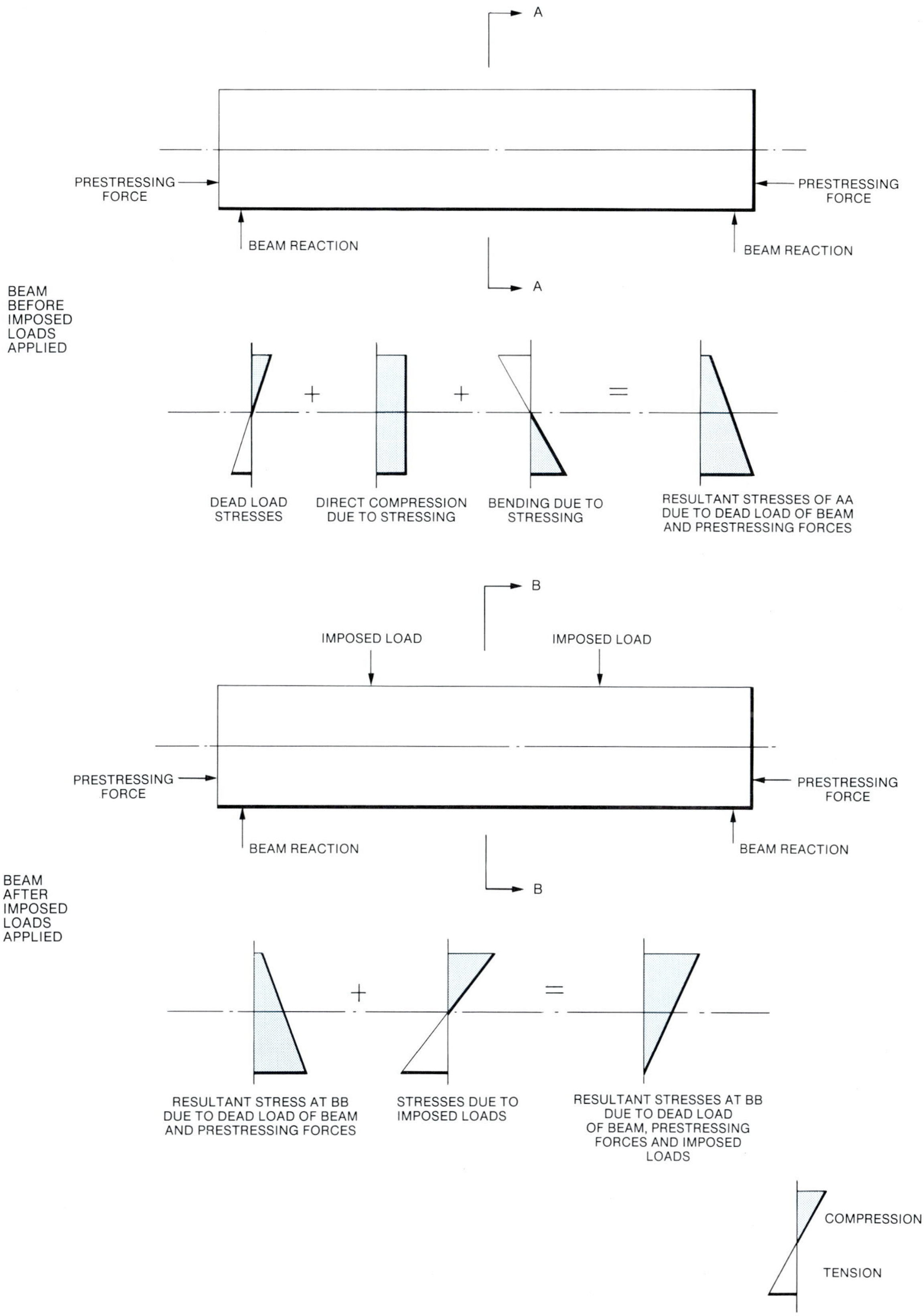

FIG. 3.44 Stresses in a precast prestressed concrete beam

Where tensioning of wires is carried out after the concrete has set, as in the case of prestressed concrete reactor pressure vessels, the method is termed *post-tensioning* and where the wires are stressed before the concrete is placed it is called *pre-tensioning*.

Post-tensioned systems are referred to as *unbonded*, when the wires or strands are protected against corrosion by specially formulated oil or grease applied prior to installation in the ducts, or by injection after installation. Bonded systems are those which are grouted-up by injection of cement grout following stressing. The grout also acts as corrosion protection provided measures are taken to ensure the absence of voids.

14.2 Prestressed piling

As described in Section 4.5 of this chapter, prestressed piles have been used in large numbers for the foundations of power stations. These piles are usually prestressed and cast in long line stressing beds, which allow several piles to be cast in one bed. Steel end plates are inserted between the piles with holes drilled to allow the uninterrupted course of the strands from the fixed achorage at one end of the bed to the stressing system at the other. Good control needs to be exerted over the increase in concrete strength so as to allow the piles to be separated and lifted from the beds as quickly as possible, and to ensure that the piles are not driven before achieving adequate strength.

The layout of strands needs to ensure that the prestress within the pile is as evenly disposed as possible. The amount of prestress is dictated not by the working load on the pile but rather by stresses imposed by lifting from the beds, storage, pitching and driving. To prevent damage to the pile head or toe during driving, the links are provided at close centres after the top and bottom length of each pile (typically 2 m to 3 m at earth end). In some cases this end reinforcement is enhanced by the addition of normal longitudinal reinforcing bars.

14.3 Prestressed concrete pressure vessels and containments

Prestressed concrete pressure vessels (PCPVs) and prestressed concrete containments (PCCs) play central roles in nuclear steam supply systems, although their design duties are somewhat different.

The PCPV is relatively thick-walled (4 m to 5 m) compared to the PCC (1 m to 1.5 m), since its primary function is to retain high gas pressure (20 bar to 40 bar) for the majority of its operational life. Typical PCPV dimensions are illustrated in Fig 3.45. In contrast, the PCC has a relatively passive function to fulfil during its operational life since it is only called upon to contain leakage at low pressure under normal reactor operating conditions. It is principally designed to retain higher pressures (typically 5 bar) which could result from low probability short duration events such as a rupture of the primary coolant circuit pipework. Typical PCCs are illustrated in Fig 3.46. Both the PCPV and the PCC have additional functions which are to provide biological shielding for the station operators and to support internal and external structures and plant with small allowable deformations under sustained temperature gradients. PCCs may also function as a missile barrier, for example, against tornado-generated missiles, turbine missiles or aircraft where applicable.

Apart from the double-barrier 1300 MW French PCCs and Canadian CANDU plants, all PCPVs and PCCs are lined with a mild steel membrane, typically 13 mm thick for the PCPV and 5 mm thick for the PCC. In both types of structure the prestressing system, in common with all prestressed concrete, is designed to resist the tensile stresses induced in the concrete by the applied loadings whether these are from mechanical loads such as internal pressure or from strain-controlled loads such as temperature cross falls.

Operating conditions require that PCPVs are equipped with thermal insulation and liner cooling water pipe systems to ensure that the liner and concrete are maintained at acceptable temperatures. These provisions are unnecessary for PCCs where internal operating temperatures are not damaging to either steel or concrete.

The design and analysis of PCPVs and PCCs has been established over the last 25 years. The applicable Standard for PCPVs is BS4975: 1973, [26]. The principles which had already been established in CEGB specifications and had been incorporated into the practical design and construction of PCPVs are reflected in this standard which is under revision. The equivalent standard used in the United States for PCPVs and PCCs is the ASME III Division 2 [27].

The service load analysis approach adopted by the CEGB for the PCPV is a *working stress* approach, based on an analysis of the vessel for a series of idealised loadings which represent the most severe combinations of load which could be applied to the vessel. The gas pressure used for design purposes is set at 10% above the normal working pressure to allow for operating transients and tolerances. The principal loading cases are as follows:

- Prestress alone at transfer force.
- Prestress plus proof test pressure at ambient temperature; proof pressure is set at 15% above the design pressure.
- Early life operating condition including prestress, plus design pressure, plus design operating temperature distribution.

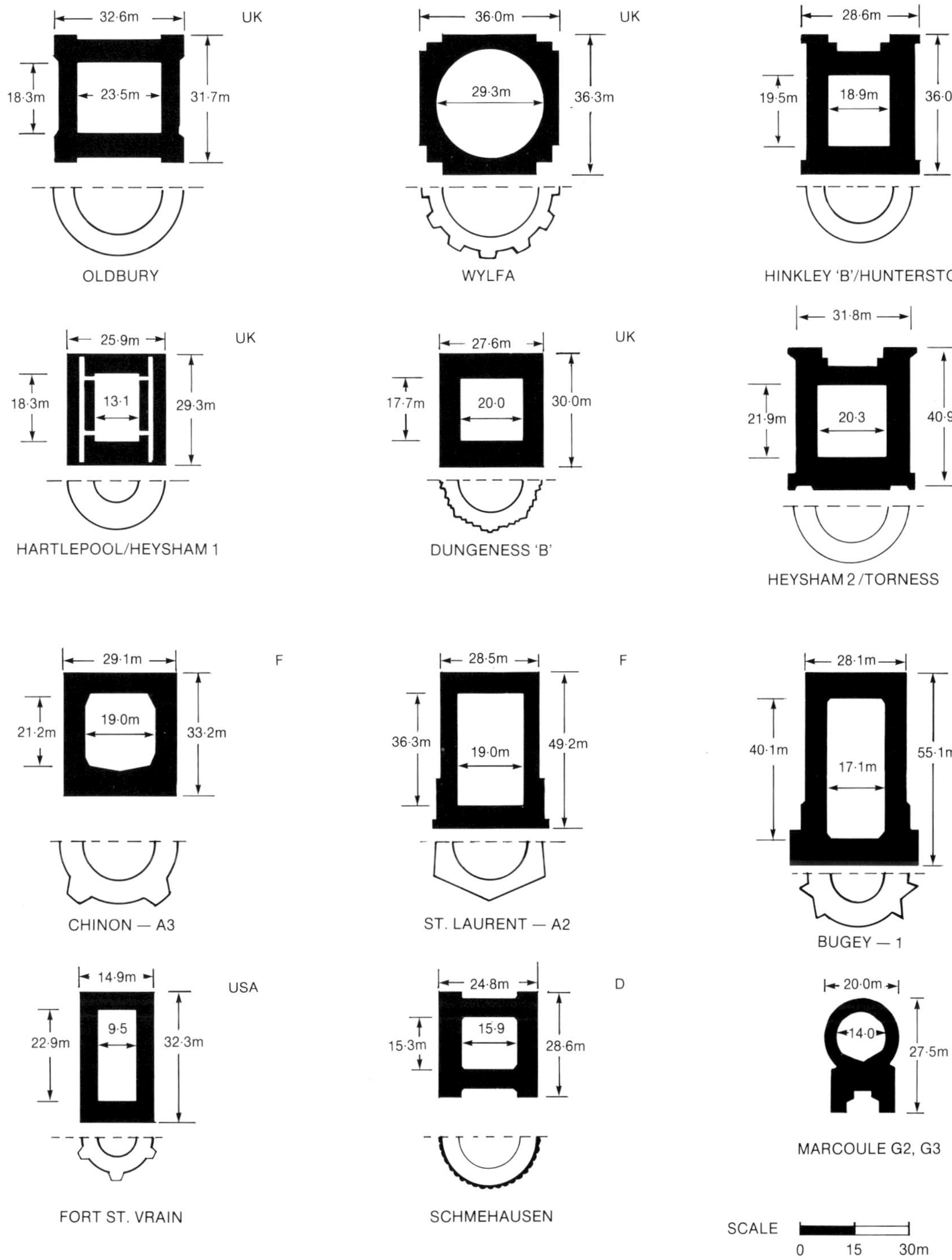

STATION	REACTOR TYPE	NO. OF UNITS	CRITICALITY DATE	WORKING PRESSURE N/mm^2	GROSS THERMAL/ NET ELEC. O/P MW
OLDBURY	MAGNOX	2	1967	2.41	750/300
WYLFA	MAGNOX	2	1969	2.64	1600/590
HINKLEY 'B'	AGR	4	1974	4.03	1400/625
HUNTERSTON	AGR	4	1974	4.03	1400/625
DUNGENESS 'B'	AGR	2	1983	3.30	1480/600
HARTLEPOOL	AGR	4	1983	4.03	1500/625
HEYSHAM 1	AGR	4	1983	4.03	1500/625
HEYSHAM 2	AGR	4	1986	4.15	1550/615
TORNESS	AGR	4	1986	4.15	1550/615
MARCOULE G2, G3	MAGNOX	2	1958	1.47	260/36
CHINON — A3	MAGNOX	1	1966	2.65	1560/480
ST. LAURENT — A2	MAGNOX	1	1971	2.85	1691/516
BUGEY — 1	MAGNOX	1	1972	4.50	1954/540
FORT ST. VRAIN	HTR	1	1974	4.86	842/330
SCHMEHAUSEN	HTR	1	1983	4.00	750/296

FIG. 3.45 Prestressed concrete pressure vessels for nuclear reactors

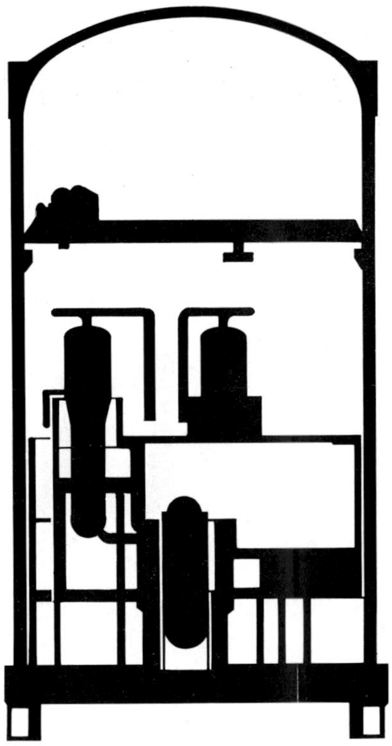

RINGHALS II WESTINGHOUSE
PWR 820MW
DRY CONTAINMENT

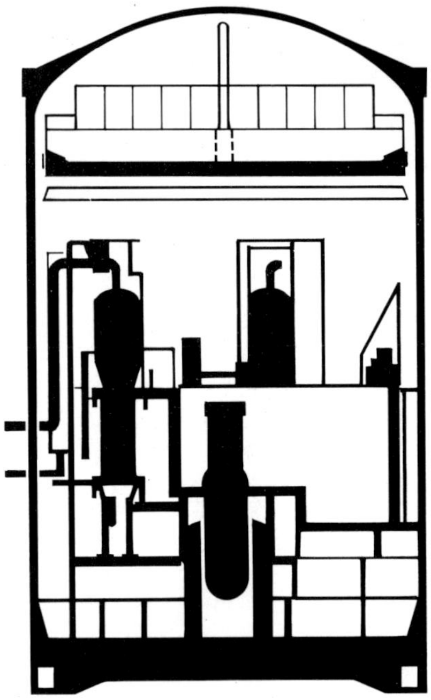

FRENCH PWR CONTAINMENTS
SINGLE BARRIER 900MW

FRENCH PWR CONTAINMENTS
DOUBLE BARRIER 1300MW

FIG. 3.46 Typical prestressed concrete containments

- Late life operating condition including prestress, plus design pressure, plus design temperature distribution.
- Start-up and shutdown conditions at early and late life including consideration of transient temperatures arising from the start-up and shutdown modes.
- Fault studies in which a number of low probability fault conditions are considered in the design. The designer may be required to show that the vessel can continue to fulfil its safety function in the event that the postulated fault occurs.

Similar conditions apply to the design of PCCs in accordance with the ASME III Division 2 specification approach although, due to the difference in duties, the emphasis is upon factored internal and external loading combinations which represent the most severe loads which the PCC may have to sustain.

In accordance with BS4975 it is an additional requirement that PCPVs have an ultimate load capacity which is generally 2.5 or 3 times the design pressure. In calculating the ultimate load factor the designer has to consider all possible modes of failure of the vessel due to internal gas pressure. This is a hypothetical loading case since the safety relief valves limit pressures under the worst credible conditions to about 30% above the design pressure. However, the ultimate load analysis is an important feature of the design requirements, since it enables the designer to check that the design will have a ductile response to pressure and that non-linear behaviour will only commence at pressures well beyond the design pressure.

In order to ensure that the mode of failure of the vessel will be close to that predicted in the ultimate load calculations, model tests are performed to determine the mode of failure and the ultimate capacity of the vessel. These model tests are normally carried out on a one tenth scale replica of the vessel. The results have shown that analytical predictions of ultimate load factor and mode of failure provide conservative estimates of the pressure that the vessel can withstand. These tests have also demonstrated the large reserves of ductility built into the vessels.

Similar tests are proposed for PCCs to be constructed by CEGB and these are required to demonstrate an ultimate load factor of at least twice design pressure.

The civil and structural construction programme for PCPVs has to be integrated into the programme for plant and reactor installation.

The milestones in vessel construction are:

- The installation and grouting of the liner base plate, sometimes carried out as a one-piece operation with the liner walls.
- The allocation and timing of bays for concrete pours to avoid heat of hydration problems in end caps and walls.
- The assembly, temporary support and concreting of the standpipe zone which carries the fuel and control rod penetrations in a closely pitched array with tight dimensional tolerances.
- The stressing of the prestressing system.

In general, the civil engineering techniques used for PCPVs and PCCs are no more complex than those in conventional structures, apart from the large scale and the need to carry out mock-up trials in advance of construction to validate the proposed methods for certain critical or complex areas.

Following construction and prestressing, PCPVs are subjected to a proof pressure test at 15% above design pressure, and unfuelled and fuelled engineering trials prior to raise power and synchronisation with the electricity grid system.

PCCs are required to pass a structural overpressure test (SOT) at 15% above design pressure and an integrated leak rate test (ILRT) at 10% above design pressure. The latter test may be repeated at intervals throughout the service life of the PCC.

The NII's nuclear site licensing conditions outlined in Section 24.5 of this chapter require that PCPVs are inspected on a regular basis and any necessary maintenance carried out. Each reactor is shut down regularly at two year intervals for maintenance. At this time the external features of the vessel are inspected by the CEGB's Appointed Examiner. The minimum programme consists of the following items:

- Prestressing system load checks to determine the residual force in the tendons.
- The condition of prestressing anchorages.
- The condition of prestressing strands or wires withdrawn from a number of tendons.
- The condition of the concrete surface.

Other items which are included in the inspections include:

- PCPV foundation settlement and tilt.
- A summary of embedded vibrating wire strain gauge readings and their correlation with theoretical predictions.
- A summary of vessel temperatures and their conformity to the operating rules for the vessel.

As a direct result of the Appointed Examiner's responsibility for regular inspection and maintenance of PCPVs, a considerable amount of information has been amassed on the performance of prestressed concrete structures.

The main conclusion that can be drawn from in-service examinations of prestressed concrete pressure vessels is that they are remarkably robust structures and that the predictions made at the design stage have been fully borne out in practice.

15 Brickwork and blockwork

15.1 General

The traditional use of facing brickwork is nowhere near so strongly rooted abroad as it is in the UK. Its use is, however, massively reduced from that adopted in the *brick temples* of pre-1939 power stations. On modern power stations, brickwork is used selectively for external and internal walling, both as infill, or panel non-load-bearing form, and in general load bearing walls; it is also used for external floor pavings.

15.2 Bricks

Clay bricks and blocks are defined and described in BS3921 [28], and are generally formed by pressing, extruding, wire cutting, and soft moulding prior to firing. Bricks are normally divided into one of the following groups:

Common bricks These are usually manufactured by wire cutting extruded lengths of clay, the bricks being dried and then burnt in a kiln normally fired by coal. Forming bricks by pressing instead of extrusion and wire cutting is also carried out at some brick yards. For internal walls, bricks are now manufactured containing up to 90% pulverised fuel ash.

Engineering bricks These are generally manufactured in two types, Class A (69.0 N/mm^2) and Class B (48.5 N/mm^2), from shale by a pressing process. They are hard, dense bricks of good strength, low permeability with good acid and frost resistance, and excellent weathering characteristics. As the name implies they are suitable for load bearing walls, piers, mass retaining walls and other demanding locations. They are also used for the internal face of walls where a high resistance to mechanical damage is required.

Facing bricks These are cut or pressed bricks finished on the exposed faces by sand blasting, or other methods of roughening, to give the required surface effect. They can be coloured before burning and typical descriptions include the words 'sand faced', 'multi-coloured' and similar descriptive terms. They are used for external walls as they are resistant to weather and have a good appearance.

Sand-lime bricks These are made to normal brick sizes from a mixture of sand, lime and water which is pressed and then subjected to high temperature steam in an autoclave which accelerates and ensures complete hydration of the lime. They are uniform in size, have a pleasant surface and colour and are used extensively unpainted for the surface of air-faced internal walls.

Concrete bricks Made from concrete, these bricks can be obtained in qualities similar to sand-lime bricks and are used in similar locations.

Lightweight bricks and blocks These are obtainable in various sizes and are manufactured from pulverised fuel ash, aerated concrete, coke breeze and similar materials. Their light weight together with high thermal and sound insulation properties make them very suitable for internal partition walls. They are not suitable for external walls except where protected on the outer face.

Extruded bricks sometimes have perforations which reduce the amount of clay in the brick and therefore its weight.

15.3 Mortar

The use of mortar in joints of appreciable thickness in brickwork enables bricks of irregular size and shape to be bonded solidly together into an element, in which the stresses from superimposed loads will be well distributed and in which there will be minimal paths for penetration of air and moisture.

The main constituents of mortar are sand, lime, cement, mixing water and other substances such as plasticisers, the proportions being varied to give a mortar suitable for the work in hand. In general it should not exceed the bricks in strength. For engineering brickwork a hard mortar composed of one part cement to three parts sand would be used whereas for normal external brickwork one part cement, one part lime and four parts of sand would be suitable. For internal block walls one part cement, two parts lime and eight parts of sand is used. Proprietary cements especially for mixing mortars are now available which can be used without additives.

15.4 Brickwork

Wire cut bricks have flat surfaces while pressed bricks have flat surfaces or frogs on one or both sides. Stronger joints are obtained in walls using bricks with frogs.

The bond used in the brickwork is most important for the appearance and strength of the finished brickwork. There are many variations on the bonds used in brick walls, but the three most common are: Stretcher Bond, English Bond and Flemish Bond.

For external wall cavity construction using two leaves of brickwork and a space usually 50 mm wide between the inner and outer skin, gives a wall with the best damp resisting properties. These walls have better thermal insulation properties than a solid brick wall of equivalent thickness.

Insulating material is frequently required to be inserted within the cavity space to increase the thermal insulation properties of the wall as a whole.

A great disadvantage with brickwork is its own deadweight. High walls without restraint at intermediate levels are subject to buckling, although these effects can be reduced by forming piers on the face of the wall at suitable intervals along its length, or by tying back to the structural columns and floors using ties which allow vertical movement.

The strength of brick walls can be increased by the use of steel reinforcement laid in the bed joints and small diameter steel bars in the vertical joints.

The correct mortar mix for the type of bricks and blocks and also the location of the work should always be used. Shrinkage and cracking can result from the use of mortar with a too high cement content.

15.5 Blocks

Blocks are categorised into one of the following forms:

- Solid blocks, which are basically solid containing no formed holes or cavities other than those inherent in the material.
- Cellular blocks, which have one or more formed cavities which do not wholly pass through the block, the closed end being laid uppermost.
- Hollow blocks, which have one or more formed holes or cavities which pass completely through the block.
- Blocks with special faces which may have a smooth, textured or profiled face, either applied as an integral part of the manufacture or applied subsequent to manufacture.

Blocks are classified according to density and compressive strength. The compressive strength of blocks in accordance with BS6073 [29] ranges from 2.8 N/mm^2 to 35 N/mm^2.

15.6 Blockwork

Concrete blockwork is a well tried form of construction offering a great many features using economic and speedy construction, and is capable of providing at one and the same time:

- The structure of a building.
- The required level of thermal and acoustic insulation, and thermal capacity.
- Good fire resistance.
- The internal and external finished surfaces.
- Adequate resistance to rain and damp penetration.
- Excellent durability.

15.7 Damp-proof courses

Cavity construction is used to prevent the horizontal movement of moisture through walls. Bitumen felt, lead or similar materials are used as impermeable horizontal layers in walls to prevent the downward movement of moisture through parapets and similar locations or the upward movement of groundwater from foundations by capillary action (see Fig 3.47). Similar vertical damp-proof courses should be employed at the sides of openings. There the inner leaf may be turned forward towards the outer leaf to close the cavity, leaving a potential capillary moisture path unless positivity sealed.

Below ground level it is necessary to prevent moisture passing through brickwork by 'tanking' the structure with asphalt or bitumastic products. This membrane may need further external support, see Fig 3.48.

15.8 Openings

Concrete lintels, in-situ or precast, steel beams or angles, proprietary pressed galvanised steel lintels (which can form the cavity damp proof tray), or brick arches are used to carry the brickwork above openings for doors, windows, ductwork, etc.

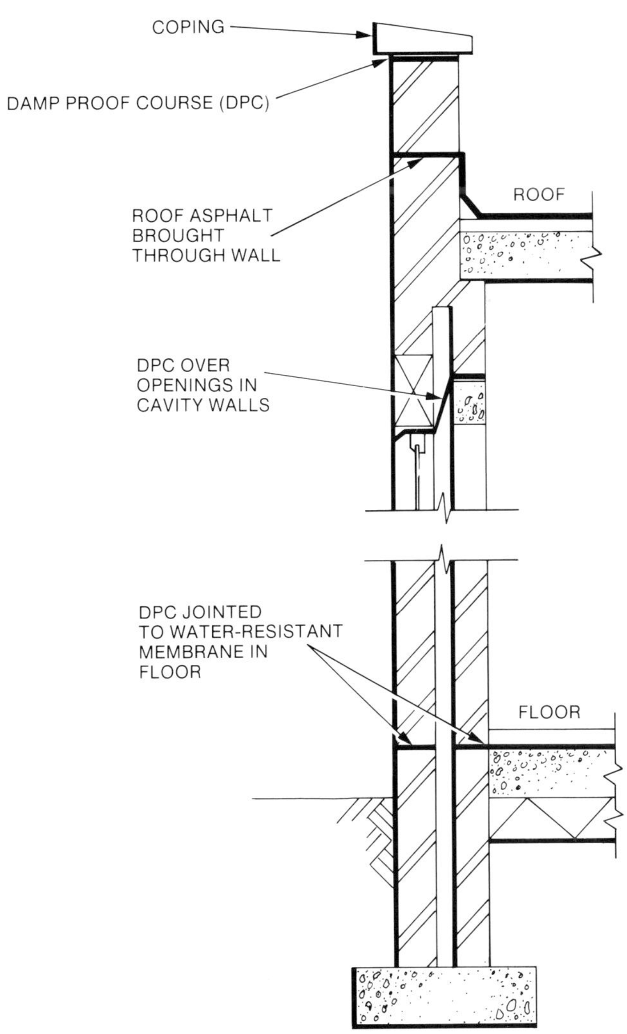

FIG. 3.47 Damp-proof courses in walls

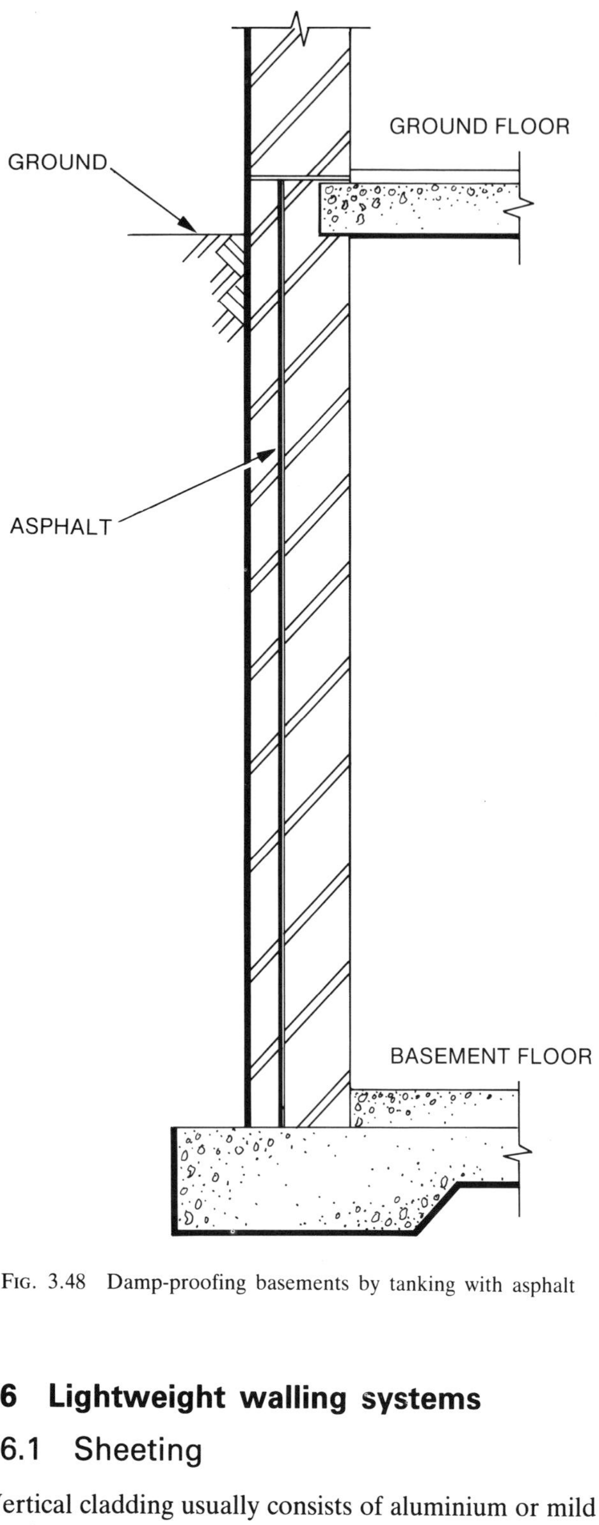

FIG. 3.48 Damp-proofing basements by tanking with asphalt

16 Lightweight walling systems

16.1 Sheeting

Vertical cladding usually consists of aluminium or mild steel galvanised sheets with a substantial protective coating system. The sheets are normally corrugated or profiled to give additional uni-directional strength. The wall cladding including aluminium is usually coated with a film of alkyd amino-based paint or other approved material. All sheets can be obtained in various colours.

Steel sheets are normally 0.8 mm thick and aluminium sheets 1.2 mm thick. The zinc coating of the steel sheets should be in accordance with BS2989 [30]. Plastic films can be applied to sheets at the works. Drying may be carried out using stove techniques or air drying processes depending on the type of coating used. The film thickness may be from approximately 25 microns for the coated finishes to films in excess of 250 microns for the laminated finishes. When comparing different materials, however, film thickness is not a measure of the relative durability characteristics of different materials.

Sheets can be cut by the manufacturers to the client's requirements and in many cases it is transport and erection difficulties which limit the length of sheet. Nevertheless, sheets up to 12 m long are commonly used for large walls to reduce the effects of horizontal joints. Cladding is also more rigid where sheets are continuous over a number of sheeting rails.

16.2 Insulation

Where exclusion of weather is the only design consideration, single sheets of cladding without a lining are sufficient. In locations where human occupancy and the necessity to preserve temperatures above those at which condensation forms (i.e., dew point temperatures) require some form of heating, then to prevent excessive heat losses a lining of thermal insulation such as plaster board or similar material is essential.

16.3 Fixings

External sheeting is normally fixed by self-tapping screws, hook bolts, clips or other suitable means directly to the angle sections forming the sheeting rails. The insulation layer may be fixed to the back of the outer sheeting or it may be supported on the inner face of the sheeting rails, thus forming a cavity which increases the thermal insulation value. Although cladding of this type can be used on a structure with a reinforced concrete frame, the spacing of the sheeting rails and other characteristics make it most suitable for use with a structural steel framework. Figure 3.49 shows a typical arrangement of insulated cladding used in conjunction with a steel frame.

To obtain natural light inside a building with this type of fabric, windows of the normal pattern or, alternatively, patent glazing consisting of wired cast-glass in aluminium or lead-covered T-shaped steel glazing bars, may be used. Another option is to use sheeting that incorporates panels of transparent material made from polyester resin reinforced with glass fibre. These sheets can be obtained with similar profiles to the main cladding material and may be incorporated directly into the cladded area. Alternatively, sheets of a different profile may be used with flashing at the top and bottom of the line of sheets.

In their application to power stations, these methods of construction have been developed to meet the need for lightweight cladding to structures. Their main

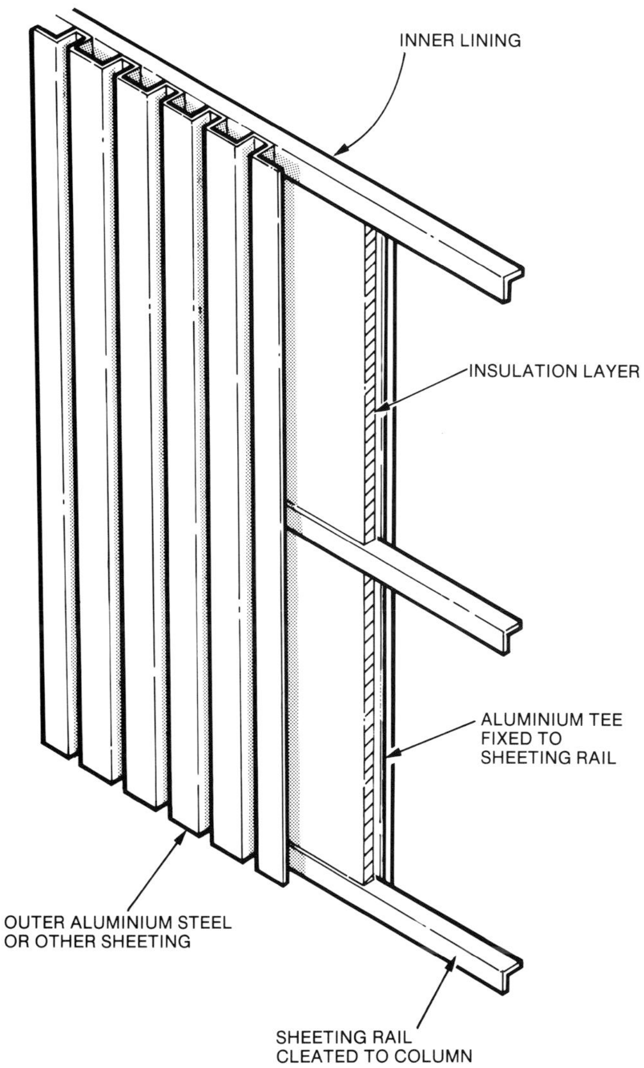

FIG. 3.49 Lightweight wall cladding

advantages over traditional materials such as brickwork, stonework or similar materials are:

- The lightness of this form of cladding reduces considerably the dead load on the structure.
- Speed of erection enables plant installation to proceed earlier in the overall programme under sheltered conditions.

16.4 Durability

Some forms of cladding, especially the plastic-faced metal sheets, are comparatively new and the development of these materials has been very rapid. Accelerated weathering tests, as developed for paint testing, have been carried out using closed carbon-arc lamps to simulate sunlight. Also exposure to ultraviolet rays to determine colour fading, and washing with artificial solutions of seawater and water containing dissolved sulphur dioxide can determine resistance to corrosion. However, the final proof of a material's durability is found only after use in a particular location.

Care must be taken when using different metals in close proximity such as sheets and fixings. If they are sufficiently different in the electrochemical series, corrosion cells will be formed unless adequate insulation and sealing of fixings and joints is used. The use of PVC tape and plastic washers together with layers of insulating board between purlins and sheets helps reduce the risk of corrosion through electrolytic action.

17 Roofing

17.1 Structural elements

These usually consist of profiled sheets of aluminium or steel protected in a similar manner to wall cladding. They are laid across steel or concrete beams to form the roof covering. The sheets are fixed together and to the roof beams by self-tapping screws, cartridge fixings or approved fixing clips.

Metal decks are designed in the same way as any other structural unit, the panels being designed as wide beams spanning between purlins. Panels with different elastic structural properties can be obtained, and it is only after considering the spans of purlins and decking that the most economic arrangement of roofing can be designed. Excessive deflection may occur due to loading which in itself will not overstress the metal decking. This distinction between elastic modulus and the material's permissible stresses in the design of roof sheeting systems may be of crucial importance. It is usual to limit deflection to one two hundred and fortieth of the span.

17.2 Insulation and weatherproofing layers

An insulating layer of Perlite (or PU) above 40 mm thick is bedded in bitumen on top of the vapour barrier and metal decking. On this is laid a weatherproofing layer consisting of two intermediate layers and a cap sheet of high performance polyester-based bitumastic felt. The joints and individual felt layers are stuck together using hot bitumen and a final surface of small white mineral chippings set in bitumen is provided to reflect heat and increase the resistance to the spread of flames in case of fire. Insulation board is kept clear of steam pipes passing through the boiler house roof, where steel plates with collars and flashings are used to exclude the weather. Glass reinforced concrete tiles are used to form the principal walkways across the roof.

A vapour barrier consisting of a sheet of hessian-reinforced aluminium foil cored felt is essential between the metal decking and the insulation board to prevent moisture being absorbed into the insulation board. Such moisture intrusion leads to both structural deterioration and reduction in insulation performance.

Purlin levels are arranged to give a minimum gradient of 40 mm in 3 m, although much higher slopes are strongly advised.

17.3 Application to power stations

Provided the felt is laid properly on this type of roof no trouble should be encountered for many years. These roofs, however, are subject to mechanical damage and care must be taken to restrict traffic to the areas which are specifically designed for it.

The advantages applicable to wall cladding regarding lightness and speed of erection apply equally to this form of roof construction. Figure 3.50 shows a typical form of roof construction. There are limitations in its use, however, for example it should not be used within 2 m of boiler escape pipes. One necessary precaution to be undertaken by the structural engineer is to provide, with one of the high steel columns, a 150 mm bore mild steel dry vertical fire main with standard fire brigade fittings at the ground floor and at turbine hall and boiler house roof levels. There should be one main to each unit. The same main can be used for boiler and tank fillings.

17.4 Durability

Roofs of this type, if constructed properly, require little maintenance when only subjected to light traffic conditions. The insulation layer, however, deteriorates rapidly if the waterproofing layer is damaged. Experience from both within and outside the power industry has shown this form of construction to be relatively very fragile and serious examples of major deterioration or failure are not uncommon.

17.5 Rainwater disposal

In view of the large areas of roofs to be drained and the consequent large amount of water to be dealt with, roof drainage and rainwater pipes require careful consideration at the design stage. If rainwater pipes are spaced at large intervals water can be running quite deep in the vicinity of the rainwater outlet during heavy storms, unless large box gutters are formed to take rainwater discharged by cross-falls and the gutters are sloped to carry the water comparatively large distances to rainwater pipes. On tall power station buildings large quantities of rainwater are precipitated directly onto the face of the building.

On some stations, water from the roof is allowed to cascade down the glazing forming the elevations of the main buildings and is collected in large gutters at operating floor level. This has the effect of washing the glazing and reduces the number of rainwater pipes.

Rainwater pipes are located preferably on the outside face of the building. Ducts are provided in welfare, offices and similar buildings to conceal internal pipes and the large pipes used for turbine and boiler houses are often fixed in the box columns of the structure. This

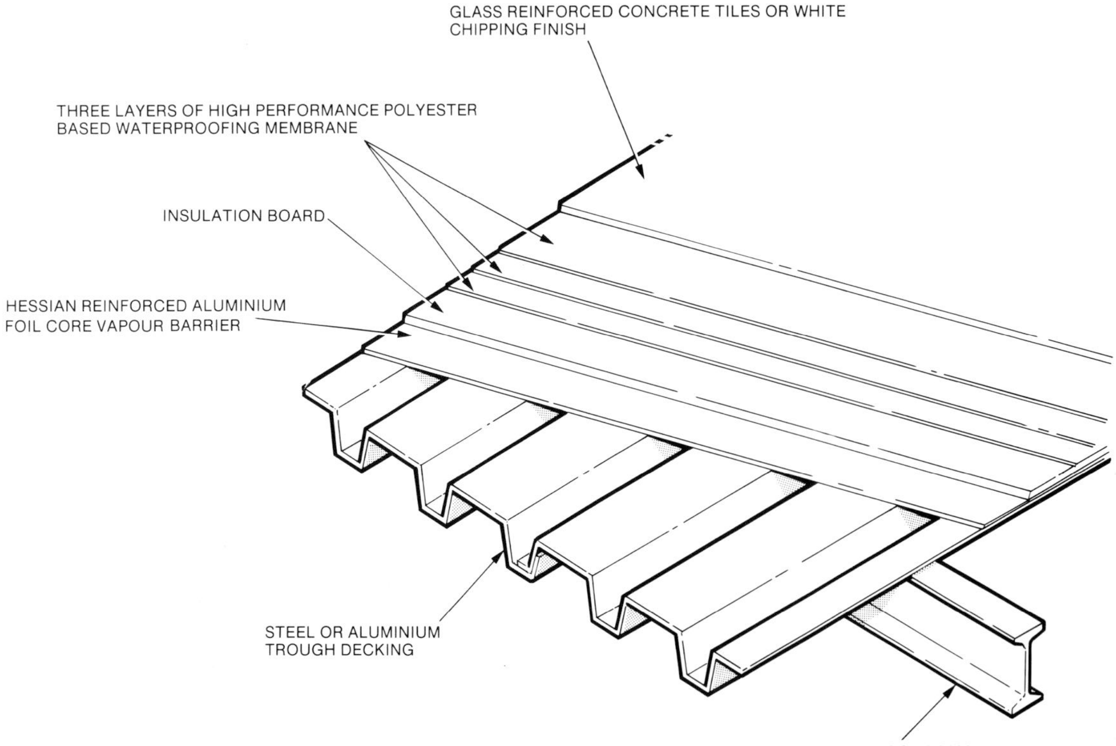

FIG. 3.50 Lightweight deck roofing

arrangement conceals the pipe and reduces the chance of damage by water due to a stopped pipe bursting at a joint. Cast iron, aluminium, and rigid plastic pipes are used for rainwater dispersal.

18 Finishes

18.1 Floor finish considerations

The following basic factors should be considered in choosing a suitable floor finish to a given area:

- Durability — the floor finish must have the ability to withstand the traffic. Maintenance appropriate to its location is essential for longevity.
- Appearance — which may be of prime importance or of little consequence depending upon location.
- Cost — capital, maintenance and the expected lifetime are the key interdependent parameters of total economic choice.
- Surface finish — highly polished surfaces will show variations in surface levelling; anti-slip properties may be important, particularly in areas of spillage, e.g., kitchens, laboratories, etc.
- Acoustic properties — sound absorption may be required in certain areas, e.g., control rooms, offices, conference rooms, corridors, etc.
- Fire resistance — in areas of high fire risk, low flame spread and low smoke emission materials may be of particular importance.
- Access to services — specially raised access floors may be required in computer rooms, cable areas, etc.
- Special requirements — floors used in conjunction with under-floor heating, or in areas of potential chemical or radiation contamination will require special materials.
- Thermal movement — suitable expansion and contraction joints may be required where differential expansion characteristics of sub-floor and floor finish may be experienced.

18.2 Types of floor finish

Having assessed the requirements of a floor finish in a particular location a selection can be made and some of the more common types available are as follows:

Trowelled concrete This is a means of obtaining a cheap, reasonable finish to a concrete slab. The concrete is trowelled smooth by hand or by machine (power floating) and the surface can be treated with a proprietary anti-dust silicate sealant. This method can provide a suitable surface for the application of other sheet or liquid resin floor finishes. Its tendency to dust under traffic can be resisted by the application of cheap penetrative sealants.

Granolithic finish This finish is applied to concrete floors and consists of a layer of concrete made from fine granite chippings and cement. The granite aggregate is graded in size from chippings down to dust. It gives a superior finish to concrete, being harder and less absorbent. This finish can be applied to set concrete, but is not recommended due to inherent problems of adhesion. It is better laid on 'green' concrete, or integrally with the concrete, to ensure a good bond and to avoid differential shrinkage of the floor and finish. The thickness laid varies from 50 mm to 75 mm and depends upon the location on the floor. The thickness can be reduced if it is laid monolithic with the concrete, which will also ensure the most positive bond with the base concrete.

Terrazzo This is an expensive finish consisting of marble chippings set in white cement. The surface is ground and polished to a flat surface after the cement has hardened. It can be laid in-situ or as tiles. This finish is very impermeable and hygienic but is so brittle that cracking occurs very easily. Expansion and contraction joints must be very carefully located. Specialist armoured terrazzo tiles are manufactured which are quite suitable for turbine hall floors and other heavy duty areas, albeit expensive.

Quarry and other vitreous tiles These are made from burnt clay and have good acid-resisting properties. Tiles can be obtained in a variety of colours with rough or smooth surfaces. Purpose-made tiles can be obtained to form coves, skirtings and other similar features. They can be set and jointed in cement or, alternatively, in a proprietary jointing compound which remains slightly plastic after setting and thus prevents failure by differential movement of the finish and slab. The fully-vitrified tiles are impermeable and are suitable for areas subject to contamination in reactor buildings.

Wood Hardwood boards or blocks are the timbers usually used for floor finishes in power stations, although softwood boards are sometimes used to provide a base for a further covering. Timbers vary and although good results can be obtained from the best hardwoods such as maple boards or oak blocks, other types of wood floors are not suitable for heavy traffic. Wood blocks laid with their grain vertical make an excellent floor finish for workshops and similar locations. The end grain gives good wearing characteristics and the nature of the surfaces means that there is a smaller chance of damage to castings and similar items placed on or moved across the floor. These blocks are often set in pitch and should not be used where the risk of fire is high.

Wood fibre blocks These are made from reconstructed timber and are very similar to wood blocks.

Sheet floorings These include linoleum, rubber and PVC. They are usually secured to a timber or cement scored base by means of a suitable adhesive. They give a good impervious hygienic finish which is not difficult to keep clean but they are not suitable for areas of heavy traffic.

Plastic tiles These are similar to sheet floors but in tileform. They are more easily laid, but have the disadvantage of many joints which are vulnerable to moisture spillage and consequent loss of adhesion, together with an inherent tendency to curl in lesser quality tiles.

In-situ composition floors These include asphalt, magnesium oxychloride, epoxide resins and many other types of finish. The quality and type of finish can be varied to resist acids, alkalis and other contaminants. Such special materials normally fall outside the architect's or civil engineer's experience and specialist advice is sought before adoption.

18.3 Finishes to walls and ceilings

For many power station buildings of brick and concrete construction a fair finish and flush pointing to the brickwork together with a good finish to the concrete is adequate. Plastering, rendering and similar treatments to walls and ceilings are called *wet processes*. Although satisfactory results are usually obtained, the time taken by application of these finishes together with the drying out periods which must be allowed before decoration can commence have led to the development of *dry finishes*. These include ceilings constructed by plastic sheet, insulation board, hardboard, metal trays and other materials which can be supported by hangers from the roof to provide a space above the ceiling for ducts and services. Insulation boards on the inside of sheet cladding and unit partitions in office blocks are typical examples of this type of finish. Many such panels can be supplied ready decorated if required.

Special finishes such as vermiculite spray can be applied in thicknesses usually between 25 mm and 50 mm to concrete, steelwork or other surfaces. Although occasionally used for insulation, the main function of these finishes is the protection of steelwork against fire.

18.4 Wall tiling and other special finishes

In kitchens, toilets, medical centres or similar locations where walls are subject to contamination by grease, dirt or corrosive substances and a high degree of hygiene is necessary, the conditions and frequent washings require the use of a better surface than is obtainable with a paint finish. In these circumstances glazed or unglazed wall tiles, in-situ terrazzo or terrazzo tiles, plastics or similar materials are used for the finish to wall surfaces. Particular attention to such decontaminable finishes is paid in the design of those areas of nuclear stations vulnerable to radioactive contamination.

18.5 Internal painting

Apart from protecting the steelwork, internal painting is used mainly for decorative reasons and is usually carried out in accordance with schemes prepared by the architects. For painting on timber, concrete, brickwork, plaster, metal or similar surfaces, the basic system is usually as follows:

(a) *Cleaning down and surface preparation* Before painting, all loose material should be removed and all surfaces thoroughly cleaned. A vacuum plant should be provided for the removal of dust. Cleanliness of the base prior to application of any protective coating is of paramount importance.

(b) *Priming coat* This seals and prepares the surface and also provides a key for subsequent coats of paint. It can also be used to protect a material during the constructional stages before the final coatings are applied near to station completion. Aluminium or conventional wood primers are normally used for timber, and alkali-resisting primers for the surfaces of new plaster, concrete, brickwork and similar surfaces. For ferrous metals a zinc phosphate or other approved rust inhibiting primer is suitable, and for galvanised surfaces an etch primer followed by a zinc phosphate primer should be used.

(c) *Undercoats* One or two coats of paint is the usual specification for undercoats, using a paint which is compatible with the primer and finishing coat.

(d) *Finishing coats* These are normally alkyd-based resin paints or similar materials which give a hard durable finish. They can be supplied to give a matt, semi-gloss or high gloss finish.

(e) *Other paints* A more economic but less durable finish can be obtained on brickwork, plaster, concrete and similar surfaces by the use of emulsion paints.

18.6 External painting

Although the decorative aspect of external painting must receive some consideration especially on the main buildings, offices, welfare blocks and similar buildings, there are many locations such as precipitator steelwork, coal gantry steelwork, external plant and other similar items where protection is of prime importance and appearance of less importance. External painting is carried out using preparations and materials similar to those used for internal work with the addition of micaceous iron oxide, chlorinated rubber, polyurethene, epoxy, acrylated rubber, vinyl acrylic and glass flake systems as required, depending on the local climate and in-service environment envisaged for the protective coating.

19 Turbine hall and boiler house construction

19.1 General

The walls in older power stations built before 1939 were usually of brickwork, built with a solid plinth course and with brick piers to carry the loads from floors, roofs, etc. Roofs were pitched and consisted of wrought iron and, later, steel roof trusses carrying timber or angle iron purlins with slated roofs. Floors were supported on cast iron joists the ends of which were supported on padstones built into the brickwork or supported on circular cast iron columns. Floors were of timber joist and board construction except for those required to carry heavy loads where cast iron plates or continuous brick arches spanning between cast iron beams were used. The introduction of rolled-steel members and later reinforced concrete enabled the main buildings to be constructed using a structural frame, in which the walls ceased to form structural elements but served mainly to insulate and exclude the weather. Flat roofs of reinforced concrete or precast concrete units covered with asphalt became common and floors were of reinforced concrete, or where heavy loads were to be carried, concrete reinforced by small section rolled-steel joists. The latter is termed *filled joist* construction.

Compound sections using one or two rolled sections with additional plates riveted to the flanges, and riveted plate girders made up from flat plates jointed by angles were introduced to carry the increased loads on columns and beams. The cladding became a plinth of brickwork to withstand mechanical damage with bitumen-coated, galvanised corrugated steel or asbestos sheeting to the walls and roofs. Lighting and ventilation were obtained by incorporating steel windows or areas of patent glazing in walls and roofs. The offices and other ancillary buildings were usually combined on the main elevation of the station, and the facing of brickwork and artificial stone on these buildings was usually extended to the whole of the main buildings on this elevation in order that a good external appearance was presented.

In the period since 1946, the standard walling construction of brickwork gave way to a number of turbine halls being built with reinforced concrete frames and roofs. Precast concrete, in-situ concrete and composite construction were used, and the use of concrete shell roofs with concrete columns and beams resulted in many excellent buildings.

Until recently it was common practice to plan the construction of a power station in two sections, the first section often being commissioned before the second was commenced. Problems presented by this form of construction included difficulties due to differential settlement of foundations owing to the settlement in the second section lagging behind that of the first. Re-levelling of crane rails was necessary and provision had to be made for movement between all interconnected parts of the building and plant. A temporary gable end had to be provided to the completed first section and this usually took the form of a steel frame covered with cladding.

Several power stations have been built in the past using a semi-outdoor construction for the boilers. A structure was not provided to house the boilers with this arrangement, although hutments and covered staircases, gantries and walkways were provided to protect operators, when working on the boilers. When these power stations were built, brick walls and concrete roofs were the accepted methods of construction for main buildings and hence considerable capital savings were effected. These savings are not so apparent when outdoor boiler construction is compared with the more modern method of boiler house construction using lightweight cladding. Outdoor construction also has the disadvantage that many repairs have to be carried out in the open and extensive scaffolding may be necessary for work at the higher levels. There are also obvious objections to this type of construction on aesthetic grounds and it is unlikely to be used in this country in the future.

Although brick and steel chimneys have been used in the past, concrete is used for chimneys on projects now under construction. For chimneys in excess of 90 m high concrete construction becomes cheaper than brick.

The first cooling towers consisted of a timber frame built over concrete cooling ponds, the outside of the frames being covered with wood boarding. Timber stacks were provided inside to ensure adequate cooling of the water. Steel was also used but corrosive conditions were so severe that in spite of maintenance painting the average life of a steel tower was only 8 years. Concrete was then adopted for cooling tower construction and the first concrete towers were similar in shape to the timber towers. These towers were superseded by the hyperbolic concrete towers, which although smaller were of the same pattern as those now being constructed.

The main buildings superstructure for a modern 2000 MW power station is unique in size and associated loads when compared with other projects carried out by building and civil engineering contractors. The design of the buildings is the result of co-ordinated efforts by civil engineers, architects, quantity surveyors, landscape architects, steelwork and reinforced concrete designers, services engineers and other specialists. Similarly the execution of the work on the site results from the combined efforts of steelwork, civil engineering and building contractors and also the many other specialist contractors and suppliers of materials.

All work on the superstructure of the main building is usually carried out under three main contracts:

- The structural steelwork contract which includes cladding and roofing.

- The painting contract.
- The building and civil engineering work contract which includes floors, brickwork, finishings, services, etc.

The programme requirements for many types of buildings are simple from the client's viewpoint, starting and finishing dates being the only limitations placed on the contractor. This type of programming is especially applicable where buildings are to be occupied by personnel who move in on completion, and can also be used for structures such as chimneys and cooling towers. The programming of contracts for the boiler house and turbine hall is, however, much more complex. Although the commencement and completion dates for the various contracts can be established from the overall project programme, many intermediate completion dates must be established to ensure that foundations and cover are available for a multitude of plant erection contractors and sub-contractors and electrical contractors to proceed.

19.2 Structural considerations

The total weight of structural steelwork required for the frame of a power station depends upon the plant layout and other factors which include the provision or omission of a basement and the use of reinforced concrete columns to the turbine hall. Hence the weight of steelwork for a 2000 MW station expressed in terms of plant capacity varies considerably from 12 t to 19.5 t per MW.

The basic dimensions of the boiler house and turbine hall are decided by the plant layout, and work cannot proceed on the design of the structures until this is established. When this stage is reached, development of the plant details and finalisation of the building design can develop together.

Because of the long beam spans and the heavy loads on columns and beams, structural steel frames are normally used, although composite construction in which reinforced concrete may be employed for turbine hall columns and for the frames of associated structures. Figures 3.51 and 3.52 show cross-sections through the steel frames for unit 1 and unit 2 respectively at Thorpe Marsh power station, which were completed in the early 1960s. The longitudinal arrangement of the turbines resulted in the turbine hall crane having a span of 24.5 m.

In stations completed before 1960, boilers were slung from the main steelwork, the top beams of the large portals of the boiler house providing support for the boilers and roof whilst the columns formed the outer wall of the boiler house. The weight of a boiler in the newer 2000 MW power stations often exceeds 13 000 tonnes, and the span of the boiler house is so large that intermediate columns are introduced to reduce the span of the steelwork carrying the boilers. The boilers are carried by slings from the overhead sling deck, supported on columns which transfer the load to the foundation. For unit 1 at Thorpe Marsh the columns on

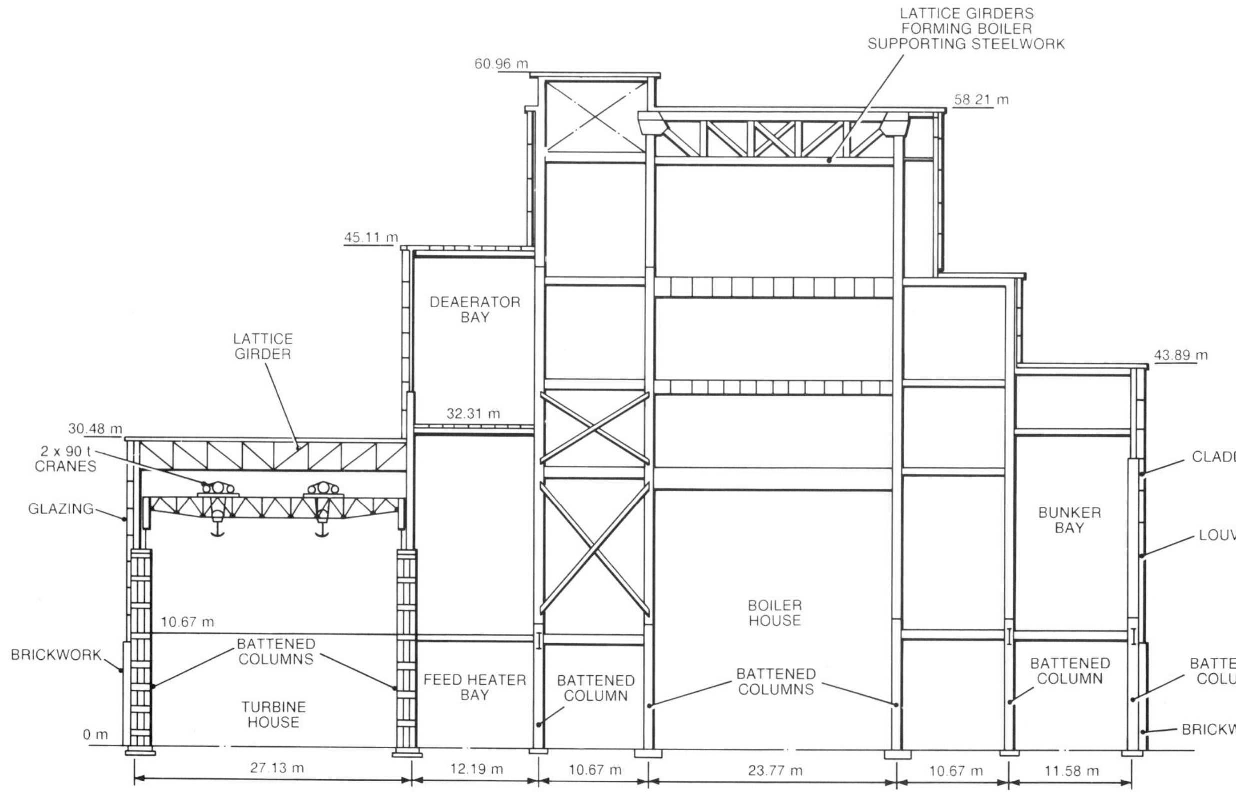

FIG. 3.51 Section through Thorpe Marsh power station unit 1

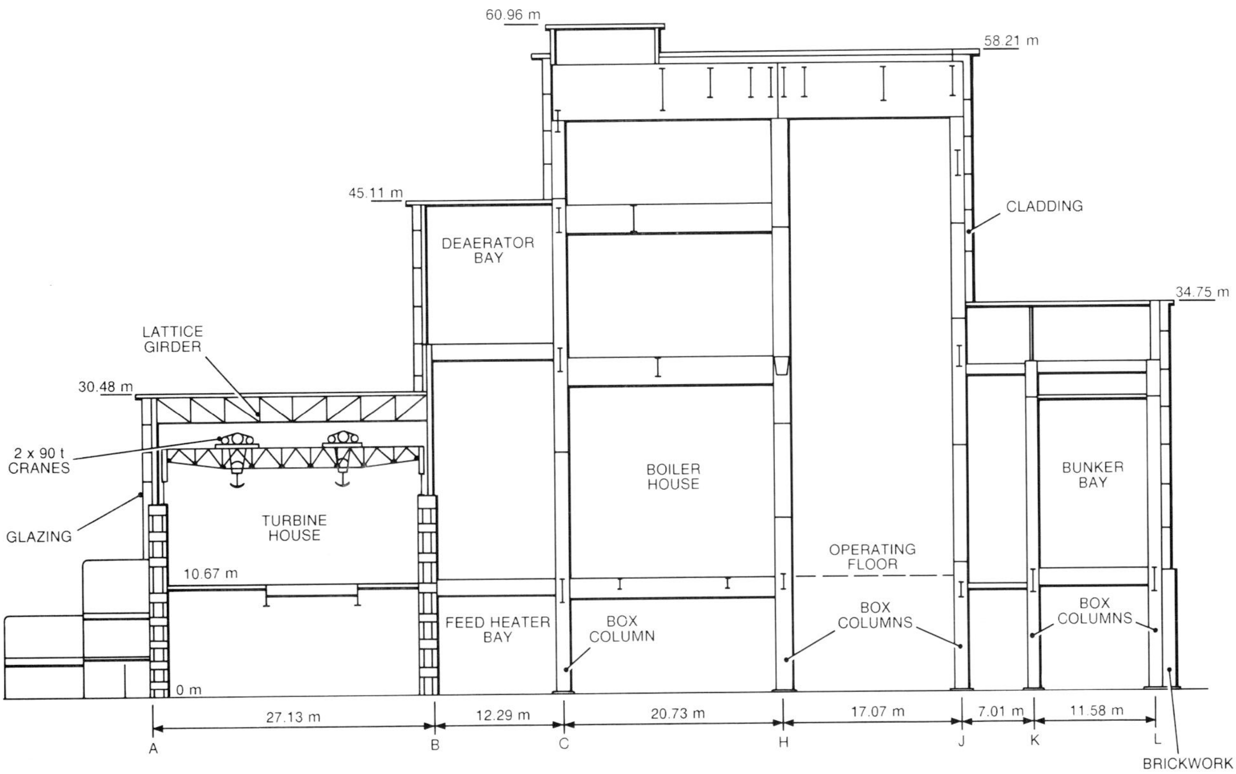

FIG. 3.52 Section through Thorpe Marsh power station unit 2

lines D and E, as shown in Fig 3.51, are used to support the boilers and for unit 2 the weight of the boilers is carried on the columns in lines C, H and J, as shown in Fig 3.52.

It is only after consultation with the steelwork contractor that decisions on the type of members to be used can be made. Certain contractors associated with plate manufacturers favour box members, whereas others can tender more economically on the basis of rolled sections. Compound members were used throughout in the frame for unit 1 at Thorpe Marsh and although similar construction methods were used for the turbine hall of unit 2, box members have been used in the boiler house construction. Lattice girders have been used to support the roof of the turbine hall for both units.

Figure 3.53 shows the cross-section through the structural steelwork for West Burton power station on the centre line of the boilers. This is a 2000 MW station with a longitudinal arrangement of the four 500 MW sets which gives a crane span in the turbine hall of 38.5 m. The boilers are supported from the columns on lines D, F and G, and all the columns in the boiler house and turbine hall, together with the main beams in the boiler house, are box members. A longitudinal arrangement of sets usually results in large spaces between boilers which in the case of West Burton have been used to accommodate the bunkers. Figure 3.54 shows the cross-section through the steelwork on the centre line of the bunkers. The roof beams used for the turbine hall at West Burton are of triangular section and lattice construction. The compression members of the roof beams are at the apex of the triangle and the tension members at the bottom, which is level with the roof, the beams thus forming a triangular shaped penthouse in which glazing and ventilation is provided.

Figure 3.55 shows the cross-section through the 2000 MW power station at Fawley. This station also has a longitudinal arrangement of sets and the span of the turbine hall crane in the station is 48.75 m. The main roof beams to the turbine hall are box girders and the roof deck is level with the underside of the main beams.

Even longer spans, up to 60 m, result from the transverse arrangement of sets on other stations. This arrangement presents problems when the heavier loads associated with larger sets are considered and the use of two crane tracks involving a further line of columns carrying two crane rails has been developed. The position of this extra line of columns is not necessarily central, its actual location depends upon the lifting arrangements for the turbine and generator. At Ferrybridge *C* power station, where this arrangement has been used, the smaller span crane is located over the generators which is advantageous owing to the greater loads on this crane. On this station, the central columns extend to the full height of the turbine hall and are used to support the roof in addition to the crane rails, thus reducing the spans of the main roof beams.

Wind pressures of 1400 N/m^2 can result in total horizontal loads up to 1500 t on the face of a boiler

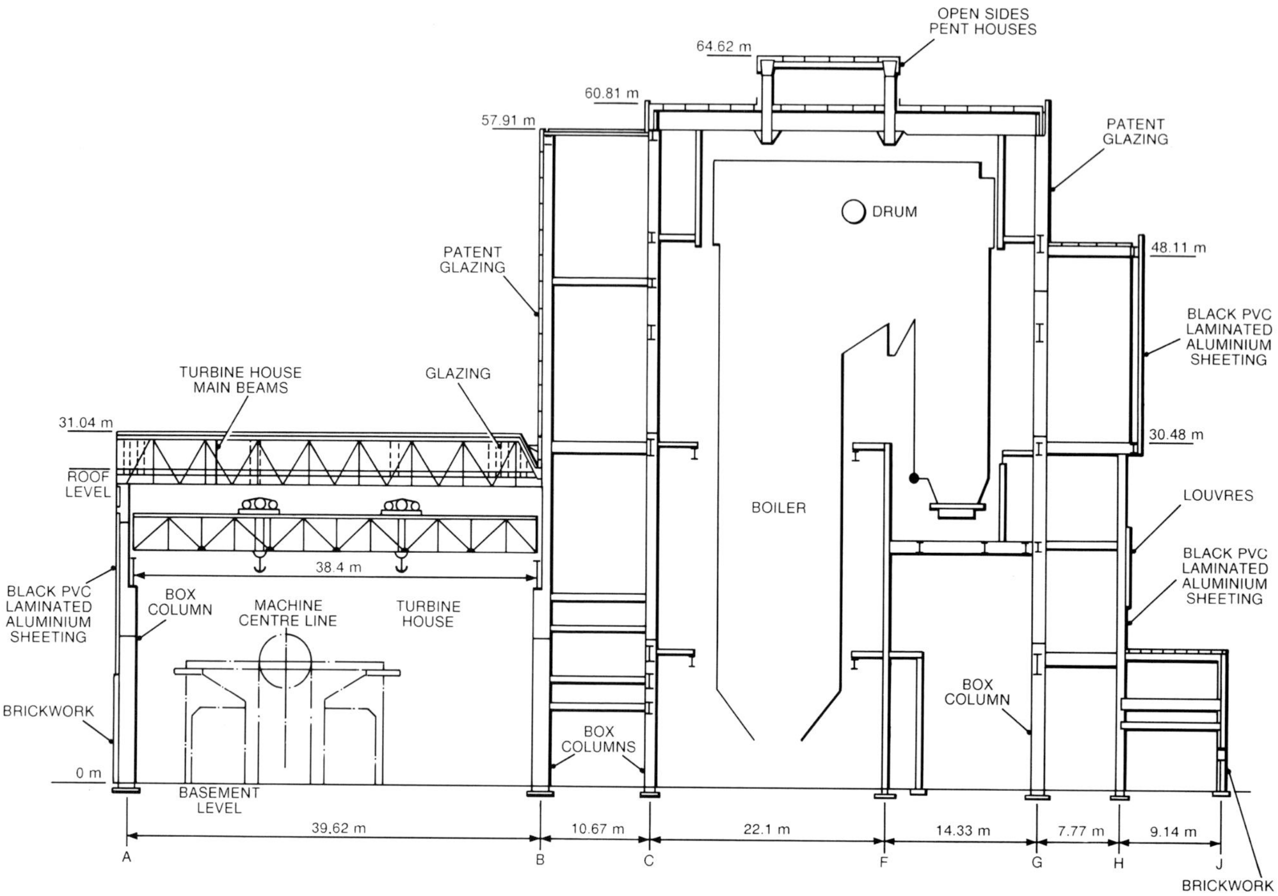

Fig. 3.53 Section through West Burton power station

house and wind girders and bracing is necessary to resist these forces. The main columns are often located at centres in excess of 12 m, and large section sheeting rails are necessary to carry the weight of the cladding and resist the wind loading on such a span. It is usual therefore to introduce intermediate columns to give a sheeting rail span usually between 2.75 m and 4.5 m.

The main roof beams in the boiler house and turbine hall support the roof structures which normally consist of lightweight decking supported on rolled section purlins. Some of the more recent stations have been provided with a roof of precast concrete units to the boiler as these have been found to be less susceptible to mechanical damage.

Beams and columns are used at intermediate levels to support floors, pipework and items of plant. Open grid flooring is normally used for floors, gantries and staircases above operating floor levels as this permits easy flow of air. The boiler house operating floor and also the floors around the sets in the turbine hall are normally of reinforced concrete construction. Solid floors are often constructed from precast concrete units or a combination of precast concrete units and in-situ concrete to facilitate speed of erection and avoid difficulties in supporting shuttering at such a height. Permanent shutterings of rigid galvanised sheet units are sometimes used.

19.3 Erection of steelwork

The design and fabrication of the steel frame is governed largely by available fabrication techniques, transport facilities and lifting equipment. Hence box columns are manufactured in lengths not usually exceeding 13 m. Sections of members such as roof trusses may be assembled into a complete member before it is lifted at the site. Columns of a cross-section 2 m by 1 m are often used and beams are up to 4 m deep. The weight of sections delivered to site is normally limited by the capacity of the works and site cranes.

Although the use of large tower type cranes is being tried out, the normal crane arrangements for erecting steelwork are shown in Fig 3.56. Derricks up to 50 t capacity are used for steelwork erection, although special erection methods involving the use of catheads on the top of columns are sometimes used for lifting special beams in excess of the crane capacity. It will be apparent from the arrangement of the cranes shown, that in many cases two cranes can be employed for a single lift as in the case of the turbine hall main beams. The derricks are usually supported on trestles known as gabbards, and the whole structure mounted on bogies travelling longitudinally along the boiler and turbine buildings. As erection for one boiler is completed, the derricks move away and are finally dismantled outside the buildings on completion of the steelwork erection.

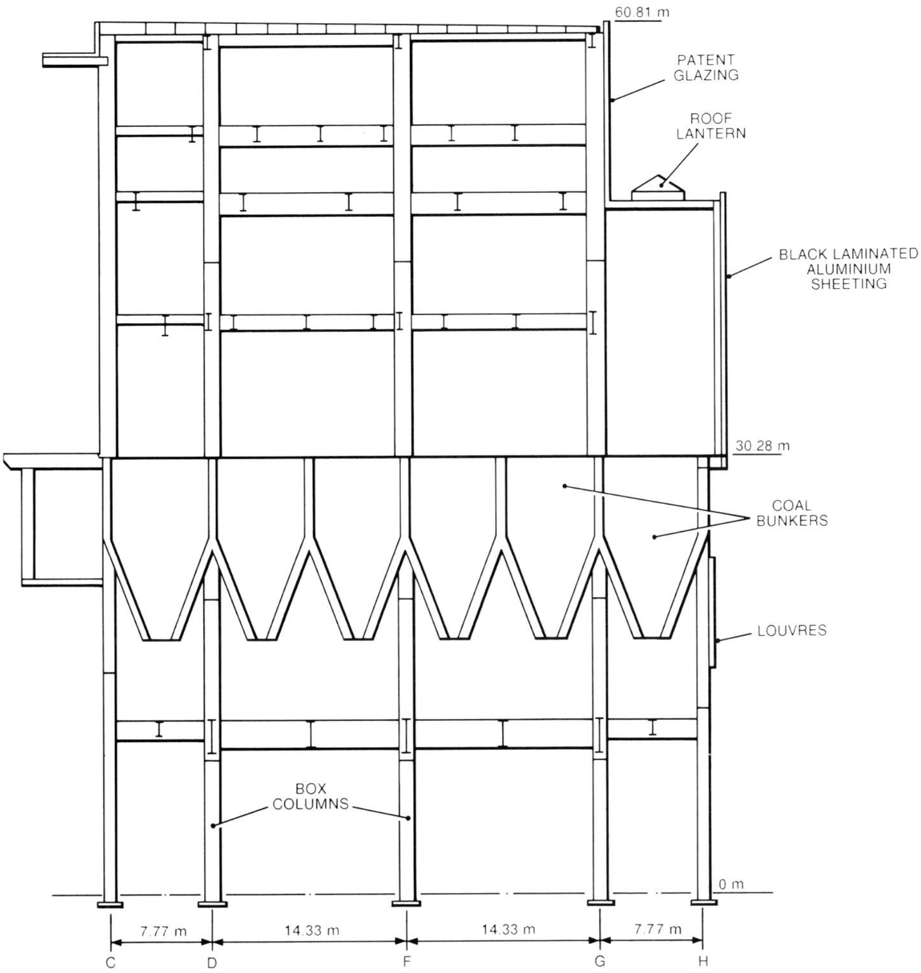

FIG. 3.54 Section through West Burton bunkers

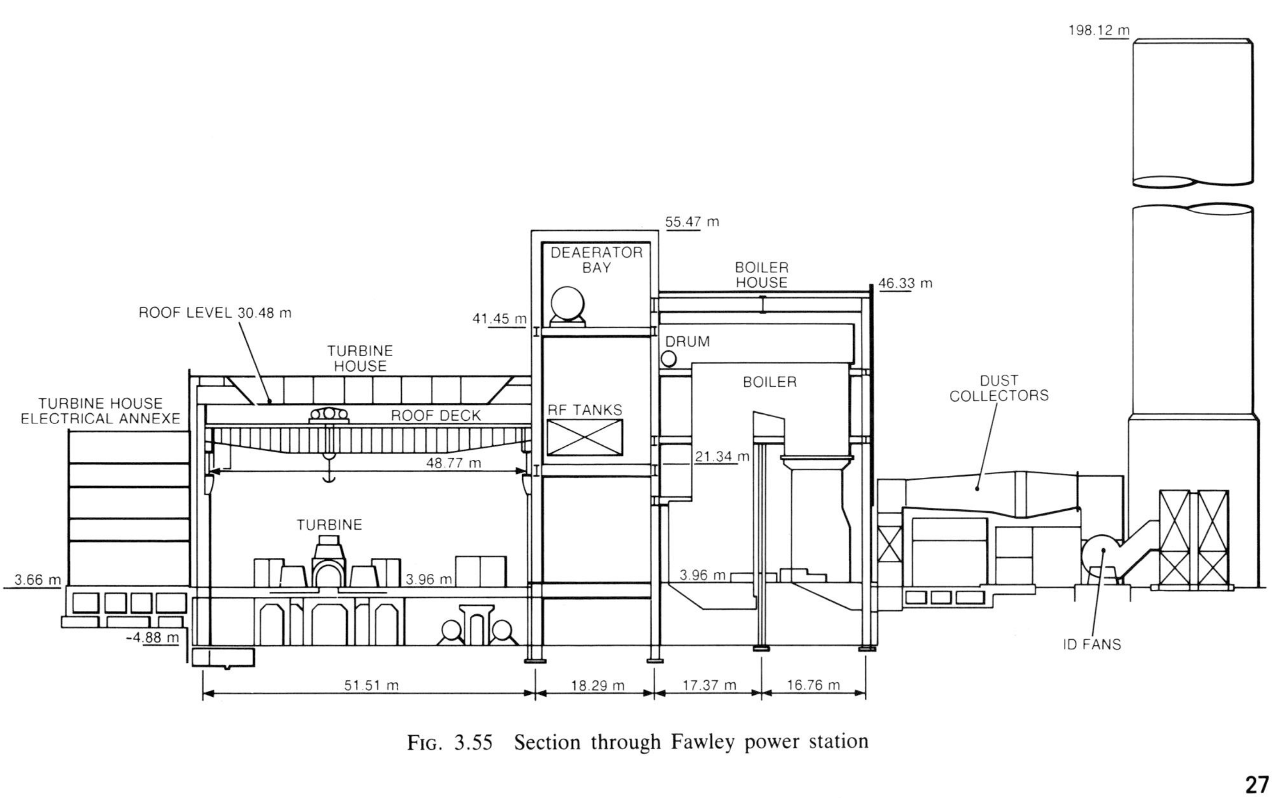

FIG. 3.55 Section through Fawley power station

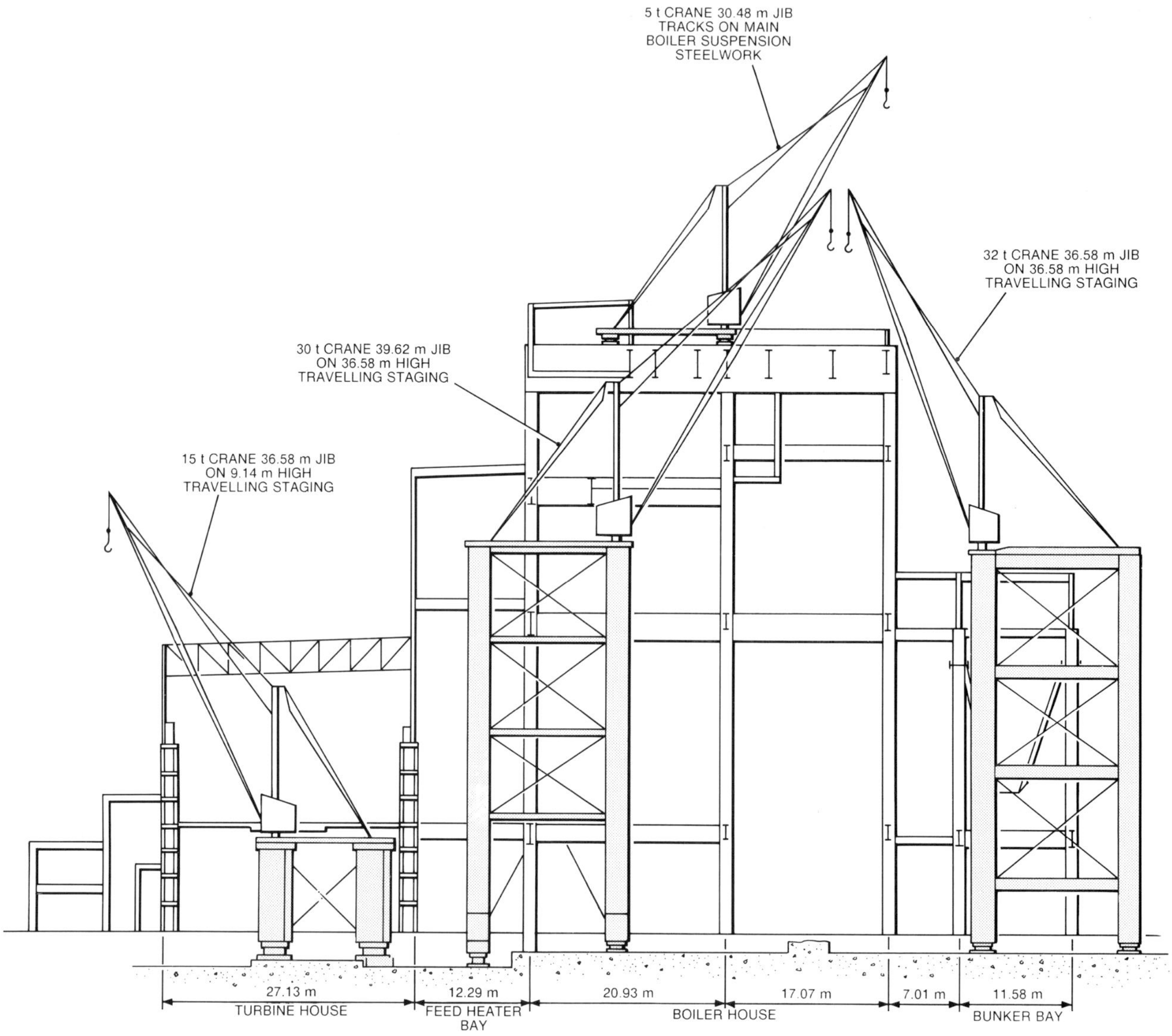

FIG. 3.56 Arrangement of cranes for erecting power station steelwork

Smaller track cranes are usually used at roof level of the boiler house to complete the lighter steelwork for the roof and intermediate floors, and also to assist in dismantling the main derricks on completion.

Use is now being made of very long-reach mobile cranes, both crawler and lorry mounted, for erection of both main and infill steel members and for the annexes and auxiliary buildings. Jib lengths exceed 60 m and loads of 100 t are possible.

The manufacture and erection of the steel frame for the turbine hall is governed by conditions very similar to those applicable to the boiler house steelwork, except that it is not usual to use track cranes at roof level due to there being very little, if any, intermediate steelwork apart from the roof purlins which are easily placed by the main derrick.

19.4 Cladding

It would be extremely difficult to use an in-situ concrete roof and brick walls for the cladding of a modern 2000 MW power station and also conform with the present programme requirements. This, together with other factors including the higher loadings which would result from the use of concrete and brickwork, has led to the adoption of metal deck roof construction and light-weight metal wall cladding.

For architectural reasons, and also to provide walls at a lower level which are more resistant to mechanical damage, it is usual to build the bottom 6 m to 12 m of the boiler house and turbine hall walls using brickwork, exposed aggregate concrete slabs or other similar material.

19.5 Ventilation

Ventilation has to be carefully considered in turbine halls. The amount of heat given off by the plant is not excessive when compared with the volume of the building and a reasonably comfortable temperature is maintained except when extreme conditions prevail externally. The build-up of humidity can, however, be a problem and air at roof level cooled below dewpoint temperature causes condensation to form resulting in degradation of the protective coatings of the building fabric and corrosion of the steelwork roof members. Condensation can be kept under control by preventing the air being cooled below dewpoint by removal of the moisture-laden air. The application of insulation to metal and the use of double glazing reduce the possibility of condensation attack. Double glazing is very expensive and in most cases ventilation to remove air before saturation point is reached is the best method of limiting or preventing condensation. This is more easily carried out if the turbine hall is open to the de-aerator bay and air from the turbine hall can be taken directly from high level into the boiler house. If this is not possible then some form of ventilation must be introduced at the highest level in the turbine hall.

Condenser design involves the provision of valves or diaphragms as protection against excessive pressure in the condensers. These valves or bursting discs are capable of discharging the full load steam flow, which for a 500 MW set is 450 m^3/s. The ventilating arrangements of the buildings must be capable of removing this quantity of steam sufficiently quickly to prevent it filling the building. To achieve this, some 40 m^2 of clear openings must be provided at high level in the turbine hall. As permanent openings of this size would result in excessive heat loss during normal operation, it follows that controlled ventilation must be provided. To function in an emergency, these ventilators must open fully in a few seconds and motor-operated louvres or opening sections of patent glazing are suitable, the motors being controlled by a pushbutton in the control room. A high degree of maintenance is necessary if ventilators of this type are to be kept in working order.

Normal ventilation of the turbine hall is provided by opening windows in the walls and roof lights.

The boiler intakes are normally located at the top levels in the boiler house and louvres are used at the lower levels in the walls (as shown for West Burton in Fig 3.53) to provide the necessary air inlet. The penthouses to allow the escape of boiler radiation losses are also shown in this figure. Approximately 100 000 m^3/min of air is required for combustion in a 2000 MW station on full load.

In addition to this air which is required for combustion, a similar amount must be allowed to pass through the boiler house to remove the boiler radiation losses making a total of 200 000 m^3/min, of which half enters the boilers, the remainder pouring out of the boiler house at the high levels. Insulation is not normally provided to the boiler house wall cladding although it is usually used on the roof decking, mainly as an element in the composite weatherproof membrane of the roof.

Figure 3.57 shows the ventilation and boiler air intakes at Thorpe Marsh power station.

Hot smoke and smoke logging can be a major problem when fires occur in turbine halls and careful consideration must be given to this matter in the ventilation design. Fire venting to reduce the temperature in the building and restrict the spread of flame must also be considered.

19.6 Floor and wall finishes

The turbine and boiler operating floors are usually finished with vitreous tiles or other high quality, hard wearing, non-absorbent floor finish. The basement floors to the turbine hall and the boiler house are provided with a quarry tile or granolithic finish and end-grain wood blocks are often used in the loading bays.

Large areas of the internal faces of walls in the turbine hall and boiler house are pointed and painted to provide a satisfactory finish. Plastering is only used in offices, toilets and other similar locations.

Insulated cladding is sometimes used for the walls of the turbine hall and large areas of windows, patent glazing or transparent sheeting are provided to give natural light.

Acoustic panelling is desirable in some areas for health and safety reasons.

Annexes to main buildings usually consist of a de-aerator bay located between the turbine hall and the boiler house and an annexe to the boiler house on the opposite side to the turbine hall, which houses ash and dust plant with associated electrical plant.

20 Reactor construction

20.1 Reactors

The first reactors used in power stations had spherical pressure vessels constructed from welded steel plate about 50 mm thick. These reactors were located in a separate large cylinder formed of concrete biological shield walls about 1.83 m thick. Integral with the walls was the concrete forming the base and pile cap.

As the development of reactors proceeded it became apparent that if the concrete shields were increased in thickness and reinforced with prestressed cables, then the steel pressure vessel could be replaced by a comparatively light steel lining about 13 mm thick to the inner face of the concrete. This design allowed for the majority of the tension forces in the concrete due to gas pressure being resisted by the prestressing cables.

Models of pressure vessels of this type were made and tested, and from estimates and assessments of performance it became apparent that the advantages of

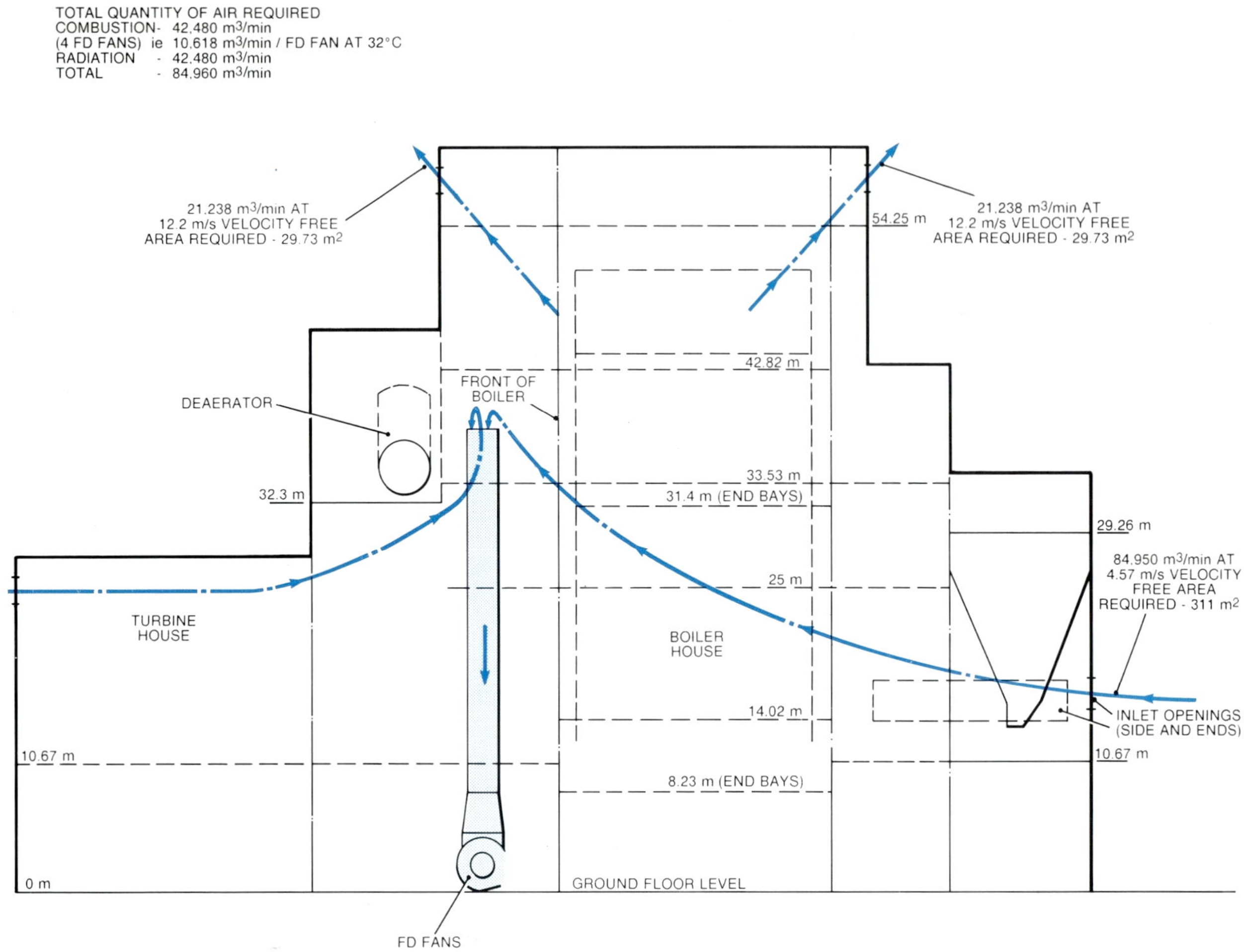

FIG. 3.57 Thorpe Marsh unit 2 ventilation and boiler air intakes

concrete pressure vessels over steel made them economically and operationally attractive. Some of the more important advantages of containing reactors in concrete pressure vessels are as follows:

- A prestressed concrete pressure vessel is cheaper than an equivalent steel vessel with a concrete biological shield.
- The concrete is subjected to its greatest compressive stresses during the prestressing operations, for when the reactor is operational, the gas pressure has the effect of reducing the compressive stresses induced in the concrete by the prestressing operations.
- The steel prestressing wires or strands which carry the tensile forces induced by the gas pressure are protected from overheating by a good thickness of concrete.
- There is relative immunity to failure from local weakness in the tensile steel wire used for the cables, as each of many tendons is anchored independently.
- The demand for skilled labour on a prestressed concrete vessel is less than for a steel vessel.
- Using prestressed concrete construction, the size of the pressure vessel and the contained pressure can be increased without the fabrication difficulties associated with thick steel plates. It therefore became possible to introduce the boilers into the pressure vessel, thereby eliminating the gas ducts and the awkward duct-to-vessel intersections.
- The load-carrying tendons can be examined during the life of the vessel and if external or ungrouted tendons are used, they can be retensioned or replaced if necessary.
- A concrete pressure vessel could not fail catastrophically if reasonably designed.

Figure 3.58 shows the arrangement of a typical cylindrical prestressed concrete pressure vessel in which the ends and walls are reinforced with prestressed tendons. There are no special requirements for the concrete used in concrete vessel construction, normal good quality concrete with a minimum crushing strength of 41.37 MN/m^2 at 28 days being suitable. The arrangement of the tendon ducts for the prestressing wires in the ends and side of the cylinder is such that the concrete will not

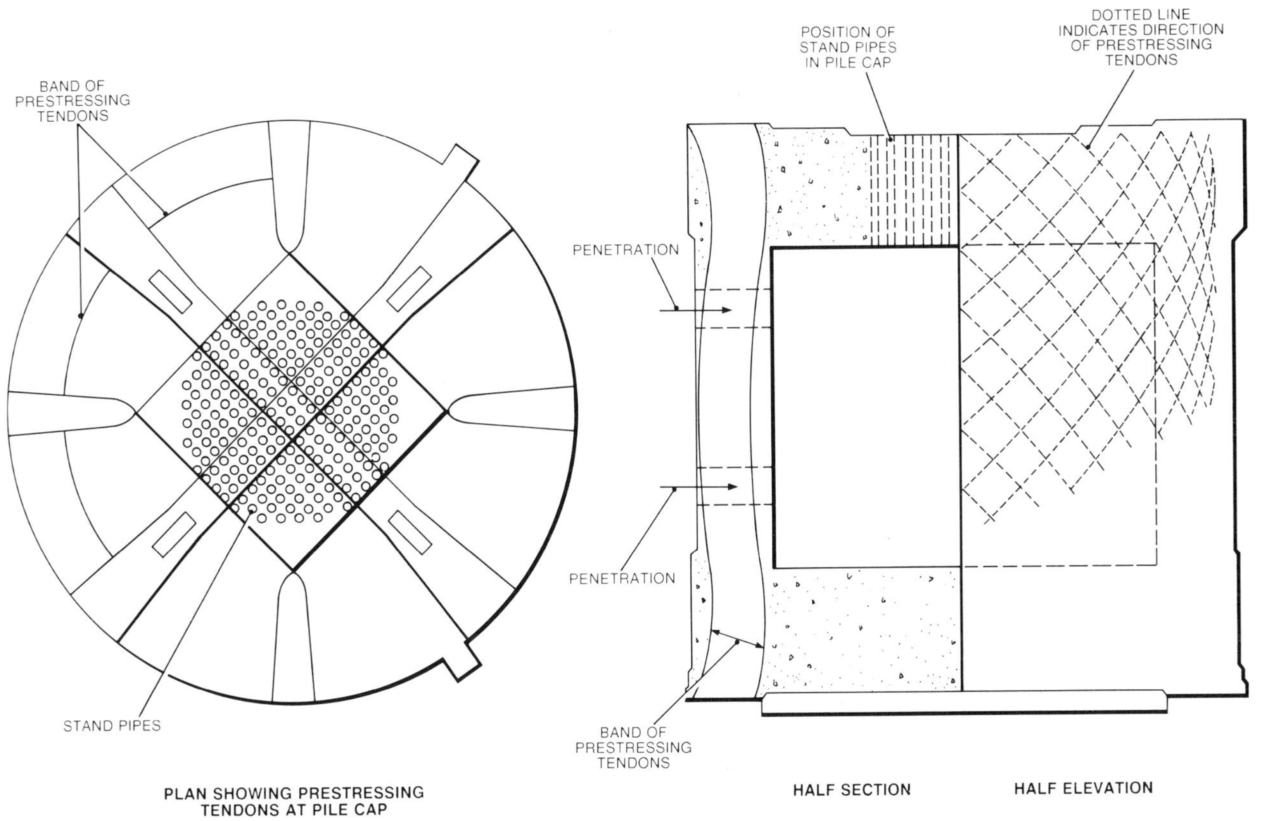

FIG. 3.58 Details of cylindrical prestressed concrete pressure vessel

be subject to cracking with gas pressures up to 1.65 times the working pressure. The precompression of the concrete due to the stressing of the cables results in compressive stresses of between 1724 kN/m^2 and 2068 kN/m^2 in the concrete, and the loading of the tendons by the jacks induces stresses up to 965 MN/m^2 in the steel. The use of the spiral layout of the cables in the walls of the cylinder ensures a uniform prestress therein, the basic angle of the spiral establishing the relationship of vertical to hoop prestressing. Alternating spiral layers are needed in both directions. Local tendon path deviations ensure there is no weakness at any of the penetrations in the walls for the entry of steam and feed pipes or control rods.

Figure 3.59 shows the arrangement of a spherical prestressed concrete reactor on which a large proportion of the prestressing tendons are fixed externally.

Over the years, the methods of prestressing pressure vessels have varied considerably from a simple three-dimensional tendon arrangement at Dungeness *B* to the use of a vertical and a wire winding system at Hartlepool and Heysham *1*. At several of the more recent stations a sophisticated variation of the spiral layout of tendons in the walls has been used that eliminates the need for end cap tendons. The spiral system has been used successfully on 50% of the UK pressure vessels. The principal advantage of this arrangement is that the anchorages of the prestressing tendons are readily accessible for load checking and replacement, if necessary, when the reactors are operating.

The arrangement of the stressing gallery for tendons in the wall of an AGR pressure vessel is shown in Fig 3.60.

20.2 Reactor buildings

Layouts of the reactor buildings and turbine halls for stations using prestressed concrete cylindrical and spherical reactors are shown in Figs 3.61 and 3.62 respectively. Generally, turbine halls are very similar to those used for coal and oil-fired stations. Although the first reactor buildings were constructed largely from concrete, structural steelwork being only used on the framing for the structures enclosing the boilers and the roof over the charge face, the more recent ones have used structural steelwork on an increasing scale for economic reasons. Figure 3.63 shows the reactor building at Oldbury power station.

The most recent stations at Heysham *2* and Torness have the turbines and reactors contained in one integrated building. The form of the main structures and the principal dimensions are shown in Figs 3.64 and 3.65. They are generally framed in reinforced concrete, with the exception of the charge hall and turbine hall

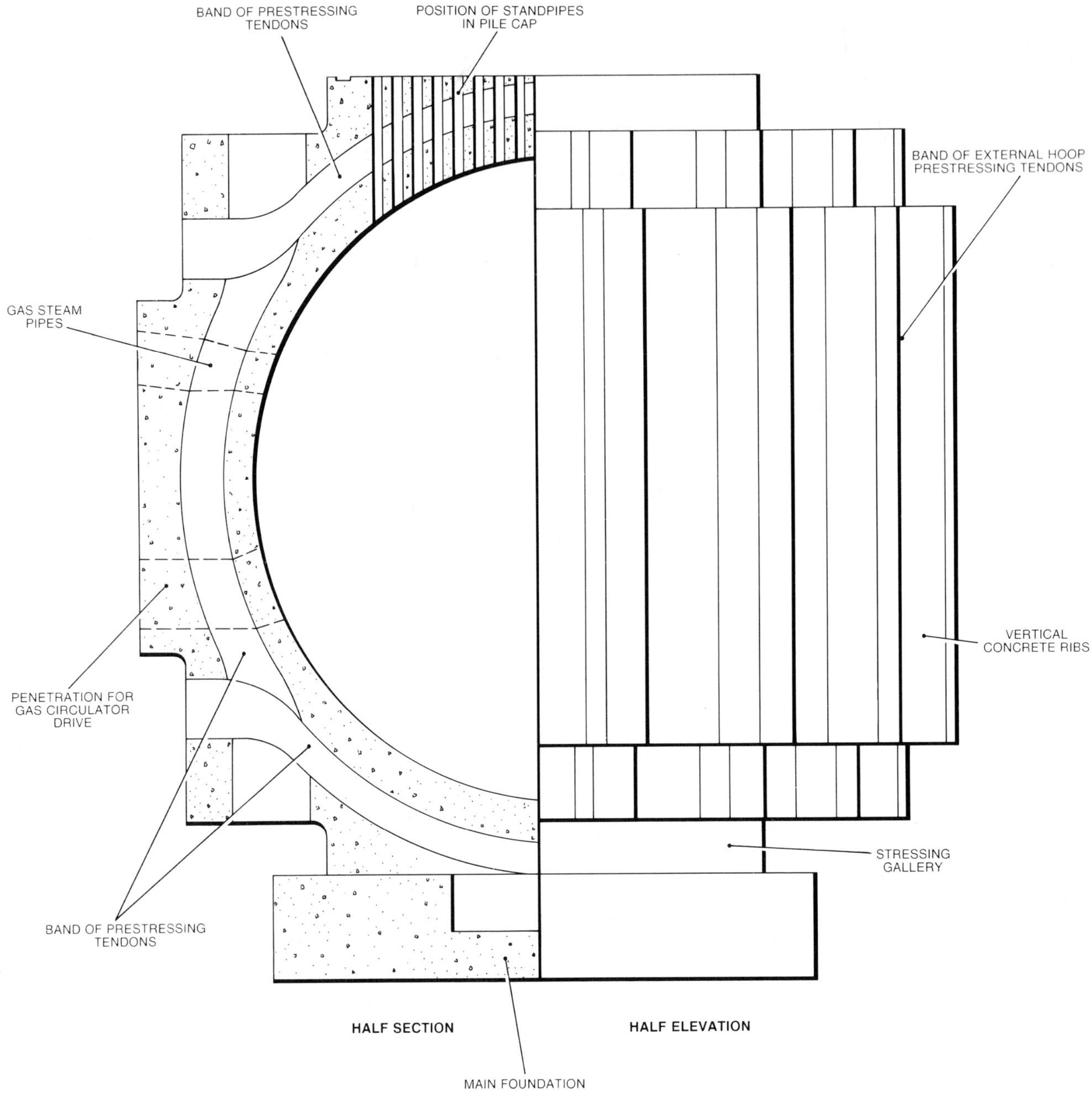

FIG. 3.59 Section through spherical prestressed concrete

above crane rail level, which are framed in structural steel.

The nuclear island buildings are all founded directly on rock, with the PCPVs, central control building and fuel handling building foundations being separate rafts, with the remainder of the structure founded on strip or isolated foundations.

21 Ancillary buildings

21.1 General

In the past there have been many views held on the desirability of including ancillary buildings as part of the main structure, grouping them in annexes to the main buildings or locating them away from the main building. The present trend is for buildings, especially those occupied by personnel for long periods, to be located away from the main buildings. This minimises noise nuisance and also enables the architect to obtain more natural lighting in the buildings.

One big disadvantage with providing accommodation for ancillary functions in the main buildings or annexes is that the steelwork must be designed in conjunction with the main building steelwork and this takes place at a very early stage in the project. Due to their size, the time of construction for ancillary buildings is less than

FIG. 3.60 Lower stressing gallery for AGR pressure vessel (see also colour photograph between pp 242 and pp 243)

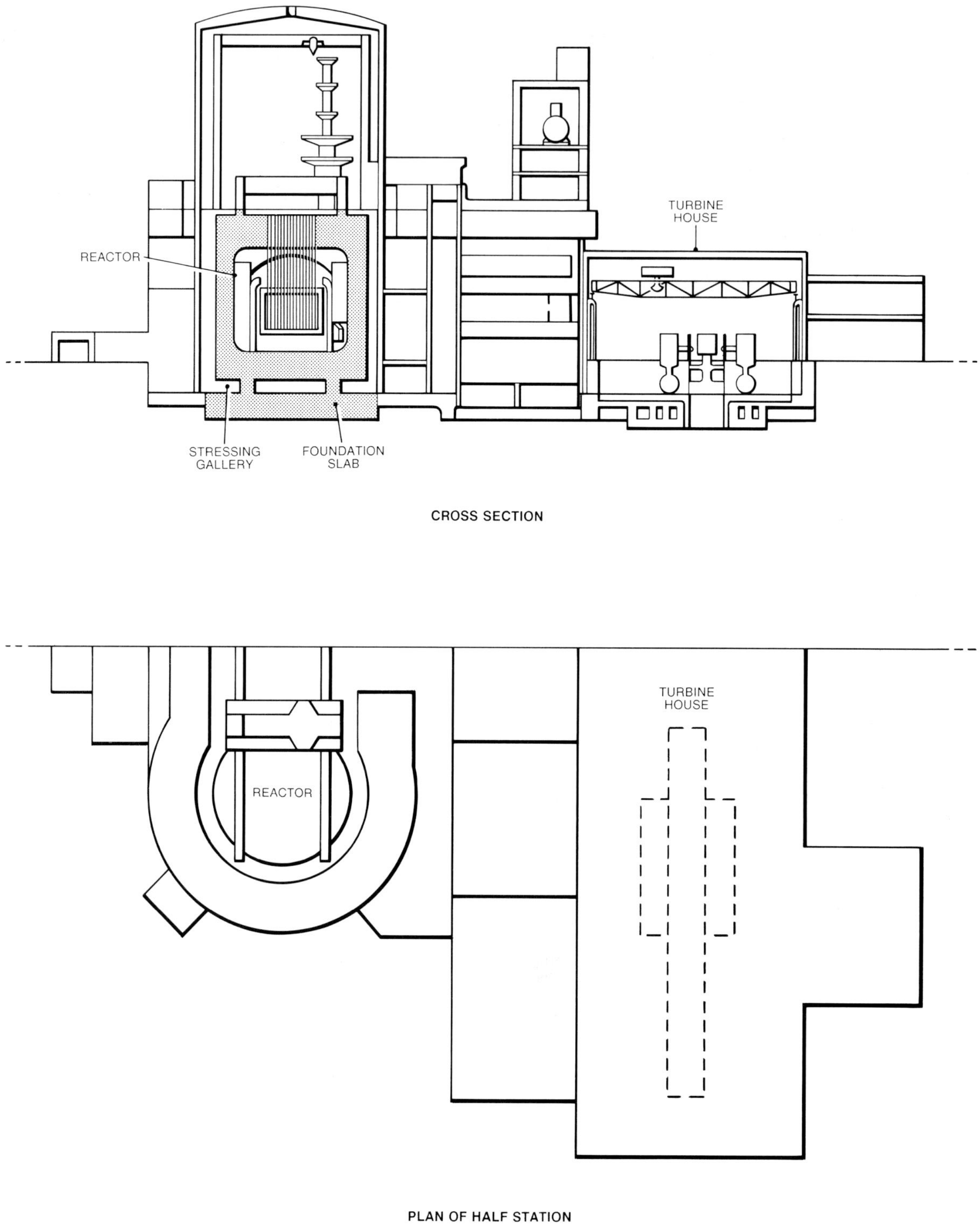

FIG. 3.61 Layout of buildings for cylindrical concrete pressure vessel

that for the main buildings and if separate buildings are used, commencement can be delayed, allowing further time for design development. In many cases this results in more economic layouts as early assessments of requirements for annexe layouts may result in excessive accommodation being provided.

21.2 Control room building

Modern control rooms house the control equipment for the complete operation of the power plant and the transmission switchgear. Some new stations have the control room centrally located in the annexe on the front of the turbine hall, others have it located at the

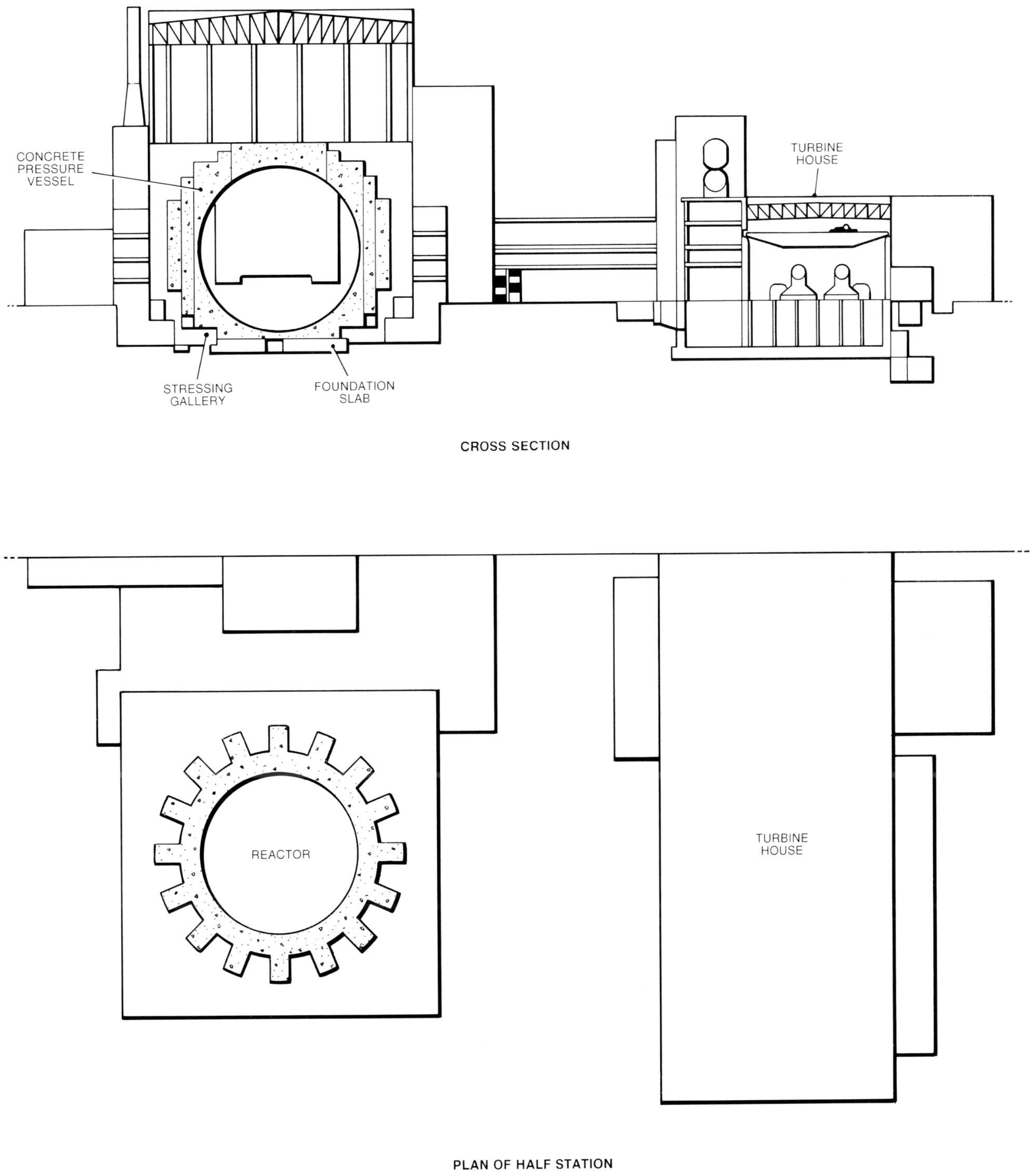

FIG. 3.62 Layout of building for spherical concrete pressure vessel

end of the turbine hall nearest to the first set to be commissioned. Others have it in a separate building either centrally located in front of or at the end of the turbine hall. Moving the control room building away from the turbine hall has the effect of reducing noise transmitted from the main buildings and a favoured position is between the turbine hall and switching compound or switch house. Another advantage of a separate control room building is that work can proceed at a very early stage and so give more time for the installation of the complex control equipment.

Although the control room building is much smaller than the main building, the programme must allow for its completion in time for commissioning the first unit. Control room buildings usually have a reinforced concrete or steel frame with brickwork, precast

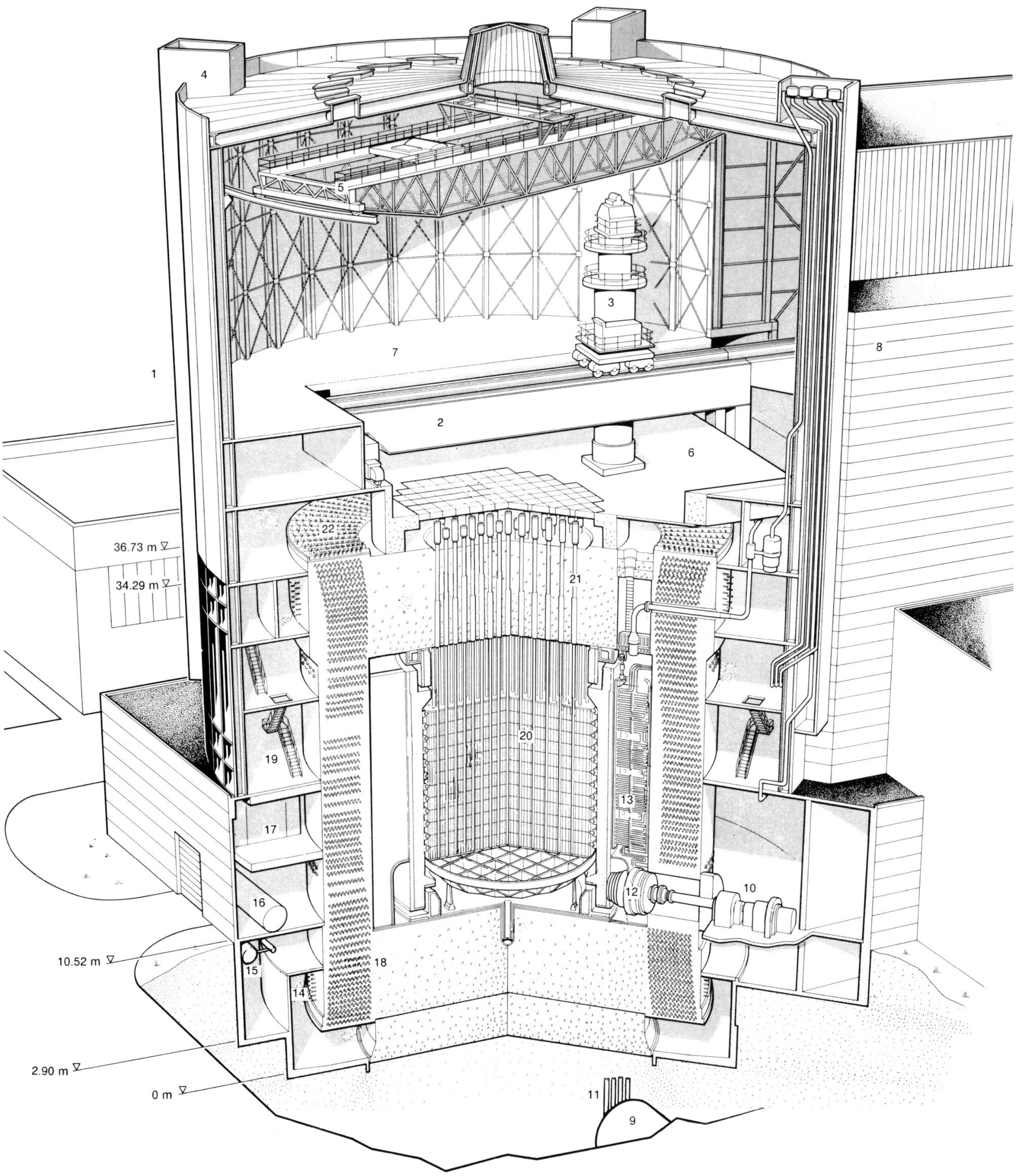

KEY

1 REACTOR 1	12 GAS CIRCULATOR
2 CHARGE MACHINE GANTRY	13 BOILER
3 CHARGE MACHINE (FUELLING MACHINE)	14 LOWER CABLE STRESSING GALLERIES
4 RELIEF VALVE OUTLETS	15 PRESSURE VESSEL COOLANT PUMPS
5 25 t CHARGE HALL CRANE	16 PRESSURE VESSEL COOLANT COOLERS
6 OPERATING FLOOR	17 FLUX MEASURING ROOM
7 CHARGE HALL (8th FLOOR)	18 CONCRETE PRESSURE VESSEL
8 CHARGE MACHINE CONTROL ROOM (10th FLOOR)	19 BOILER ANNEXE
9 TUNNEL	20 REACTOR CORE
10 GAS CIRCULATOR DRIVE AND AUXILIARIES	21 CHARGE AND CONTROL / BCD STANDPIPES
11 STEAM AND FEED WATER PIPES TO TURBINE HALL	22 UPPER CABLE STRESSING GALLERIES

Fig. 3.63 Reactor building at Oldbury magnox power station

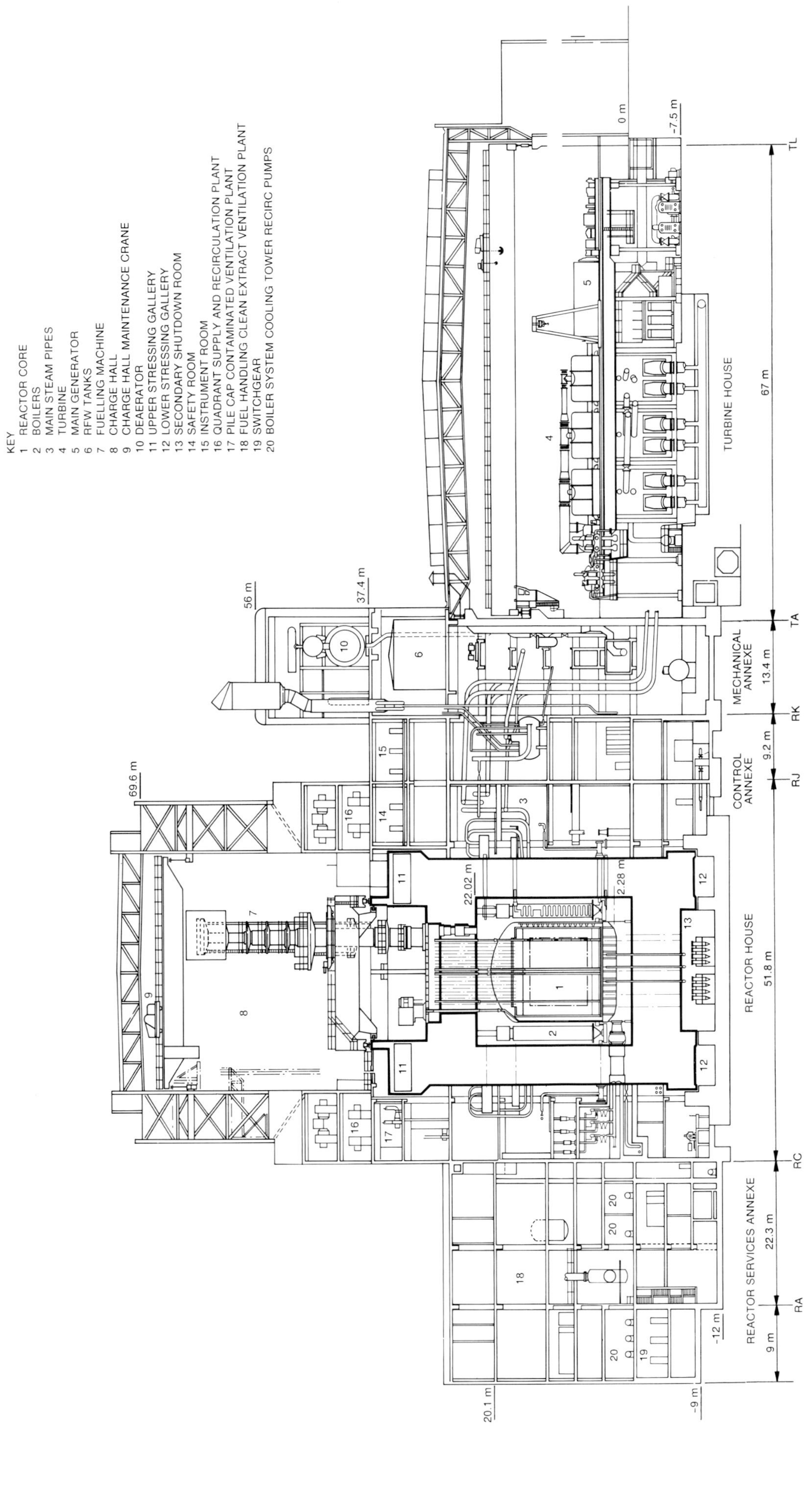

FIG. 3.64 Section through station — Heysham 2

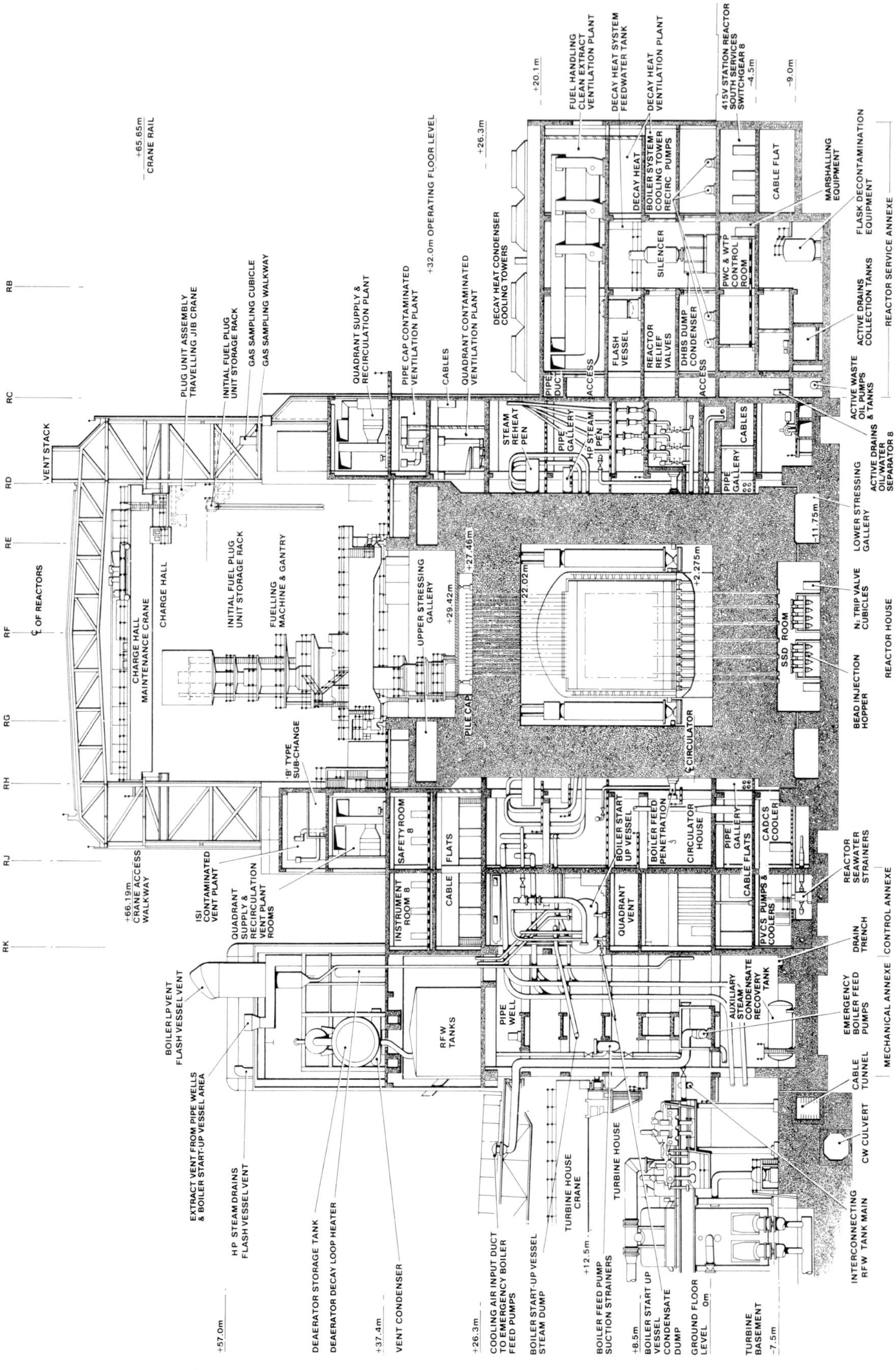

FIG. 3.65 Section through reactor — Heysham 2

concrete or similar cladding. They are usually several storeys in height, cable basements being provided under the control room and switchgear rooms. Due to the large number of cable openings, trenches, plinths and similar arrangements, floors are usually of in-situ concrete construction as is the roof of the building.

The control room building has finishings of a fairly high standard in the entrance hall, staircases and the control room itself. Terrazzo or similar floors are frequently used in entrance halls and staircases and rubber or similar noise-absorbent floor finishes in the control room. Panelling or similar finishes are used on the walls in these rooms, and suspended ceilings using acoustic tiles or other removable panels are used to conceal ventilating ducts and other services located at high level. Cable basements usually have granolithic floors, fair faced brick or concrete walls and concrete ceilings.

To keep the air free from dust and to maintain a low humidity a forced or balanced plenum air conditioning system is installed and doors are located in such a manner that air locks are provided to minimise heat loss. The necessity for accurate reading of instruments on the panels means that the lighting of the control room including the arrangement of windows requires special treatment. Diffused ceiling lighting usually gives good results.

21.3 Gas turbine house

Special features to be considered during the design of this building are the large filters and ducts for combustion air to the sets and cooling air to and from the electrical generators and oil coolers, built onto the structural foundations. Preferably the gas turbine house should be independent of the main buildings and located in an area away from offices and similar buildings because of the noise factor. A steel or reinforced concrete frame is normally used as provision must be made for an overhead crane of about 15 t capacity. To give good sound insulation 340 mm or 460 mm brick or similar construction walls are used together with a precast or in-situ concrete roof. Natural lighting is best obtained by the use of high-level windows or preferably by roof lighting.

The ventilation of the gas turbine houses is usually by means of louvres and as the turbines operate for comparatively short periods and are not manned, the louvres can be electrically operated with control thermostats or similar equipment.

The turbine hall, filter room and other ancillary plant rooms usually have a granolithic floor finish with fair faced brick walls and concrete ceilings. The control rooms have plastered walls and ceilings together with thermoplastic tile or similar floor finishes. Heating of the control room is normally by means of electric tubular or similar heaters.

21.4 CW pumphouse

The CW pumps on some power stations are located in open pits, unprotected from the weather. If outdoor equipment is not used they are provided with a low superstructure.

Other arrangements of pumphouses require superstructures up to 15.24 m high, these having structural steel or reinforced concrete frames which provide support for overhead cranes and carry the walls and roofs. Lightweight metal deck roofing and lightweight metal wall panels are the usual forms of cladding.

Circular pumphouses have been constructed for hydraulic reasons and because they have certain structural advantages for the foundation design. They usually have a radial crane for handling pump components.

To prevent mechanical damage to the walls at low level it is usual to have a plinth of brick or precast concrete panels 3 m to 3.7 m high and this is normally finished with a fair face internally. The cladding is not usually lined and large areas of glazing are provided. Ventilation is by manually controlled opening windows. Heating is not normally provided. Floors generally have a granolithic finish.

Should it be necessary to provide accommodation for personnel or control panels, then brick or similar materials are used for the walls to partition off the section required, the ceiling being of wood wool slabs or similar insulating materials. Finishes are used to suit the requirements of the occupied area with heating and ventilation being provided accordingly.

21.5 Coal, dust and ash plant buildings

These buildings, apart from the amenities for personnel operating the plant, usually consist of steel frames cladded in a similar manner to CW pumphouses. The cladding improves the appearance of the structures and acts as a weatherproof covering to protect plant and personnel. In addition to supporting the cladding the steel frame is used to support pipes, ductwork and plant items. It also supports cranes and lifting beams necessary for plant repair and maintenance.

Some of the smaller buildings, such as ash pumphouses and the superstructures of transfer houses located mainly below ground level which do not warrant a steel frame, are constructed from load bearing brickwork with in-situ or precast concrete roofs. These buildings are not normally heated and ventilation is by means of opening windows. Provision for personnel, control panels or similar equipment is also made in a similar manner to that used for CW pumphouses.

21.6 Workshops and stores

The main workshops and stores are usually steel framed buildings with insulated lightweight cladding for walls and lightweight metal deck roofs. Overhead travelling cranes are provided in the main workshops and the heavy stores. Natural lighting is normally provided by roof lights and glazing in the walls. It is normal for walls to have a plinth 3 m to 3.7 m high of brickwork or precast concrete panels to help resist mechanical damage. A typical layout of stores and workshops is shown in Fig 3.66.

Ventilation is usually by manually operated opening windows or roof ventilators in the workshops and stores. Hot water or steam unit heaters at high level are the most satisfactory forms of heating for these buildings. Heating of the stores office is usually done in the same manner as that used for the main office block. Walls in the workshops are usually fair faced brickwork to the plinth, and floors have a granolithic finish except for the machine shop and other special areas where end grain wood block is sometimes used.

Special attention must be paid to garages, paint shops, joiners' shops, blacksmiths' shops and other similar locations where the fire risk is high. One satisfactory method of dealing with this problem is to locate them in brick buildings with concrete roofs, separate from, or as annexes to the workshops. Some form of extract ventilation is provided in the joiners' and paint shops, especially if woodworking machinery is used and spray painting is carried out.

21.7 Offices, welfare blocks, laboratories and similar buildings

These may be separate buildings or they may be combined in a central block. If single-storey construction is used then the buildings are often of load bearing wall construction. If they consist of more than one storey then a frame of steel or reinforced concrete is used.

Heating of these buildings is usually by means of hot water or steam and there are many attractive heating

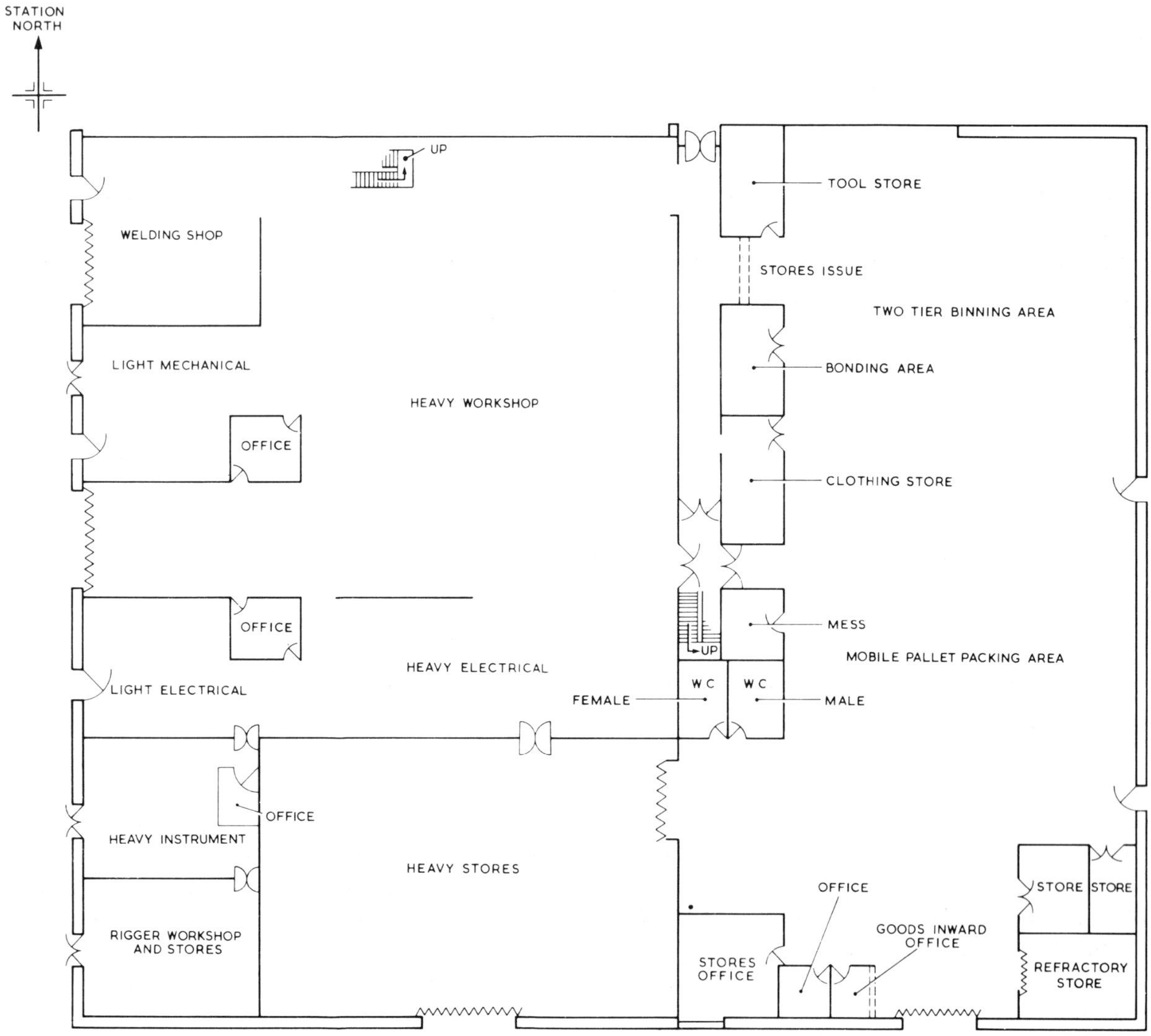

Fig. 3.66 Arrangement of stores and workshops

units now used in place of the conventional radiators. Most heaters consist of a bank of finned tubes located behind a pressed metal cover with inlet and outlet grilles. Circulation of air through this type of heater may be controlled by a simple damper type of operating gear or, for locations where a greater flexibility of control is required, fans can be incorporated to speed the air flow over the heater tubes thus increasing the rate of heat transfer.

Ventilation to individual rooms is normally by means of opening windows except in special locations in the laboratory. In the canteen kitchen extract fans are used to create an induced plenum system to prevent the spread of air contaminated by odours. Special inlets may be provided in the kitchen area to provide the comparatively large quantity of air necessary to give the required air changes. If these inlets cannot be located in positions to ensure reasonably clean air, then some kind of filter should be incorporated in the inlet duct.

The standard of finishings in these buildings is usually fairly high, with plastered walls and ceilings, and thermoplastic or wood block floors. In offices it is usual to complete floor finishings between outer walls before partitions are erected. As these are normally supplied in sections, they can be easily taken down and re-erected to a different layout if required.

Welfare amenities are provided as necessary throughout the station, separate facilities being provided for coal gangs and other personnel working in isolated locations. Toilet and washing facilities are also provided at central points in the station buildings but the main locker and ablution facilities are normally provided in a central block adjacent to the gate house and clocking arrangements. Quarry tiles are the most commonly used floor finish in the welfare blocks, walls being plastered or tiled as necessary.

A typical layout of offices, welfare, laboratory and canteen facilities is shown in Fig 3.67.

There are many other buildings located in different parts of the station which require special treatment. Amongst these are the water treatment plant house where special consideration must be given to the floor and wall finishes to resist acids and alkalis.

The main switching station consists of large lattice frames cladded with lightweight wall cladding and roofing. A brick plinth wall is normally provided and the floor finish is usually bitumen macadam. The steelwork is usually designed to support electrical equipment in addition to the wind, snow and dead load of the structure.

Heating and ventilation is not normally provided in these buildings, and a small structure is usually constructed from load bearing brickwork with a concrete roof to house the office, stores and welfare facilities for the switching station. A typical arrangement of a 132 kV air blast switch house is shown in Fig 3.68.

22 Chimneys, cooling towers and precipitators

22.1 Chimneys

Although single flue chimneys have been used on 2000 MW power stations the present trends are to use multi-flue stacks for all future stations of this type. For power stations with four boilers, four flues, one to each boiler, are contained in a single concrete cylinder which acts as a windshield. Figure 3.69 shows a typical arrangement of this type of chimney. Although this arrangement is suitable for 2000 MW stations, other arrangements of the flues in the windshield including discharging the flue gases from two boilers into one flue may be necessary for larger stations. The outer cylinder of concrete is constructed to withstand the greater part of the wind forces and its own dead load. The flues are built off the lowest intermediate floor or foundation slab and an acid resisting brickwork or other suitable lining is provided. There is no structural connection between the flues and the windshield, the flues being permitted to expand or contract freely according to temperature variations. The space between the windshield and the flues is accessible at all times, including the period when the boilers are operating. Access to the various levels is provided by a concrete or steel staircase and in some chimneys of this type, a lift is also provided.

Other types of multi-flue chimneys have windshields supporting floors at 9 m to 12 m intervals, the flues consisting of brick cylinders with external insulation built off each of these floors to form continuous gas ducts.

On the first type of chimney the windshield and flues are of reinforced concrete not normally exceeding 305 mm thick, apart from the lower section below the floor supporting the concrete flues. Brick linings to the flues are normally only one half brick thick and are self-supporting for heights up to 12 m. Concrete corbels, therefore, are cast on the inner face of the flue to carry each successive lift of brickwork. The use of corbels at these intervals has the advantage that in addition to restricting the height of brick lining sections it also provides joints for accommodating differential expansion between the lining and chimney, the joint being packed with glass wool and covered with a lead flashing. Brickwork is built from best quality engineering bricks jointed in acid resisting compound with a cavity between the concrete and brickwork. On single-flue chimneys the cavity is filled with vermiculite or glass wool insulation or is ventilated by holes in the concrete, and on multi-flue chimneys the cavities are either sealed or filled with insulation as for single flue chimneys.

In the second type of chimney the construction of the intermediate floors governs the construction programme and therefore it is advantageous to reduce the number of floors. With the development of tongued

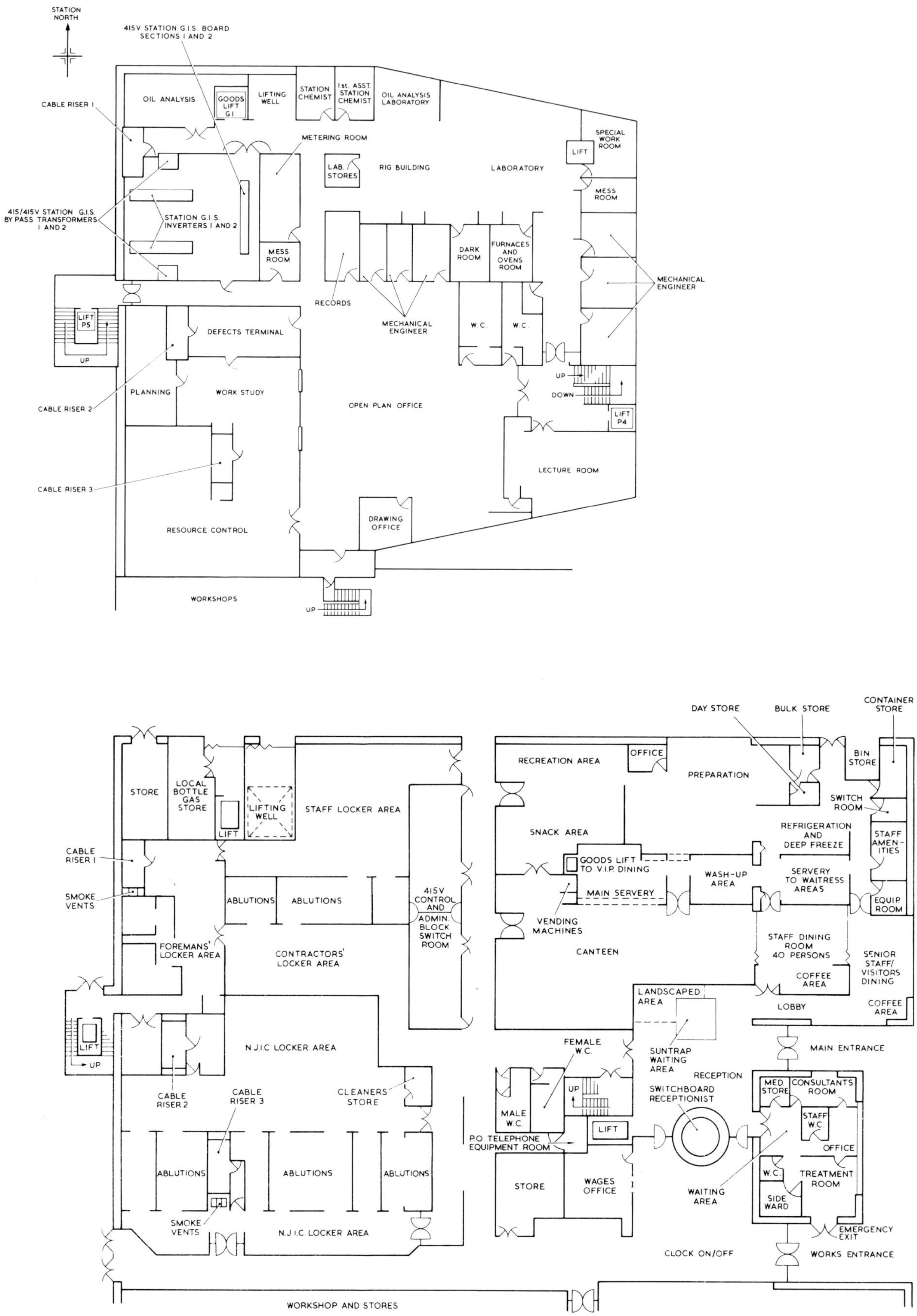

FIG. 3.67 Arrangement of offices, laboratories and welfare facilities

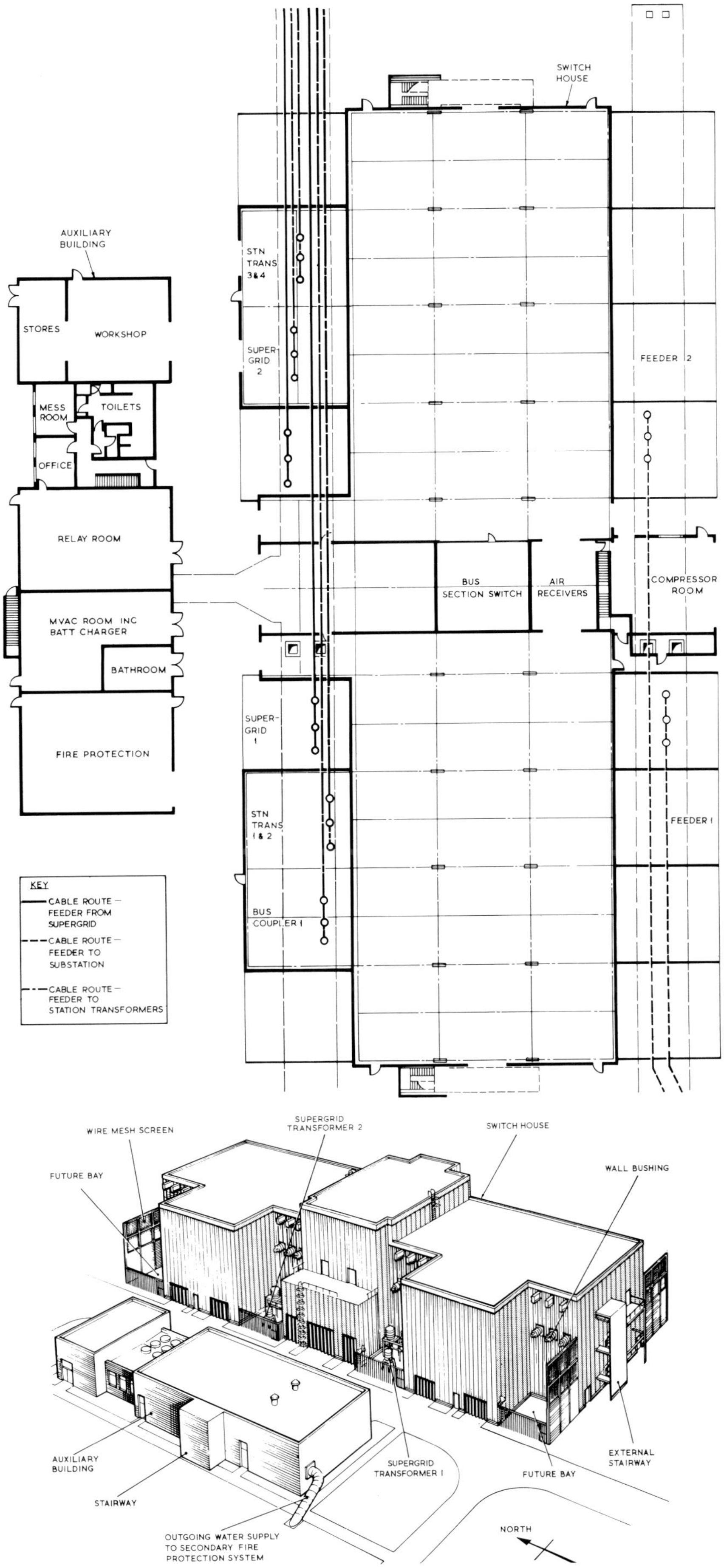

FIG. 3.68 132 kV substation

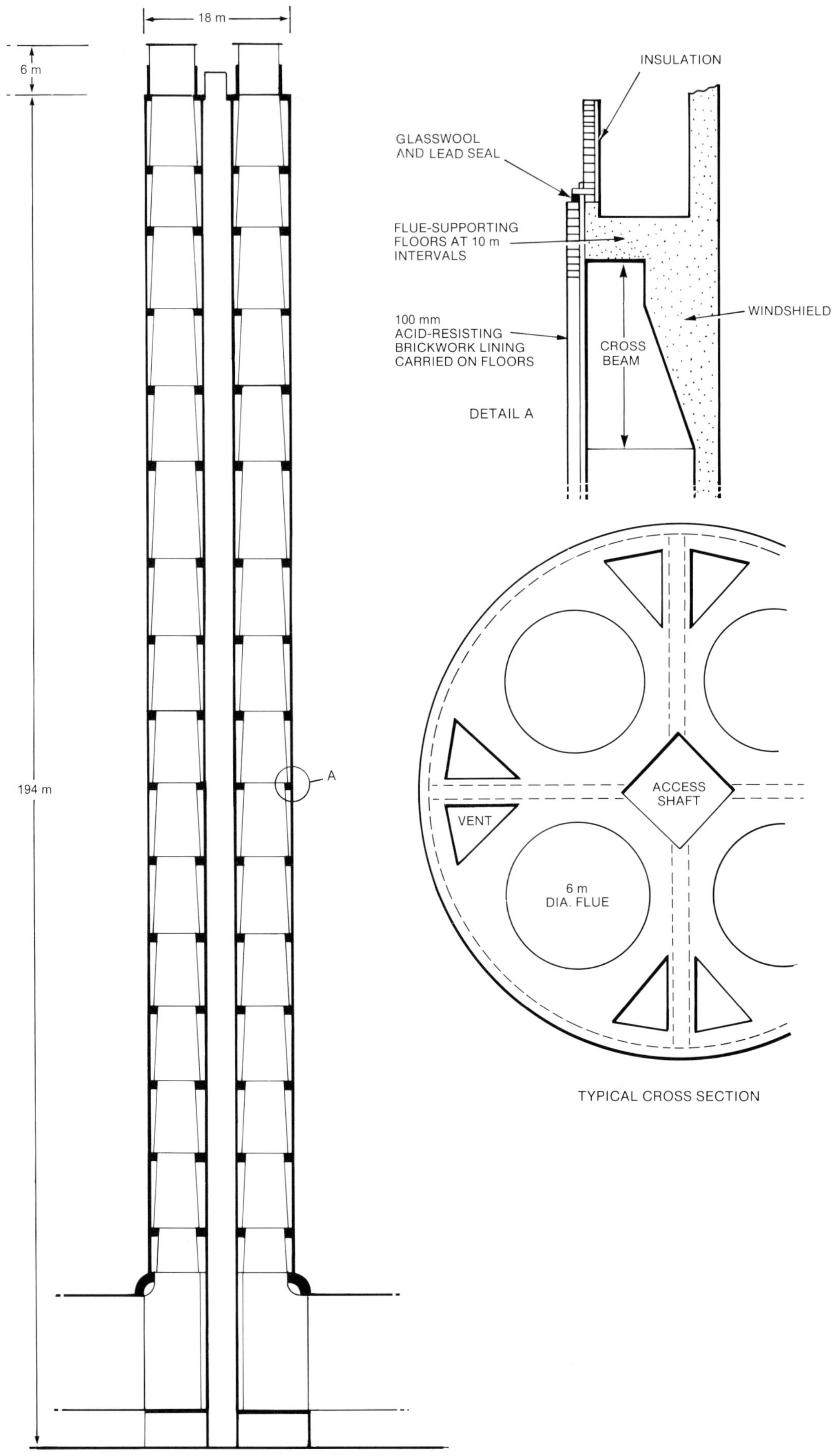

FIG. 3.69 Multi-flue chimney

and grooved blockwork in lieu of plain no-frog bricks it is feasible to space these floors at intervals up to 30 m.

Protection against acid attack can be given to the concrete at the top of chimneys by the use of engineering brickwork or protective coatings applied to the outer surface of the concrete windshield and flues. Quarry or similar tiles are used on the flat surface on the top of the windshield and cast iron segments are used on the top of the flues.

If flue gas is to be prevented from penetrating the chimney lining and entering the cavity between the lining and the concrete, a negative pressure must be maintained inside the chimney. The difference in pressure caused by the relative densities of the gases inside the chimney and the atmosphere must exceed the losses caused by the bend at the flue entry, the head loss due to friction inside the chimney and the head necessary to give the gas the required exit velocity of about 25 m/s.

A negative pressure inside the flue is more important in chimneys where flues are built in brickwork off intermediate floors, because a gas leakage in a chimney of this type means that gas would enter the occupied zone inside the windshield.

The shafts of single-flue chimneys and the windshields of multi-flue chimneys must be designed to withstand wind and dead loads and temperature stresses. Building Code of Practice CP3: Chapter V: Part 2 [16] requires that structures whose greatest lateral or vertical dimension exceeds 50 m shall be designed for a 15 second gust wind speed, but it is prudent to apply a factor to allow for dynamic effects in the preliminary design. The basic design, as a cantilever resisting overturning under wind forces considered as static loading, may be based on any of several well documented procedures. However, windshields have aspect ratios (i.e., height/mean diameter) in the range 10 to 12, and it is necessary (particularly for a windshield enclosing free-standing shafts) to investigate the ovalling stresses caused by the varying pressure distribution around the windshield, which result in positive and negative bending moments in the horizontal plane. Generally, these two aspects of design are considered separately and this is probably adequate for a ratio of mean diameter/shell thickness up to 50.

The distribution of pressure around the windshield has in the past been based on wind tunnel measurements at values of Reynolds number somewhat lower than those that actually occur, but the CEGB has carried out full-scale measurements to determine a realistic pressure distribution. Less is known about the internal pressure on the windshield; the presence of ventilation louvres at top and bottom of the windshield will cause the internal pressure to vary between them.

In designing the floors inside the windshield, their effect as stiffening diaphragms should be considered, otherwise the windshield could be of uneconomic thickness. The floor design must also include areas of open-mesh flooring to allow sufficient upflow of air to cool the interspace, in which the temperature should not normally exceed 38°C.

Although it is usual to provide an expansion gap between the floors and free-standing concrete shafts in a windshield, the floors may be brought into contact with the shafts and load transferred laterally owing to horizontal deflection of the windshield. Hence the shafts must be designed to withstand a proportion of the total wind load based on the relative stiffness of shafts and windshield.

The design of the windshield is based on an elastic analysis for a 15 second gust wind speed. The sections should be checked using a load factor analysis for overturning moment resulting from a wind speed of 1.5 times the design wind speed.

As previously stated, it is desirable in chimney design to apply a factor to the wind forces which will adequately allow for dynamic effects. This factor for a single-flue chimney is related to the natural frequency of the chimney, but for multi-flue chimneys a full investigation is required. Excessive oscillations may occur in steel chimneys owing to vortex shedding and buffeting and, whilst no significant vibration has been noticed in concrete chimneys, it is important to determine the conditions under which such vibrations could occur. In the last decade, the effect of dynamic forces has been studied in greater detail and guidelines for design purposes are now incorporated in a 'Model Code for Concrete Chimneys' [31] prepared by Comité International des Cheminées Industrielles (CICIND). In addition the latest edition of DIN 1056 'Free Standing Chimneys' [32] incorporates requirements for the consideration of the effects of oscillation, considered in the form of a static load equivalent.

Temperature stresses have been traditionally calculated on the temperature differential which exists across the flue walls and which causes tensile strain on the cooler face. However, the presence of long vertical cracks in several tall chimneys built for the CEGB since 1960 suggests that temperature stresses have been underestimated and that an empirical approach based on experience would provide better answers.

In accordance with Civil Aviation Authority requirements, chimneys are provided with aircraft warning lights and the usual arrangement consists of lights located at the top of the chimney and at 50 m vertical intervals. Three fittings at 120° positions or four fittings at 90° positions depending upon the arrangement of flues are provided at each level. In the past, fittings have each had four tungsten bulbs with individual red shades, but a recently-developed aircraft warning beacon has a discharge lamp in a glass fibre casing with a vertical transparent plastic front. This latter fitting has an anticipated lamp life several times greater than that for a tungsten fitting, and all future chimneys will be equipped with discharge fittings. On multi-flue chimneys, the lights are usually fixed to doors in the outer face of the windshield in such a manner that when the door opens, the light is brought into the chimney where maintenance can be easily carried out. The floors provided inside the windshield are usually located at levels which coincide with the aircraft warning light doors. On chimneys without internal access, i.e., single-flue chimneys, lamp replacement and maintenance must be carried out by steeplejacks, and because of this duplication of fittings is required.

A lightning protection system is essential on such high structures as chimneys. Air terminals are provided on the top of the windshield and flues, and coronary bands are provided at various levels to intercept strikes below the top of the chimney. If a strike occurs, the down conductors are at a high potential and to avoid the risk of a flashover of other metal structures inside the chimney all reinforcement and steelwork are tied in to the down conductors to ensure that no difference of potential occurs.

BS6651 — Code of Practice for Protection of Structures Against Lightning [33] — permits the use of steel reinforcement to concrete as down conductors, provided the reinforcement cage is adequately earthed and tests on completion show that there is adequate continuity. Cast iron cappings, handrails or other air terminations must be bonded to the reinforcement.

Although access sockets in which steeplejacks may screw their ladder fixing hooks are provided on the outer face of single-flue chimneys, sockets are not provided on multi-flue chimneys.

22.2 Cooling towers

If a piled foundation is used for a cooling tower (Fig 3.70), the column loads are transmitted direct to the pile caps and the pond wall serves only to retain water. However, if ground conditions permit, a strip footing is provided round the periphery of the pond and the wall provides a rib thus forming an inverted T-beam.

The design of the hyperboloidal shell is based upon BS4485 Part 4 [34], but as towers increase in size (165 m), the dynamic response of the tower shell causing an enhancement of the stress pattern in the shell has been studied in both model and full scale tests and revised design requirements form the basis of a new draft of BS4485. Within the tower from just above the open column area down to pond level is the 'cascading' structure. The water from the condensers is pumped through the CW culverts up to the top of this structure or 'stack'. At this level it enters a large box culvert running diametrically across the top of the stack and into pipes connected at intervals along both sides of this culvert, which feed the sprayers arranged over the top of the stack. The stack is constructed independently of the tower shell so as not to throw any load upon it in the event of differential settlement, and is usually constructed from precast concrete columns supporting perforated precast concrete beams. Laced through the perforations of these beams are triangular section timber fillets at varying levels within the cascade structure to ensure a diverse path for the water falling from the spray nozzles. The damp warm conditions in which the timber is located make it very susceptible to wet rot attack. A high degree of protection is given by subjecting the timber to a pressure impregnation process of salt solution using mercury, copper or arsenical salts. Asbestos cement sheets are also used in cooling stacks instead of timber and in view of their anticipated longer life their use may be economic. Before deciding the type of stack some assessment of durability must be made after analysing the cooling water as this may affect the life and economics of a particular form of construction.

Since the early 1980s a new type of packing comprising closely spaced corrugated PVC sheeting formed into modules for ease of handling has begun to replace these long established forms of packing. Its use shows a marked increase in efficiency and its light weight allows the use of a pack support structure with wider spaced columns (9 m) thus improving the air flow within the tower.

This type of packing has shown a tendency in certain areas of the UK to biofouling, i.e., the build up of algae type growths in the packing matrix due to the warm humid conditions. Massive load increases can result, exacerbated by silt deposits and sometimes by carbonate deposits in hard water areas. Hence the predicted build-up of loading must be allowed for in the structural design of the pack support structure.

Eliminators are usually constructed in a similar manner to the cooling stack but at a higher level.

Lightning protection is not normally provided on cooling towers as the shape of the tower and the reinforcement contained in the concrete are considered adequate protection.

22.3 Precipitators

Precipitators (Fig 3.71) are essentially civil engineering work, although they are designed by the boiler contractor and are included as part of the boiler contract. Occasionally steel precipitators are used but the common form of housing for the dust extraction plant is a concrete structure. The effects of high temperature on the concrete, the consequent differential thermal movement between hot and cold inner and outer faces of concrete walls, thermal expansion of the structure and the formation of acid condensates inside the precipitators are the major problems associated with precipitator design. Concrete which contains a high percentage of pulverised fuel ash is particularly suited to precipitator casing construction owing to its good resistance to acid and high temperature. Hair cracking of the finished concrete is also reduced owing to the slower hardening of this type of concrete. In some cases exterior insulation is employed to reduce condensation.

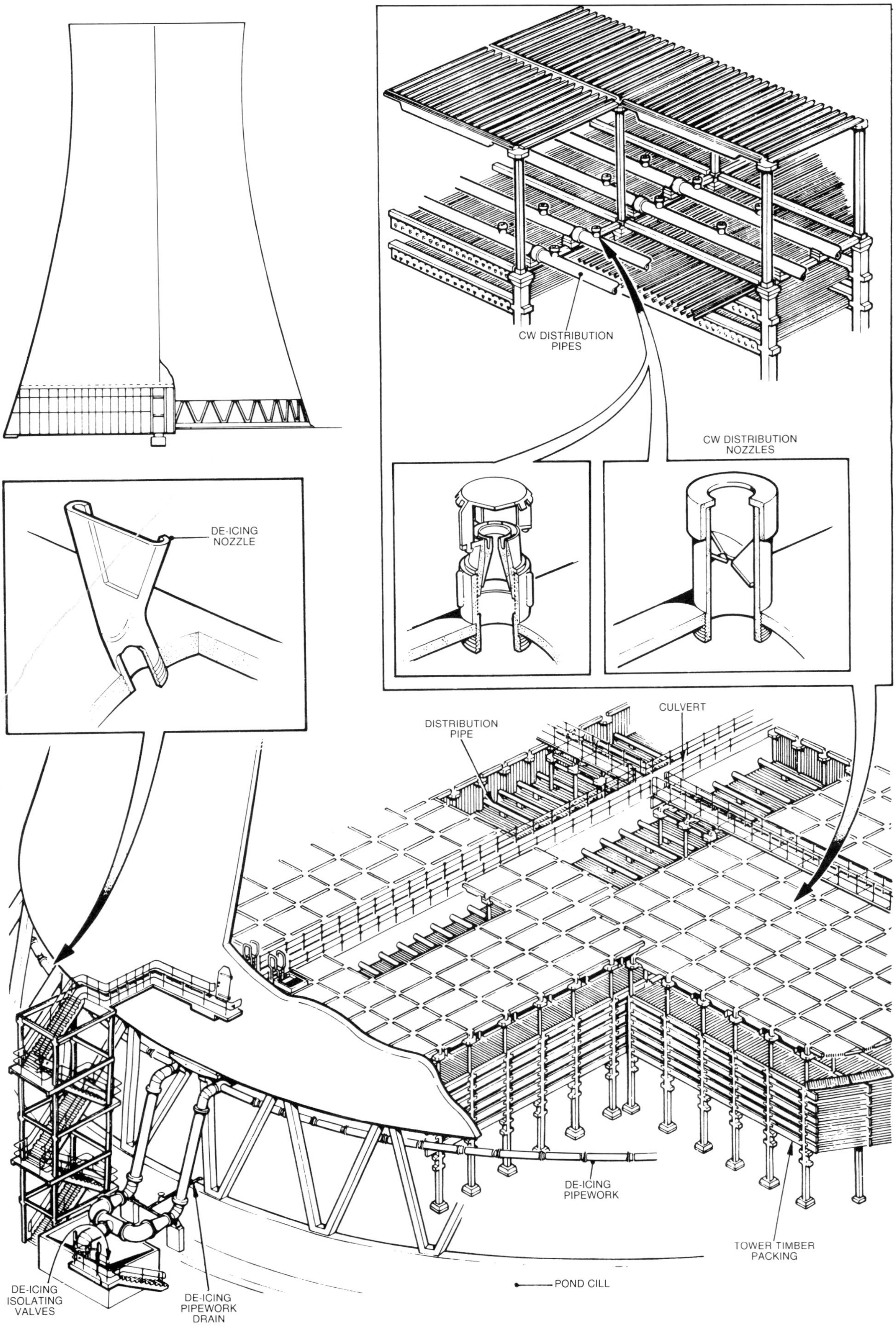

FIG. 3.70 Cooling tower construction

FIG. 3.71 Precipitators
(see also colour photograph between pp 242 and pp 243)

23 Architecture and landscape

23.1 General power station architecture

The CEGB, under the Electricity Act 1957, was required to pay attention to the appearance of new power stations, both in architectural concept and in matters of detail for its environmental amenity suitability. In so doing the executive architect would seek to exploit the engineering form of the plant and its supporting structures. The prime architectural objective would be to ensure the best possible appearance of the project, including buildings, structures and plant as seen from such viewpoints as are predominent in the public's perception of a power station, and to present a confident and consistent image as part of the CEGB's corporate design policy.

The executive architect plays a major role throughout the planning stages of the project, hopefully influencing both client and critic for the common good. In the implementation phase he is required to contribute to the design of both the plant-dominated structures and the buildings that are primarily occupied by people.

The landscape architect will study and assess the effect of the project upon the landscape. He will identify the landscape objectives and prepare a strategy plan showing the key landscape elements. He will design the various elements in his plan and will participate in their implementation by specialist contractors. Long term maintenance of the landscape plan differs from virtually all other power station maintenance commitments simply because the schemes mature and possibly wither, naturally.

23.2 Landscape considerations

Power generation is now largely a rural activity, with the burden of the visual and operational effects falling on the countryside. Although the cost of the necessary work is small by comparison with the overall project budget, the significance of environmental issues in the planning and consent procedure is large, and growing. The best power station design is the one for which consent can be obtained, and to achieve this economically in landscape terms, a range of expertise and activities is required. The proposed site must be visually analysed by the appointed landscape architect, who will seek to minimise adverse effects of the project. His skill will identify features to be emphasised or hidden, and his landscape scheme should pay due regard to the quality of the locality and its inhabitants.

Any existing woodlands both on and adjacent to site should be examined by a forester, and his recommendations incorporated into the landscape plan. Forest plantations may need to be converted from timber cropping to permanent woodland regimes if their screening value is important. Ecological and wildlife considerations must also be adequately reported on by a specialist, and provision made for the protection of important or unique features of the proposed site.

Early commissioning of environmental specialists is vital if basic site layout decisions are to reflect the broad context of the site. In addition, wide ranging consultation with local authorities and such bodies as the Royal Fine Art Commission is required. This may be perceived as increasing the time and effort required to reach a final solution, but a better and more acceptable scheme will result. Landscape work outside the site boundary should also be planned for where particularly sensitive views are identified.

23.3 Preparatory works

Preparatory landscape work is work to be undertaken early in the project in order to safeguard or enhance final results. This consists of the identification and adequate protection of existing features to be kept (trees and woods, watercourses, rare habitats) and the conservation of topsoil by the proper design of storage mounds. Topsoil stacks should be kept as shallow as possible in order to minimise degradation and loss of soil life. Stacks may stand for several years before re-use, and must be sown with grass and kept mown in order to reduce weed propagation.

Excavation and grading to achieve the desired optimal land form is the main part of the preliminary works stage. Storage or disposal of topsoil and subsoil should be integrated into the scheme design, for example, by the construction of earth mounds, banks or terraces to final profiles from surplus materials. Double handling later in the project should be avoided wherever possible. Where soil is to remain undisturbed, early tree planting should be considered in order to gain growth in the early years. Suitable sites are the main access roads and junctions, car parks, off-site planting, and the outer fringes of contractors' working and storage areas. Particular attention should be paid to buildings which are inhabited from the start of the project (such as hostels, construction offices, visitor reception facilities). This is partly because a good standard of external environment in such places is a counterbalance to harsh working environments elsewhere, and partly to demonstrate the standard which it is intended to achieve at the completion of the project. Although landscape provided around temporary buildings may be sacrificed at the end of the job, it may be fully justified for its humanising effects during the years it exists.

23.4 Landscape layout

Landscape work within the scope and ability of a power station construction can rarely be used to hide major structures. Hence the idea that the purpose of landscaping is to screen ugliness is true only in a very small degree. It is best perceived as a base setting from which the main buildings are seen to rise. Generally, power station buildings become simpler and bolder in outline and massing in their higher parts, whereas the scale of trees is fixed, and relates best to small or intermediate structures. Thus trees become ground cover planting in the scale of a modern power station. Station sites contain many small elements up to about 15 metres (vehicles, fences, small buildings, storage tanks) and at this level some screening by trees is both possible and usually desirable. Even then it will not be achieved until a station is well into its lifetime.

Landscape is both hard and soft, so paving, walls and other features must be given appropriate consideration. The context or climate of a development may require more hard finishes and less planting, resulting in higher initial cost but lower maintenance. Areas around the gatehouse, administration building and along the access road may justify more expensive finishes, the planting of larger trees and possibly additional water features in order to demonstrate a high standard of site operation.

Other elements which must be considered in the overall design are fences (type and position), lighting, road layout and detailing, direction signs, and the siting and operation of a permanent contractors' area. Sports facilities are likely to be required by staff on non-operational land, and may be incorporated into the original site landscape work.

The implementation of landscape contracts tends to be concentrated at the end of the project, as areas are completed or vacated by contractors. The reclamation of contractors' storage areas, often virtually derelict and with no topsoil, forms the largest element in the workload, and for it to be successful, specialist advice must be sought. To ensure that promises given publicly at the outset are met, it is vital to safeguard the budget for land reclamation until the end of the job.

Landscape is dynamic, and changes must be planned for during the scheme's development. Pioneer planting on poor ground may need to be replaced when soil conditions have improved.

Trees will require thinning and eventually replacing, so second generation planting should be included in the original designs. Provision for maintenance and remedial work must be made, both by allocation of finance, and by the employment of suitable staff. On coal-fired stations, ash disposal is an important activity having environmental impact, and comprehensive landscape schemes will be required in order to secure consent for such works. The final completion of an ash scheme can of course only be achieved at the end of the station's life.

Finally, since less than half of a power station site may be occupied by operational buildings, many opportunities exist for the beneficial use of the remainder. Woodland is a valuable feature in its own right for its effect in providing shelter and a wildlife habitat, but grassland, marsh or water bodies may also contribute to the local environment. In general, economic considerations militate against the provision of high maintenance, park type planting on any large scale. Grass mowing in large areas is acceptable but again the use of small manicured lawn areas should be restricted to focal points if used at all.

Figure 3.72 shows an example of a nuclear station layout showing elements of the landscape scheme and development of the land for various uses after construction.

24 Regulations

24.1 Government instruments

Amongst the many Government Regulations which apply to power station construction and operation are:

(a) The Factories Act, 1961 which applies to health, safety and welfare of the personnel working in the power station when commissioned, relates to:

The Building (Safety, Health and Welfare) Regulations, 1958.

The Construction (General Provisions) Regulations, 1961.

The Construction (Lifting Operations) Regulations, 1961.

(b) The Offices, Shops and Railway Premises Act, 1963 which lays down the minimum standards for office accommodation.

(c) The Health and Safety at Work Act, 1974.

24.2 Factories Act

Only the aspects of this legislation which relate to the layout of buildings are dealt with here.

The Factories Act, 1961, applies to power stations when commissioned, and clauses relating to the following topics must be borne in mind in the design and construction stages:

Safe means of access to place of work This covers items such as ladders, walkways, doorways, lifts, etc., and during the commissioning of a power station the Inspector of Factories will inspect these items. The Act requires adequate access and fencing in all working platforms and footways and the Inspector of Factories will state his requirements regarding the measures to be adopted. It would be very difficult to advise the Factory Inspector of the developments in design and to obtain his requirements at an early stage in the project. For this reason and also to obviate any tendency to remove the responsibility for implementing the Act from the employer to the inspector, inspections do not normally take place until construction is well advanced.

Means of escape in case of fire Sections of the Act lay down the provision which must be made for exit doors, their marking, warning systems, etc. The provision must be certified by the local Fire Authority as reasonable.

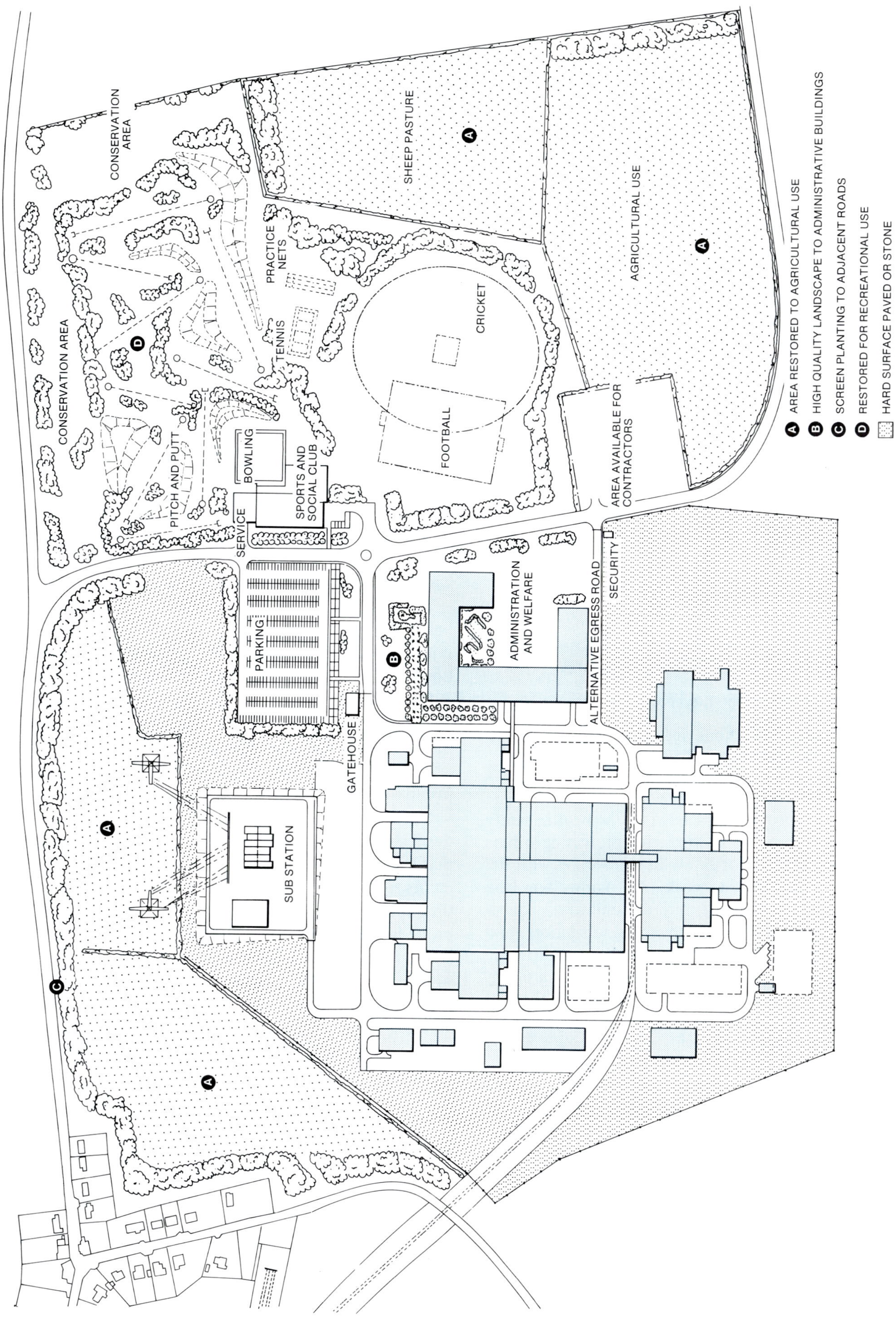

FIG. 3.72 Station landscaping

Fixed lifting equipment All lifting tackle, blocks, cranes, etc., are tested by the manufacturers, marked with the safe working load and test certificates are issued to the CEGB when they are taken over. There are, however, many lifting installations which rely on the attachment to and strength of building frame in addition to the lifting equipment and runway beams or crane rails and these can only be tested after installation. Proof loads in excess of the working load must be traversed over the whole operating area of the crane before the certificate is issued and the lifting equipment marked with the safe working load.

Health (general provisions) This part of the Act sets out the minimum requirements relating to drinking water supplies, washing facilities, sanitary accommodation, accommodation for clothing, first aid, ventilation and heating of working places, notification of accidents and other similar matters. The CEGB has issued its own Safety Rules and Standards which provide accommodation with facilities in excess of the minimum required by the Act.

24.3 Health and Safety at Work Act, 1974

The Act is divided into four parts:

Part 1 Health, safety and welfare in connection with work, and control of dangerous substances and certain emissions into the atmosphere.

Part 2 The employment medical advisory service.

Part 3 Building regulations and amendment of building (Scotland) Act 1959.

Part 4 Miscellaneous and general.

The Act makes provision for securing the health, safety and welfare of persons at work, for protecting others against risks to health or safety in connection with the activities of persons at work, for controlling the keeping and use and preventing the unlawful acquisition, possession and use of dangerous substances, and for controlling certain emissions into the atmosphere.

A novel feature of this Act was to move such provisions from the civil to the criminal code, thereby emphasising the responsibility of the employer to his workforce and of the engineer for the safety of his design and construction proposals. As with all such legislation the proof of default rests heavily on the court's interpretation of 'negligence' and 'reasonable practicability'.

24.4 Building regulations

Compliance with Building Regulations 1985 is mandatory. The classes of building which are exempt are listed in Schedule 3 Building Regulation 1985. The precise position of the CEGB remains undetermined, however, as the Act allows a 'self certification of compliance' concept for major utilities which has still to be worked out in detail.

The main purpose of the Building Regulations is to ensure the health and safety of people in or about the building. They are also concerned with energy conservation and access to buildings for the disabled.

24.5 Nuclear station licensing

The main legislation governing the safety of nuclear installations is the Health and Safety at Work Act 1974 and the associated relevant statutory provisions of the Nuclear Installations Act 1965. Under these Acts any site proposed for a commercial nuclear installation requires the granting of a nuclear site licence to the owner and operator by the Health and Safety Executive (HSE). HM Nuclear Installations Inspectorate (NII) is the part of the HSE responsible for administering this licensing function.

The Acts lay down general requirements for the safety of nuclear power plants. Specific requirements are a matter for the NII to formulate and apply. These may include safety cases relating to different stages of the licensing process or conditions, binding in law, which may be attached to the site licence.

The Nuclear Installations Act places an absolute liability upon the CEGB as licensee with regard to injury to persons or damage to property. This is additional to the CEGB responsibility under the Health and Safety at Work Act for the safe design and operation of its nuclear power stations so as to ensure the health and safety of its employees and other persons.

The licensing process for a given site and proposed installation begins with the applicant consulting the NII about the information required prior to formal procedures being implemented. When the NII is satisfied that the proposed installation should meet its requirements for licensing, the applicant is advised of this and then makes formal planning application to the appropriate Secretary of State. The procedure to be followed is laid down in the Electrical Lighting Act 1909 under which consent is required before any power station can be built or extended. If the application is successful, and this may involve Public Inquiry proceedings, a licence to install and operate the nuclear installation is required to be granted following submission of a satisfactory safety case to the NII.

Formal application requires the preparation of a Preliminary Safety Report (PSR) on completion of the design of the main features of the buildings and plant, followed by a more detailed Pre-Construction Safety Report (PCSR) based on the contract design. Given a satisfactory outcome to the NII review of the PCSR and subject to Consent having been obtained under the Electric Lighting Act, a nuclear site licence may be granted.

The construction and commissioning of the plant are further subject to a number of stage controls imposed on the licensee by the NII. The first control or hold point is usually a requirement for a formal NII consent to commence the pouring of permanent concrete for the foundations. This consent allows construction to proceed until the next controlling hold point is reached and so on. Finally consent is required before fuel can be loaded.

The operation of the completed plant is governed by licensing conditions which require that a maintenance schedule, covering inspection, testing and maintenance of the plant, is prepared by the licensee and approved by the NII. This schedule of work has to be completed satisfactorily, largely during each of the obligatory biennial shutdowns for maintenance inspection which every reactor has to undergo, before the NII will give consent for the next start-up and continued operation. In the case of stations with prestressed concrete pressure vessels (PCPVs) the civil engineering features of the PCPVs are subjected to inspections and tests under this licensing procedure. The submission of a structural assessment report to the NII prepared by the Appointed Examiner of the CEGB's Civil Engineering Branch, gives the NII the assurance of the PCPV's fitness for further operation. The PCPV schedule includes examinations and texts on the prestressing system and concrete structure together with precise surveying of the PCPV deformations and foundation settlement. Further details of these statutory examinations are given in Section 14.3 of this chapter.

24.6 Offices, Shops and Railway Premises Act

The parts of this Act which affect the design of power stations are those relating to the standards required for office accommodation. Premises must be registered and the Act lays down that a minimum standard of cleanliness must be maintained. Although it is sometimes misleading to quote requirements out of context, a minimum area of 4 m^2 must be allotted for each person in a room, or if the ceiling is lower than 3 m then a volume of 11.5 m^3 must be substituted. For accurate information the Act should be consulted as circumstances can affect interpretation. It also lays down that a minimum temperature of 20°C must be maintained and that adequate lighting and ventilation shall be provided. BS6465 Part 1 [35] regarding minimum provision of sanitary conveniences and washing facilities and adequate supplies of drinking water, must be complied with. Arrangements must also be made for accommodating and, where necessary, drying clothes.

The Act also covers seating requirements, eating facilities and also the construction of floors, passages and stairs, especially finishes and handrails.

The part of the Act relating to fire precautions covers provision of escape routes, fire fighting appliances and fire alarms. It also covers escape drills for employees and the procedure for issuing fire certificates.

The Act also covers procedures for notification of accidents and the division of responsibilities if the occupier is not the owner of the office building.

25 Civil engineering contracts

25.1 Special aspects of civil engineering contracts

In general the type of contract chosen for civil engineering work depends on the allocation of risk between the employer and the contractor. These range from the turnkey contracts where the contractor has 100% of the risk, to the cost reimbursable contract where the employer has 100% of the risk. The various types of contract between these two limits can best be illustrated as shown in Fig 3.73.

The decision on the type of contract to be used largely depends on the availability and accuracy of the information available at the time of the tender.

25.2 Forms of contract

25.2.1 Turnkey contracts

This type of contract is not in general use for civil engineering contracts. It is used where the employer issues a performance specification and will accept the contractor's design and construction standards as meeting that performance specification.

25.2.2 Lump sum contracts

In this type of contract the contractor tenders a lump sum for works designed by the engineer. The contractor accepts the full risks of all construction costs and the price is only subject to change due to variations to the tender design or other modifications ordered by the engineer.

25.2.3 Remeasurement contracts

For conventional civil engineering the remeasurement form of contract is generally used. For this type of contract the engineer, whether he is a consulting engineer or an in-house member of staff, prepares tender designs of the proposed Works. From these the quantity surveyor prepares a Bill of Approximate Quantities for the anticipated requirements and these are issued to contractors who price the various items and submit a tender. The design of the works is frequently further developed by the engineer during the contract and the contractor constructs the works in accordance with the final design. The construction drawings are normally issued in stages as the design is completed. Measurement for the final account is car-

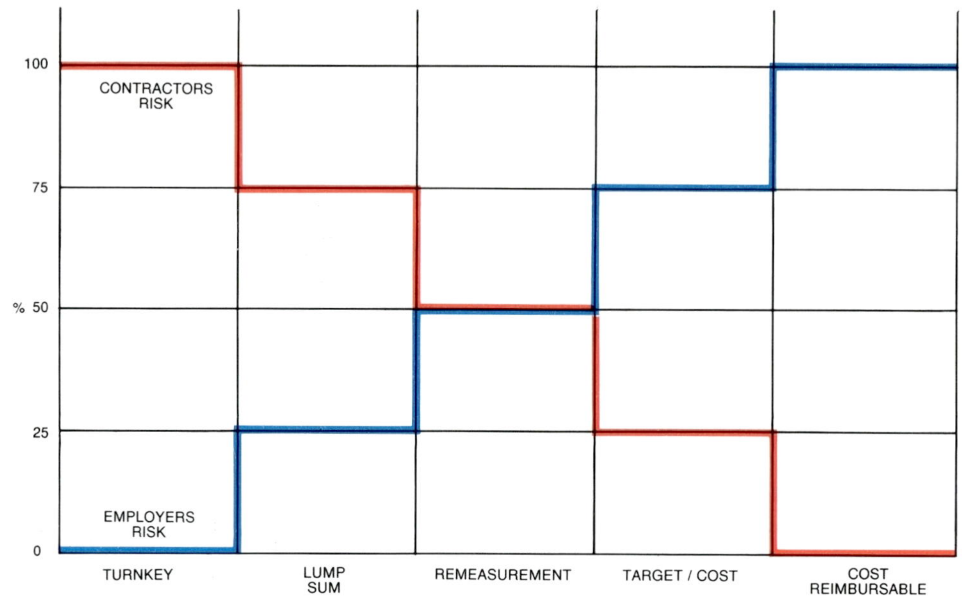

FIG. 3.73 Risk factors of contract type

ried out by the quantity surveyor from the construction drawings and site instructions and priced at the tender or analogous rates or on a day-work basis to arrive at the contract sum. The employer accepts the risks due to the changes in design between the tender design and the 'as constructed' design. The contractor accepts the risk of labour, material and plant requirements for constructing the works to the engineer's design. Owing to unexpected changes in ground conditions, programme modifications, delays in issuing the engineer's designs and such like, additional costs may arise and these are subject to claims by the contractor.

25.2.4 Target/cost contracts

Where, owing to programme restrictions it is not possible for the engineer to prepare a tender design which reasonably reflects the final requirements, or where it is not possible to carry out sufficient ground investigations to determine the inherent risks, or where it is unreasonable for any other reason (for example, works involving innovation or exceptional technical complexity) for the contractor to accept the risks in a remeasurement type contract, a target/cost contract is used. For this contract the target price is established by remeasurement and pricing of the 'as constructed' work, in accordance with rates agreed at tender time. The actual costs are fully recorded and the final payment is based on the relationship between the target and actual costs. This ensures that the contractor has a reasonable profit for the work and the employer is not embarrassed by an influx of additional cost claims.

One advantage of the target type of contract is that with a final price depending on the relationship between target price and actual cost, the contractor is encouraged to carry out the work in the most economical manner in order to maximise the differential in his favour.

25.2.5 Cost reimbursable contracts

This type of contract is used where it is not possible to define the works and the employer undertakes to pay the full costs of construction. This form is often liable to a progressive loss of financial control. However there are various control mechanisms that can be employed, for example, management-type contracts.

25.3 Contract strategy

25.3.1 Extent of civil engineering work

The first civil contractor starts when the initial investigations are commenced for a site and the last civil contractor finishes with the handing over of a complete and landscaped station. Between these dates various contractors will carry out foundations, structural steelwork and buildings under different contracts and at different times to meet the overall station programme.

25.3.2 Type of contract

As the civil engineering final design of power station works is invariably developed by the engineer during the currency of the contract, a firm design cannot be prepared at tender time. Consequently remeasurement contracts are generally adopted.

25.3.3 Extent of contract

The value, programme and type of work required should be taken into consideration when determining the extent of individual contracts. Of these three criteria, the type of work is of major importance. Though many contractors do not specialise in any particular field, some forms of construction do require specialised expertise.

The contract programme is determined by the completion dates required to meet the overall station programme and the essential design information receipt dates for tender and construction drawings preparation. From these dates the construction period and the value of the work should be determined to ensure that the creation of a viable contract is possible. For a power station the different types of work and their suggested grouping within contracts together with reasons and type of contract are given in Table 3.6. Note that some ancillary work will be required in all contracts but is mentioned in only one contract where the main work is centred, e.g., drainage roadworks; and not all sections are included.

25.4 Contract placing

25.4.1 Programme

From the overall programme for the station the necessary completion date for each section of the work can be determined. By suitable grouping of activities and sections of the work an issue of enquiry date is established for each contract.

25.4.2 Invitation to tender

An invitation to tender should be issued to selected contractors (who may have pre-qualified if the extent of contracts warrants it). This invitation should give all tenderers identical information and request a bid compliant with the invitation. Normally not more than six tenders are sought.

25.4.3 Tender documentation

The basic tender documentation should consist of six major sections:

- Conditions of contract.
- Programme requirements.
- Specification for the work, including materials and standards.
- Bills of quantities.
- Form of tender.
- Drawings.

All documents should be mutually explanatory and consistent. The bills of quantities for lump sum contracts will detail the information on the drawings. For remeasurement contracts they will be an interpretation of the work outlined on the drawings, to provide sufficient information to enable the contractor to submit a tender and price.

25.4.4 Return of tenders

Tenders should be returned by a specified time and date. They are opened by a selected panel and all prices recorded at the opening.

25.4.5 Tender appraisal

All tenders are checked to ensure that they comply with the enquiry requirements and are arithmetically correct. A financial assessment of any qualification or alternative bid is made and a formal paper prepared recommending the tender most economic to the employer.

25.4.6 Tender acceptance

The successful tender should be accepted within the programme period and all qualifications or ambiguities should be cleared before the formal award of the contract.

25.5 Contract administration

25.5.1 General

Good contract administration ensures that the works are constructed to the specified quality, within the programme time and within the approved budget. In accordance with the conditions of contract the engineer's representative is delegated specific duties to perform on behalf of the engineer. At the start of each contract these duties should be detailed in writing by the engineer to the contractor. The engineer's representative must record all approvals, condemnations and instructions, ensuring that copies are circulated to all staff responsible for the contract administration.

25.5.2 Specific quality

Civil engineering works are a site-based activity and require detail supervision. To ensure that the works are constructed as designed and specified and in accordance with quality assurance procedures the engineer's representative and associated staff supervise each phase of construction, obtain test certificates for the materials, carry out any test specified in the specification or standards and check all work for line and level.

25.5.3 Programme requirements

To ensure that the works are constructed to programme it is essential that the contractor has full details of the

TABLE 3.6

Suggested extent of civil contracts for a conventional power station

CONTRACT Section of Work	REASON FOR INCLUSION IN CONTRACT	TYPE OF CONTRACT
SITE INVESTIGATION		
Site investigation	Specialist work	Remeasurement
ACCESS ROAD		
Off-site access road	Early requirement specialist work	Remeasurement
PRELIMINARY WORKS		
Site clearance — levelling Access roads Drainage water supply Contractors hardstandings Temporary offices and buildings	Early requirements	Remeasurement
MAIN FOUNDATIONS		
1 Foundations for main building ancillary buildings culverts pumphouse chimney	Large civil engineering works of a similar nature	Remeasurement
2 Coal handling plant Ash handling plant Flue gas desulphurisation plant	Ditto — but could be combined with 1 depending on interconnections and programme	Remeasurement
3 Railway works	Ditto — could be combined with 1 or 2 or treated as specialist contract	Remeasurement
STRUCTURAL STEELWORKS		
Main building — steelwork Main building — cladding Main building — roofing Ancillary building — steelwork	Specialist contractor) combined Specialist sub-contractor) due to Specialist sub-contractor) programme	Remeasurement
SUPERSTRUCTURE		
Main building — above foundation slab Ancillary buildings	Major civil engineering work normally requiring separation from the main foundations due to programme and interface connections	Remeasurement
CHIMNEY		
Main chimney	Specialist contractor	Lump sum depending on design/tender information
COOLING WATER SYSTEM		
Off-shore tunnels or submerged tube	Specialist contractor — large contract warrants individual management	Remeasurement or target/ cost or turnkey dependent on design/tender information and ground conditions
or		
Cooling tower — foundations shells	Specialist contractor	Remeasurement
— packing and supports	Specialist sub-contractor Either design by engineer or design by contractor to performance specification	Remeasurement/Turnkey
LANDSCAPING		
Landscaping	Specialist contractor	Remeasurement
PILING		
All foundations as required	Engineer to determine loads and numbers of piles. Specialist contractor to design piles for specified loads	Remeasurement

engineer's requirements in sufficient time to enable him to plan and organise the work and his resources of labour, plant and material.

To monitor the performance under each contract the engineer convenes regular (monthly) meetings with the contractor's senior management and the respective staffs to consider and act on the information required, the progress to date and the programme for completion.

Intermediate meetings between the engineer's representative and the contractor's site staff are held to ensure that any undertakings given or instructions issued at or following the engineer's meeting are carried out. All meetings are minuted and agreed by both parties.

25.6 Budgetary approval and control

25.6.1 General

If budgetary control is to be achieved, constant and detailed analysis of the developing design and the engineer's instructions is required. Changes in cost fall under four main headings:

- Changes in design.
- Design development.
- Site instructions/daywork.
- Additional costs.

25.6.2 Changes in design

Changes in design arise from alterations in plant or electrical requirements (i.e., loadings or layout) and variations in ground conditions. If costs arising from these changes are not noted in good time they can have a devastating effect on budgetary control. It is recommended that a quantity surveyor is located in the engineer's design office and monitors the changes between the tender design/bills of quantities and the design currently being prepared. The quantity surveyor should estimate the cost of these changes and notify them to the employer and the engineer so that corrective measures are taken or adequate financial provision made.

25.6.3 Design development

The financial effects of the progressive small changes due to design development are more difficult to detect than the larger design change. This is due to the fact that individually they are of a minor nature but their cumulative effect can be large. Budgetary allowances should be made for design development and the quantity surveyor can assess the actual changes by maintaining the remeasurement of the works closely behind the issue of working drawings, costing this remeasurement and comparing it with the budgetary allowance. The cumulative effect and comparison can be notified as previously stated.

25.6.4 Site instructions

Copies of all site instructions and daywork orders should be passed to the quantity surveyor who ensures that any changes in cost are recorded and notified, as previously stated.

25.6.5 Additional costs

In accordance with the conditions of contract the contractor is entitled to additional costs where the employer or the engineer fail to meet their obligations under the contract. The contractor must give notice and details of his assessment of such additional costs. The engineer shall, on receipt of such notice, inform the employer and quantity surveyor. The employer and engineers should take corrective measures to reduce the impact of the costs. A preliminary assessment can be made and notified by the quantity surveyor so that financial provision can be made.

26 References

[1] BS5930: Code of practice for site investigations: 1981

[2] BS8004: Code of practice for foundations: 1986

[3] BS8110: Structural use of concrete: 1985
Part 1: Code of practice for design and construction
Part 2: Code of practice for special circumstances
Part 3: Design charts for singly reinforced beams, doubly reinforced beams and rectangular columns

[4] Hadjian, A.H.: Seismic isolation of nuclear plants: Nuclear Engineering and Design, Vol. 84 No. 3: 1985

[5] BS882: Specification for aggregates from natural resources for concrete: 1983

[6] BS12: Specification for ordinary and rapid-hardening Portland cement: 1978

[7] BS4027: Specification for sulphate-resisting Portland cement: 1980

[8] BS3148: Methods of test for water for making concrete: 1980

[9] BS3892: Pulverised fuel ash:
Part 1: 1982 — Specification for pulverised fuel ash for use as a cementitious component in structural concrete
Part 2: 1984 — Specification for pulverised fuel ash for use in grouts and for miscellaneous uses in concrete

[10] BS5075: Concrete admixtures:
Part 1: 1982 — Specification for accelerating admixtures, retarding admixtures and water reducing admixtures
Part 2: 1982 — Specification for air-entering admixtures
Part 3: 1985 — Specification for superplasticising admixtures

[11] BS5572: Code of practice for sanitary pipework: 1978

[12] BS8301: Code of practice for building drainage: 1985

[13] Reservoirs Act 1975

[14] Factories Act: The work in compressed air — special regulations, : SI1958 No. 61: SI1960 No. 1307: SI1973 No. 36: 1958

[15] BS6399: Loading for buildings: 1984
Part 1: Code of practice for dead and imposed loads

[16] CP3: Chapter V: Loading
Part 2: 1972 Wind loads

[17] BS4360: Specification for weldable structural steels: 1986

[18] BS5950: Structural use of steelwork in building: 1985
Part 1: Code of practice for design in simple and continuous construction: hot rolled sections
Part 2: Specification for materials, fabrication and erection: hot rolled sections
Part 4: Code of practice for design of floors with profiled steel sheeting
Part 5: Code of practice for design of cold formed sections

[19] Constrado — Steelwork design guide to BS5950: Part 1: Volume 1, Section properties and member capacities: 1985

[20] Pal, D.C. and Parker, J.V.: The aseismic design of a reactor building for the advanced gas cooled reactor power plant: Proc. Conf. (ICE) Earthquake Engineering in Britain, University of East Anglia: 18–19 April 1985

[21] Smith, C.R.: Seismic design approach for the Sizewell *B* nuclear power plant: Proc. Conf. (ICE) Earthquake Engineering in Britain, University of East Anglia: 18–19 April 1985

[22] ACI 349–80: Code requirements for nuclear safety related concrete structures and commentary — ACI 349R–80: American Concrete Institute, Detroit, Michigan, USA

[23] ACI 318M–83: Building code requirements for reinforced concrete structures: American Concrete Institute, Detroit, Michigan, USA

[24] ASCE: Manual and Report on Engineering Practice — No. 58: Structural analysis and design of nuclear plant facilities: American Society of Civil Engineers: 1980

[25] Park, R. and Paulay, T.: Reinforced concrete structures: John Wiley: 1975

[26] BS4975: Specification for prestressed concrete pressure vessels for nuclear reactors: 1973

[27] ASME III Division 2: Specification for prestressed concrete reactor vessels and containments: ASME, New York, USA

[28] BS3921: Specification for clay bricks: 1985

[29] BS6073: Precast concrete masonry units: 1981
Part 1: Specification for precast concrete masonry units
Part 2: Method for specifying precast concrete masonry units

[30] BS2989: Specification for continuously hot-dip zinc-coated and iron-zinc alloy coated steel: wide strip, sheet/plate and silt wide strip: 1982

[31] Model code for concrete chimneys; Part A; The shell: Comité International des Cheminées Industrielles: 1984

[32] DIN 1056: Free standing chimneys: Deutsches Institut fur Normung, E.V. Berlin

[33] BS6651: Code of practice for protection of structures against lightning: 1985

[34] BS4485: Specification for water cooling towers:
Part 1: 1969 (1982) — Glossary of terms
Part 2: 1969 — Methods of test and acceptance testing
Part 3: 1977 — Thermal and functional design of cooling towers Addendum No. 1 (1978) to Part 3 — Factory prefabricated cooling towers
Part 4: 1975 — Structural design of cooling towers

[35] BS6465: Sanitary installations: 1984
Part 1: Code of practice for scale of provision, selection and installation of sanitary appliances

[36] BS8004: Code of practice for foundations: 1986

Appendix A

Estimation of the carrying capacity of piles

A1 Driven pile in non-cohesive soil

The best known of the dynamic formulae is the Hiley formula. This is based on the impact of elastic bodies and equates the energy of the hammer blow to the resistance of the ground to the penetration of the pile. Allowances are made for loss of energy due to elastic contractions of the pile, dolly and subsoil as well as the losses due to the inertia of the pile. As originally proposed the formula is as follows:

$$R = \frac{Wh\eta}{S + c/2}$$

where R = ultimate driving resistance in tons

W = weight of hammer in tons

h = height of free fall of hammer in inches

S = final set or penetration per blow in inches

c = sum of temporary elastic compressions in inches of pile; dolly and ground; these values are also dependent on whether driving is easy, medium or hard, and are obtained from tables [36]

η = efficiency of blow; this value is dependent on the ratio of the weight of the pile to the hammer and the dolly type. This value may be obtained from tables [36]

The factor of safety to be applied to the ultimate driving resistance obtained by the formula should never be less than 2. However, if there is any suggestion of time-dependent relaxation of resistance this formula should be used with extra care and higher factors for safety used. An increase in these values should in any case be made for structures sensitive to settlement.

A2 Driven and bored piles in cohesive soil

The carrying capacity may be estimated when the shear strength of the soil is known. The first term in the following example formula is that due to skin friction and the second term is due to end bearing support.

For a bored pile in London Clay the estimated safe load is:

$$\Pi dL \times \frac{0.45C_s}{F} + \frac{\Pi d^2}{4} + \frac{9w_{C_B}}{F} \text{ kN}$$

where d = diameter of pile in metres

L = effective length of pile in metres

C_s = average undrained cohesion over the length of the pile shaft in kN/m^2

C_B = typical undrained cohesion at base of pile in kN/m^2

w = a factor which depends on the diameter of the pile

F = factor of safety, about 2 on skin friction and 2.5 for end bearing

SUBJECT INDEX